Surface Modification of Magnesium and its Alloys for Biomedical Applications

Related titles

Surface modification of biomaterials
(ISBN 978-1-84569-640-5)

Coatings for biomedical applications
(ISBN 978-1-84569-568-2)

Metals for biomedical devices
(ISBN 978-1-84569-434-0)

Woodhead Publishing Series in Biomaterials: Number 89

Surface Modification of Magnesium and its Alloys for Biomedical Applications

Volume 1: Biological Interactions, Mechanical Properties and Testing

Edited by

T. S. N. Sankara Narayanan, Il-Song Park and Min-Ho Lee

ELSEVIER

AMSTERDAM • BOSTON • CAMBRIDGE • HEIDELBERG
LONDON • NEW YORK • OXFORD • PARIS • SAN DIEGO
SAN FRANCISCO • SINGAPORE • SYDNEY • TOKYO
Woodhead Publishing is an imprint of Elsevier

WP WOODHEAD PUBLISHING

Woodhead Publishing is an imprint of Elsevier
80 High Street, Sawston, Cambridge, CB22 3HJ, UK
225 Wyman Street, Waltham, MA 02451, USA
Langford Lane, Kidlington, OX5 1GB, UK

British Library Cataloguing in Publication Data
A catalogue record for this book is available from the British Library.

Library of Congress Control Number: 2014955924

ISBN 978-1-78242-077-4 (print)
ISBN 978-1-78242-082-8 (online)
ISBN 978-1-78242-498-7 (two-volume set – print)
ISBN 978-1-78242-499-4 (two-volume set – online)

For information on all Woodhead Publishing publications visit our website at http://store.elsevier.com/

Typeset by TNQ Books and Journals
www.tnq.co.in

Printed and bound in the United Kingdom

Contents

List of contributors

R.A. Antunes Universidade Federal do ABC, São Paulo, Brazil

Andrejs Atrens The University of Queensland, St Lucia, QLD, Australia

Lokesh Choudhary Monash University, VIC, Australia

P.K. Chu City University of Hong Kong, Hong Kong, China

M.C.L. de Oliveira Electrocell Ind. Com. Equip. Elet. LTDA, Technology, Entrepreneurship and Innovation Center (CIETEC), São Paulo, Brazil

Martin Durisin Medical University of Hannover, Hannover, Germany

T. Imwinkelried Formerly Synthes GmbH, Oberdorf, Switzerland (Presently at Robert Mathys Foundation)

Nicholas Travis Kirkland Nagasaki University, Nagasaki, Japan

Min-Ho Lee Chonbuk National University, Jeonju, Republic of Korea

Xiao Lin Chinese Academy of Sciences, Shenyang, China

Anneke Loos Hannover Medical School, Hannover, Germany

T.S.N. Sankara Narayanan Chonbuk National University, Jeonju, Republic of Korea

Il-Song Park Chonbuk National University, Jeonju, Republic of Korea

R.K. Singh Raman Monash University, VIC, Australia

Jay Waterman University of Canterbury, Christchurch, New Zealand

G.S. Wu City University of Hong Kong, Hong Kong, China

Ke Yang Chinese Academy of Sciences, Shenyang, China

Y.F. Zheng Peking University, Beijing, China

W.R. Zhou Peking University, Beijing, China

Woodhead Publishing Series in Biomaterials

1 **Sterilisation of tissues using ionising radiations**
Edited by J. F. Kennedy, G. O. Phillips and P. A. Williams
2 **Surfaces and interfaces for biomaterials**
Edited by P. Vadgama
3 **Molecular interfacial phenomena of polymers and biopolymers**
Edited by C. Chen
4 **Biomaterials, artificial organs and tissue engineering**
Edited by L. Hench and J. Jones
5 **Medical modelling**
R. Bibb
6 **Artificial cells, cell engineering and therapy**
Edited by S. Prakash
7 **Biomedical polymers**
Edited by M. Jenkins
8 **Tissue engineering using ceramics and polymers**
Edited by A. R. Boccaccini and J. Gough
9 **Bioceramics and their clinical applications**
Edited by T. Kokubo
10 **Dental biomaterials**
Edited by R. V. Curtis and T. F. Watson
11 **Joint replacement technology**
Edited by P. A. Revell
12 **Natural-based polymers for biomedical applications**
Edited by R. L. Reiss et al
13 **Degradation rate of bioresorbable materials**
Edited by F. J. Buchanan
14 **Orthopaedic bone cements**
Edited by S. Deb
15 **Shape memory alloys for biomedical applications**
Edited by T. Yoneyama and S. Miyazaki
16 **Cellular response to biomaterials**
Edited by L. Di Silvio
17 **Biomaterials for treating skin loss**
Edited by D. P. Orgill and C. Blanco
18 **Biomaterials and tissue engineering in urology**
Edited by J. Denstedt and A. Atala

19 **Materials science for dentistry**
B. W. Darvell
20 **Bone repair biomaterials**
Edited by J. A. Planell, S. M. Best, D. Lacroix and A. Merolli
21 **Biomedical composites**
Edited by L. Ambrosio
22 **Drug–device combination products**
Edited by A. Lewis
23 **Biomaterials and regenerative medicine in ophthalmology**
Edited by T. V. Chirila
24 **Regenerative medicine and biomaterials for the repair of connective tissues**
Edited by C. Archer and J. Ralphs
25 **Metals for biomedical devices**
Edited by M. Ninomi
26 **Biointegration of medical implant materials: Science and design**
Edited by C. P. Sharma
27 **Biomaterials and devices for the circulatory system**
Edited by T. Gourlay and R. Black
28 **Surface modification of biomaterials: Methods analysis and applications**
Edited by R. Williams
29 **Biomaterials for artificial organs**
Edited by M. Lysaght and T. Webster
30 **Injectable biomaterials: Science and applications**
Edited by B. Vernon
31 **Biomedical hydrogels: Biochemistry, manufacture and medical applications**
Edited by S. Rimmer
32 **Preprosthetic and maxillofacial surgery: Biomaterials, bone grafting and tissue engineering**
Edited by J. Ferri and E. Hunziker
33 **Bioactive materials in medicine: Design and applications**
Edited by X. Zhao, J. M. Courtney and H. Qian
34 **Advanced wound repair therapies**
Edited by D. Farrar
35 **Electrospinning for tissue regeneration**
Edited by L. Bosworth and S. Downes
36 **Bioactive glasses: Materials, properties and applications**
Edited by H. O. Ylänen
37 **Coatings for biomedical applications**
Edited by M. Driver
38 **Progenitor and stem cell technologies and therapies**
Edited by A. Atala
39 **Biomaterials for spinal surgery**
Edited by L. Ambrosio and E. Tanner
40 **Minimized cardiopulmonary bypass techniques and technologies**
Edited by T. Gourlay and S. Gunaydin
41 **Wear of orthopaedic implants and artificial joints**
Edited by S. Affatato
42 **Biomaterials in plastic surgery: Breast implants**
Edited by W. Peters, H. Brandon, K. L. Jerina, C. Wolf and V. L. Young

43 **MEMS for biomedical applications**
Edited by S. Bhansali and A. Vasudev
44 **Durability and reliability of medical polymers**
Edited by M. Jenkins and A. Stamboulis
45 **Biosensors for medical applications**
Edited by S. Higson
46 **Sterilisation of biomaterials and medical devices**
Edited by S. Lerouge and A. Simmons
47 **The hip resurfacing handbook: A practical guide to the use and management of modern hip resurfacings**
Edited by K. De Smet, P. Campbell and C. Van Der Straeten
48 **Developments in tissue engineered and regenerative medicine products**
J. Basu and J. W. Ludlow
49 **Nanomedicine: Technologies and applications**
Edited by T. J. Webster
50 **Biocompatibility and performance of medical devices**
Edited by J-P. Boutrand
51 **Medical robotics: Minimally invasive surgery**
Edited by P. Gomes
52 **Implantable sensor systems for medical applications**
Edited by A. Inmann and D. Hodgins
53 **Non-metallic biomaterials for tooth repair and replacement**
Edited by P. Vallittu
54 **Joining and assembly of medical materials and devices**
Edited by Y. (Norman) Zhou and M. D. Breyen
55 **Diamond-based materials for biomedical applications**
Edited by R. Narayan
56 **Nanomaterials in tissue engineering: Fabrication and applications**
Edited by A. K. Gaharwar, S. Sant, M. J. Hancock and S. A. Hacking
57 **Biomimetic biomaterials: Structure and applications**
Edited by A. J. Ruys
58 **Standardisation in cell and tissue engineering: Methods and protocols**
Edited by V. Salih
59 **Inhaler devices: Fundamentals, design and drug delivery**
Edited by P. Prokopovich
60 **Bio-tribocorrosion in biomaterials and medical implants**
Edited by Y. Yan
61 **Microfluidic devices for biomedical applications**
Edited by X-J. James Li and Y. Zhou
62 **Decontamination in hospitals and healthcare**
Edited by J. T. Walker
63 **Biomedical imaging: Applications and advances**
Edited by P. Morris
64 **Characterization of biomaterials**
Edited by M. Jaffe, W. Hammond, P. Tolias and T. Arinzeh
65 **Biomaterials and medical tribology**
Edited by J. Paolo Davim
66 **Biomaterials for cancer therapeutics: Diagnosis, prevention and therapy**
Edited by K. Park

67 **New functional biomaterials for medicine and healthcare**
E. P. Ivanova, K. Bazaka and R. J. Crawford
68 **Porous silicon for biomedical applications**
Edited by H. A. Santos
69 **A practical approach to spinal trauma**
Edited by H. N. Bajaj and S. Katoch
70 **Rapid prototyping of biomaterials: Principles and applications**
Edited by R. Narayan
71 **Cardiac regeneration and repair Volume 1: Pathology and therapies**
Edited by R-K. Li and R. D. Weisel
72 **Cardiac regeneration and repair Volume 2: Biomaterials and tissue engineering**
Edited by R-K. Li and R. D. Weisel
73 **Semiconducting silicon nanowires for biomedical applications**
Edited by J. L. Coffer
74 **Silk biomaterials for tissue engineering and regenerative medicine**
Edited by S. Kundu
75 **Biomaterials for bone regeneration: Novel techniques and applications**
Edited by P. Dubruel and S. Van Vlierberghe
76 **Biomedical foams for tissue engineering applications**
Edited by P. Netti
77 **Precious metals for biomedical applications**
Edited by N. Baltzer and T. Copponnex
78 **Bone substitute biomaterials**
Edited by K. Mallick
79 **Regulatory affairs for biomaterials and medical devices**
Edited by S. F. Amato and R. Ezzell
80 **Joint replacement technology Second edition**
Edited by P. A. Revell
81 **Computational modelling of biomechanics and biotribology in the musculoskeletal system: Biomaterials and tissues**
Edited by Z. Jin
82 **Biophotonics for medical applications**
Edited by I. Meglinski
83 **Modelling degradation of bioresorbable polymeric medical devices**
Edited by J. Pan
84 **Perspectives in total hip arthroplasty: Advances in biomaterials and their tribological interactions**
S. Affatato
85 **Tissue engineering using ceramics and polymers Second edition**
Edited by A. R. Boccaccini and P. X. Ma
86 **Biomaterials and medical device associated infections**
Edited by L. Barnes and I. Cooper
87 **Surgical techniques in total knee arthroplasty (TKA) and alternative procedures**
Edited by S. Affatato
88 **Lanthanide oxide nanoparticles for molecular imaging and therapeutics**
G. H. Lee
89 **Surface modification of magnesium and its alloys for biomedical applications Volume 1: Biological interactions, mechanical properties and testing**
Edited by T. S. N. Sankara Narayanan, Il-Song Park and Min-Ho Lee

90 **Surface modification of magnesium and its alloys for biomedical applications Volume 2: Modification and coating techniques**
Edited by T. S. N. Sankara Narayanan, Il-Song Park and Min-Ho Lee

91 **Medical modelling: The application of Advanced Design and Rapid Prototyping techniques in medicine Second Edition**
Edited by R. Bibb, D. Eggbeer and A. Paterson

Part One

Fundamentals of surface-modified magnesium and magnesium alloys for biomedical applications

Revolutionising biodegradable biomaterials – significance of magnesium and its alloys

1

Andrejs Atrens
The University of Queensland, St Lucia, QLD, Australia

1.1 Introduction

Biodegradable medical implant applications are a relatively new area for magnesium (Mg) alloys (Atrens, Liu, & Zainal Abidin, 2011; Atrens, Liu, Zainal Abidin, & Song, 2011; Carboneras, Garcia-Alonso, & Escudero, 2011; Hänzi, Gerber, Schinhammer, Löffler, & Uggowitzer, 2010; Kalb, Rzany, & Hensel, 2012; Kraus et al., 2012; Lei Tang, Cai, Feng, & Li, 2012; Remennik, Bartsch, Willbold, & Witte, 2011; Salunke, Shanov, & Witte, 2011; Virtanen, 2011; Witte, 2010; Zainal Abidin, Atrens, Martin, & Atrens, 2011; Zainal Abidin, Martin, & Atrens, 2011; Zainal Abidin et al., 2013; Zberg, Uggowitzer, & Loeffler, 2009; Zhou et al., 2011). There is a considerable recent literature on this topic (Bobe et al., in press; Bornapour, Muja, Shum-Tim, Cerruti, & Pekguleryuz, 2013; Brar, Ball, Bergland, Allen, & Manuel, 2013; Brar, Wong, & Manuel, 2012; Carboneras et al., 2011; Castellani et al., 2011; Choudhary, 2012; Erdmann et al., 2011; Gastaldi et al., 2011; Ge, Dellasega, Demir, & Vedani, in press; González et al., 2012; Gu, Xie, Li, Zheng, & Qin, 2012; Jang, Collins, Sankar, & Yun, in press; Kraus et al., 2012; Kubásek, Vojtěch, Lipov, & Ruml, 2013; Kuhlmann et al., in press; Li & Zheng, in press; Li, Shih, Parish, & Atrens, 2012; Li et al., in press; Mao et al., 2012; Qiu et al., 2013; Rad, Idris, Kadir, & Farahany, 2012; Salunke et al., 2011; Scheideler et al., in press; Tan, Yu, Wan, & Yang, in press; Vojtěch, Kubásek, Šerák, & Novák, 2011;Wang et al., 2011; Willbold, Kaya, Kaya, Beckmann & Witte, 2011; Yang et al., in press; Zhang et al., 2012; Zhang, Wang, Tuan, & Xue, 2012; Zhao, Shi, & Xu, 2013; Zong et al., 2012). The implant biodegrades after it has completed its function as an implant. Key to this application is an understanding of the corrosion behaviour of Mg alloys. This understanding should make use of the extensive existing research, which has been carried out and is being carried out, for the use of Mg alloys in other applications, particularly transport (auto construction, aerospace) (Song & Atrens, 2003).

It is vital for researchers in this area to understand key aspects about Mg metallurgy and Mg corrosion. There are unique features concerning Mg that make the study of Mg corrosion fascinating, but can lead to misleading or erroneous conclusions if appropriate care is not taken. Clearly, care must be taken in all research; however, Mg behaves differently than other metals, so there are more traps for the unwary.

Surface Modification of Magnesium and its Alloys for Biomedical Applications. http://dx.doi.org/10.1016/B978-1-78242-077-4.00001-2

Researchers interested in Mg for biodegradable applications are urged to consult the prior literature, particularly the recent reviews (Atrens, Liu, Zainal Abidin, 2011; Atrens, Liu, Zainal Abidin, & Song, 2011). It is not the intention to repeat (or even summarise) herein all the information contained therein. Researchers should also consult our other prior reviews on Mg corrosion (Atrens, Winzer, Dietzel, Srinivasan, & Song, 2011; Atrens, Dietzel, Srinivasan, Winzer, & Kannan, 2011, Atrens, Shi, & Song, 2011; Song & Atrens, 1999; Song & Atrens, 2000; Song & Atrens, 2003; Winzer et al., 2005), papers on measurement of Mg corrosion (Liu, Qiu, Zhao, Song, & Atrens, 2008; Liu et al., 2009; Shi & Atrens, 2011; Shi, Liu & Atrens, 2010, Shi, Prasad, & Atrens, 2012a; Song, Atrens, & StJohn, 2001), the Mg corrosion mechanism (Atrens & Dietzel, 2007; Atrens, 2013; Cao et al., 2013a; Cao, Shi, Song, Liu, & Atrens, 2013b; Liu et al., 2008, Liu, Schmutz, Uggowitzer, Song, & Atrens et al., 2010; Qiao, Shi, Hort, Zainal Abidin, & Atrens, 2012; Shi & Atrens, 2013a; Shi et al., 2012; Shi, Jia, & Atrens, 2012b; Shi, Cao, Song, & Liu, 2013b; Song, Atrens, St. John, Nairn, & Lang, 1997; Song, Atrens, StJohn, Wu, Nairn, 1997; Song, Atrens, Wu, & Zhang, 1998; Song, Atrens, & Dargusch, 1999; Zhao, Liu, Song, & Atrens et al., 2008a; Zhao, Liu, Song, & Atrens, 2008b), galvanic corrosion (Atrens et al., 2011; Jia et al., 2004; Jia, Atrens, Song, & Muster 2005; Jia, Song, & Atrens, 2005; Jia, Song, & Atrens 2006; Jia, Song, & Atrens, 2007; Shi et al., 2012c, Winzer et al., 2005), stress corrosion cracking (SCC) (Atrens et al., 2011; Atrens et al., 2011; Bobby Kannan, Dietzel, Blawert, Atrens, & Lyon, 2008; Shi et al., 2012b; Song, Blawert, Dietzel, & Atrens., 2005; Winzer, Atrens, Dietzel, Song, & Kainer, 2007a,b,c; Winzer, Atrens, Dietzel, Song, & Kainer, 2008a,b,c; Winzer, et al., 2008), surface films (Liu et al., 2009; Liu et al., 2010; Seyeux et al., 2009), and flammability (Liu et al., 2012; Prasad, Shi, & Atrens, 2012a,b). There is also much recent literature (Alvarez-Lopez et al., 2010; Bakhsheshi-Rad, Abdul-Kadir, Idris, & Farahany et al., 2012; Čapek & Vojtěch, 2013; Choudhary & Singh Raman, 2013; Han et al., 2012; Kalb et al., 2012; Liu et al., 2010; Minárik et al., in press; Pu et al., 2012; Sun, Li, & Fang et al., 2011; Yang et al., 2012; Zhou, Shen, & Aung et al., 2010) on these topics, and there are also much valuable data in the ASM Handbook (Shaw & Wolfe, 2005).

This chapter focuses on recent developments, under the following subheadings: (1) corrosion of Mg alloys, (2) ultra-pure Mg, (3) measurement of Mg corrosion, (4) *in vitro* and *in vivo* comparison, (5) Mg corrosion mechanism, and (6) galvanic corrosion. The important points are summarised in the final section.

1.2 Corrosion of magnesium alloys

Mg is the most active of the engineering materials. Mg has a high driving force for corrosion, weakly checked by the poorly protective corrosion product films that typically form on the surface of Mg in most corrosion situations (Song & Atrens, 1999, 2000, 2003).

Mg alloys corrode faster than high-purity (HP) Mg because of the microgalvanic acceleration caused by second phases. In this context, HP Mg is defined as Mg with an impurity level lower than the tolerance limit (Liu et al., 2009). Thus, purification may produce an alloy with a corrosion rate lower than that of an impure alloy. Nevertheless, the HP Mg alloy has a corrosion rate greater than that of HP Mg.

Particularly problematic are second phases associated with the impurity elements, particularly iron (Fe). Fe is easily picked up during melt processing in Fe containers or hot or cold working with Fe tools. These Fe-rich phases, and particles, can increase the corrosion rate by orders of magnitude, even for concentrations of Fe in the Mg alloy in the parts per million (ppm) range (Liu et al., 2008, 2009). This is clear from Figure 1.1 (Atrens et al., 2011; Liu et al., 2009), in which the corrosion of low-purity Mg (280 ppm Fe) and HP Mg (45 ppm) is characterised using hydrogen evolution during an immersion test. The corrosion rate is high if the Fe content is above the impurity limit, whereas the corrosion rate is low for HP Mg.

One molecule of hydrogen is produced, during Mg corrosion in an immersion test, by each corroded Mg atom, as is apparent from the overall corrosion reaction:

$$Mg + H^+ + H_2 \rightarrow Mg^{++} + OH^- + H_2 \quad (1.1)$$

The measurement of the evolved hydrogen is an easy and convenient means to characterise Mg corrosion (Song et al., 2001), and measurement is possible of (1) the instantaneous corrosion rate, (2) how the instantaneous corrosion rate changes with time, and (3) the average corrosion rate over the whole exposure period. Measurement of the corrosion rate using hydrogen evolution is much easier than measurement of the corrosion rate (1) by weight loss or (2) using electrochemical measurements. Weight loss measurements require considerable care and experimental skill to ensure (1) that all the corrosion products have been removed after the corrosion exposure without the

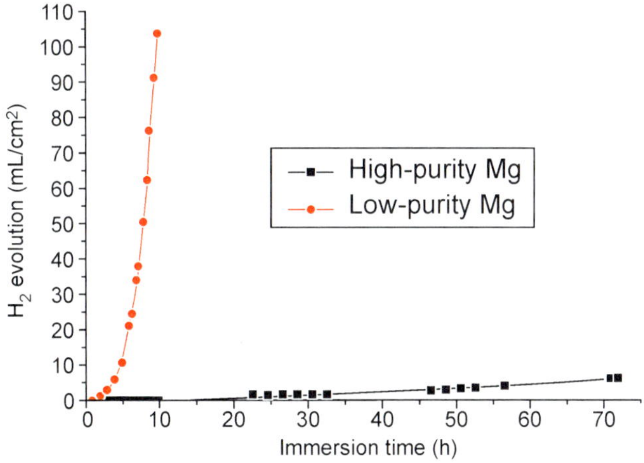

Figure 1.1 The corrosion of low purity (280 ppm Fe) and HP (45 ppm) Mg was characterised using hydrogen evolution. One molecule of hydrogen is produced by each corroded Mg, as is apparent from the overall corrosion reaction. The corrosion rate is high if the Fe content is above the impurity limit, whereas the corrosion rate is low for HP Mg. Liu et al. (2009).

removal of any Mg metal, and (2) that the amount of moisture on the specimen is similar when the specimen is weighed before and after corrosion exposure. Electrochemical measurements require a different skill set, and they also require the appropriate electrochemical apparatus. Electrochemical methods include Tafel extrapolation of polarisation curves, electrochemical impedance spectroscopy (EIS), and polarisation resistance. Furthermore, anyone using electrochemical methods of measurement for Mg corrosion should be aware of the known limitations of these techniques, which are unique and specific to Mg (Atrens et al., 2011; Cao et al., 2013a,b; Shi & Atrens, 2011; Shi et al., 2010, 2012a; Song & Atrens, 2003). Hydrogen evolution measurements also need care and may lead to corrosion rates lower than the actual corrosion rate, particularly for (1) HP Mg and (2) Mg alloy corrosion in synthetic body fluids (Zainal Abidin et al., 2011; Zainal Abidin et al., 2013). In these cases, not all the evolved hydrogen is collected because some hydrogen dissolves in the Mg metal.

Nevertheless, it is best practice to use a number on independent measurements of the Mg corrosion rate and to convert each independent measurement to the same units so that they can be easily compared. The independent measurements should all have the same quantitative value if they are measuring the same quantity.

The Fe tolerance limit can be understood by reference to the Mg-Fe phase diagram, as illustrated in Figure 1.2 (Atrens et al., 2011; Liu et al., 2009). The Mg-Fe phase diagram is a eutectic diagram. The eutectic composition corresponds to 180 ppm (0.018 wt%) Fe. This eutectic composition corresponds to the tolerance limit for as-cast Mg. The microstructure contains the body centred cubic (BCC) Fe rich phase for the solidification of a casting with a Fe content greater than the eutectic Fe content, and consequently the corrosion rate is high. The solidification of a casting containing less than 180 ppm is expected, under normal solidification, to result in alpha-Mg containing Fe in solid solution, and consequently the corrosion rate is low under these conditions, as is evident from Figure 1.1. However, Figure 1.2 indicates that such a casting is in a metastable state with respect to the Fe in solid solution if the Fe content is greater than a few ppm. Heat treatment at elevated temperatures leads to higher corrosion rates due to precipitation of Fe rich phases, as shown in Figure 1.3. This implies that corrosion rates are expected to be high for Mg-X alloys solution heat treated to dissolve the alloying elements (X), because the solution heat treatment causes the precipitation of the Fe rich phases.

Furthermore, Figure 1.2 indicates that a casting with a Fe content less than the Fe tolerance limit for as-cast Mg can also be of low purity and can have high corrosion rate, if the Fe rich phase forms during solidification, despite the low Fe content. This is clear from the data in Figure 1.4 (Qiao et al., 2012). Figure 1.4 presents data for HP (40 ppm) Mg characterised using hydrogen evolution, specimens 26 and 27 compared with corrosion of specimens 19 and 28, which came from a Helmholtz Zentrum Geesthacht (HZG) ingot (30 ppm). Specimens 26 and 27 were HP Mg. In contrast, specimens 19 and 28 did not behave consistent with the behaviour of HP Mg, despite their Fe content being below the Fe tolerance limit for Mg castings. Moreover, the corrosion morphology was also not consistent with the corrosion of HP Mg as is evident from Figure 1.5.

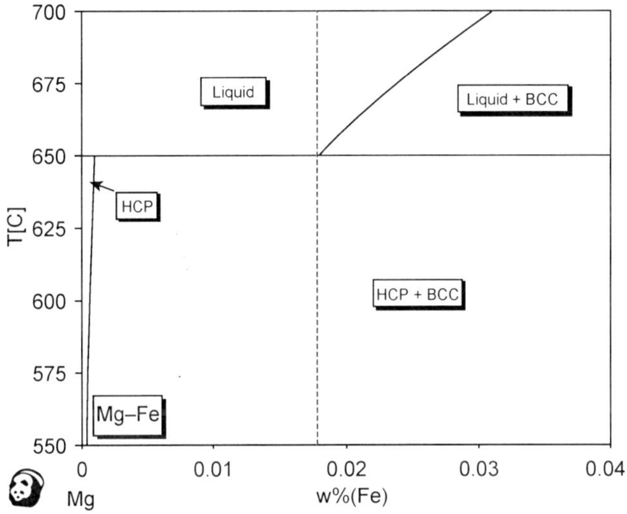

Figure 1.2 The Mg-Fe phase diagram is a eutectic phase diagram with the eutectic composition corresponding to 180 ppm Fe. This eutectic composition corresponds to the tolerance limit for as-cast Mg. The microstructure contains the BCC Fe rich phase for the solidification of a Mg casting with a Fe content greater than the eutectic Fe content, and consequently the corrosion rate is high. The solidification of a casting containing less than 180 ppm is expected under normal solidification to result in alpha-Mg containing the Fe in solid solution, and consequently the corrosion rate is low under these conditions, as is evident from Figure 1.1. However, Figure 1.2 indicates that such a casting is in a metastable state with respect to the Fe in solid solution. Heat treatment at elevated temperatures leads to higher corrosion rates due to precipitation of Fe rich phases, as shown in Figure 1.3. This implies that corrosion rates are expected to be high for Mg-X alloys solution heat treated to dissolve the alloying elements, because the solution heat treatment causes the precipitation of the Fe rich phases. Liu et al. (2009).

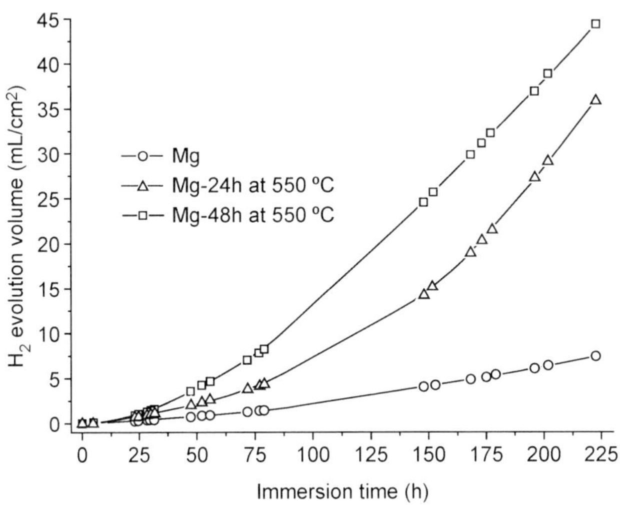

Figure 1.3 The corrosion of high-purity (45 ppm) Mg characterised using hydrogen evolution. After heat treatment the corrosion rate has increased significantly. Liu et al. (2009).

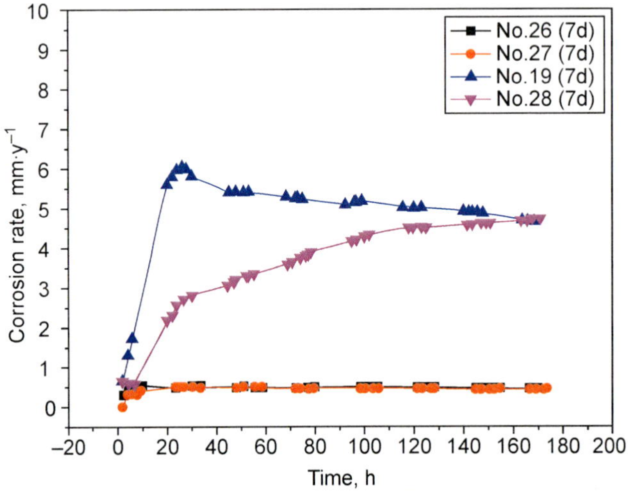

Figure 1.4 The corrosion of HP (40 ppm) Mg characterised using hydrogen evolution, specimens 26 and 27 compared with corrosion of specimens 19 and 28, which came from an HZG ingot (30 ppm). Specimens 19 and 28 did not behave consistent with the behaviour of HP Mg, despite their Fe content being below the normal Fe tolerance limit for Mg castings. Moreover, the corrosion morphology was also not consistent with the corrosion of HP Mg, as is evident from Figure 1.5. Qiao et al. (2012).

Figure 1.5 Typical corrosion morphology for specimens from the HZG Mg ingot (30 ppm) was not consistent with the behaviour of HP Mg, despite their Fe content being below the normal Fe tolerance limit for Mg castings. Qiao et al. (2012).

Figure 1.1 illustrates a further important point (Atrens et al., 2011; Liu et al., 2009). The data for low-purity Mg clearly shows that the corrosion rate is not linear with time. This is common for Mg alloys. The corrosion rate is small initially, then the corrosion rate accelerates to a steady-state corrosion rate. This is a common observation, both *in vitro* in laboratory studies and *in vivo* (Remennik et al., 2011). Figure 1.4 shows a similar behaviour for specimens 19 and 28 (Qiao et al., 2012). The corrosion rate was initially small, then accelerated and achieved a steady-state corrosion rate that was an order of magnitude greater than the initial corrosion rate.

1.3 Ultra-pure magnesium

The corrosion of Mg alloys is currently too fast for applications as a biodegradable implant (Atrens et al., 2011; Atrens et al., 2011; Song & Atrens, 2003). Furthermore, the corrosion rate of all existing Mg alloys is faster than that of ultra-HP Mg (Atrens et al., 2011;u Atrens et al., 2011; Song & Atrens, 2003). There are two possible future approaches: (1) development of Mg alloys with corrosion rates lower than that of HP Mg or (2) surface modification of existing Mg alloys to reduce the corrosion rate. This second approach is the theme of many of the contributions to this book. This section is related to the development of Mg alloys with corrosion rates lower than that of HP Mg.

Our recent work discovered how to produce ultra-HP Mg alloys, when ultra-pure Mg alloys are defined as having a Fe content less than 2 ppm. Figure 1.6 shows how molten Mg can be purified with zirconium (Zr) (Prasad et al., 2012c) to produce ultra-HP (UP) Mg-X alloys. Zr is added to the molten Mg alloy containing Fe. Zr reacts with Fe, and forms $FeZr_x$ particles. These particles settle to the bottom of the melt, as also shown by Qian et al. (2001), because their density is significantly higher than molten Mg. The melt composition attains the equilibrium composition corresponding to the maximum solubility of Fe and Zr in the molten Mg at the reaction temperature. Figure 1.6(a) shows that 2 ppm Fe (0.0002% Fe) is in equilibrium with 0.2% Zr and $FeZr_2$ in molten Mg at 660 °C. Thermodynamics indicate that Zr similarly removes aluminium (Al), manganese (Mn), silicon (Si), nickel (Ni), cobalt (Co), tin (Sn), and antimony (Sb) from molten Mg (Prasad et al., 2012). Our preliminary research (Prasad et al., 2012c) showed that (1) the melt treatment with Zr works in purifying molten Mg and (2) the resultant melt composition is in good agreement with theoretical expectations as shown in Figure 1.6(b). The isothermal sections in Figure 1.6(a) and (b) were calculated using the Pandat thermodynamic package computherm.

Ultra-pure Mg-X alloys (where X may be elements such as dysprosium [Dy] and gandolinium [Gd]) may be stainless after solution heat treatment (SHT) to dissolve the alloying element into solid solution. Dy and Gd have high solubilities in Mg and may produce surface films (Atrens et al., 2011; Atrens et al., 2011; Hort et al., 2010; Song & Atrens, 2003; Yang et al., 2012) on these Mg-X alloys that are substantially more protective than those that form naturally on UP Mg. This is the mechanism that produces stainless steels. A highly protective surface passive film forms on stainless steels for a chromium (Cr) content greater than 10.5% (Bruesch et al., 1984, 1985;

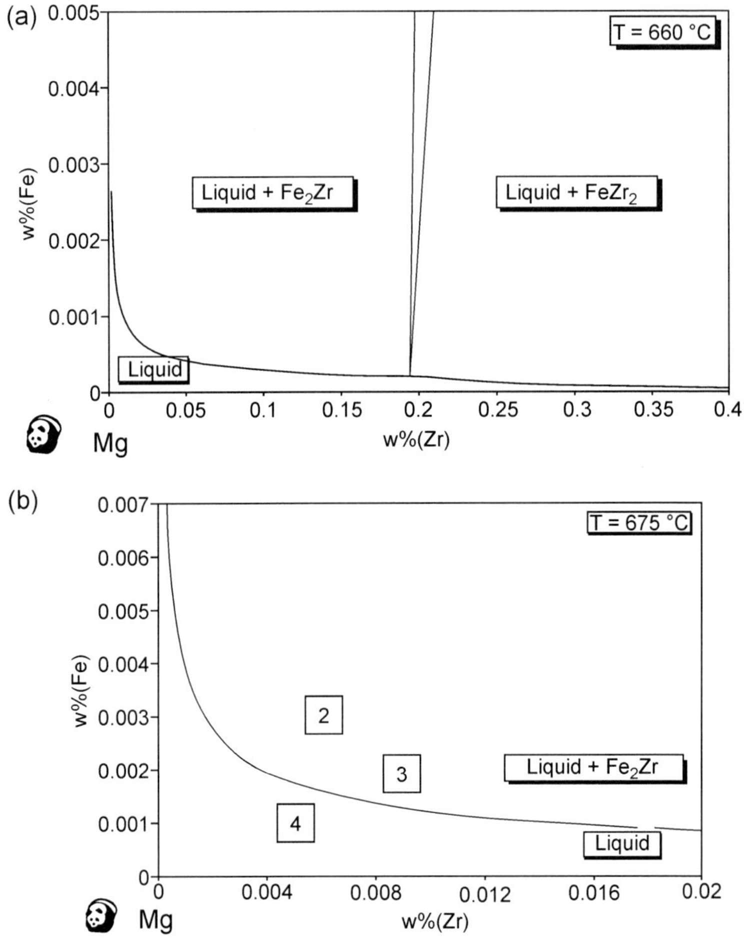

Figure 1.6 (a) Isothermal section through the Mg-Fe-Zr phase diagram at 660 °C showing the maximum solubility of Fe and Zr in liquid Mg at 660 °C, (b) comparison of experimental measurements with the expectation from the phase diagram at 675 °C There was good agreement between experiment and the theoretical thermodynamic predictions. The isothermal sections in (a) and (b) were calculated using the Pandat thermodynamic package. Prasad et al. (2012).

Heine & Kirchheim, 1990; Jin & Atrens, 1987, 1988; Kirchheim et al., 1989; Lim & Atrens, 1991, 1992a,b,c; Olefjord & Elfstrom, 1982; Schneider et al., 1990). In comparison, red rust (iron oxide) is not protective for a lower Cr content. On this basis, we suggested $MgCr_y$ (Song & Atrens, 2003) in 2003. However, extensive efforts using rapid solidification by Uhlenhaut (2008) were not successful, because molten Cr and Mg formed immiscible liquids (like oil and water). Now our recent work (Prasad et al., 2012c) opened up the possibility to produce stainless Mg using conventional metallurgy.

Ultra-HP Mg-X alloys (X = Zn, Sn, Pb) are expected to have low corrosion rates, significantly lower than those of ultra-pure Mg, because the electrochemical properties of the Mg-X alloys may be changed compared with the ultra-HP Mg sufficiently by making the cathodic hydrogen evolution reaction much more difficult than on ultra-HP Mg (Frankel et al., 2013; Pourbaix, 1966). This is an entirely new approach to the development of Mg alloys with corrosion rates lower than those of ultra-HP Mg.

Previously, the intrinsic corrosion behaviour of these Mg-X alloys could not be assessed. Assessment of their intrinsic corrosion behaviour requires Mg-X alloy production (not possible previously in ultra-HP form) and solution heat treatment so that all the alloying elements are in solid solution. But the intrinsic corrosion behaviour is masked by a Fe impurity concentration above the Fe tolerance limit (Liu et al., 2008, 2009; Song & Atrens, 2003). Above the Fe tolerance limit, the corrosion rate is high, because the corrosion behaviour is dominated by the presence of Fe-rich particles. Our recent research (Liu et al., 2009) showed that the Fe tolerance limit is ~2 ppm after heat treatment of the type needed to bring all the alloying elements into solid solution. That means that the Fe impurity element dominates the corrosion behaviour, and the corrosion rate is high, if the Fe concentration is greater than ~2 ppm for a Mg-X alloy, heat-treated to bring all the alloying element into solid solution. And such low Fe concentrations had not been achieved for these alloys previously. These alloys in the as-cast condition were available in HP form, because Fe concentrations between 40 ppm and 180 ppm for cast alloys (Liu et al., 2009) are readily achieved in industrial practice. However, their corrosion rate is high in the cast condition because of galvanic acceleration by the second phase in the two-phase cast microstructure. And after heat treatment of these alloys with Fe contents greater than ~2 ppm, the corrosion behaviour is dominated by the Fe impurity element.

There is no alternative method to produce ultra-high purity (UP) Mg-X alloys, with X = Dy and Gd. The alloying elements (particularly Dy and Gd) are not available in sufficient purity to allow production of UP Mg-X alloys by conventional melting with UP Mg produced by vacuum distillation as produced by Uggowitzer (2012) or Hort (2013). The Fe content is typically over 1000 ppm for master alloys containing Dy and Gd. Furthermore, vacuum distillation cannot produce Mg-X alloys because of the high melting point of the alloying elements.

Mg-X alloys (X = Dy, Gd) have been shown to be biocompatible (Feyerabend et al., 2010; Hort et al., 2010; Yang et al., 2012).

The logical and feasible next step is to produce ultra-pure Mg-X alloys and to study their corrosion behaviour with the expectation that these alloys will produce Mg alloys that have corrosion rates lower than or equal to that of HP Mg.

1.4 Measurement of magnesium corrosion

Weight-loss experiments, properly carried out, are routinely used to measure long-term corrosion rates. Moreover, weight loss can be used to measure both *in vitro* corrosion rates and *in vivo* corrosion rates. The best and most reliable comparison would be if these weight-loss measurements, both *in vivo* and *in vitro*, were carried out by the same experimenter or by the same research group. Then, it would be expected that there would be similar errors in both the in vitro and the *in vivo* measurements, for measurements carried out by the same experimenter. It is more difficult to carry out comparisons between groups.

In contrast, Tafel extrapolation of polarisation curves (typically measured soon after specimen immersion in the solution) (Atrens et al., 2011; Shi et al., 2010; Song & Atrens, 2003) has not yielded values of corrosion rate for Mg that have agreed with other independent measurements of corrosion rate, like weight loss or hydrogen evolution. This is despite the fact that Tafel extrapolation is routinely used for the measurement of the corrosion rate for most corroding metals (Jones, 1992). There are some well-known reasons why Tafel extrapolation has not yielded good measurements for Mg corrosion (Shi & Atrens, 2011; Shi et al., 2012b). In the present circumstances, it must be considered the responsibility of any researcher who wishes to use Tafel extrapolation, to be aware of the known limitation and to provide supporting data from at least one of the other methods for the measurement of the corrosion rate of Mg alloys. Furthermore, any research that provides Mg corrosion data based only on Tafel extrapolation must be viewed critically.

One reason for the lack of agreement between Tafel extrapolation and other measurements of corrosion for Mg alloys is that Tafel extrapolation is often carried out for polarisation curves measured soon after specimen immersion in the solution (Kirkland et al., 2012). The reason for the problem is obvious from the data in Figure 1.7 (Zainal Abidin et al., 2011). The open symbols show the corrosion

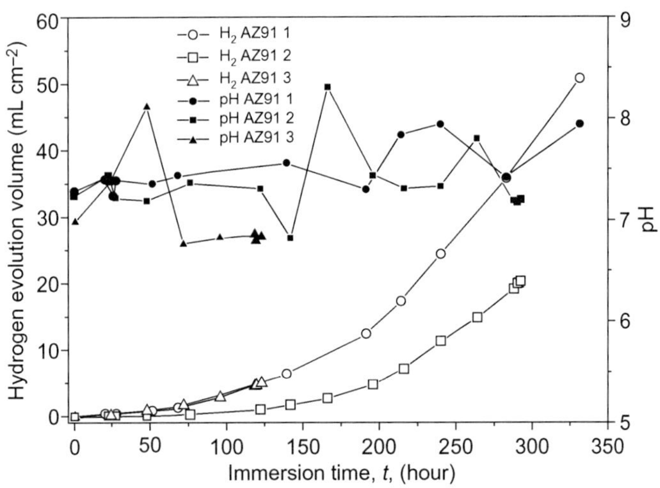

Figure 1.7 Corrosion behaviour as characterised by hydrogen evolution (open symbols), as well as the solution pH (corresponding closed symbols) during the immersion test for AZ91 in Nor's solution (CO_2 – bicarbonate buffered Hank's solution, with the CO_2 partial pressure equal to 0.009 atm). Zainal Abidin et al. (2011).

of AZ91, as characterised by hydrogen evolution, of AZ91 in Nor's solution (CO_2 – bicarbonate buffered Hank's solution, with the CO_2 partial pressure equal to 0.009 atm). This shows that the corrosion rate of AZ91 was initially low during an incubation period, whereas afterwards the corrosion rate accelerated to a steady state value of ~18 mm y^{-1}. Corrosion rates varied by a factor of two for identical specimens tested under identical conditions. This variability is common in the study of Mg corrosion (Atrens et al., 2011; Shi & Atrens, 2011; Zhao et al., 2008a). In cases such as that illustrated in Figure 1.7, there is little relationship between the initial corrosion rate and the steady-state corrosion rate.

Similarly, it has been reported that the corrosion rate increases *in vivo*. Witte et al. (2005) found the *in vivo* corrosion rate for AZ91, implanted into the femora (bone) of guinea pigs, was P_{r6w} E= 0.43 mm y^{-1} (over 6 weeks) and P_{r18w} = 1.3 mm y^{-1} (over 18 weeks; where the subscript "*r*" indicates the corrosion rate was evaluated from the decrease in size of the implant). The corrosion rate increased with increasing exposure time. Similarly, Xu et al. (2007), for as-cast Mg-1.2Mn-1Zn implanted into rat femora, reported corrosion rates of P_{r9w} = 0.3 mm y^{-1} (over 9 weeks) and P_{r18w} = 0.7 mm y^{-1} (over 18 weeks), consistent with those of Witte et al. (2005), which also increased with increasing exposure time.

Similarly, the corrosion rate accelerated for the data for low-purity Mg in Figure 1.1, the HP Mg after heat treatment in Figure 1.3, specimens 19 and 28 in Figure 1.4. In all these cases, the measurement of corrosion rate soon after specimen immersion does not provide a good estimate of the long-term corrosion, which is what is needed if the service life is to be predicted.

It is for this reason that the advice of Kirkland et al. (2012) is not appropriate that the corrosion rate (by Tafel extrapolation of polarisation curves) should be assessed from polarisation curves measured soon after specimen immersion in the solution.

Fishing-line specimens and plug-in specimens were developed (Shi & Atrens, 2011) in response to the known issues (Atrens et al., 2011; Shi et al., 2010; Song & Atrens, 2003) with the measurement of Mg corrosion using electrochemical methods. Fishing-line specimens were designed as the most minimalist possible specimen mount and identified the issues of crevice corrosion for Mg during immersion tests (Shi & Atrens, 2011).

Plug-in specimens were subsequently designed, Figure 1.8 (Shi & Atrens, 2011), to have no crevice and to allow reliable polarisation curves to be measured. Three reasons were identified (Shi & Atrens, 2011) why Tafel extrapolation had previously not yielded corrosion rate measurements in agreement with independent measurements: (1) crevice corrosion can occur in the specimen mount when the specimen is mounted in a metallurgical mount or similar as widely used in Mg corrosion studies, Figure 1.9; (2) Tafel extrapolation has often been used to measure corrosion behaviour soon after specimen immersion. This initial corrosion behaviour may not correlate with steady state corrosion behaviour, either *in vivo* or *in vitro*, see previous section also; (3) Figure 1.10 shows that at times longer than 14 days, the corrosion rate increased significantly, but this increase in corrosion rate was not measured by the Tafel

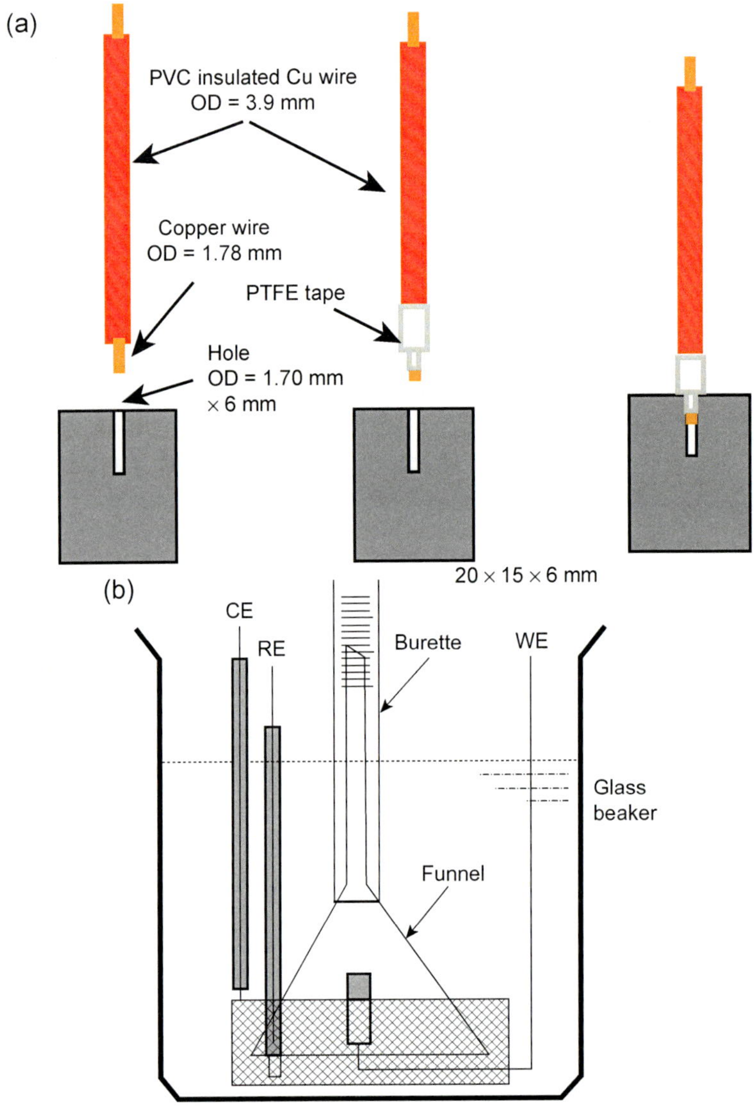

Figure 1.8 (a) The assembly of the plug-in specimen involves the insertion of a slightly oversized bared copper wire into the hole in the side of the specimen. The absence of a crevice is ensured by the fact that the end of the square cut polyvinyl chloride (PVC) insulation is pressed firmly against the specimen side and the use of the PTFE tape as shown. After the immersion test, the average corrosion rate can be measured by weight loss. During the immersion test the instantaneous corrosion rate can be measured using hydrogen evolution, Tafel extrapolation of polarisation curves, and EIS, using the electrochemical cell as illustrated in (b). The average corrosion rate can also be measured from the total volume of evolved hydrogen. Shi and Atrens (2011). (b) The electrochemical cell. The plug-in specimen is the working electrode. The counter electrode is a Pt mesh. Hydrogen evolved from the specimen is collected by the burette, which is initially full of solution. Shi and Atrens (2011).

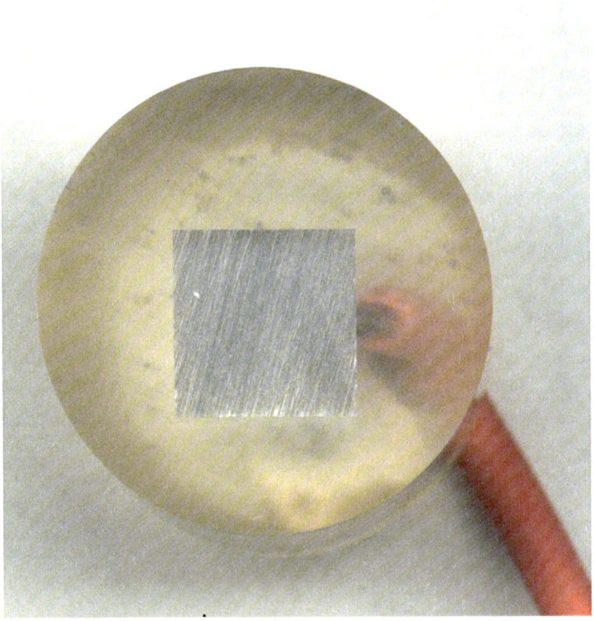

Figure 1.9 Typical mounted Mg specimen. The Mg is mounted in a metallurgical resin, with the electrical connection to the back of the Mg specimen, electrically insulated by the resin. The specimen is quick and easy to prepare. It is easy to prepare the exposed surface to any desired finish. Shi and Atrens (2011).

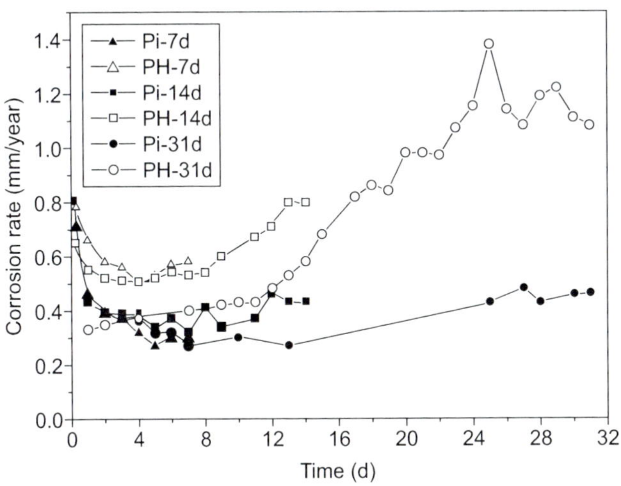

Figure 1.10 Instantaneous corrosion rate evaluated from hydrogen evolution, P_H, data for three plug-in specimens for HP Mg immersed for up to 31 days in 3.5% NaCl saturated with $Mg(OH)_2$. Also presented are the corrosion rate evaluated by Tafel extrapolation of polarisation curves, P_i. Shi and Atrens (2011).

extrapolation. This was attributed to the evolving hydrogen and corrosion products causing a decoupling of the corrosion and electrochemical measurements.

Plug-in specimens enabled the measurement of reliable cathodic polarisation curves for HP Mg in 3.5% NaCl saturated with $Mg(OH)_2$, for an immersion period less than 10 days (Shi & Atrens, 2011). Over the first 10 days, the apparent valence for Mg was 1.45 in support of the Mg corrosion mechanism involving the uni-positive Mg^+ ion (Atrens & Dietzel, 2007; Atrens et al., 2011; Atrens et al., 2011; Song & Atrens, 1999, 2003) and close to the value of 1.5 measured by Petty et al. (1954). The corrosion rate for HP Mg in 3.5% NaCl saturated with $Mg(OH)_2$ was ~0.3 mm y^{-1}, compared with ~1.0 mm y^{-1}, which had previously been measured with mounted specimens (Song & Atrens, 1999, 2000, 2003). The plug-in specimens allowed measurement of a lower corrosion rate due to the absence of crevice corrosion.

The build-up of surface corrosion products and evolved hydrogen for longer immersion times may have caused a decoupling of the corrosion electrochemistry at the corroding surface and that which was measured by the polarisation curve (Shi & Atrens, 2011). There were similar observations (Zainal Abidin et al., 2011) for Mg alloys in Hank's solution at 37 °C.

1.5 *In vitro* and *in vivo* comparison

There are several papers (Staiger et al., 2011; Witte et al., 2006) that point to a lack correspondence between the corrosion rates measured *in vitro* tests in the laboratory and the behaviour *in vivo*.

This issue was considered in part by Zainal Abidin et al. (2011), and their comparison of their data with the literature led to the view that a correspondence between *in vitro* and *in vivo* is only to be expected if the corrosion mechanism is the same in the laboratory tests *in vitro* and *in vivo* in the body. This is the same principle that is valid in all cases when a short-term laboratory test is designed to predict long-term behaviour in service (ASM International, 1987; Baboian, 1995; Jones, 1992). This means that the test method needs to predict long-term or steady-state corrosion rates, and this is particularly an issue for Mg corrosion, in which it is generally found that the corrosion of Mg alloys increases with exposure time to a steady-state corrosion rate (Zainal Abidin et al., 2011). Please also see previous section. It is also necessary that the chemical composition of the *in vitro* testing solution is appropriate. It is particularly important that Mg corrosion is sufficiently understood, so that the critical aspects of the environment are identified and controlled. Our present state of knowledge indicates that it is vital to control the following: (1) temperature (37 °C), (2) chloride concentration, (3) pH (7.4), and (4) buffer system. For tests in which these are controlled to be similar to those in the body, there is a reasonable correspondence between *in vitro* and *in vivo* (Remennik et al., 2011; Walker et al., 2012; Zainal Abidin &, 2012; Zainal Abidin et al., 2013); however, it is clear that more research is needed to understand the corrosion biomechanism, particularly the controlling factors.

1.6 Magnesium corrosion mechanism

The latest critical review of the Mg corrosion mechanism was carried out by Shi et al. (2012b), who concluded that all the data are consistent with the Mg corrosion mechanism involving the uni-positive Mg^+ ion. Subsequently, Kirkland et al. (2012) analysed data on Mg corrosion and proposed that the data were not consistent with the existence of the uni-positive Mg^+ ion, but a simple recalculation of their published data does not support this proposal. More recently, Frankel et al. (2013) proposed that the cathodic reaction (hydrogen evolution) is speeded up on anodic polarisation. However, there was no substantiation for this proposal. See also the recent papers (Atrens, 2013; Cao et al., 2013a,b; Shi et al., 2013b).

1.7 Galvanic corrosion

Figure 1.11(a) presents a typical morphology of galvanic corrosion (Shi et al., 2012b). In this case, an HP Mg plate with a steel insert was placed horizontally in 3.5% NaCl solution saturated with $Mg(OH)_2$. The galvanic corrosion consisted of general corrosion on which was superimposed heterogeneous corrosion. The corrosion was deepest in the Mg at the interface with the steel and decreased with distance from the interface. The rate of galvanic corrosion (expressed as a current density in Figure 1.11(b) (Shi et al., 2012b) decreased as a distance from the interface.

This is typical of galvanic corrosion of Mg (Atrens et al., 2011; Jia et al., 2004, 2005, 2006, 2007; Shi et al., 2012b,c). Galvanic corrosion occurs at a macro-scale when Mg is connected to a less reactive metal in an electrolyte. Mg is the most reactive engineering material, so corrosion of Mg is always accelerated by galvanic coupling.

Galvanic corrosion is always to be considered whenever corrosion protection of a Mg component is by a metallic layer, such as a chromium or stainless steel layer on the Mg component. High corrosion rates can occur in the Mg component if there are any defects or holidays in the coating, particularly as the large area of the coating provides a massive cathode to accelerate the corrosion of the Mg at the coating defect. Such galvanic corrosion will also occur if the coating is damaged in service.

Galvanic corrosion also occurs on a microscale for multiphase Mg alloys, which is one reason why Mg alloys have corrosion rates faster than that of HP Mg. The second phases typically accelerate the corrosion of the alpha–Mg matrix. Such galvanic corrosion can also be caused by Fe rich phases and particles in the microstructure (Cao et al., 2013b; Shi et al., 2013b).

1.8 Summary of important points

1.1 It is vital for researchers in this area to understand key aspects about Mg metallurgy and Mg corrosion. There are unique features concerning Mg that can lead to misleading or erroneous conclusions if appropriate care is not taken. Mg has behaviour that is different from other metals, so there are more traps for the unwary.

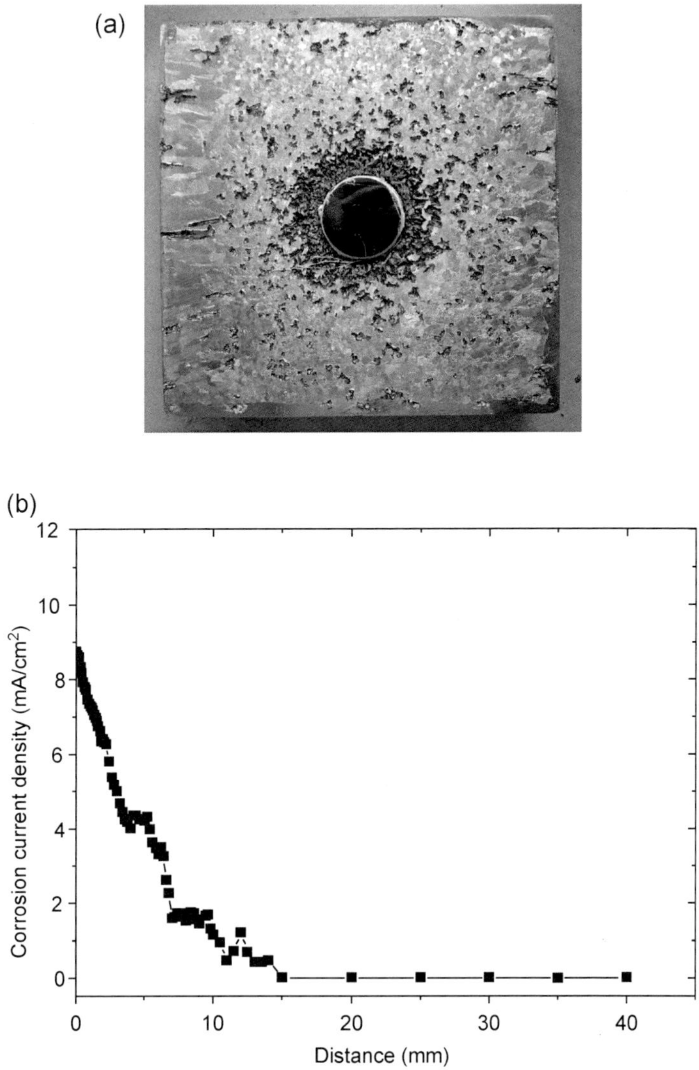

Figure 1.11 (a) Typical morphology of galvanic corrosion of HP Mg due to a circular carbon steel insert, after immersion horizontally in 3.5%NaCl solution saturated with $Mg(OH)_2$ solution. (b) The speed of the galvanic corrosion (as indicated by the corrosion current density) is fastest in the Mg at the interface with the steel, and decreases with distance from the interface Shi et al. (2012).

2.1 Mg has a high-driving force for corrosion weakly checked by the poorly protective corrosion product films that typically form on the surface of Mg in most corrosion situations.
2.2 Mg alloys corrode faster than HP Mg because of the micro-galvanic acceleration caused by second phases.
2.3 HP Mg is Mg with an impurity content below the tolerance limit.
2.4 Purification may produce a Mg alloy with lower corrosion rate than an impure alloy; nevertheless, the HP Mg alloy has a corrosion rate greater than that of HP Mg.

2.5 The measurement of the evolved hydrogen is an easy and convenient means to characterise Mg corrosion.

2.6 It is best practice to use a number on independent measurements of the Mg corrosion rate (hydrogen evolution, weight loss, electrochemical methods [including Tafel extrapolation, electrochemical impedance spectroscopy (EIS), and polarisation resistance]), and to convert each independent measurement to the same units so that they can be easily compared, (mm/y is suggested). The independent measurements should all have the same quantitative value if they are measuring the same quantity.

2.7 The corrosion rate is high for Mg alloys containing impurity elements above their tolerance limit.

2.8 Fe is easily picked up in molten Mg by handling in Fe containers, and Fe is easily picked up during hot working from Fe containing tools (rolling, equal channel angular pressing (ECAP)).

2.9 The Fe tolerance limit for Mg castings is 180 ppm, provided that no Fe rich phases form during solidification.

2.10 If Fe rich phases form during solidification, the corrosion rate of cast Mg can be high, even if the Fe content is 30 ppm.

2.11 The Fe tolerance limit for Mg alloys after heat treatment can be as low as 2 ppm.

3.1 Ultra-HP Mg alloys may have corrosion rates significantly better than the best available current Mg alloys that all have corrosion rates equal to or higher than that of HP Mg.

4.1 Weight-loss measurements are routinely used to measure long-term corrosion rates and can be used to measure both *in vitro* corrosion rates and *in vivo* corrosion rates.

4.2 Tafel extrapolation has not yielded values of corrosion rate for Mg that have agreed with other independent measurements of corrosion rate, like weight loss or hydrogen evolution. This is despite the fact that Tafel extrapolation is routinely used for the measurement of the corrosion rate for many corroding metal systems.

4.3 In the present circumstances, it must be considered the responsibility of any researcher who wishes to use Tafel extrapolation to be aware of the known limitation and to provide supporting data from at least one of the other methods for the measurement of the corrosion rate of Mg alloys.

4.4 Any research that provides Mg corrosion data based only on Tafel extrapolation must be viewed critically.

4.5 One reason for the lack of agreement between Tafel extrapolation and other measurements of corrosion for Mg alloys is that Tafel extrapolation is often carried out for polarisation curves measured soon after specimen immersion in the solution. However, for many cases of Mg corrosion, there is little relationship between the initial corrosion rate and the steady-state corrosion rate.

4.6 A good estimate of long-term corrosion is needed to be able to predict the service life of an implant in the body.

4.7 It is for these reasons that the advice of Kirkland et al. (2012) is not appropriate that the corrosion rate (by Tafel extrapolation of polarisation curves) should be assessed from polarisation curves measured soon after specimen immersion in the solution.

4.8 Plug-in specimens allow measurement of the corrosion rate in a number of independent ways. After the immersion test, the average corrosion rate can be measured by weight loss. During the immersion test, the instantaneous corrosion rate can be measured using hydrogen evolution, Tafel extrapolation of polarisation curves, and electrochemical impedance spectroscopy. The average corrosion rate can also be measured from the total volume of evolved hydrogen.

4.9 The corrosion rate measured using electrochemical methods (Tafel extrapolation, EIS) typically is less than that measured using hydrogen evolution or weight loss, consistent with the Mg corrosion mechanism involving the uni-positive Mg^+ ion.

4.10 The build-up of surface corrosion products and evolved hydrogen may cause a decoupling of the corrosion electrochemistry at the corroding surface and that, which was measured by the polarisation curve.

5.1 A correspondence between *in vitro* and *in vivo* is only to be expected if the corrosion mechanism is the same in the laboratory tests *in vitro*, and *in vivo* in the body.

5.2 Our present state of knowledge indicates that it is vital for an *in vitro* test to control the following: (1) temperature (37 °C), (2) chloride concentration, (3) pH (7.4), and (4) buffer system.

5.3 For tests in which these are controlled to be similar to those in the body, there is a reasonable correspondence between *in vitro* and *in vivo*.

6.1 All existing data are consistent with the Mg corrosion mechanism involving the uni-positive Mg^+ ion.

7.1 Galvanic corrosion of Mg typically consists of general corrosion on which is superimposed heterogeneous corrosion. The galvanic corrosion is typically deepest in the Mg at the interface, and decreased with distance from the interface.

Acknowledgements

This research is supported by the Australian Research Council Centre of Excellence Design of Light Alloys.

References

Alvarez-Lopez, M., Pereda, M. D., del Valle, J. A., Fernandez-Lorenzo, M., Garcia-Alonso, M. C., Ruano, O. A., et al. (2010). Corrosion behaviour of AZ31 magnesium alloy with different grain sizes in simulated biological fluids. *Acta Biomaterialia, 6*, 1763–1771.

Atrens, A. (2013). Overview of the Mg Corrosion Mechanism, in Corrosion, Passivity, and Energy: A Symposium in Honor of Digby D. Macdonald, ECS Transactions, *50*(31), 335–344.

Atrens, A., & Dietzel, W. (2007). The negative difference effect and unipositive Mg^+. *Advanced Engineering Materials, 9*, 292–297.

Atrens, A., Dietzel, W., Srinivasan, P. B., Winzer, N., & Kannan, M. B. (2011). Stress corrosion cracking (SCC) in magnesium alloys. In V. S. Raja, & T. Shoji (Eds.), *Stress corrosion cracking: Mechanisms, materials and application to industrial problems* (pp. 341–380). (Chapter 9) Cambridge: Woodhead.

Atrens, A., Liu, M., & Zainal Abidin, N. I. (2011). Corrosion mechanism applicable to biodegradable magnesium implants. *Materials Science and Engineering B, 176*, 1609–1636.

Atrens, A., Liu, M., Zainal Abidin, N. I., & Song, G. (2011). Corrosion of magnesium (Mg) alloys and metallurgical influence. In G. L. Song (Ed.), *Corrosion of magnesium alloys* (pp. 117–165). (Chapter 3) Cambridge: Woodhead.

Atrens, A., Shi, Z., & Song, G. (2011). Numerical modelling of galvanic corrosion of magnesium (Mg) alloys. In G. L. Song (Ed.), *Corrosion of magnesium alloys* (pp. 455–483). (Chapter 12) Cambridge: Woodhead.

Atrens, A., Winzer, N., & Dietzel, W. (2011). Stress corrosion cracking of magnesium alloys. *Advanced Engineering Materials, 13*, 11–18.

Atrens, A., Winzer, N., Dietzel, W., Srinivasan, P. B., & Song, G. (2011). Stress corrosion cracking of magnesium (Mg) alloys. In G. L. Song (Ed.), *Corrosion of magnesium alloys* (pp. 299–364). (Chapter 8) Cambridge: Woodhead.

ASM International. (1987). *Metals handbook, Ninth Edition, volume 13 corrosion*. ASM International.

Baboian, R. (Ed.). (1995). *Corrosion tests and standards: application and interpretation*. ASTM.

Bakhsheshi-Rad, H. R., Abdul-Kadir, M. R., Idris, M. H., & Farahany, S. (2012). Relationship between the corrosion behavior and the thermal characteristics and microstructure of Mg–0.5Ca–xZn alloys. *Corrosion Science, 64*, 184–197.

Bobby Kannan, M., Dietzel, W., Blawert, C., Atrens, A., & Lyon, P. (2008). Stress corrosion cracking of rare-earth-containing magnesium alloys ZE41, QE22, and Elektron 21 (EV31A) compared with AZ80. *Materials Science and Engineering A, 480*, 529–553.

Bobe, K., Willbold, E., Morgenthal, I., Andersen, O., Studnitzky, T., Nellesen, J. et al. In vitro and in vivo evaluation of biodegradable, open-porous scaffolds made of sintered magnesium W4 short fibres. Acta Biomaterialia, Uncorrected Proof, Available online 28 March 2013 (in press)

Bornapour, M., Muja, N., Shum-Tim, D., Cerruti, M., & Pekguleryuz, M. (2013). Biocompatibility and biodegradability of Mg–Sr alloys: the formation of Sr-substituted hydroxyapatite. *Acta Biomaterialia, 9*, 5319–5330.

Brar, H. S., Ball, J. P., Berglund, I. S., Allen, J. B., & Manuel, M. V. (2013). A study of a biodegradable Mg–3Sc–3Y alloy and the effect of self-passivation on the in vitro degradation. *Acta Biomaterialia, 9*, 5331–5340.

Brar, H. S., Wong, J., & Manuel, M. V. (2012). Investigation of the mechanical and degradation properties of Mg–Sr and Mg–Zn–Sr alloys for use as potential biodegradable implant materials. *Journal of the Mechanical Behavior of Biomedical Materials, 7*, 87–95.

Bruesch, P., Atrens, A., Muller, K., & Neff, H. (1984). Corrosion of stainless steels in chloride solutions—an examination of the passive film by XPS. *Fresenius Zeitschrift fur Analytische Chemie, 319*, 812–821.

Bruesch, P., Muller, K., Atrens, A., & Neff, H. (1985). Corrosion of stainless steels in chloride solution: an XPS investigation of passive films. *Applied Physics A, 38*, 1–18.

Cao, F., Shi, Z., Hofstetter, J., Uggowitzer, P. J., Song, G., Liu, M., & Atrens, A. (2013a). Corrosion of ultra-high-purity Mg in 3.5% NaCl solution saturated with $Mg(OH)_2$. *Corrosion Science, 76*, 78–99.

Cao, F., Shi, Z., Song, G. L., Liu, M., & Atrens, A. (2013b). Corrosion behaviour in salt spray and in 3.5% NaCl solution saturated with $Mg(OH)_2$ of as-cast and solution heat-treated binary Mg-X alloys: X = Mn, Sn, Ca, Zn, Al, Zr, Si, Sr. *Corrosion Science, 76*, 60–97.

Čapek, J., & Vojtěch, D. (2013). Properties of porous magnesium prepared by powder metallurgy. *Materials Science and Engineering: C, 33*, 564–569.

Carboneras, M., Garcia-Alonso, M. C., & Escudero, M. L. (2011). Biodegradation kinetics of modified magnesium based materials in cell culture medium. *Corrosion Science, 53*, 1433–1439.

Castellani, C., Lindtner, R. A., Hausbrandt, P., Tschegg, E., Stanzl-Tschegg, S. E., Zanoni, G., et al. (2011). A Weinberg bone–implant interface strength and osseointegration: biodegradable magnesium alloy versus standard titanium control. *Acta Biomaterialia, 7*, 432–440.

Choudhary, L. (2012). RK Singh Raman magnesium alloys as body implants: fracture mechanism under dynamic and static loadings in a physiological environment. *Acta Biomaterialia, 8*, 916–923.

Choudhary, L., & Singh Raman, R. K. (2013). Mechanical integrity of magnesium alloys in a physiological environment: slow strain rate testing based study. *Engineering Fracture Mechanics, 103*, 94–102.

Erdmann, N., Angrisani, N., Reifenrath, J., Lucas, A., Thorey, F., Bormann, D., et al. (2011). Biomechanical testing and degradation analysis of MgCa0.8 alloy screws: a comparative in vivo study in rabbits. *Acta Biomaterialia, 7*, 1421–1428.

Feyerabend, F., Fischer, J., Holtz, J., Witte, F., Willumeit, R., Druker, H., et al. (2010). Evaluation of short-term effects of rare earth and other elements used in magnesium alloys on primary cell lines. *Acta Biomaterialia, 6*, 1834.

Frankel, G. S., Samaniego, A., & Birbilis, N. (2013). Evolution of hydrogen at dissolving magnesium surfaces. *Corrosion Science*. http://dx.doi.org/10.1016/j.corsci.2013.01.017.

Gastaldi, D., Sassi, V., Petrini, L., Vedani, M., Trasatti, S., & Migliavacca, F. (2011). Continuum damage model for bioresorbable magnesium alloy devices—application to coronary stents. *Journal of the Mechanical Behavior of Biomedical Materials, 4*, 352–365.

Ge, Q., Dellasega, D., Demir, A.G., Vedani, M. The processing of ultrafine-grained Mg tubes for biodegradable stents. Acta Biomaterialia, Corrected Proof, Available online 17 January 2013 (in press)

González, S., Pellicer, E., Fornell, J., Blanquer, A., Barrios, L., Ibáñez, E., et al. (2012). Improved mechanical performance and delayed corrosion phenomena in biodegradable Mg–Zn–Ca alloys through Pd-alloying. *Journal of the Mechanical Behavior of Biomedical Materials, 6*, 53–62.

Gu, X. N., Xie, X. H., Li, N., Zheng, Y. F., & Qin, L. (2012). In vitro and in vivo studies on a Mg–Sr binary alloy system developed as a new kind of biodegradable metal. *Acta Biomaterialia, 8*, 2360–2374.

Han, G., Lee, J. Y., Kim, Y. C., Park, J. H., Kim, D. I., Han, H. S., et al. (2012). Preferred crystallographic pitting corrosion of pure magnesium in Hanks' solution. *Corrosion Science, 63*, 316–322.

Hänzi, A. C., Gerber, I., Schinhammer, M., Löffler, J. F., & Uggowitzer, P. J. (2010). On the in vitro and in vivo degradation performance and biological response of new biodegradable Mg–Y–Zn alloys. *Acta Biomaterials, 6*, 1824–1833.

Heine, B., & Kirchheim, R. (1990). Dissolution rates of iron and chromium and Fe-Cr-alloys in the passive state. *Corrosion Science, 31*, 533–538.

Hort, N. Group leader, Helmholtz Zentrum Geeschacht, German National Lab, private communication 2013.

Hort, N., Uang, Y., Fechner, D., Stormer, M., Blawert, C., Witte, F., et al. (2010). Magnesium alloys as implant materials—principles of property design for Mg-RE alloys. *Acta Biomaterialia, 6*, 1714.

CompuTherm http://www.computherm.com.

Jang, Y., Collins, B. Sankar, J., Yun, Y. Effect of biologically relevant ions on the corrosion products formed on alloy AZ31B: an improved understanding of magnesium corrosion. Acta Biomaterialia, Uncorrected Proof, Available online 25 March 2013 (in press)

Jia, J. X., Atrens, A., Song, G., & Muster, T. (2005). Simulation of galvanic corrosion of magnesium coupled to a steel fastener in NaCl solution. *Materials and Corrosion, 56*, 468–474.

Jia, J. X., Song, G., & Atrens, A. (2005). Boundary element predictions of the influence of the electrolyte on the galvanic corrosion of AZ91D coupled to steel. *Materials and Corrosion, 56*, 259–270.

Jia, J. X., Song, G. L., & Atrens, A. (2006). Influence of geometry on galvanic corrosion of AZ91D coupled to steel. *Corrosion Science, 48*, 2133–2153.

Jia, J. X., Song, G., & Atrens, A. (2007). Experimental measurement and computer simulation of galvanic corrosion of magnesium coupled to steel. *Advanced Engineering Materials, 9*, 65–74.

Jia, J. X., Song, G., Atrens, A., StJohn, D., Baynham, J., & Chandler, G. (2004). Evaluation of the BEASY program using linear and picewise linear approaches for the boundary conditions. *Materials and Corrosion, 55*, 845–852.

Jin, S., & Atrens, A. (1987). ESCA—studies of the structure and composition of the passive film formed on stainless steels by various immersion times in 0.1M NaCl solution. *Applied Physics A, 42*, 149–165.

Jin, S., & Atrens, A. (1988). ESCA—studies of the surface film formed on stainless steels by exposure to 0.1M NaCl solution at various controlled potentials. *Applied Physics A, 46*, 51–65.

Jones, D. A. (1992). *Principles and prevention of corrosion*. Prentice Hall.

Kalb, H., Rzany, A., & Hensel, B. (2012). Impact of microgalvanic corrosion on the degradation morphology of WE43 and pure magnesium under exposure to simulated body fluid. *Corrosion Science, 57*, 122–130.

Kirchheim, R., Heine, B., Fischmeister, H., Hofmann, S., Knote, H., & Stolz, U. (1989). The Passivity of iron-chromium alloys. *Corros. Sci, 29*, 899–917.

Kirkland, N. T., Birbilis, N., & Staiger, M. P. (2012). Assessing the corrosion of biodegradable magnesium implants: a critical review of current methodologies and their limitations. *Acta Biomaterialia, 8*, 925–936.

Kirkland, N. T., Williams, G., & Birbilis, N. (2012). Observations of the galvanostatic dissolution of pure magnesium. *Corrosion Science, 65*, 5–9.

Kraus, T., Fischerauer, S. F., Harenzi, A. C., Uggowitzer, P. J., Löffler, J. F., & Weinberg, A. M. (2012). Magnesium alloys for temporary implants in osteosynthesis: in vivo studies of their degradation and interaction with bone. *Acta Biomaterialia, 8*, 1230–1238.

Kuhlmann, J., Bartsch, I., Willbold, E., Schuchardt, S., Holz, O., Hort, N., et al. Fast escape of hydrogen from gas cavities around corroding magnesium implants. Acta Biomaterialia, Corrected Proof, Available online 12 October 2012 (in press)

Kubásek, J., Vojtěch, D., Lipov, J., & Ruml, T. (May 2013). Structure, mechanical properties, corrosion behavior and cytotoxicity of biodegradable Mg–X (X = Sn, Ga, In) alloys. *Materials Science and Engineering: C, 33*, 2421–2432.

Lei, T., Tang, W., Cai, S. H., Feng, F. F., & Li, N. F. (2012). On the corrosion behavior of newly developed biodegradable Mg-based metal matrix composites produced by in situ reaction. *Corrosion Science, 54*, 270–277.

Li, Y., Wen, C., Mushahary, D., Sravanthi, R., Harishankar, N., Pande, G., et al. (2012). Mg–Zr–Sr alloys as biodegradable implant materials. *Acta Biomaterialia, 8*, 3177–3188.

Li, H.F., Xie, X.H., Zhao, K., Wang, Y.B., Zheng, Y.F., Wang, W.H., et al. In vitro and in vivo studies on biodegradable CaMgZnSrYb high-entropy bulk metallic glass, Acta Biomaterialia. Corrected Proof, Available online 1 February 2013 (in press)

Li, N., Zheng, Y. Novel magnesium alloys developed for biomedical application: a review, Journal of Materials Science & Technology, Corrected Proof, Available online 9 February 2013 (in press)

Lim, A. S., & Atrens, A. (1991). ESCA studies of Si-Fe. *Applied Physics A, 53*, 273–281.

Lim, A. S., & Atrens, A. (1992a). ESCA studies of Fe-Ti. *Applied Physics A, 54*, 500–507.

Lim, A. S., & Atrens, A. (1992b). ESCA studies of Ni-Cr. *Applied Physics A, 54*, 343–349.

Lim, A. S., & Atrens, A. (1992c). ESCA studies of Cr-Co. *Applied Physics A, 54*, 270–278.

Liu, M., Qiu, D., Zhao, M. C., Song, G., & Atrens, A. (2008). The effect of crystallographic orientation on the active corrosion of pure magnesium. *Scripta Materialia, 58*, 421–424.

Liu, M., Schmutz, P., Zanna, S., Seyeux, A., Ardelean, H., Song, G., et al. (2010). Electrochemical reactivity, surface composition and corrosion mechanisms of the complex metallic alloy Al_3Mg_2. *Corrosion Science, 52*, 562–578.

Liu, M., Schmutz, P., Uggowitzer, P. J., Song, G., & Atrens, A. (2010). The influence of Yttrium (Y) on the corrosion of Mg-Y binary alloys. *Corrosion Science, 52*, 3687–3701.

Liu, M., Shih, D. S., Parish, C., & Atrens, A. (2012). The ignition temperature of Mg alloys WE43, AZ31 and AZ91. *Corrosion Science, 54*, 139–142.

Liu, M., Uggowitzer, P. J., Nagasekhar, A. V., Schmutz, P., Easton, M., Song, G., et al. (2009). Calculated phase diagrams and the corrosion of die-cast Mg-Al alloys. *Corrosion Science, 51*, 602–619.

Liu, M., Uggowitzer, P. J., Schmutz, P., & Atrens, A. (2008). Calculated phase diagrams, iron tolerance limit and corrosion of Mg-Al alloys. *JOM, 60*(12), 39–44.

Liu, C. L., Wang, Y. J., Zeng, R. C., Zhang, X. M., Huang, W. J., & Chu, P. K. (2010c). In vitro corrosion degradation behaviour of Mg–Ca alloy in the presence of albumin. *Corrosion Science, 52*, 3341–3347.

Liu, M., Zanna, S., Ardelean, H., Frateur, I., Schmutz, P., Song, G., et al. (2009). A first quantitative XPS study of the surface films formed, by exposure to water, on Mg and on the Mg-Al intermetallics: Al_3Mg_2 and $Mg_{17}Al_{12}$. *Corrosion Science, 51*, 1115–1127.

Mao, L., Yuan, G., Wang, S., Niu, J., Wu, G., & Ding, W. (2012). A novel biodegradable Mg–Nd–Zn–Zr alloy with uniform corrosion behavior in artificial plasma. *Materials Letters, 88*, 1–4.

Minárik P., Král., R., Janeček., M. Effect of ECAP processing on corrosion resistance of AE21 and AE42 magnesium alloys, Applied Surface Science, Corrected Proof, Available online 27 December 2012 (in press)

Olefjord, I., & Elfstrom, B.-O. (1982). The composition of the surface during passivation of stainless steels. *Corros, 38*, 47–52.

Petty, R. L., Davidson, A. W., & Kleinberg, J. (1954). The anodic oxidation of magnesium metal: evidence for the existence of unipositive magnesium. *Journal of the American Chemical Society, 76*, 363–366.

Pourbaix, M. (1966). Atlas of electrochemical equilibria in aqueous solutions. *NACE, Cebelcor.*

Prasad, A., Shi, Z., & Atrens, A. (2012a). Flammability of Mg-X binary alloys. *Advanced Engineering Materials, 14*, 772–784.

Prasad, A., Shi, Z., & Atrens, A. (2012b). Influence of Al and Y on the ignition and flammability of Mg alloys. *Corrosion Science, 54*, 153–163.

Prasad, A., Uggowitzer, P. J., Shi, Z., & Atrens, A. (2012c). Production of high purity magnesium alloys by melt purification with Zr. *Advanced Engineering Materials, 14*, 477–490.

Pu, Z., Song, G. L., Yang, S., Outeiro, J. C., Dillon, O. W., Jr, Puleo, D. A., et al. (2012). Grain refined and basal textured surface produced by burnishing for improved corrosion performance of AZ31B Mg alloy. *Corrosion Science, 57*, 192–201.

Qian, M., Zheng, L., Graham, D., Frost, M. T., & St. John, D. H. (2001). Settling of undissolved zirconium particles in pure magnesium melts. *J Light Metals, 1*, 157–165.

Qiao, Z., Shi, Z., Hort, N., Zainal Abidin, N., & Atrens, A. (2012). Corrosion behaviour of a nominally high purity Mg ingot produced by permanent mould direct chill casting. *Corrosion Science, 61*, 185–207.

Qiu, J. F. X., Niu, X., Tian, Z., Sun, W., Liu, X., Li, Y., et al. (2013). Microstructure, mechanical properties, in vitro degradation and cytotoxicity evaluations of Mg–1.5Y–1.2Zn–0.44Zr alloys for biodegradable metallic implants. *Materials Science and Engineering: C, 33*, 2345–2352.

Rad, H. R. B., Idris, M. H., Kadir, M. R. A., & Farahany, S. (2012). Microstructure analysis and corrosion behavior of biodegradable Mg–Ca implant alloys. *Materials & Design, 33*, 88–97.

Remennik, S., Bartsch, I., Willbold, E., & Witte, F. (2011). New, fast corroding high-ductility Mg-Bi-Ca and Mg-Bi-Si alloys, with no clinically observable gas formation in bone implants. *Materials Science and Engineering B, 176*, 1653–1659.

Salunke, P., Shanov, V., & Witte, F. (2011). High purity biodegradable magnesium coating for implant application. *Materials Science and Engineering: B, 176*, 1711–1717.

Scheideler, L., Füger, C., Schille, C., Rupp, F., Wendel, H. P., Hort, N. et al, Comparison of different in vitro tests for biocompatibility screening of Mg alloys. Acta Biomaterialia, Corrected Proof, Available online 18 February 2013 (in press)

Schneider, A., Kuron, D., Hofmann, S., & Kirchheim, R. (1990). AES analysis of pits and passive films formed on Fe-Cr Fe-Mo and Fe-Cr-Mo alloys. *Corrosion Science, 31*, 191–196.

Seyeux, A., Liu, M., Schmutz, P., Song, G., Atrens, A., & Marcus, P. (2009). ToF-SIMS depth profile of the surface film on pure magnesium formed by immersion in pure water and the identification of magnesium hydride. *Corrosion Science, 51*, 1883–1886.

Shaw, B. A., & Wolfe, R. C. (2005). Corrosion of magnesium and magnesium-base alloys. *Corrosion: Materials, 13B*, 205–227. ASM Handbook, ASM International.

Shi, Z., Liu, M., & Atrens, A. (2010). Measurement of the corrosion rate of magnesium alloys using Tafel extrapolation. *Corrosion Science, 52*, 579–588.

Shi, Z., & Atrens, A. (2013a). Comments on the paper entitled "Observations of the galvanostatic dissolution of pure magnesium" by N.T. Kirkland, G. Williams and N. *Birbilis. Corrosion Science, 77*, 403–406.

Shi, Z., & Atrens, A. (2011). An innovative specimen configuration for the study of Mg Corrosion. *Corrosion Science, 53*, 226–246.

Shi, Z., Cao, F., Song, G. L., Liu, M., & Atrens, A. (2013b). Corrosion behaviour in salt spray and in 3.5% NaCl solution saturated with $Mg(OH)_2$ of as-cast and solution heat-treated binary Mg-RE alloys: RE = Ce, La, Nd, Y, Gd. *Corrosion Science, 76*, 98–118.

Shi, Z., Jia, J. X., & Atrens, A. (2012c). Galvanostatic anodic polarisation curves and galvanic corrosion of AZ31B in 0.01M Na_2SO_4 saturated with $Mg(OH)_2$. *Advanced Engineering Materials, 14*, 324–334.

Shi, Z., Jia, J. X., & Atrens, A. (2012b). Galvanostatic anodic polarisation curves and galvanic corrosion of high purity Mg in 3.5%NaCl saturated with $Mg(OH)_2$. *Corrosion Science, 60*, 296–308.

Shi, Z., Prasad, A., & Atrens, A. (2012a). Plug-in specimens for the measurement of the corrosion rate of Mg alloys. *JOM, 64*(6), 657–663.

Song, G., & Atrens, A. (2000). Chapter 4 corrosion of non-ferrous alloys. III. Magnesium alloys. In M. Schütze (Ed.), *Corrosion and Environmental Degradation* (pp. 131–171). Wiley.

Song, G., & Atrens, A. (1999). Corrosion mechanisms of magnesium alloys. *Advanced Engineering Materials, 1*, 11–33.

Song, G., & Atrens, A. (2003). (Invited review) Understanding magnesium corrosion mechanism: a framework for improved alloy performance. *Advanced Engineering Materials, 5*, 837–858.

Song, G. L., Atrens, A., & Dargusch, M. (1999). Influence of microstructure on the corrosion of diecast AZ91D. *Corrosion Science, 41*, 249–273.

Song, G., Atrens, A., & St. John, D. (2001). An hydrogen evolution method for the estimation of the corrosion rate of magnesium alloys. In J. N. Hryn (Ed.), *Magnesium Technology 2001* (pp. 255–262). Wiley.

Song, G. L., Atrens, A., St. John, D. H., Nairn, J., & Lang, Y. (1997). Electrochemical corrosion of pure magnesium in 1N NaCl. *Corrosion Science, 39*, 855–875.

Song, G. L., Atrens, A., St. John, D., Wu, X., & Nairn, J. (1997). The anodic dissolution of magnesium in chloride and sulphate solutions. *Corrosion Science, 39*, 1981–2004.

Song, G. L., Atrens, A., Wu, X., & Zhang, B. (1998). Corrosion behaviour of AZ21, AZ501 and AZ91 in sodium chloride. *Corrosion Science, 40*, 1769–1791.

Song, R. G., Blawert, C., Dietzel, W., & Atrens, A. (2005). A study of the stress corrosion cracking and hydrogen embrittlement of AZ31 magnesium alloy. *Materials Science and Engineering A, 399*, 308–317.

Staiger, M. P., Feyerabend, F., Willumeit, R., Sfeir, G. S., Zheng, Y. F., Virtanen, S., et al. (2011). Summary of the panel discussions at the 2nd symposium on biodegradable metals, Maratea, Italy, 2010. *Mater. Sci. Eng. B, 176*, 1596–1599.

Sun, H., Li, C., & Fang, W. (2011). Corrosion behavior of extrusion-drawn pure Mg wire immersed in simulated body fluid. *Transactions of Nonferrous Metals Society of China, 21*, s258–s261.

Tan, L., Yu, X., Wan, P., Yang, K. Biodegradable materials for bone repairs: a review. Journal of Materials Science & Technology, Corrected Proof, Available online 21 March 2013 (in press)

Uggowitzer, P. (2012). Prof at ETHZ, patent application EP.6.

Uhlenhaut, D.I. (2008) PhD ETHZ Nr 17958 (2003–2008), private communication.

Virtanen, S. (2011). Biodegradable Mg and Mg alloys: corrosion and biocompatibility. *Materials Science and Engineering B, 176*, 1600–1608.

Vojtěch, D., Kubásek, J., Šerák, J., & Novák, P. (2011). Mechanical and corrosion properties of newly developed biodegradable Zn-based alloys for bone fixation. *Acta Biomaterialia, 7*, 3515–3522.

Walker, J., Shadanbaz, S., Kirkland, N. T., Stace, E., Woodfield, T., Staiger, M. P., et al. (2012). Magnesium alloys: predicting in vivo corrosion with in vitro immersion testing. *J Biomedical Materials Research B: Applied Biomaterials, 100b*(4), 1134–1141.

Wang, Y., Lim, C. S., Lim, C. V., Yong, M. S., Teo, E. K., & Moh, L. N. (April 2011). In vitro degradation behavior of M1A magnesium alloy in protein-containing simulated body fluid. *Materials Science and Engineering: C, 31*, 579–587.

Willbold, E., Kaya, A. A., Kaya, R. A., Beckmann, F., & Witte, F. (2011). Corrosion of magnesium alloy AZ31 screws is dependent on the implantation site. *Materials Science and Engineering: B, 176*, 1835–1840.

Winzer, N., Atrens, A., Dietzel, W., Raja, V. S., Song, G., & Kainer, K. U. (2008). Characterisation of stress corrosion cracking (SCC) of Mg-Al alloys. *Materials Science and Engineering A, 488*, 339–351.

Winzer, N., Atrens, A., Dietzel, W., Song, G., & Kainer, K. U. (2008c). Comparison of the linearly increasing stress test and the constant extension rate test in the evaluation of transgranular stress corrosion cracking of magnesium. *Materials Science and Engineering A, 472*, 97–106.

Winzer, N., Atrens, A., Dietzel, W., Song, G., & Kainer, K. U. (2007a). Evaluation of the delayed hydride cracking mechanism for transgranular stress corrosion cracking of magnesium alloys. *Materials Science and Engineering A, 466*, 18–31.

Winzer, N., Atrens, A., Dietzel, W., Song, G., & Kainer, K. U. (2008b). The fractography of stress corrosion cracking (SCC) of Mg-Al alloys. *Metallurgical and Materials Transactions A, 39*, 1157.

Winzer, N., Atrens, A., Dietzel, W., Song, G., & Kainer, K. U. (2007b). Magnesium stress corrosion cracking. *Transactions of Nonferrous Metals Society of China, 17*, S150–S155. Part A Sp. Iss. 1 November 2007.

Winzer, N., Atrens, A., Dietzel, W., Song, G., & Kainer, K. U. (2007c). Stress corrosion cracking in magnesium alloys: characterisation and prevention. *JOM, 59*(8), 49–53.

Winzer, N., Atrens, A., Dietzel, W., Song, G., & Kainer, K. U. (2008a). Stress corrosion cracking (SCC) in Mg-Al alloys studied using compact specimens. *Advanced Engineering Materials, 10*, 453–458.

Winzer, N., Atrens, A., Song, G., Ghali; E., Dietzel, W., Kainer, K. U., et al. (2005). A critical review of the stress corrosion cracking (SCC) of magnesium alloys. *Advanced Engineering Materials, 7*, 659–693.

Witte, F. (2010). The history of biodegradable magnesium implants: a review. *Acta Biomaterial, 6*, 1680–1692.

Witte, F., Fischer, J., Nellesen, J., Crostack, H. A., Kaese, V., Pisch, A., et al. (2006). In vitro and in vivo corrosion measurements of magnesium alloys. *Biomater, 27*, 1013–1018.

Witte, F., Kaese, V., Haferkamp, H., Switzer, E., Meyer-Lindenberg, A., Wirth, C. J., et al. (2005). In vivo corrosion of four magnesium alloys and the associated bone response. *Biomaterials, 26*, 3557–3563.

Xu, L., Yu, G., Zhang, E., Pan, F., & Yang, K. (2007). In vivo corrosion behaviour of Mg-Mn-Zn alloy for bone implant application. *J. Biomed. Mater. Res, A 83*, 703–711.

Yang, L., Hort, N., Laipple, D., Hoeche, D., Huang, Y., Kainer, K. U., et al. (2012a). Element distribution in the corrosion layer and cytotoxicity of alloy Mg-10Dy during in vitro biodegradation. *Acta Biomaterialia*. http://dx.doi.org/10.1016/j.actbio.2012.10.001.

Yang, L., Huang, Y., Feyerabend, F., Willumeit, R., Mendis, C., Kainer, K. U., et al. Microstructure, mechanical and corrosion properties of Mg–Dy–Gd–Zr alloys for medical applications. Acta Biomaterialia, Corrected Proof, Available online 20 March 2013 (in press)

Yang, L., Huang, Y., Feyerabend, F., Willumeit, R., Kainer, K. U., & Hort, N. (2012b). Influence of ageing treatment on microstructure, mechanical and bio-corrosion properties of Mg–Dy alloys. *Journal of the Mechanical Behavior of Biomedical Materials, 13*, 36–44.

N. I. Zainal Abidin, PhD. Thesis, The University of Queensland, 2012.

Zainal Abidin, N. I., Atrens, A. D., Martin, D., & Atrens, A. (2011). The corrosion of high purity Mg, Mg2Zn0.2Mn, ZE41 and AZ91 in Hank's solution at 37 °C. *Corrosion Science, 53*, 3542–3556.

Zainal Abidin, N. I., Martin, D., & Atrens, A. (2011). Corrosion of high purity Mg, AZ91, ZE41, and Mg2Zn0.2Mn in Hank's solution at room temperature. *Corrosion Science, 53*, 862–872.

Zainal Abidin, N. I., Rolfe, B., Owen, H., Malisano, J., Martin, D., Hofstetter, J., Uggowitzer, P. J., & Atrens, A. (2013). The in vivo and in vitro corrosion of high-purity magnesium and magnesium alloys wz21 and az91. *Corrosion Science, 75*, 354–366.

Zberg, B., Uggowitzer, P. J., & Loeffler, J. F. (2009). MgZnCa glasses without clinically observable hydrogen for biodegradable implants. *Nature Material, 8*, 887–891.

Zhang, X., Wang, Z., Yuan, G., & Xue, Y. (2012). Improvement of mechanical properties and corrosion resistance of biodegradable Mg–Nd–Zn–Zr alloys by double extrusion. *Materials Science and Engineering: B, 177*, 1113–1119.

Zhang, X., Yuan, G., Mao, L., Niu, J., Fu, P., & Ding, W. (2012). Effects of extrusion and heat treatment on the mechanical properties and biocorrosion behaviors of a Mg–Nd–Zn–Zr alloy. *Journal of the Mechanical Behavior of Biomedical Materials, 7*, 77–86.

Zhao, M. C., Liu, M., Song, G. L., & Atrens, A. (2008b). Influence of pH and chloride ion concentration on the corrosion of Mg alloy ZE41. *Corrosion Science, 50*, 3168–3178.

Zhao, M. C., Liu, M., Song, G., & Atrens, A. (2008a). Influence of the-phase morphology on the corrosion of the Mg alloy AZ91. *Corrosion Science, 50*, 1939–1953.

Zhao, X., Shi, L., & Xu, J. (2013). Biodegradable Mg–Zn–Y alloys with long-period stacking ordered structure: optimization for mechanical properties. *Journal of the Mechanical Behavior of Biomedical Materials, 18*, 181–190.

Zhou, Y. L., Luo, D. M., Hu, W. Y., Li, Y., Hodgson, P., & Wen, C. (2011). Microstructural characterization and mechanical properties of Mg-Zr-Ca alloys prepared by hot-extrusion for biomedical applications. *Advance Science Letters, 4*, 2860–2863.

Zhou, W., Shen, T., & Aung, N. N. (2010). Effect of heat treatment on corrosion behaviour of magnesium alloy AZ91D in simulated body fluid. *Corrosion Science, 52*, 1035–1041.

Zong, Y., Yuan, G., Zhang, X., Mao, L., Niu, J., & Ding, W. (2012). Comparison of biodegradable behaviors of AZ31 and Mg–Nd–Zn–Zr alloys in Hank's physiological solution. *Materials Science and Engineering: B, 177*, 395–401.

Surface modification of magnesium and its alloys for biomedical applications: opportunities and challenges

2

T.S.N. Sankara Narayanan, Il-Song Park, Min-Ho Lee
Chonbuk National University, Jeonju, Republic of Korea

2.1 Introduction

The development of biodegradable implants is indeed fascinating, and among the various types of materials used in this regard, magnesium (Mg) and its alloys assume significance (Song & Song, 2007; Staiger, Pietak, Huadmai, & Dias., 2006; Virtanen, 2011; Witte, 2010; Witte et al., 2008; Zeng, Dietzel, Witte, Hort, & Blawert, 2008). Its biodegradable property in combination with its nontoxic nature makes Mg and its alloys potential candidate materials for biodegradable implants. However, the rapid corrosion, generation of a large volume of hydrogen gas, accumulation of hydrogen bubbles in gas pockets adjacent to the implant, and an increase in the local pH of body fluid are the most critical limitations in using Mg and its alloys as implant materials. A detailed account of revolutionizing Mg-based biomaterials is addressed in Chapter 1 of this book. If Mg has to be used as a biodegradable implant material, then it should meet the following requirements.

- It should maintain sufficient mechanical strength and integrity until the affected part of the body is healed.
- It should exhibit resistance to corrosion in body fluid during the initial periods of implantation and subsequently corrode in a controlled and uniform fashion.
- Its corrosion products should not exceed the acceptable absorption level of the human body.

Reducing the rate of corrosion of Mg is the most appropriate strategy because a low corrosion rate of an Mg implant implies a decrease in the extent of hydrogen evolution and alkalization, which enable the human body to gradually absorb or consume the corrosion products. Alloying and surface treatment/coatings were explored to reduce the corrosion rate and to improve the biocompatibility of Mg and its alloys. Development of Mg alloys with better corrosion resistance, mechanical integrity, and biocompatibility is a challenging task. The ability to modify surface properties to achieve enhanced corrosion resistance and biocompatibility, while retaining the bulk properties of materials, is the unique advantage of surface modification methods. The importance of engineering implant surfaces to make them more biocompatible was addressed recently by Bauer, Schmuki, Mark, and Park (2013). The details of various

Surface Modification of Magnesium and its Alloys for Biomedical Applications. http://dx.doi.org/10.1016/B978-1-78242-077-4.00002-4

surface modification methods and their importance for biomedical applications of Mg alloys are addressed in recent reviews (Hornberger, Virtanen, & Boccaccini, 2012; Wang, Tang et al., 2012; Yang, Cui, & Lee, 2011). This chapter aims to address the role of various methods available for engineering the surface of Mg and its alloys toward the development of degradable biomaterials, with a focus on their opportunities and limitations.

2.2 Thermal, hydrothermal, and alkaline heat treatment-based surface modifications

2.2.1 Self-passivation

The protective nature of the oxide film that naturally forms on metallic materials and its adherence to the respective substrate can be understood from the Pilling–Bedworth (PB) ratio of the corresponding oxide. Because of the low PB ratio, magnesium oxide (MgO), which is the oxide layer that naturally forms on Mg, is not protective and has a low stability in aqueous solutions. In the case of Mg alloys, if the alloying elements possess sufficient driving force for diffusion to the surface and thermodynamic stability, then their selective oxidation could be an effective way to develop a protective oxide film on the surface of Mg alloys. This concept forms the basis of imparting self-passivation to Mg alloys. Among the various alloying elements, yttrium and scandium oxides possess high thermodynamic stability. Brar, Ball, Berglund, Allen, and Manuel (2013) studied the protective ability of the oxide film formed on an Mg-3Sc-3Y alloy after thermal oxidation at 500 °C for various durations, from 0.5 to 25 h. The oxide layer formed on this alloy was mainly composed of scandium oxide than yttrium oxide, with a higher volume fraction of scandium oxide. Oxidation of the Mg-3Sc-3Y alloy for 24 h enabled a decrease in the rate of degradation by almost 100-fold and a negligible degradation rate in Hank's balanced salt solution at 37 °C for up to 23 days (Figure 2.1). The degradation products of the Mg-3Sc-3Y alloy showed no significant toxicity to osteoblastic cells and promoted cell attachment and proliferation. The findings of Brar et al. (2013) confirmed that selective oxidation through alloying Mg with scandium and yttrium is indeed an effective method to control the rate of degradation of Mg alloys. The absence of any significant toxicity and the promotion of cell attachment and proliferation makes this type of modification more promising. Brar et al. (2013) suggested that this approach is suitable for applications in which the implant geometry restricts the use of coating technologies. This approach, however, could not be extended to all types of Mg alloys.

Mg alloys containing aluminum (Al) and zinc (Zn), such as the AZ91 Mg alloy, contain a primary α-phase, coarse β-phase particles, and a fine ($\alpha + \beta$) eutectic phase. If the volume fraction of β-phase particles is small, then they serve as a galvanic cathode to accelerate the corrosion of the α-phase. However, if a continuous β-phase network is formed on the surface, then the β-phase acts as a barrier, resisting the corrosion of the Mg alloy. This conceptual approach of using a microstructural modification

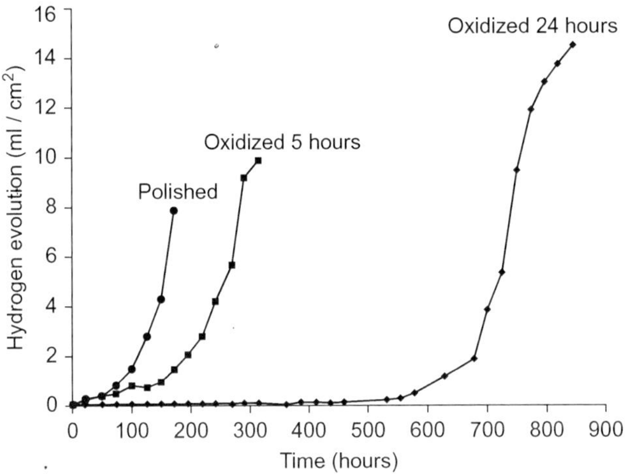

Figure 2.1 The hydrogen evolution behavior of a polished and oxidized Mg-3Sc-3Y alloy in Hank's balanced salt solution at 37 °C (electrolyte volume-to-sample surface area ratio is ~150).
Adapted from Brar et al. (2013) with permission from Elsevier.

of the AZ91 Mg alloy as the basis for imparting a self-passivation mechanism after the corrosion process was explored by Zhao et al. (2012). According to them, the microstructural modification of the AZ91 Mg alloy could be achieved by heat treatment at 410 °C for 24 h, followed by furnace cooling up to 213 °C for about 90 min and subsequent quenching in water at 10 °C. Heat treatment prompted dissolution of the fine lamellar ($\alpha + \beta$) eutectic phase and the coarse β-phase particles in the α-phase matrix, resulting in the formation of an Al-containing, supersaturated α-phase matrix, while subsequent cooling in the furnace induced precipitation of Al atoms from the supersaturated matrix to finally form fine, continuous β precipitates in the fine lamellar ($\alpha + \beta$) microconstituent. This microstructural modification decreased the corrosion current density of the AZ91 Mg alloy from 2×10^{-4} to 5.4×10^{-5} A/cm^2 in simulated body fluid (SBF) and from 1.9×10^{-6} to 7×10^{-7} A/cm^2 in cell culture medium. In addition, this modification increased the number of MC3T3-E1 preosteoblasts after 5 h of cell culture, indicating its ability to promote higher initial cell adhesion during the early stages of cell culturing. The larger number of adherent cells on the modified AZ91 Mg alloy surface after 1, 3, and 7 days of cell culture confirmed its improved cell viability and cell proliferation.

Wu, Zhao, Zhang, Ibrahim, and Chu (2013) applied the similar concept for altering the microstructure of a Mg-7.5Al-0.8Zn-0.2Mn alloy by a suitable combination of heat treatment and quenching, performed in two stages, to induce the self-passivation mechanism after corrosion in cell culture medium. The redistribution of the β-phase enabled quicker formation of galvanic cells, which increased the rate of corrosion of the Mg alloy during the initial stages of immersion in cell culture medium. Although the primary corrosion product of magnesium hydroxide ($Mg(OH)_2$) is not protective,

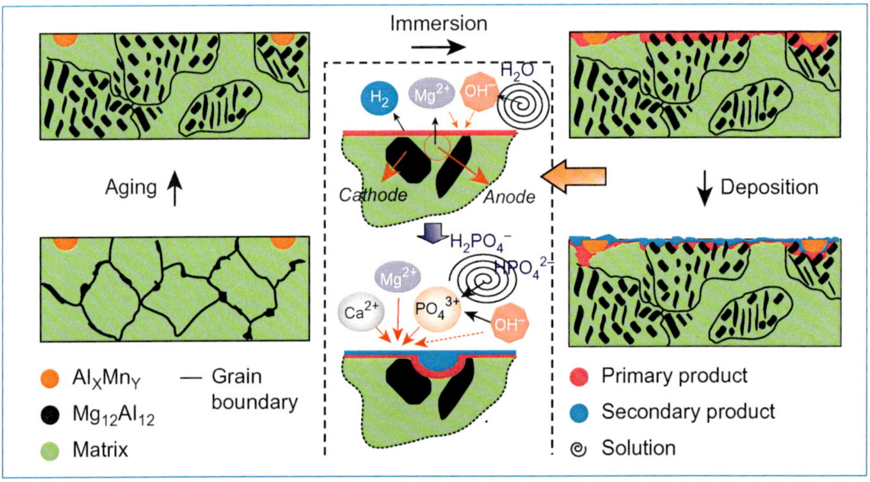

Figure 2.2 Schematic diagram of the self-protection mechanism of an Mg-7.5Al-0.8Zn-0.2Mn alloy induced by a suitable combination of heat treatment and quenching. The Mg alloy samples were heated from 23 to 440 °C at a rate of 10 °C/min, soaked at 440 °C for 24 h in air, and quenched in water at 13 °C. Subsequently, they were heated from 23 to 200 °C, soaked at 200 °C for 48 h, and quenched in water at 13 °C.
Adapted from Wu, Zhao et al. (2013) with permission from Elsevier.

precipitation of the secondary corrosion products such as calcium phosphate (or $Ca_{10}(PO_4)_6(OH)_2$) and $Mg_3(PO_4)_2$ over the entire surface of the Mg alloy provided good surface coverage and reduced the electrochemical activity of the alloy in the cell culture medium. In addition, the modified Mg alloy promoted attachment of osteoblasts on the surface and better cell spreading after 1, 3, and 7 days of cell culture. The schematic of the self-protection mechanism of the Mg-7.5Al-0.8Zn-0.2Mn alloy induced by a suitable combination of heat treatment and quenching is depicted in Figure 2.2.

2.2.2 Hydrothermal treatment

Modification of the surface of pure Mg and the AZ91 Mg alloy by hydrothermal treatment using either deionized water or sodium hydroxide (NaOH) was explored by many researchers (Feng et al., 2013; Gupta, Mensah-Darkwa, & Kumar. 2014; Zhu, Wu, Zhang, & Zhao, 2011; Zhu, Zhao, Zhang, & Wu, 2012). In general, a uniform, compact, and adherent surface layer composed of $Mg(OH)_2$ was formed after hydrothermal treatment. Fourier-transform infrared spectra and X-ray diffraction measurements confirmed the chemical nature of the coating. The thickness of the $Mg(OH)_2$ coating was largely a function of the pH and the treatment time: the higher the pH and longer the treatment time, the larger the thickness. The morphological features of the $Mg(OH)_2$ coating were highly dependent on the pH of the medium and treatment time (Figure 2.3). The $Mg(OH)_2$ coating formed on Mg and the AZ91 Mg alloy by hydrothermal treatment was hydrophilic, showed strong adherence to the Mg alloy,

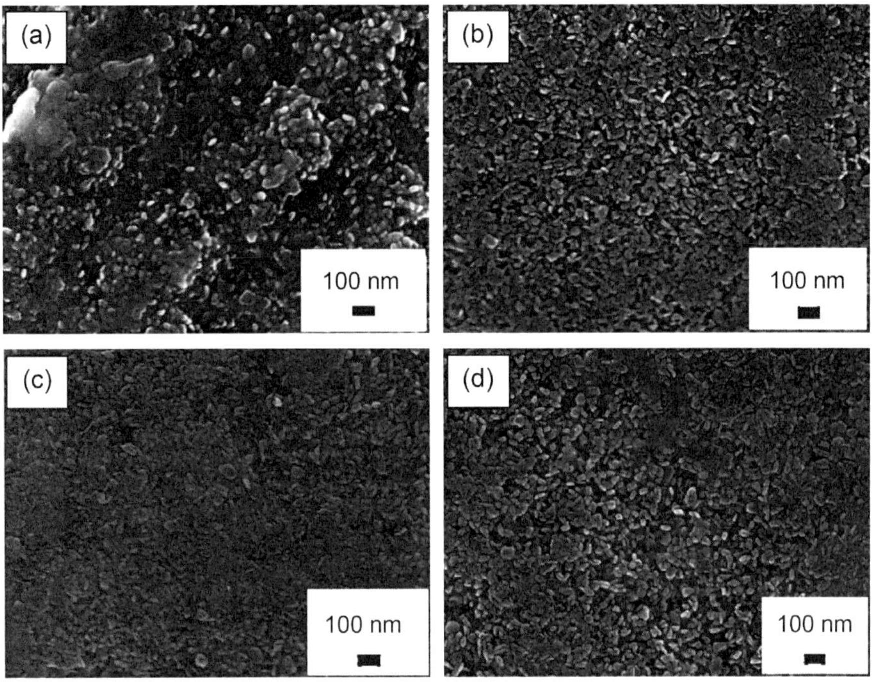

Figure 2.3 Surface morphology of the magnesium hydroxide coating deposited on an AZ31 magnesium alloy after hydrothermal treatment using 5.66 wt% sodium hydroxide at 160 °C for 1 h (a), 2 h (b), 3 h (c), and 4 h (d).
Adapted from Zhu et al. (2011) with permission from Elsevier.

and offered higher corrosion resistance than its untreated counterpart in 3.5% sodium chloride (NaCl), phosphate-buffered saline (PBS), and Hank's solution. However, the ability of $Mg(OH)_2$ to afford long-term protection against corrosion for the Mg and its alloys in SBF is a matter of concern.

2.2.3 *Alkaline heat treatment*

The effect of alkaline heat treatment (AHT) on the corrosion behavior and cytotoxicity of pure Mg (99.9%) in SBF was studied by Li, Gao, and Wang (2004). Pure Mg was soaked in supersaturated sodium bicarbonate ($NaHCO_3$)—magnesium carbonate solution (starting pH of 9.3) for 24 h followed by heat treatment at 773 K for 10 h. After AHT, Mg showed an improved corrosion resistance in SBF, as evidenced by the lower loss of mass after 14 days of immersion and by a slow change in pH of the SBF observed during the first 6 days of immersion. In addition, after AHT, Mg showed growth of calcium (Ca) phosphate-based apatite with a molar ratio of 1.858 after 14 days of immersion in SBF and no signs of morphological changes on cells or inhibitory effect on cell growth.

Gu, Zheng, Cheng, and Zheng (2009) studied the effect of AHT on the corrosion behavior of Mg—Ca alloy (1.4 wt% Ca). They soaked the Mg—Ca alloy in three different types of alkaline solutions, namely, Na_2HPO_4, Na_2CO_3, and $NaHCO_3$ for 24 h followed by heat treatment at 773 K for 12 h. During the initial alkaline treatment, depending on the type of alkaline solution, compounds such as magnesium carbonate, calcium carbonate, magnesium sodium phosphate, and calcium phosphate were deposited on the surface. These deposits were loose and defective and accelerated the corrosion rate of the Mg–Ca alloy. However, during the subsequent heat treatment, these loose surface layers were fused and consolidated, which was accompanied by the oxidation of the substrate, resulting in the formation of a compact MgO layer on the surface of the Mg–Ca alloy (Figure 2.4). The average thickness of the MgO layer formed after AHT in Na_2HPO_4, Na_2CO_3, and $NaHCO_3$ are 13, 9, and 26 μm, respectively. After AHT, the surface roughness of the Mg–Ca alloy increased. AHT in all the three solutions improved the corrosion resistance of the Mg–Ca alloys; based on the extent of protection against corrosion, they were ranked as follows: AHT-$NaHCO_3$ > AHT-Na_2HPO_4 > AHT-Na_2CO_3. The increase in surface area of the sample after AHT has enabled the formation of calcium phosphate. The cytotoxicity evaluation showed that AHT of the Mg–Ca alloy did not induce toxicity to L-929 cells.

Nan, Yangmei, and Bangcheng (2013) soaked commercially pure Mg (99.95% purity) in 5 mL of 1 M $NaHCO_3$ at 60 °C for 24 h followed by heat treatment in air at 773 K for 10 h. According to them, the AHT reduced the rate of Mg degradation, supported by the slow change in the pH of SBF and smaller volume of hydrogen gas released during corrosion. After immersion in SBF for 24 h, the AHT led to the formation of $Ca_2P_2O_7$. According to El Kady, Mohamed, and El-Bassyouni (2009), $Ca_2P_2O_7$ is one of the intermediate products of the biomineralization process, which has the potential to regulate the onset of calcification and could promote mineralization in the body. The AHT Mg samples pretreated in serum cell culture medium for 24 h showed a smaller extent of adhesion and spreading of MG63 cells after 4 and 6 h of cell culture but exhibited more spreading with sphere-like characteristics after 8 h of cell culture.

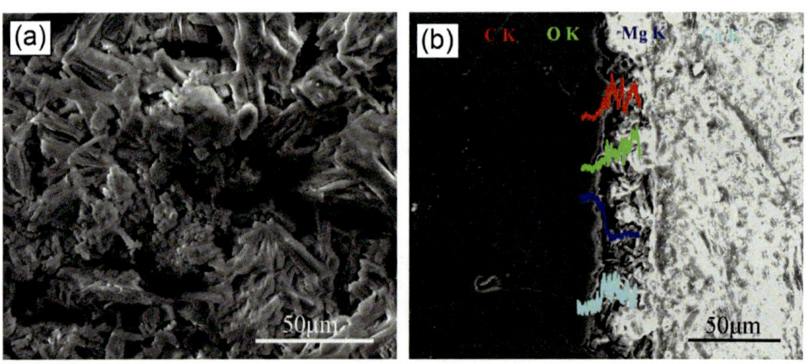

Figure 2.4 Scanning electron micrographs taken at the surface (a) and at the cross-section (b) of a magnesium–calcium alloy (1.4 wt% calcium) after alkaline treatment in sodium bicarbonate for 24 h followed by subsequent heat treatment at 773 K for 12 h.
Adapted from Gu, Zheng, Cheng et al. (2009) with permission from Elsevier.

The findings of Li et al. (2004), Gu, Zheng, Cheng et al. (2009), and Nan et al. (2013) clearly show that AHT decreased the rate of degradation of Mg and its alloys in SBF, and the formation of an MgO layer is responsible for the observed improvement in resistance against corrosion. Deposition of MgO on the surface of Mg and its alloys could be accomplished by heat treatment. However, the amount and crystallinity of MgO formed exclusively by heat treatment were smaller than those formed by AHT. The MgO layers prepared only by heat treatment are porous and nonprotective, whereas those prepared by AHT are relatively dense and compact. The dense and compact nature and the amount and crystallinity of MgO are decisive factors in controlling the rate of degradation of Mg and its alloys in SBF.

2.3 Chemical surface modifications

2.3.1 Chemical passivation

Passivation of the surface of Mg is the simplest approach in surface modification. Passivation of commercially pure Mg (99.9% purity) by soaking in 1 M NaOH for 24 h or in modified SBF (m-SBF) at 37 °C for 5 days was explored by Lorenz et al. (2009) as a possible means of surface modification of Mg. The passive layer formed on Mg after immersion in 1 M NaOH was very thin (on the order of a few nanometers) and composed of $MgO/Mg(OH)_2$. The passive film formed on Mg after immersion in m-SBF was relatively thicker, on the order of tens of micrometers, and composed of amorphous carbonated calcium and magnesium phosphate. Both types of passive films, however, were not highly protective in m-SBF, which was attributed to the insufficient thickness and compactness of the passive films formed in 1 M NaOH and the higher porosity and roughness of the passive film formed in m-SBF. Both types of passive films showed a significantly larger number of human HeLa and mouse fibroblasts cells on their surfaces than those observed on the polished Mg surface. The passive film formed on Mg in 1 M NaOH showed a significant increase in the cell survival rate. The passive film formed on Mg in m-SBF seems to promote a large number of cell nuclei during initial cell adhesion, whereas it does not support cell spreading or favor the formation of actin cytoskeleton after 24 h of cell culture. The inability of the passive film formed on Mg in m-SBF to control the rate of corrosion resulted in a large increase in pH that led to cell death. Hence, it is evident that chemical passivation using NaOH and m-SBF has limited potential in the surface modification of Mg and its alloys toward the development of degradable implants.

2.3.2 Reaction with ionic liquids

Ionic liquids (ILs) are salts that exist as liquid at temperatures below 100 °C, and they are usually composed of a combination of organic cations and organic/inorganic anions. The interaction between ILs such as trihexyl(tetradecyl)phosphonium diphenylphosphate ([P6,6,6,14][DPP]), trihexyl(tetradecyl)phosphonium bis(trifluoromethanesulfonyl) amide ([P66614][NTf2]), and trihexyl(tetradecyl)phosphonium bis

(2,4,4-trimethylpentyl)phosphinate ([P66614][$(iC_8)_2PO_2$]) and Mg alloys, namely, ZE41 and AZ31, was studied by various groups (Efthimiadis et al., 2010; Forsyth et al., 2009; Howlett et al., 2010; Huang, P. et al., 2013; Latham, Howlett, MacFarlane, & Forsyth, 2011). The findings of their studies indicate that the interaction between ILs and Mg alloys is not a simple physisorption of the anions of the ILs on the surface of Mg alloys; instead, it most likely involves a chemical or electrochemical process. The nonreversible nature of the cyclic voltammetric scans of AZ31 Mg alloy in ([P66614] [$(iC_8)_2PO_2$) further confirms this attribute (Latham et al., 2011). The surface films formed on Mg alloys have a thickness of a few tens of nanometers, in which the inner layer is dominated by metal oxides/hydroxides while the outer layer mainly consists of organics, most likely entrapped breakdown products and components of ILs. These surface films offered reasonable protection for the Mg alloys against corrosion in a chloride-containing environment. However, there are many challenges to overcome before realizing the potential utility of ILs for protection of Mg alloys against corrosion. The intrinsic heterogeneity of Mg alloys poses a major challenge in generating a uniform surface film. Acid pickling pretreatment and application of a potential bias during treatment with ILs are some possible means of achieving better surface homogeneity of Mg alloys, whereas the addition of suitable amounts of water in ILs enhances its reactivity (Murase et al., 2013). In spite of these modifications, the slow reaction kinetics warrant a longer treatment time (24 h in most cases), and the resultant surface films are by no means free of defects. The current status of the role of ILs for corrosion protection of Mg alloys recently was reviewed by Huang et al. (2013). According to them, much research still is needed to understand the possible means of generating robust passivating films on reactive metal surfaces such as Mg in the presence of ILs. Until then, the possibility of using ILs for controlling the rate of corrosion of Mg alloys toward the development of degradable implants is limited.

2.3.3 Self-assembled monolayers

The spontaneous formation of a highly ordered monolayer structure of certain organic compounds, particularly those with a long alkyl chain, on the surface of various substrates is referred as a self-assembled monolayer (SAM). The head groups of these organic compounds anchor to the surface, while their hydrophobic tail prevents the approach of water molecules toward the surface. The interfacial properties of the SAM are largely determined by the type of the head group and the length of the tail groups. SAMs form by a simple chemisorption process, making them strongly adhere to the surface. The ability of SAMs to establish conformal contact regardless of the shape of the substrates and the capability to control their thickness and composition offer a wide window of opportunity to explore their potential applications in many areas. The kinetics and mechanism of formation of SAMs was reviewed by Schwartz (2001). Liu, Yu, Zhou, and Wu (2006) successfully formed SAMs of sodium dodecanoate ($C_{12}H_{23}O_2Na$), sodium palmitate ($C_{16}H_{31}O_2Na$), and sodium stearate ($C_{18}H_{35}O_2Na$) on an AZ91D Mg alloy in ethanol and showed that these SAMs could offer a corrosion protection efficiency of up to 98.5% in 0.2 M NaCl. A distinct trend observed in the three SAMs is that the longer the alkyl chains of alkylcarboxylate, the better the extent of corrosion protection. This is because of the increase in the thickness of the SAMs with an increase in the length of the alkyl chains of these compounds. An increase in treatment time of up to 24 h also showed an increase in corrosion protection efficiency, probably because of a better alignment and packing of the SAMs. The schematic structure of the n-alkylcarboxylate monolayer on the surface of the Mg alloy with C_{18} carboxylate is depicted in Figure 2.5. Gao, Qiao, and Xin (2010) formed SAMs of stearic acid (SA) on heat-treated (773 K for 10 h) 4N-Mg. According to them, a combination of heat treatment and SAMs decreased the rate of degradation of Mg from 0.11 to 0.05 mm/year and induced calcium phosphate deposition in SBF. The improvement in corrosion resistance following this treatment also was evidenced by the marginal decrease in the diameter of the Mg sample by 0.2 mm after 12 weeks of implantation in rabbits when compared with the uncoated sample, which exhibited a larger decrease in diameter of 0.93 mm. In addition, when the Mg samples modified by the combination of heat treatment and SAMs were implanted into the rabbits, no infections occurred, and the Mg^{2+} concentration in blood after 2, 6, and 12 weeks remained in the natural range.

Ishizaki, Okido, Masuda, Saito, and Sakamoto (2011) and Ishizaki, Teshima, Masuda, and Sakamoto (2011) formed SAMs of alkanoic and phosphonic acids on an AZ31 Mg alloy using liquid phase and vapor phase methods. Accordingly, a phosphonic acid-derived SAM-modified AZ31 Mg alloy offered better resistance to corrosion than the unmodified alkanoic acid-derived alloys. They also reported that

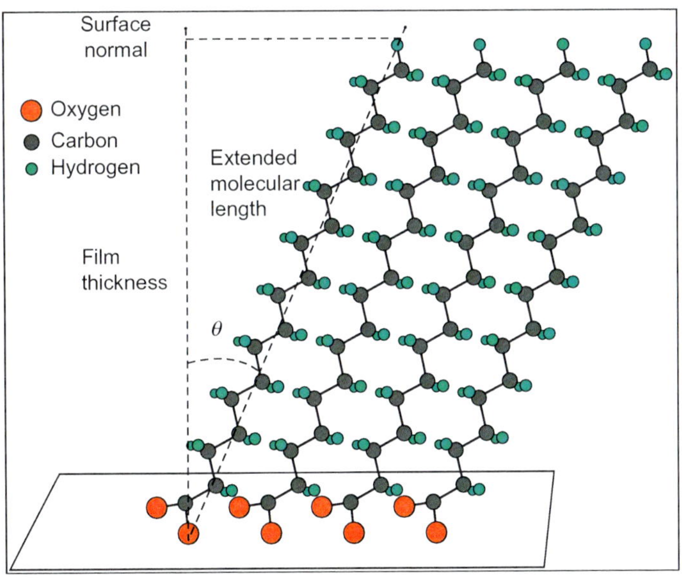

Figure 2.5 Schematic structure of the n-alkylcarboxylate monolayer on the surface of a magnesium alloy with C_{18} carboxylate. The monodentate bonding was estimated assuming a C–O bond normal to the substrate surface.
Adapted from Liu et al. (2006) with permission from Elsevier.

the adhesion and chemical stability of SAMs of alkyl phosphonic acid prepared by a simple immersion method was very low when compared with those prepared by the vapor phase method. These inferences clearly point out that SAMs should be well-ordered and closely packed, with their hydrophobic terminal methyl group oriented outward, and that there exists a strong van der Waals interaction between the alkyl chains, thus endowing them with a high homogeneity, good order, and better chemical stability to achieve a better protection of Mg alloys against corrosion. It has been reported that water contact angles could be a good indicator of the nature of SAMs; in general, those that exhibit high hydrophobicity offer better protection against corrosion. Salman and Okido (2012) reported that the SAMs of oleic and stearic acids on AZ31 Mg alloy formed in ethanol showed larger contact angles than those formed using acetone and hexane, probably because of the formation of magnesium ethoxide on the Mg alloy's surface. Hence it is evident that selection of a suitable solvent also is a key in the formation of well-ordered and homogeneous SAMs.

Grubac et al. (2012) suggested that SAMs of palmitic and stearic acids have much potential for designing specific interface architectures in degradable implant technology. They formed SAMs of palmitic and stearic acids on bare as well as anodized AZ91D alloy and evaluated their corrosion resistance in Hank's solution at 37 °C. Their findings further confirmed that the longer the hydrocarbon chains of alkylcarboxylate, the better the corrosion protection of the SAMs formed by it. In addition, they showed that the anodic oxide film provided an excellent base for the chemisorption of carboxylic acids, resulting in a significant decrease in the rate of degradation of the Mg alloy. Grubac et al. (2013) studied the effect of SAMs of dodecylphosphonic acid and octadecyl-phosphonic acid, in both as-deposited and heat-treated (120 °C for 5.5 and 24 h) conditions, on the corrosion resistance of AZ91D Mg alloy in Hank's solution at 37 °C. Accordingly, in the as-deposited condition, the intermolecular interaction (Van der Waals and hydrogen bonding) was apparently stronger than the substrate–molecule interaction, whereas after heat treatment the transformation of the phosphonic acid monolayer to a phosphonate enabled a strong chemical bond with the surface, thus leading to a significant increase in protection against corrosion. Based on the findings of the role of SAMs for protection of Mg alloys against corrosion, it is evident that the formation of well-ordered and closely packed SAMs with a high homogeneity, good order, and better chemical stability is the key step in achieving better corrosion protection of Mg alloys toward the development of degradable Mg-based biomaterials. The rapid decrease in the contact angle of some of the SAM-modified Mg alloy surfaces in aqueous solutions within a short duration of time has raised questions on their long-term chemical stability in human body fluid.

2.3.4 Chemical conversion coatings

2.3.4.1 Cerium-, nicotinic acid-, titanate-, and sodium bicarbonate-based conversion coatings

Cui et al. (2011) studied the formation of cerium conversion coatings on an AZ31 Mg alloy. The cerium conversion coating was dense, composed of CeO_2, CeO, Ce_2O_3,

MgO, $Mg(OH)_2$, and aluminum oxide (Al_2O_3), effectively reduced the rate of the AZ31 Mg alloy when compared with the untreated alloy after 24 h of immersion in Hank's solution, and showed excellent biocompatibility. Deposition of the nicotinic acid-based conversion coatings improved the resistance of the AZ31 Mg alloy to corrosion, and it did not indicate any cytotoxic activity in a normal adult stem cell culture, suggesting its good biocompatibility in cell culture *in vitro* (Bikulčius, Ručinskienė, Bukelskienė, Selskis, & Grigucevičienė, 2012). Yang, Tsai, Huang, and Lin (2012) studied the deposition of a titanate conversion coating on AZ31 Mg alloy and evaluated the corrosion behavior of the coated Mg alloy in 0.05 M NaCl, 0.1 M sodium sulfate, and after 24 h of the salt spray test (ASTM B117). The titanate conversion-coated Mg alloy offered a better corrosion resistance than the uncoated alloy. The effect of deposition of a chemical conversion coating by immersion in saturated aqueous $NaHCO_3$ solution on the corrosion resistance of AZ31 and AZ61 Mg alloys in 0.6 M NaCl was studied by Feliu, Samaniego, El-Hadad, and Llorente (2013). The resultant coating was composed of carbonates of magnesium and sodium and aluminum hydroxide and offered reasonable protection when compared with its untreated counterpart. The formation of a typical "mud crack" pattern, possibly caused by the release of hydrogen gas and/or dehydration of the coating after deposition, limits the extent of protection against corrosion offered by these coatings.

2.3.4.2 Phosphate conversion coatings

Deposition of phosphate conversion coatings has been explored by many researchers as a surface modification method to improve the corrosion resistance of Mg and its alloys (Hu, Meng, Chen, & Wang, 2012; Li, Lian, Niu, & Jiang, 2006; Van Phuong, Lee, Chang, Kim et al., 2013; Van Phuong, Lee, Chang, & Moon, 2013; Xu, Zhang, & Yang, 2009). Deposition of a Zn phosphate coating is one of the most well-studied systems and was recently reviewed by Van Phuong, Lee, Chang, Kim et al. (2013). It has been reported that the experimental conditions should be optimized carefully to achieve a good-quality deposit. Being a conversion coating, the resultant phosphate coating is porous in nature, which limits its fullest potential in improving the corrosion resistance of Mg alloys (Hu et al., 2012; Li et al., 2006; Van Phuong, Lee, Chang, Kim et al., 2013; Van Phuong, Lee, Chang, & Moon, 2013). Xu et al. (2009) used $Ca(H_2PO_4)_2 \cdot H_2O$ and $Zn(H_2PO_4)_2 \cdot 2H_2O$ as main components of a phosphating bath and deposited brushite coating with small amounts of Zn on the surface of an Mg–Mn–Zn alloy. The brushite coating provided good protection to the Mg alloy against corrosion in SBF. However, in spite of its ability to reduce the corrosion rate, it fails to completely stop the degradation of Mg in SBF. As a surface coating, the brushite coating can improve the surface biocompatibility of Mg alloys because it can be transformed into hydroxyapatite (HA) phase in SBF. A detailed account of phosphate conversion treatment of Mg alloys for biomedical applications is presented by Zhang in Chapter 2 of Volume 2 of this book. In addition, the deposition of strontium phosphate and manganese phosphate coatings on Mg alloys and their possible implications in biomedical applications are addressed by Chen et al. in Chapter 15 of Volume 2 of this book. Deposition of $Mg(OH)_2$-HA and HA coatings also was explored using the chemical conversion method (Chen, Birbilis, & Abbott, 2011; Wang, Huang et al., 2013). Deposition of calcium phosphates and HA assumed significance because of their bioactive properties. More details on the various methods of deposition and characteristics of calcium phosphates and HA are covered in Chapters 3, 7, and 10 of Volume 2 of this book.

2.3.4.3 Phytic acid-based conversion coatings

Phytic acid-based conversion coatings assumed significance because of their nontoxic and biocompatible nature as well as the chelating ability of phytic acid with Mg and its alloys. Jianrui, Yina, and Weidong (2006) studied the deposition of phytic acid-based conversion coatings on an AZ91D Mg alloy as a function of the concentration of phytic acid, pH, temperature, and immersion time. Accordingly, phytic acid coatings deposited under optimum conditions (concentration of phytic acid: ~0.5–1%; temperature: ~25–60 °C, pH ~3–5, and treatment time: ~30–60 min) offered a reasonably good protection of the AZ91D Mg alloy against corrosion in 3.5% NaCl. Cui et al. (2008) studied the mechanism of formation of phytic acid-based conversion coatings on AZ91D Mg alloy using 5 g/L of phytic acid at 20 °C as a function of time. Accordingly, in an aqueous solution, the phosphate group of the phytic acid is ionized, forming phytic acid radicals with a different number of phosphate radicals, phosphate hydrogen radicals, and hydroxyl ions. The phytic acid ion combined with the surface Mg ions to form a magnesium phytate complex on the surface of the Mg alloy. The phytic acid conversion coating exhibited a good bonding strength (5B as per ASTM D 3359-90) and offered reasonable protection against corrosion in 3.5% NaCl. Ye, Zheng, Wang, Xi, and Li (2012) pointed out that phytic acid coatings deposited on WE43 Mg alloys at pH 3 and 5 were relatively thicker than those coated at pH 8 and 10; among them, those coated at pH 5 offered a marginal increase in resistance against corrosion in SBF (Figure 2.6), better cell viability, and an acceptable hemolysis rate. The findings of Jianrui et al. (2006), Cui et al. (2008), and Ye, Zheng et al. (2012) clearly indicate that in spite of its complexing ability, nontoxicity, biocompatibility, cell viability, and hemolytic ability, the presence of cracks is a serious limitation in terms of achieving higher corrosion resistance with phytic acid-based conversion coatings.

Gupta, Mensah-Darkwa, and Kumar (2013), Gupta, Mensah-Darkwa, Sankar, and Kumar (2013), and Chen, Huang, and Peng (2013) made some modifications in the methodology of deposition to overcome this limitation. In one of the approaches, Gupta, Mensah-Darkwa, Sankar et al. (2013) deposited the phytic acid conversion coating by immersing Mg in a phytic acid solution (50% water) at room temperature for 3 h, drying at 60 °C for 3 h, followed by immersion in an SA solution (1 mg/mL in ethanol) at 80 °C for 2 h, with a thickness of ~2.3 μm. The posttreatment in an SA solution filled the microcracks present in the phytic acid conversion coating, improved the smoothness of the coating, and reduced the corrosion current density of phytic acid conversion-coated Mg in PBS from 1.16×10^{-6} to 1.03×10^{-6} A/cm^2 (Gupta, Mensah-Darkwa, Sankar et al., 2013). In another approach, the phytic acid-coated Mg was subjected to a heat treatment at 95 °C for 1 h under a vacuum. In spite of a very slight shrinkage (~1.4%) in the film thickness, the heat treatment helped to annihilate the microcracks

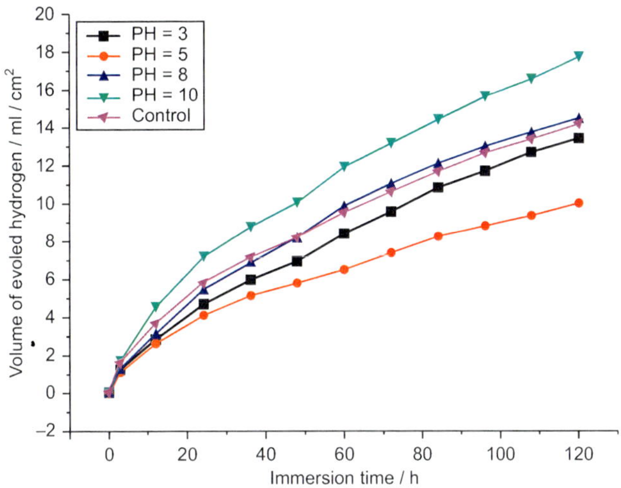

Figure 2.6 Comparison of the volume of hydrogen that evolved during the immersion of an untreated and phytic acid conversion–coated WE43 magnesium alloy (at various pHs) in simulated body fluid as a function of immersion time.
Adapted from Ye, Zheng et al. (2012) with permission from Elsevier.

present in the phytic acid conversion-coated layer, with no formation of oxides. The heat treatment after coating reduced the corrosion current density of phytic acid conversion-coated Mg in PBS from 1.16×10^{-6} to 9.27×10^{-7} A/cm^2 (Gupta, Mensah-Darkwa, & Kumar, 2013). Chen, Wan et al. (2013) suggested that pretreatment of Mg with alkaline in 3 M NaOH at 60 °C for about 24 h helped to immobilize the phytic acid layer, with uniform distribution on the surface following the establishment of a covalent bond via the hydrolysis/neutralization reaction between the OH^- on Mg and H^-O_3PO-R (myo-inositol-1,2,3,4,5,6-hexakisphosphate) on the phytic acid molecule. The formation of a dense and homogenous phytic acid conversion coating by this methodology decreased the rate of degradation of Mg in PBS. The improved bonding of the phytic acid molecules and the clinically acceptable hemolysis ratio of the coated Mg suggests that it has potential application in the development of degradable Mg-based cardiovascular stents.

2.3.4.4 Fluoride conversion coatings

Deposition of fluoride conversion coatings on Mg and its alloys has been studied by many researchers, particularly because of their suitability for biomedical applications and their ability to serve as a pretreatment for bioactive coatings. A detailed account of fluoride conversion coatings of Mg alloys is presented by da Conceicao in Chapter 1 of Volume 2 of this book, and their suitability for biomedical applications is addressed here. Fluoride coatings offer many advantages.

- Fluoride conversion coatings reduced the corrosion rate of Mg and its alloys under both *in vitro* and *in vivo* conditions (Figure 2.7), confirming their ability to control the localized increase in pH and accumulation of hydrogen gas (Carboneras, García-Alonso, & Escudero,

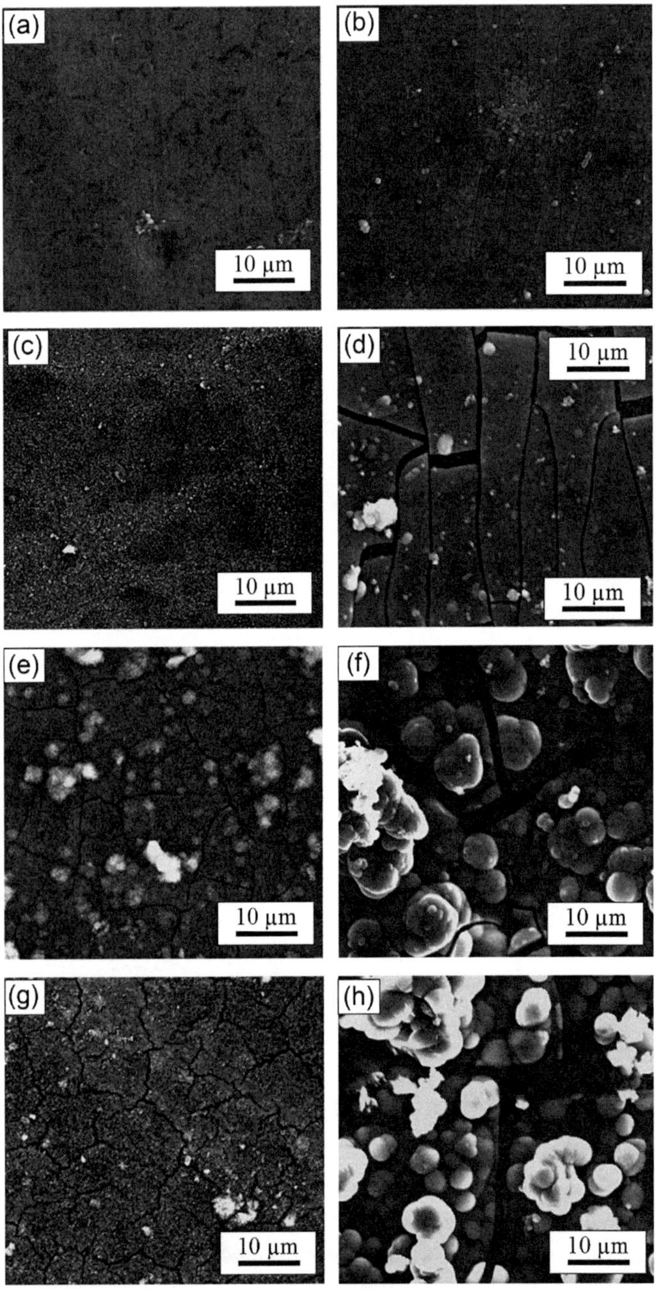

Figure 2.7 Surface morphologies of a fluoride-treated (a, c, e, g) and bare (b, d, f, h) AZ31B magnesium alloy after immersion in simulated blood plasma for different durations: 3 days (a, b); 7 days (c, d); 14 days (e, f); and 30 days (g, h).
Adapted from Yan et al. (2010) with permission from Elsevier.

2011; Chiu, Wong, Cheng, & Man, 2007; da Conceicao, Scharnagl, Blawert, Dietzel, & Kainer, 2010; da Conceicao, Scharnagl, Dietzel, & Kainer, 2012; Drynda et al., 2010; Drynda, Seibt, Hassel, Bach, & Peuster, 2013; Witte et al., 2010; Yan et al., 2010).

- It offered a good biocompatibility (Carboneras et al., 2011; Drynda et al., 2010, 2013; Thomann et al., 2010; Witte et al., 2010), demonstrated good clinical tolerance in a rabbit model (Thomann et al., 2010), revealed no adverse effects during surgical implantation (Drynda et al., 2013), and showed no significant signs of gas accumulation at the implantation site (Drynda et al., 2013), suggesting its better tissue compatibility.
- It ensured a gradual degradation, which helped the Mg alloy to maintain a better mechanical property for as long as 45 days in SBF, thus meeting one of the important requirements of a biodegradable material (Pereda et al., 2010; Yan et al., 2010).
- It offered an improved cell response for MG63 cells and MC3T3-E1 cells after 72 h of cell culture (Li, Li et al., 2013) and promoted the adhesion and proliferation of 3T3 cells (Ye, Chen, You, & Liu, 2010).
- It enhanced the deposition of calcium phosphate, suggesting its ability to accelerate the rate of mineral deposition and promote bone healing (Bakhsheshi-Rad, Idris, Abdul-kadir, & Daroonparvar, 2013; Li, Li et al., 2013).
- It demonstrated a significant increase in bonelike fraction on Mg alloy sponges, as evidenced by more rapid bone adaptation, better tolerance, higher vascularization, and good osseointegration (Lalk et al., 2013).
- The low solubility of Mg fluoride (MgF_2) coating in aqueous physiological solutions (0.13 g/L) decreased its leaching rate, as evidenced by the absence of an elevated fluoride concentration in regions adjacent to the bone (Drynda et al., 2010; Witte et al., 2010).
- The fluoride ions released from MgF_2 coating did not show any cytotoxic effects (Grillo, Alvarez, & Fernández Lorenzo de Mele, 2011; Pereda, Alonso, Gamero, del Valle, & Fernández Lorenzo de Mele, 2011).
- The MgF_2 coating effectively restricted biofilm formation against two common nosocomial biofilm-forming pathogens: *Escherichia coli* and *Staphylococcus aureus* (Lellouche, Kahana, Elias, Gedanken, & Banin, 2009).
- It exhibited no harm to the red blood cells (Ye et al., 2010), while it could enhance the lysosomal activity under specific conditions (Grillo et al., 2011).
- It served as a good pretreatment for the deposition of bioactive coatings such as HA and dicalcium phosphate dihydrate (DCPD) obtained by electrodeposition (ED) and electrophoretic deposition (EPD) methods (Bakhsheshi-Rad, Idris, & Abdul-kadir, 2013; Jo et al., 2011; Rojaee, Fathi, & Raeissi, 2013a), and it increased the adhesion and bonding strength of these coatings (Bakhsheshi-Rad, Idris, & Abdul-kadir, 2013).

Based on these characteristics, MgF_2-coated Mg and its alloys is considered suitable for cardiovascular stents (Bakhsheshi-Rad, Idris, Abdul-kadir et al., 2013; Drynda et al., 2010) and for biodegradable implants in endoprothesic applications (Yan et al., 2010). In spite of these numerous advantages, the acceptance of MgF_2-coated Mg alloys as candidate materials for degradable implants still remains limited because the MgF_2 coating prepared by chemical conversion method (immersion in 20−50 wt% HF at 300 K for 6−168 h) is usually thin ($\sim$1−3 μm), porous, and could offer only a limited increase in the corrosion resistance in SBF (Bakhsheshi-Rad, Idris, Abdul-kadir et al., 2013; Carboneras et al., 2011; Chiu et al., 2007; da Conceicao et al., 2010, 2012; Drynda et al., 2010; Pereda et al., 2010, 2011; Thomann et al., 2010; Witte et al., 2010; Yan et al., 2010; Ye et al., 2010). Chen, Song, Zhang, Li,

Wang et al. (2011) showed that these coatings were severely cracked and delaminated after 4 h of dynamic tests, questioning their endurance for stent applications. Hence, it is generally felt that they can be used mainly as a pretreatment.

The MgF_2 coatings having a larger thickness (15–20 μm) when prepared by an alternate methodology that involves immersion of the Mg alloy in 200 g/L NaOH for 3 h followed by conversion of $Mg(OH)_2$ to MgF_2 through subsequent immersion in 40% HF for 96 h, which also failed to show a significant improvement in performance (Drynda et al., 2010, 2013; Witte et al., 2010). In spite of its larger thickness (15–20 μm), Drynda et al. (2010) observed H_2 evolution after 8–40 h. Witte et al. (2010) showed that the MgF_2 coating deposited on an LAE 442 Mg alloy vanished after 4 weeks of implantation. Drynda et al. (2013) observed a large reduction in the area of the MgF_2-coated Mg-0.6% Ca alloy after 3 months of implantation. Accordingly, for the MgF_2 coatings prepared by this methodology, the outer layer (only 5 μm from the top layer) is rich in fluorine, and it decreased significantly along its thickness from the top surface layer to the inner layers, which are rich in oxygen (Drynda et al., 2013). This raises the question of whether this methodology allows the complete conversion of $Mg(OH)_2$ to MgF_2. According to Drynda et al. (2010), an increase in thickness of the MgF_2 coating beyond 20 μm resulted in the formation of capillary-branched cracks over the whole surface. This would jeopardize the protectiveness of the MgF_2 coating.

The mechanism of deposition of the MgF_2 coating using the chemical conversion method is not completely understood. The kinetics of coating formation is very slow. It is presumed that the coated layer consists of a mixture of MgF_2 and $Mg(OH)_{2-x}F_x$, and the latter compound gets converted to MgF_2 when x becomes 2. The passivating nature of MgF_2 is considered one of the reasons for the slow kinetics of deposition. The slow conversion of $Mg(OH)_{2-x}F_x$ to MgF_2 by its reaction with HF could also account for the poor kinetics. The exact reasons for the slow kinetics are not yet fully ascertained. The formation of MgF_2 coating by immersion in NaOH followed by the subsequent conversion of $Mg(OH)_2$ to MgF_2 by immersion in HF could also involve intermediates such as $Mg(OH)_{2-x}F_x$. The influence of the volume fraction of $Mg(OH)_{2-x}F_x$ on the corrosion resistance of the resultant coatings obtained by these methodologies is not yet fully understood. A correlation of the corrosion resistance of fluoride conversion coatings with their composition is not established. Because the characteristics of MgF_2-coated Mg and its alloys are much suitable for biomedical applications, modification of the existing methodologies to prepare MgF_2 coatings with desirable characteristics will be beneficial. Its ability to serve as a pretreatment for the subsequent deposition of bioactive coatings suggests that the MgF_2 coating is a promising surface modification method.

2.3.5 *Sol-gel coatings*

2.3.5.1 *Titania, HA, and calcium phosphate glass ceramic coatings*

Amaravathy, Rose, Sathiyanarayanan, and Rajendran (2012) prepared a titanium dioxide (TiO_2) coating on an AZ31 Mg alloy through the sol-gel method using the dip coating technique. The as-deposited TiO_2 coatings were dried in an oven at 60 °C for 1 h for gelation and subsequently sintered at 120, 240, and 360 °C for 1 h. Deposition of the TiO_2 coating increased the root mean square roughness of the AZ31 Mg alloy from 0.1053 to 0.1334 nm, decreased the contact angle from 68° to 23°, provided better cell adhesion, and provided a higher degree of cell spreading and attachment. However, when immersed in SBF, severe local collapses of the TiO_2 coating occurred on the seventh day, which indicates its inability to offer good long-term protection against corrosion for the AZ31 Mg alloy. This is because of the presence of uniformly distributed pores and cracks in the TiO_2 coating, which are formed during the sintering process.

Rojaee, Fathi, and Raeissi (2013b) prepared a sol-gel-derived nanostructured HA coating on an AZ91 Mg alloy using the dip coating technique. The as-deposited coatings were dried at room temperature for 24 h, followed by drying at 60 °C for 24 h and sintering at 400 °C for 6 h. The resultant coatings were dense, homogeneous, and crack free and exhibited no obvious delamination or gaps between the coating and the substrate, with an average thickness of 6.3 ± 1.1 μm. Because of the low sintering temperature (400 °C), the degree of crystallinity of the nanostructured HA coating was only about 45%. The nanostructured HA coating promoted biomineralization around the wounded tissue and decreased the extent of biocorrosion and the rate of release of Mg^{2+} ions, while the relatively slow increase in pH of the SBF provided enhanced cell growth and proliferation. However, the major limitation is the poor bonding strength (4.2 ± 0.3 MPa) of the nanostructured HA coating prepared by this method, resulting in both cohesive and adhesive failures. In general, an increase in sintering temperature is a viable option to improve adhesive strength. As Mg alloys possess a relatively low melting point and poor heat resistance, the possibility of improving the adhesive strength of a nanostructured HA coating by increasing the sintering temperature is rather limited.

Tang, Xin, and Wang (2013) inserted a titania buffer layer using the sol-gel dip coating method to improve the bonding strength between a calcium phosphate layer and an AZ31 Mg alloy. The as-deposited TiO_2 coatings were dried at 60 °C for 30 min and subsequently sintered at 400 °C for 30 min for two cycles. The calcium phosphate coating was deposited over the TiO_2 coating, followed by drying at 60 °C for 30 min and sintering at 400 °C for 30 min for two cycles. The TiO_2 coatings exhibited the presence of microcracks and micropores, which were randomly distributed on their surface and formed due to the thermal mismatch between the coating and the AZ31 Mg alloy. Amaravathy et al. (2012) also reported the formation of cracks and pores in TiO_2 coatings prepared using sol-gel dip coating method. However, the cracks present in the TiO_2 coatings were filled in and completely covered by the calcium phosphate coating deposited over them, which resulted in the formation of a dense and compact coating. The insertion of the TiO_2 coating significantly improved the bonding strength of the calcium phosphate layer to the AZ31 Mg alloy. Owing to the low sintering temperature (400 °C), the degree of crystallinity of the calcium phosphate coating was poor.

Ren et al. (2013) prepared calcium phosphate glass ceramic coatings on an AZ31 Mg alloy using the sol-gel dip coating method. The as-deposited coatings were dried at 60 °C for 2 h, followed by heat treatment at 400, 450, and 500 °C for 2 h, followed by furnace cooling. The calcium phosphate glass ceramic coatings subjected to heat treatment at 400 °C were smooth, compact, and homogeneous, were composed entirely of glass phase, and showed a smooth transition from the AZ31 Mg alloy to the coating, with an average thickness of ~1.2 μm. However, coatings subjected to heat treatment at 450 and 500 °C were not as compact as the one treated at 400 °C and showed the generation of secondary pores following the emergence of a crystalline phase of the calcium phosphate glass. Coatings treated at 500 °C showed a random distribution of crystalline particles composed of $Ca_2P_2O_7$ and $Ca_4P_6O_{19}$ phases. The calcium phosphate glass ceramic coatings prepared by heat treatment at 400 °C offered improved resistance against corrosion in SBF than those prepared at 450 and 500 °C, which is evidenced by lower i_{corr} values in the potentiodynamic polarization test and a slow increase in the pH of the SBF up to 7 days in an immersion test. Moreover, after immersion in SBF for 7 days, coatings heat treated at 400 °C showed a Ca-to-phosphate atomic ratio of 1.58, whereas coatings heat treated at 450 and 500 °C showed a lower Ca-to-phosphate atomic ratio of 1.02.

2.3.5.2 Bioactive glass coatings

Bioactive glass is considered an ideal materials for use in bone tissue engineering applications because of its excellent osteoconductivity, bioactivity, and controllable biodegradability (Hench, 1997). Moreover, bioactive glass such as 45S5 Bioglass® is widely accepted for use in medical devices and is approved by the US Food and Drug Administration (Best, Porter, Thian, & Huang, 2008). Hence, the deposition of bioactive glass coatings could offer many benefits in terms of improving the corrosion resistance of Mg and its alloys, in addition to providing a bioactive surface for better cell growth. The deposition of the bioactive 45S5 glass–ceramic coating (45S5 BGC) on an AZ31 Mg alloy by sol-gel dip coating method was explored by Ye, Cai et al. (2012) and Dou et al. (2013). These coatings were deposited in cycles, each of which involved immersion of the Mg alloy in the sol followed by its withdrawal at a speed of 0.5 mm/s, aging at room temperature for 24 h, drying at 60 °C for 1 h, and calcination at 500 °C for 1.5 h. Subsequent treatment cycles were performed after the previous one dried (Dou et al., 2013; Ye, Cai et al., 2012). The deposited coatings were homogeneous, dense, continuous, and crack free only up to three cycles, with a thickness of 0.48–1.00 μm. The high viscosity of the precursor sol and evolution of in-plane tensile residual stress during densification of the coating are considered responsible for the formation of cracks (Dou et al., 2013). The as-deposited 45S5 BGC was calcinated at 500 °C to relieve residual stress either by structural relaxation or by softening the glassy phase (Dou et al., 2013). The 45S5 BGC is composed of an amorphous phase and a crystalline $Na_2Ca_2Si_3O_9$ phase. It enabled an anodic shift in E_{corr} from −1.60 to −1.48 V versus SCE and a decrease in i_{corr} from 4.48 to 0.16 μA/cm^2 in SBF, which also was reflected in an immersion test during the first 7 days of immersion in SBF (Ye, Cai et al., 2012). The slow increase in pH following the decrease in the extent of dissolution of the coated Mg alloy could provide a suitable environment for cell survival (Ye, Cai et al., 2012).

A decrease in mass loss of the coated AZ31 Mg alloy, from 78.04% to 2.31%, also was observed after 7 days of immersion in culture media (Dou et al., 2013). The formation of calcium-deficient HA indicates its ability to promote biomineralization in m-SBF (Dou et al., 2013). However, after 14 days of immersion, the corrosion resistance of the coated samples decreased because of the cracking of the glass–ceramic coating (Figure 2.8) (Dou et al., 2013; Ye, Cai et al., 2012), suggesting its inability to offer long-term protection against corrosion. Huang et al. (2013) recently prepared mesoporous 45S5 bioactive glass-ceramic (45S5 MBGC) coatings on an AZ31 magnesium alloy using the sol-gel dip coating method and an evaporation-induced self-assembly process. Accordingly, unlike the 45S5 BGC, the 45S5 MBGC coatings possess a higher volume fraction of the crystalline $Na_2Ca_2Si_3O_9$ phase, a low elastic modulus, low susceptibility for cracking, higher surface roughness, a large specific surface area, large pore volume, and increased wettability (Huang et al., 2013). The 45S5 MBGC coatings offered a significant decrease in i_{corr} from 1.12×10^{-5} to 6.24×10^{-7} A/cm^2. The transformation of the crystalline $Na_2Ca_2Si_3O_9$ into amorphous apatite in SBF makes the coating more bioactive (Huang et al., 2013). However, the ability of the coating to offer long-term protection against corrosion is yet to be

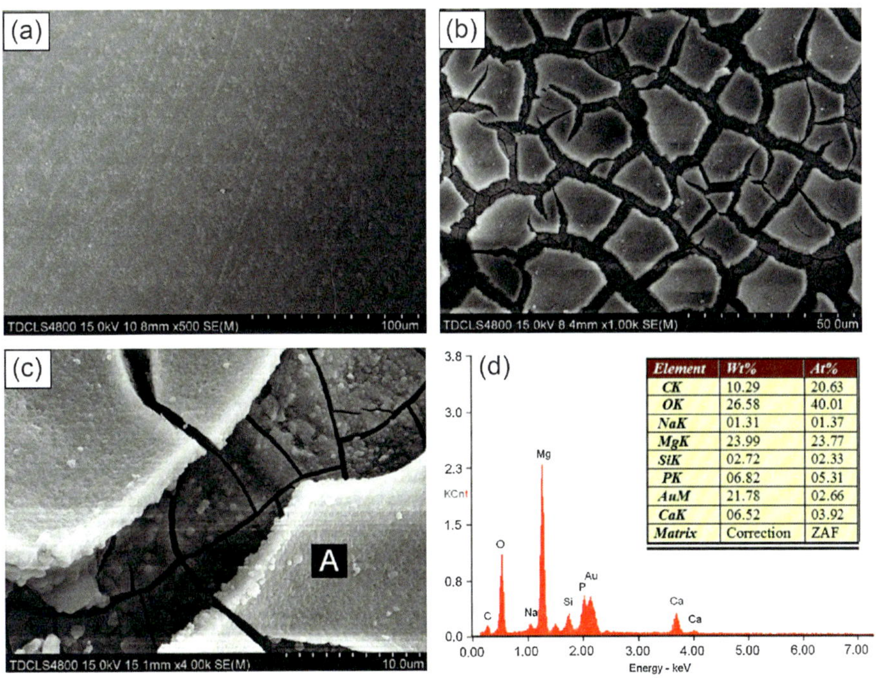

Figure 2.8 Surface morphology of the bioactive glass–ceramic-coated AZ31 magnesium alloy after immersion in simulated body fluid for 3 days (a) and 14 days (b). (c) An enlarged view of (b). (d) Energy-dispersive X-ray spectroscopy analysis performed on the area marked as "A" in (c).
Adapted from Ye, Cai et al. (2012) with permission from Elsevier.

ascertained. From the findings of Ye, Cai et al. (2012), Dou et al. (2013), and Huang et al. (2013), it can be inferred that 45S5 BGC and 45S5 MBGC coatings could slow down the corrosion rate of the AZ31 Mg alloy, exhibit better bioactivity, and promote biomineralization. However, the susceptibility of the coating to cracking after immersion in SBF is the major drawback, which limits its acceptance as a suitable surface coating for Mg alloys in the development of degradable implants.

2.3.6 Silane coatings

Silanes are a group of silicon-based organic–inorganic materials, which are biocompatible. In aqueous solution, silanes readily hydrolyze and form silanol groups (SiOH) that enable their attachment to any hydrated metal surface (metal-OH) through the formation of silicon (Si)–oxygen–metal bonds (Plueddemann, 1991). In addition, the silanol groups could undergo self-cross-linking via siloxane bonds (Si–O–Si), leading to the formation of a protective layer chemically bound to metallic substrates (van Ooij et al., 2000). Furthermore, silanes could provide functional moieties that facilitate the attachment of bioactive molecules and promote interfacial interactions between the metal implants and the surrounding cells and tissue (Weetall, 1993). The effect of a silane coating on the resistance of Mg alloys to corrosion has been studied by many researchers (Liu et al., 2013; Pinto, Carmezim, Ferreira, & Montemor, 2010; Xue et al., 2012; Zomorodian et al., 2012). According to Pinto et al. (2010), the formation of a silane layer on an WE54 Mg alloy increased its impedance by two orders of magnitude in 0.005 M NaCl, and the impedance values remained in around 1 $M\Omega\ cm^2$ for several hours. However, the impedance value dropped to 0.2 $M\Omega\ cm^2$ after 3 days of immersion and became similar to that of the untreated Mg alloy after 1 week of immersion. Zomorodian et al. (2012) also observed a stable impedance value for a silane-coated AZ31 Mg alloy for a duration of only 48 h of immersion in Hank's solution. With a further increase in immersion time, an abrupt decrease in impedance value and the formation of large defects in the silane coating were observed. Accumulation of the corrosion products in these defects also was observed (Figure 2.9). The inferences of Pinto et al. (2010) and Zomorodian et al. (2012) indicate that the silane coating offers only a short-term protection against corrosion for Mg alloys. With an increase in immersion time, penetration of the corrosive medium through the silane layer reaches the interface between the silane coating and the Mg alloy, initiating corrosion of the Mg alloy. Alkalinization promotes swelling and decomposition of the silicon dioxide network, whereas accumulation of corrosion products promotes delamination and blistering of the coating with the formation of defects, thus exposing the Mg alloy to further degradation. Hence, increasing the bond strength and hydrolytic stability of silane coatings becomes critical in achieving reasonably good protection of Mg alloys against corrosion. Pinto et al. (2010) attempted a hybrid treatment that involves the formation of an anodic oxide film on the surface of an WE54 Mg alloy by immersing it in 0.1 M NaOH for 24 h, with an impressed potential of 0.5 V as the first step, followed by coating it with bis-[triethoxysilylpropyl]tetrasulfide silane as the second step. A synergistic effect of this hybrid treatment has been reported to

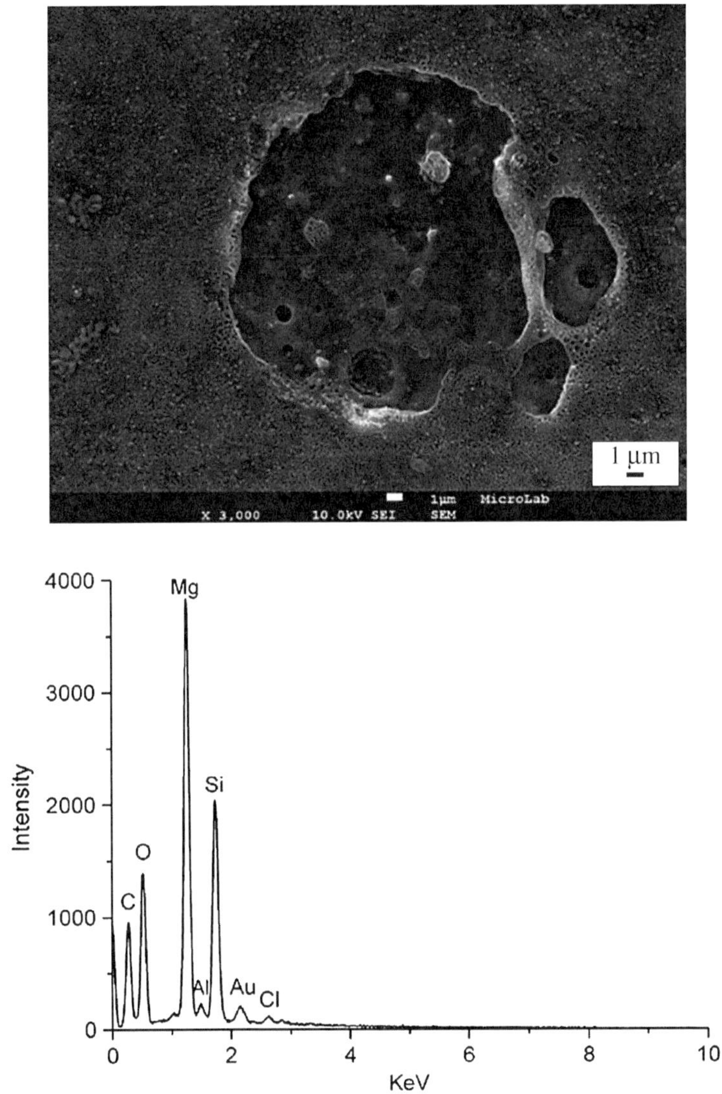

Figure 2.9 Surface morphology of the silane coating deposited on an AZ31 magnesium alloy after immersion in Hank's solution for 48 h and the EDX analysis performed on the corrosion products containing holes.
Adapted from Zomorodian et al. (2012) with permission from Elsevier.

offer some improvement in resistance against corrosion. However, Zomorodian et al. (2012) reported that direct current polarization treatment of an AZ31 Mg alloy in 0.1 M NaOH has no relevant effect on the protective performance of the glycidoxypropyltrimethoxysilane coating. Xue et al. (2012) combined bis-[triethoxysilyl] ethane (BTSE) and epoxy resin to modify the surface of an Mg–4Y alloy. This epoxy-modified BTSE silane coating successfully increased the resistance of the Mg−4Y alloy to corrosion at the initial stage of implantation, whereas it fails to offer long-term corrosion resistance in 0.9 wt% NaCl. The formation of cracks in the epoxy-modified BTSE silane coating, caused by the high temperature (100 °C) and shorter time (15 min) used during the curing process, is considered responsible for this behavior. Another limitation of this type of coating is that the epoxy resin alone is not biodegradable in a short time period. Liu et al. (2013) suggested a two-step procedure to introduce multifunctional, anticorrosive silane coatings on AZ31 Mg alloys. According to their methodology, in the first step, the Mg alloy is activated by immersion in NaOH, followed by treatment with BTSE to immobilize a layer of densely cross-linked silane coating. In the second step, amine functionality is imparted to the surface by treating the modified surface of the Mg alloy with 3-amino-propyltrimethoxysilane. In addition, they also showed that it is possible to functionalize the silane-modified surface of the Mg alloy with heparin to render the coating hemocompatible. The silane coatings act like a physical barrier and offered an improvement in corrosion resistance of the AZ31 Mg alloy. Functionalization of the silane-coated AZ31 Mg alloy with heparin indeed decreased the extent of protection against corrosion offered by the silane coating, while they significantly lower platelet adhesion. The ability of the silane coatings to provide a suitable base to impart biofunctional properties is beneficial. However, the bond strength and hydrolytic stability of silane coatings need to be improved for their acceptance as a surface modification method for Mg alloys toward the development of Mg-based degradable biomaterials.

2.3.7 Coating with biomolecules

Gelatin is a hydrolyzed collagen that is relatively more stable and easy to preserve than collagen. Nevertheless, it retains good bioactivity to promote cell attachment. Chan, Chian, and Tan (2013) deposited a gelatin coating on an amorphous $Mg_{67}Zn_{28}Ca_5$ alloy by electrospinning. The solvent used and the process parameters of electrospinning were optimized in such a way to render sufficient time for the solvent to evaporate before the gelatin fibers were deposited on the Mg alloy. Electrospinning provided a viable option to generate a nanofibrous structure that closely resembled tissue architecture and offered improved biocompatibility. Because gelatin is water soluble, the as-deposited gelatin coating was subjected to cross-linking by a dehydrothermal method. The as-deposited porous nanofibrous nature of the coating was quite similar to conventional electrospun structures, and it was not altered after cross-linking by the dehydrothermal method (Figure 2.10). The methodology offered unique advantages of controlling the thickness and cross-linking by varying the duration of electrospinning and dehydrothermal treatment. The porous/fibrous structure of the gelatin coating deposited on the amorphous $Mg_{67}Zn_{28}Ca_5$ alloy reduced the area of the alloy exposed to the external environment. The gelatin-coated amorphous $Mg_{67}Zn_{28}Ca_5$ alloy exhibited no indirect cytotoxicity, and it supported attachment of L929 and MG63 cell lines with high viability. Detachment of the gelatin coating is a matter of great concern, and its long-term stability needs to be ascertained.

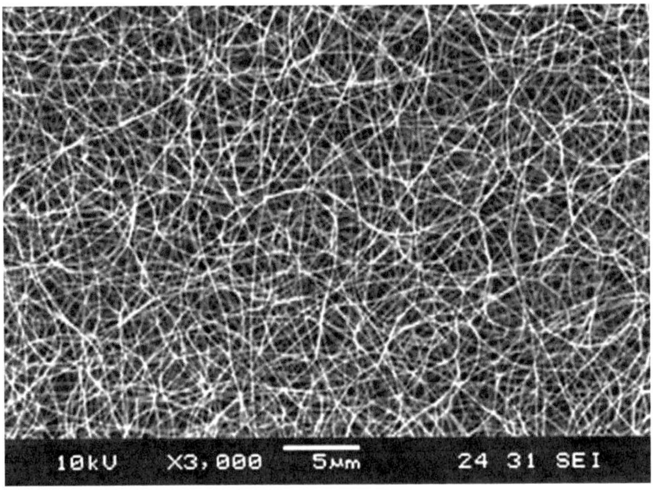

Figure 2.10 Scanning electron micrograph of an amorphous $Mg_{67}Zn_{28}Ca_5$ alloy coated with gelatin by electrospinning. A 10% gelatin solution in 1,1,1,3,3,3-hexafluro-2-propanol was electrospun at 25 kV for 1 h with a tip-to-alloy distance of 12 cm, freeze-dried overnight, followed by dehydrothermally cross-linking at 140 °C under vacuum pressure of 10^{-3} mm Hg for 5 days.
Adapted from Chan et al. (2013) with permission from Elsevier.

Chitosan is the N-deacetylated product of chitin, a linear polysaccharide derived from the shells of crustaceans such as crab and shrimp, the cuticles of insects, and the cell walls of fungi. Chitosan coatings assumed significance as biocompatible coatings following evidence of their immunological activity, bacteriostatic properties, biocompatibility, biodegradability, and low toxicity. In addition, they can be used to deliver drugs and promote wound healing. Fekry, Ghoneim, and Ameer (2014) deposited a chitosan coating on an AZ91E Mg alloy using the dip coating method. The Mg alloy was immersed in solutions containing 5%, 10%, and 15% chitosan dissolved in 1% acetic acid for 3 h followed by drying at 80 °C for 2 h. The chitosan coating was adherent and decreased the corrosion rate of the Mg alloy in a weakly acidic synthetic sweat medium. Among them, those coated using 15% chitosan offered better resistance against corrosion. Gu, Zheng, Lan et al. (2009) deposited a chitosan coating on an Mg-1Ca alloy using four different types of chitosan with varying degrees of deacetylation, intrinsic viscosity, and molecular weight. These coatings were deposited using the dip coating method at a withdrawal speed of 1 cm/min for 1, 3, 6, and 9 cycles, followed by drying at 60 °C for 20 min. Accordingly, the characteristics of the chitosan coating showed a strong dependence on its molecular weight, which increased the viscosity of the solution used for deposition. Only coatings deposited using chitosan with a molecular weight of 2.7×10^5 for 6 cycles offered a better resistance against corrosion in SBF. A detailed account of surface modification of Mg alloys by chitosan coatings is presented by Zheng and coworkers in Chapter 11 of Volume 2 of this book.

Albumin is the most abundant blood protein and is biocompatible (Peters, 1996). Albumin-coated surfaces have demonstrated antithrombogenic properties (Yamazoe, Oyane, Nashima, & Ito, 2010; Yamazoe & Tanabe, 2010;). However, attempts to coat the surface of Mg and its alloys with albumin have resulted in the formation of a rough and inhomogeneous coating often incorporated with $Mg(OH)_2$, phosphates, or other salts. Killian, Wagener, Schmuki, and Virtanen (2010) and Wagener, Killian, Turhan, and Virtanen (2013) showed that it is possible to functionalize Mg with protein layers via linker molecules. Initially, the Mg was passivated in 1 M boiling NaOH for 20 min. Next, three different linker molecules, namely, aminopropyl-triethoxysilane (APTES) plus ascorbic acid (VitC), carbonyldiimidazole (CDI), and SA were pre-coated over the passivated Mg surface. They were subsequently soaked in an aqueous albumin solution for different durations of time (from 0.25 h up to 24 h). The quality of the resultant coating depended on the uniformity and roughness of the precoating and the interaction between the linker molecules and albumin. Among the linker molecule precoatings, the CDI coating was very rough and inhomogeneous. Though relatively smooth, the SA coating lacked homogeneity. Among the three, only the APTES plus VitC coating was relatively smooth and homogeneous. The CDI coating was hydrophilic, whereas both the APTES plus VitC and SA coatings were hydrophobic. Because protein adsorption is preferred on hydrophobic surfaces, the interaction between protein and the APTES plus VitC and SA coatings were good. The Nyquist plots recorded during the albumin coating deposited for different durations of time (from 0.25 h up to 24 h) show an inductive loop in the low-frequency region, which indicates that all the three types of linker molecule precoatings were not able to prevent the dissolution of Mg. In addition, a reduction in peak intensities of the protein signals was observed for longer treatment times used during albumin coating. These inferences confirm that the linker molecule precoatings could not form a completely dense layer on the surface. Moreover, delamination of the linker molecule precoating might have occurred with an increase in treatment time.

2.3.8 Polymer coatings

Surface modification of Mg and its alloys by polymeric coatings is indeed an interesting approach, and in this respect biodegradable polymers assume significance toward the development of Mg-based degradable biomaterials. In fact, biodegradable polymers such as poly(L-lactic acid) (PLLA), poly(-caprolactone) (PCL), and poly(-glycolic acid) have been approved for human clinical uses including small load-bearing bone implants and cardiovascular interventions (O'Brien & Carroll, 2009). Many researchers have studied the deposition of PCL, PLLA, poly(glycolic acid), and others on Mg and its alloys using dip and spin coating methods (Alabbasi, Liyanaarachchi, & Kannan, 2012; Degner, Singer, Cordero, Boccaccini, & Virtanen, 2013; Kim et al., 2013; Ostrowski, Lee, Roy, Ramanathan, & Kumta, 2013; Wong et al., 2010; Xu & Yamamato, 2012). The findings of their studies showed that these coatings reduce the degradation rate of Mg without much accumulation of hydrogen gas during the initial stages of degradation, exhibit no inflammation or necrosis, provide good cytocompatibility, and promote larger volumes of new bone formation.

However, they also have some serious limitations that restrict their widespread clinical acceptance. Bobby Kannan Mathan has addressed the surface modification of Mg alloys by biodegradable polymers in Chapter 13 of Volume 2 of this book; only some interesting facts about them are presented in this section.

Chen, Song, Zhang, Li, Zhao et al. (2011) reported that PLA coating deposited on pure Mg using the dip coating method is thick, nonuniform, and possesses insufficient adhesion to offer reasonably good protection against corrosion. Park et al. (2013) pointed out the evolution of porosity in a PCL coating prepared by the dip coating method following solvent removal. They noticed rapid water absorption at the early stage of corrosion, resulting in detachment of the PCL coating. An inhomogeneous durability of the coating with the formation of gas pockets, leading to an eventual detachment of poly(lactic-*co*-glycolic acid) deposited by the dip coating method, was observed by Ostrowski et al. (2013). It is important to produce a homogeneous, nonporous coating with good adhesion to achieve better performance. According to Xu and Yamamato (2012) and Alabbasi et al. (2012), polymer coatings produced by spin coating are thin, homogeneous, and relatively more adherent than those prepared using the dip coating method. However, poor adhesion of PLLA, PCL, and poly(ether imide) coatings deposited by the spin coating method also was observed by many researchers (Alabbasi et al., 2012; Degner et al., 2013; Kim et al., 2013). Hence, in spite of their ability to show a reduced degradation rate of Mg during the initial periods of immersion and good cytocompatibility, the degradable polymer coatings deposited by both the dip and spin coating methods could not offer a long-term protection against corrosion. A careful analysis of the mode of failure of the coatings revealed detachment of coatings induced by bulk erosion/degradation, resulting in the formation of many micro-/macropores and cracks in early stages of the degradation (Tamada & Langer, 1993). It has been reported that many biodegradable polymer coatings, such as PCL, could actually enhance the corrosion rate of Mg alloys since they degrade by the hydrolysis of their ester bonds and release carboxylic acids into the local environment (Chen, Song, Zhang, Li, Zhao et al., 2011). In addition, once an Mg alloy is exposed to a corrosive environment, the formation of $Mg(OH)_2$ could accelerate the hydrolysis rate and weaken the stability of the PCL coating. Moreover, the acidic products generated following the hydrolysis of PCL could react with the substrate under aqueous conditions (Wang, He et al., 2013). This leads to the question of whether deposition of degradable polymer coatings is useful for improving the performance of Mg-based biomaterials. If so, what are the requirements of such coatings?

Wang, He et al. (2013) developed a surface-eroding coating of poly(1,3-trimethylene carbonate) (PTMC) on an Mg alloy and compared its performance with the bulk-eroding PCL coating. Accordingly, the PTMC coating (1) exhibited uniform surface erosion against the nonuniform bulk erosion shown by the PCL coating; (2) effectively protected the corrosion of the Mg alloy under dynamic degradation test conditions; (3) reduced the corrosion current density by three orders and one order of magnitude when compared with an uncoated and PCL-coated Mg alloy, respectively; (4) decreased platelet adhesion and erythrocyte attachment; (5) reduced the percentage of hemolysis; (6) showed no excessive inflammation, necrosis, or hydrogen gas accumulation; and (7) exhibited less volume reduction and fewer corrosion products compared with the controls after 52 weeks under *in vivo* conditions. The homogeneous surface erosion of the PTMC coating from exterior to interior is one of the beneficial attributes for the development of fully biodegradable cardiovascular stents. A schematic of the degradation of polymer coatings by surface and bulk erosion is depicted in Figure 2.11.

Turhan, Weiser, Jha, and Virtanen (2011) explored deposition of polypyrrolle for the protection of an AZ91D Mg alloy against corrosion by electropolymerization. Accordingly, the choice of potential range and scan rate during deposition is critical in determining the uniformity and adhesion of the coating to achieve a better corrosion resistance. Srinivasan, Ranjani, and Rajendran (2013) evaluated the performance of polypyrrolle coatings electropolymerized on an AZ31 Mg alloy in SBF. Accordingly, in addition to the potential range, the monomer (pyrrolle) concentration is also important; only at a suitable concentration were the resultant polypyrrolle coatings uniform and crack free, with a distinct cauliflower-like morphology.

Deposition of conducting polymer coatings on Mg alloys assumes significance because of their ability to load anti-inflammatory drugs, which can be subsequently released upon electric stimulation. Poly(3,4-ethylenedioxythiophene) (PEDOT) is a good candidate material for such applications. PEDOT can be electrochemically deposited on many metallic surfaces. However, the rapid corrosion of Mg in many electrolyte mediums poses a serious problem in the electrochemical deposition of PEDOT on Mg and its alloys. Luo and Cui (2011) considered ILs as a suitable medium for the deposition of PEDOT because ILs are highly conductive and stable, with a very wide electrochemical window, and PEDOT coatings could be directly electrodeposited on Mg. The PEDOT coatings electrodeposited on Mg were uniform and reduced the corrosion current density of Mg by about 50%. The ability of the methodology to load an anti-inflammatory drug, namely, dexamethasone, in the PEDOT coating during ED and its subsequent release by electrical stimulation confirmed its potential application in delivering anti-inflammatory drugs. Sebba et al. deposited PEDOT coatings on Mg alloys for neural implant applications (Sebaa, Dhillon, & Liu, 2013). The details about the method of deposition, characteristics of the coatings, and their performance are addressed in Chapter 12 of Volume 2 of this book. The PEDOT coatings electrodeposited using an IL as the electrolyte medium

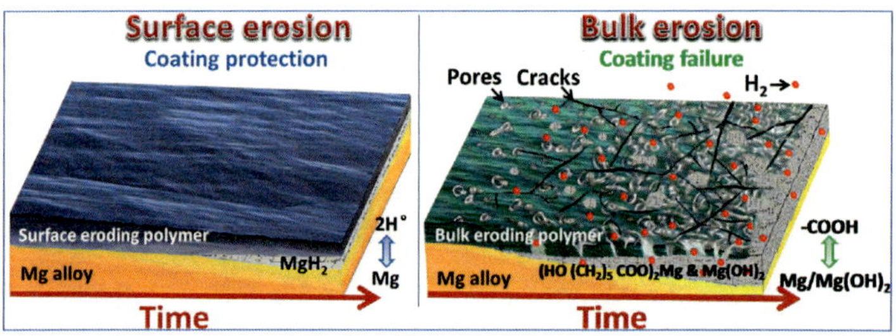

Figure 2.11 Schematic of the degradation behavior of polymer coatings by surface erosion and bulk erosion mechanisms.
Adapted from Wang, He et al. (2013) with permission from Elsevier.

also has some limitations. In spite of its ability to reduce the corrosion rate by about 50%, the PEDOT coating could not totally prevent the Mg from corrosion. In addition, after multiple electrical stimulations for drug release, the PEDOT coating could partly detach from the substrate, and the anions of the IL released during the process should be biocompatible and nontoxic. It has been reported that PEDOT is biocompatible, and its mode of degradation under *in vivo* conditions has not yet been fully studied.

2.4 Electrochemical surface modification

2.4.1 Anodizing and microarc oxidation

Anodizing is a well-known electrochemical oxidation process for producing thick and stable oxide film on metals and alloys to improve their hardness and resistance against both corrosion and wear for engineering applications. The electrochemically formed oxide layer also has the ability to serve as an excellent base to improve the adhesion of paint coatings and to impregnate dyes for esthetic or architectural applications. In addition, these oxide coatings possess higher thermal stability, thermal shock resistance, and heat resistance; high dielectric strength; and good optical properties (Blawert, Dietzel, Ghali, & Song, 2006). Anodizing is widely used to treat Al, Mg, titanium, tantalum, niobium, zirconium (Zr), and their alloys. The anodizing behavior of Mg alloys is strongly influenced by the applied voltage or current, the type of electrolytes, the nature and concentration of the alloying elements, and the microstructure of the alloy (Blawert et al., 2006; Hiromoto & Yamamoto, 2010; de Oliveira, Pereira, Correa, & Antunes, 2013; Xue, Yun, Schulz, & Shanov, 2011). In general, different passive and active states were found at different anodizing voltage or current regions. Sparking, microarcing or arcing were usually observed at high voltages (above 50 V), and they have been referred to as microarc oxidation (MAO) or plasma electrolytic oxidation (PEO) processes (Blawert et al., 2006; Ghasemi, Raja, Blawert, Dietzel, & Kainer, 2008; Hussein, Zhang, Nie, Xia, & Northwood, 2011; Malayoglu, Tekin, & Shrestha, 2010). Above the breakdown potential, porous films were formed. The breakdown potential could vary with the type of Mg alloy and the electrolytes used. A major limitation of oxide films formed electrochemically by anodizing was their adverse effect on fatigue properties, caused by oxidation-induced surface tensile stress, structural defects in the oxide layer, and softening of the substrate with age following the heat associated with oxide film formation, which restricts their widespread acceptance for many applications. For Mg and its alloys, the combination of these factors seemed to be particularly disadvantageous because magnesia has both high specific heat during formation and a substantial lattice misfit with the metal. Conventional anodizing failed to minimize the risk of premature fatigue failure of Mg alloys. Because fatigue strength is critical for load-bearing implants, the suitability of this method for surface engineering of Mg and its alloys becomes questionable. In this respect, PEO is considered an option to reduce this risk of fatigue failure (Yerokhin et al., 2004).

Details on the mechanism of deposition of MAO and the characteristics of coatings are covered in Chapter 8 of Volume 2 of this book. The architecture of MAO coatings on Mg and its alloys has a three-layer structure: a porous outer layer with several large, deep pores/cavities, a middle layer with less porosity, and a thin barrier layer. The pores and cracks formed in MAO coatings help to release residual stress and significantly improve the mechanical interlocking effect, resulting in a higher bond strength. However, the presence of a higher pore density on the surface of the MAO coatings of Mg and its alloys increases the effective surface area and thus the tendency of the corrosive medium to be adsorbed by and concentrate in these pores. This facilitates quicker infiltration of the corrosive medium into the inner regions of the coating and subsequently down to the substrate, thus deteriorating the resistance of the coating to corrosion by changing its local pH. The pore density, distribution of pores, and interconnectivity of the pores with the substrate are the critical factors that decide its ability to protect against corrosion. The various strategies adopted to improve the corrosion resistance of MAO-coated Mg alloys is addressed in Chapter 9 of Volume 2 of this book. It is believed that with the advent of new strategies to achieve a better resistance against corrosion for MAO-coated Mg alloys, this method of surface modification would assume significance in the development of degradable Mg-based biomaterials.

2.4.2 Cathodic plasma electrolytic deposition

Paulmier, Bell, and Fredericks (2007) developed a novel cathodic plasma electrolytic deposition method that combines galvanic processes and plasma–chemical phenomena and operates at atmospheric pressure. The basic principle of this method is the application of a high voltage between a cathode and an anode (with a cathodic-to-anodic area ratio of 1:0.08) immersed in an electrolyte solution. The high electric field near the cathode induces a strong Joule heating in its vicinity and triggers the formation of a vapor sheath around it. Because the electric conductivity of this vapor sheath is much lower than that of the liquid solution, the voltage drop occurs primarily within the vapor sheath, thus inducing a strong electric field within the separated bubbles. If the applied voltage is sufficiently high, a plasma glow discharge is generated within the bubble, resulting in dissociation of the vaporized electrolyte and formation of a continuous coating on the cathode. High deposition rate, ability to produce uniform and thicker coatings, operation at atmospheric pressure, and direct production of crystalline coatings without the need for annealing after deposition are the major advantages of this method. Liu, Pan, Yang, Cai, and Chen (2012) deposited Al_2O_3–ZrO_2 ceramic coatings on an WE43 Mg alloy by cathodic plasma electrolytic deposition using a solution mixture of $Al(NO_3)_3$ (20 g/L), $Zr(NO_3)_4$ (4 g/L) and ethanol at 400 V for 1 h, with a pulse frequency of 100 Hz and a duty cycle of 30%. The resultant coatings composed of t-ZrO_2, α-Al_2O_3, and γ-Al_2O_3 phases enabled an anodic shift in E_{corr} by 300 mV and a 50-fold decrease in i_{corr} in SBF. The biocompatible nature of the ceramic oxide phases and the significant improvement in corrosion resistance are certainly encouraging. However, the surface roughness and presence of many irregularly distributed open macropores in the resultant coatings are the major concerns in achieving a long-term protection against corrosion.

2.4.3 Electrodeposition

Electrodeposition (ED) is a versatile and cost-effective method for depositing metals, alloys, metallic oxides, ceramics, and composites. It offers the unique advantages of controlling the thickness and chemical composition of the coatings by suitably varying the deposition potential/current. Use of a pulsed current mode instead of the conventional direct current mode extends the window of opportunity of this methodology to prepare good-quality coatings. ED is widely used to modify the surface of Mg and its alloys, and among them deposition of HA has assumed significance (Jamesh, Kumar, Sankara Narayanan, 2012; Song, Shan, & Han, 2008). The electrolyte solution used for the deposition of HA contains 0.1 M $Ca(NO_3)_2$, 0.06 M $(NH_4)_3PO_4$, and 10 mL/L of 30 vol% hydrogen peroxide (pH 4.2; temperature, 27 °C). An increase in temperature has enabled an increase in the reactivity and rate of deposition. The as-deposited coating mainly consisted of DCPD (also referred as brushite), which converts to HA after subsequent immersion in 1 M NaOH at 80 °C for 2 h (Jamesh et al., 2012; Song et al., 2008). The ED HA coating helped to improve the corrosion resistance of Mg alloys, in addition to providing better bioactivity.

In recent years, the use of pulsed current mode for ED of HA and ED of substituted or doped HA coatings has received considerable attention. Among them ED is a versatile and cost-effective, calcium-deficient HA and F-, Sr-, Zn-, and Si-doped HA are considered beneficial because of their ability to provide an increase in bioactivity in SBF (Drevet & Benhayoune, 2013; Huang et al., 2013; Meng et al., 2011; Qiu, Wan, Tan, Fan, & Yang, 2014; Wang et al., 2011; Wang, Guan, Wang, Ren, & Wang, 2010). One of the major impediments of the ED HA coating is the poor bond strength (about 4–6 MPa) (Ban & Hasegawa, 2002), which could lead the coating to peel off after implantation. Because MAO coatings have a porous network, they could be exploited to pin HA within the pores. This concept was effectively used by many researchers (Chen et al., 2012; Gao et al., 2011; Shi, Qi, Chen, & Shi, 2011; Zhao, Guo et al., 2013) to deposit DCPD, HA, and rodlike nano HA by ED. This approach helped to seal the pores in the MAO coating, increased the bonding strength of the HA coating, and increased the bioactivity and corrosion resistance of the resultant composite coating.

Cathodic electrochemical deposition of metal hydroxides following electrochemical reduction of the corresponding nitrates and their subsequent conversion to the corresponding metal oxides by thermal treatment is a well-established method. This methodology was used by Li, Wang, Ho, Li, and Yen (2011) to deposit MgO on an AZ91D Mg alloy. In spite of its ability to improve the corrosion resistance of the Mg alloy, the adhesion strength (10 MPa) is much less than that required (35 MPa) for its successful use in orthopedic implant applications. Lei, Ouyang, Tang, Li, and Zhou (2010) explored anodic electrochemical deposition of MgO on an Mg-Zn-Ca alloy. According to them, the resultant dense MgO coatings offered an improvement in corrosion resistance in SBF. A detailed account of the anodic ED of MgO is presented by Lei in Chapter 6 of Volume 2 of this book.

2.4.4 Electrophoretic deposition

EPD is a versatile and cost-effective material processing technique to produce homogeneous and dense ceramic, polymer, and composite coatings for biomedical applications. It offers excellent control of the thickness, stoichiometry, and microstructure of the deposited layers by suitably adjusting the applied voltage and treatment time. The fundamentals, mechanism, and kinetics of EPD were addressed in many reviews (Besra & Liu, 2007; Boccaccini, Cho, Subhani, Kaya, & Kaya 2010; Boccaccini, Van der Biest, & Talbot, 2002; Boccaccini & Zhitomirsky, 2002; Corni, Ryan, & Boccaccini, 2008; Farrokhi-Rad, Loghmani, Shahrabi, & Khanmohammadi, 2014; Heavens, 1990; Mohan, Durgalakshmi, Geetha, Sankara Narayanan, & Asokamani, 2012; Sarkar & Nicholson, 1996; Van der Biest & Vandeperre, 1999; Zhitomirsky, Roether, Boccaccini, & Zhitomirsky, 2009). EPD has been widely used for deposition of inorganic particles (particularly HA), bioactive glass, and their composites on stainless steel, titanium alloys, and shape memory alloys for biomedical applications. Adhesion of the EPD coatings in their as-deposited conditions is often considered a limitation. Hence, after deposition, a heat-treatment step is usually adopted to increase the adhesion and density of the coating. Because Mg has a relatively low melting point, heat treatment of EPD coatings after being deposited on Mg and its alloys at higher temperatures is a difficult proposition. Many researchers have considered MAO coatings deposited on Mg and its alloys as a pretreatment for deposition of a variety of inorganic materials by EPD (Razavi, Fathi, Savabi, Beni et al., 2013, 2014; Razavi, Fathi, Savabi, Mohammad Razavi et al., 2013, 2014; Rojaee et al., 2013a; Wu, Wen, Dai, Lu, & Yang, 2010; Zhang, Dai, Wei, & Wen, 2012). The MAO coatings are highly adherent to the base metal and are porous. It is believed that the porous nature of the MAO coating could help to pin the inorganic materials deposited by EPD in the subsequent stage. Razavi, Fathi, Savabi, Beni et al. (2013, 2014), and Razavi, Fathi, Savabi, Mohammad Razavi et al. (2013, 2014) deposited nanostructured bredigite ($Ca_7MgSi_4O_{16}$), diopside ($CaMgSi_2O_6$), and akermanite ($Ca_2MgSi_2O_7$) coatings by EPD over an MAO-coated AZ91 Mg alloy and showed that these coatings increased the resistance against corrosion and improved the *in vitro* bioactivity. Wu et al. (2010) deposited a calcium phosphate/chitosan composite coating by EPD on an MAO-coated AZ91D Mg alloy followed by a conversion coating process in PBS. Zhang et al. (2012) confirmed that the binding strength of the calcium phosphate/chitosan composite coating deposited by EPD over an MAO-coated AZ91D Mg alloy is good. Rojaee et al. (2013a) compared the ability of a fluoride conversion coating and an MAO coating deposited on an AZ91 Mg alloy as a pretreatment for the subsequent deposition of nanostructured HA coating by EPD. Accordingly, the nanostructured HA coatings deposited over an MAO-coated Mg alloy offered better resistance against corrosion and better bioactivity. Hence, it is clear that to overcome the limitations due to adhesion of the as-deposited coatings by EPD, MAO coatings could be used as a pretreatment to improve the adhesive strength and to achieve a better corrosion resistance and bioactivity of the composite coatings deposited by the combination of MAO and EPD.

2.5 Plasma surface modification

A variety of plasma surface modification methods, which include physical vapor deposition (PVD) processes (evaporation, magnetron sputtering, ion plating, and ion beam-assisted deposition), ion implantation, plasma spraying, and others, have been explored to modify the surface of Mg alloys. The role of plasma surface modification of magnesium alloys for biomedical application has been reviewed by Yang, Cui, Lee, and Wang (2010). Stephen Abela presents a detailed account of PVD on Mg alloys for biomedical applications in Chapter 4 of Volume 2 of this book.

2.5.1 Physical vapor deposition

Chromium nitride, titanium nitride, TiAlN, diamond-like carbon film, and others were successfully deposited on Mg alloys using PVD (Chen, Huang et al., 2013; Zhang, Yang, Cui, Lee, & Lee, 2010). The major challenges encountered in PVD of various types of coatings on Mg alloys were the control of deposition temperature and achieving good adhesion of the coatings. The application of a pulsed bias voltage during deposition could help reduce the deposition temperature; preventing oxidation of Mg alloys during deposition helped to improve the adhesion of the PVD coatings. In spite of these advancements in the deposition process, the major limitation of coatings deposited by PVD is the presence of pores and pinholes, which in most cases are impossible to eliminate completely (Hoche, Groß, Foerster, Schmidt, & Adamitzki, 2009). Because these coatings are very thin (on the order of only a few microns), the defects in them allow quick permeation of the corrosive medium through to the surface, leading to the formation of a galvanic cell, resulting in pitting corrosion of the Mg alloy (Figure 2.12). The inferior quality and poor adhesion of the coatings prepared using a conventional PVD technique are considered the direct consequences of the low energy of the atoms arriving at the substrate during film growth. It implies that if the energy of the incident atoms or ions is increased, then the quality and adhesion of the coatings could be improved. This led to the development of the filtered cathodic arc deposition method in which the high kinetic energy and ionization rate (close to 100%) lead to the deposition of coatings with superior properties, including excellent adhesion and high density (Tay, Zhao, & Chua, 2006).

Xin, Liu, Zhang, Jiang et al. (2008), Xin, Liu, Zhang, Huo et al. (2008), and Xin et al. (2009) deposited Al/Al_2O_3, Zr/ZrO_2, and Zr/ZrN bilayer coatings on an AZ91D Mg alloy by filtered cathodic arc deposition. These coatings are ~1.5 μm thick. However, they are uniform, dense, and have no visible pores and cracks. The deposition of the transition layer (Al in case of Al/Al_2O_3 and Zr in case of Zr/ZrO_2 and Zr/ZrN) helped to improve the adhesion of the subsequently deposited Al_2O_3, ZrO_2, and ZrN layers. In addition, the transition layer buffers the stress developed

Figure 2.12 Comparison of the degradation behavior of naked and diamond-like carbon–coated AZ31 and AM50 magnesium alloys after 5, 10, and 15 days of immersion in simulated body fluid (SBF) at 37 °C. Inset: Scanning electron micrographs and energy-dispersive X-ray spectroscopy analysis performed at the surface of the samples after immersion in SBF for 15 days. Scale bar = 10 μm.
Adapted from Zhang, Yang et al. (2010) with permission from Elsevier.

due to the mismatch between the Mg alloy and the Al_2O_3, ZrO_2, and ZrN layers. The Al/Al_2O_3, Zr/ZrO_2, and Zr/ZrN bilayer coatings deposited by filtered cathodic arc deposition offered a significant improvement in the resistance of the AZ91D Mg alloy to corrosion in SBF. However, the coatings could not completely prevent penetration of the electrolyte, which led to deterioration of their protective properties after long-term exposure in SBF.

2.5.2 Plasma-enhanced chemical vapor deposition

Li et al. (2012) and Li, Cheng et al. (2013) deposited amorphous Si and silicon carbide (SiC) films on an WE43 Mg alloy using PECVD. Accordingly, these films effectively slowed the degradation rate and alleviated local alkalization of the WE43 alloy in SBF at 37 °C (Figure 2.13). In addition, the extraction medium of the Si- and SiC-coated WE43

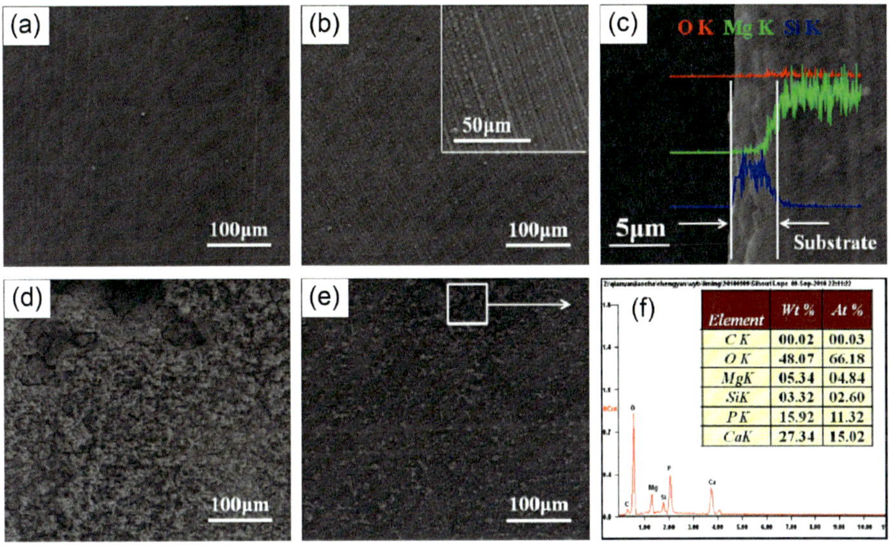

Figure 2.13 Scanning electron microscopy images of the uncoated (a, d) and plasma-enhanced chemical vapor deposited (PECVD) silicon (Si)-coated (b, e) WE43 magnesium (Mg) alloy before immersion (a, b) and after 240 h of immersion (d, e) in simulated body fluid. (c) Cross-sectional morphologies and energy-dispersive X-ray spectroscopy (EDS) line scan of a PECVD Si-coated WE43 Mg alloy. (f) Energy-dispersive X-ray spectroscopy analysis of the corrosion products on an Si-coated WE43 alloy.
Adapted from Li, Cheng et al. (2013) with permission from Elsevier.

Mg alloy exhibited no inhibitory effect on L929 cell growth and enabled higher cell viabilities than the uncoated alloys. Moreover, the hemolysis rate of the Si- and SiC-coated WE43 alloy is considerably decreased. Levy and Aghion (2013) deposited a 1-μm-thick neodymium (Nd) coating on an Mg−1.2%Nd−0.5%Y−0.5%Zr−0.4%Ca alloy (EW10X04) by PVD using an electron gun evaporator and further subjected it to diffusion heat treatment at 320 °C for 3 h in a protective inert atmosphere of nitrogen and 0.2% sulfur hexafluoride at high vacuum. The resultant coatings were homogeneous, adherent, and offered better corrosion resistance for the Mg alloy in simulated physiological electrolyte. The formation of a continuous network of passive intermetallic at grain boundaries in the form of $Mg_{41}Nd_5$ and enrichment of the oxide film with Nd and Nd oxides such as Nd_2O_3 and Nd_6O_{11} are considered responsible for the improved corrosion behavior.

2.5.3 Ion plating

Zhang, Xu, and Yang (2005) explored the use of ion plating to deposit a well-adherent, 10-μm-thick, dense, and pore- and crack-free Ti coating on pure Mg with good interfacial bonding between the Ti coating and the Mg substrate. The resultant Ti coating decreased the corrosion current density by one order of magnitude lower than that of the uncoated Mg.

2.5.4 Sputtering

Sputtering is one of the well-known methods for the deposition of thin films to modify the surface properties of materials, imparting them with, for example, better resistance against corrosion, improved biocompatibility, enhanced bioactivity (Liu, Chu, & Ding, 2004). The ability to deposit thin films at relatively higher rates over a larger deposition area with low substrate heating makes magnetron sputtering an effective method for the surface modification of Mg alloys. Wu (2007) used multi-magnetron sputtering to prepare an Al coating and an Al/Ti multi-layer coating on an AZ31 Mg alloy, and both offered improved resistance against corrosion. Tacikowski, Banaszek, and Smolik (2014) recently used a hybrid method that combines magnetron sputtering, PVD, and a chemical treatment to prepare an Al/Ti/TiN coating on an AZ91D Mg alloy. The coated layer consists of an outer TiN coating produced by PVD with an Al subcoating adjacent to the substrate and a thin transition Ti subcoating, both prepared by magnetron sputtering. The outer TiN coating is sealed using a boiling aqueous solution for 15−120 min. The presence of the Al sublayer and its diffusion bonding with the Mg alloy during deposition of the TiN_2TiAl_{10} composite coating effectively separates the Mg alloy from the corrosive environment, which is evidenced by the anodic shift in E_{corr} of the Mg alloy by 500 mV.

Plasma surface modification processes also have been explored for the deposition of calcium phosphate coatings. Surmenev (2011) reviewed the various plasma-assisted methods of depositing calcium phosphate-based coatings. Among them, plasma spraying, radiofrequency magnetron sputtering, pulsed laser deposition, and ion beam-assisted deposition assume significance.

2.5.5 Plasma immersion ion implantation

Plasma immersion ion implantation (PIII) has emerged as a key surface modification method in biomedical engineering because of its ability to overcome the line-of-sight limitations encountered by conventional beam-line ion implantation coupled with a high process efficiency and flexibility in treating components with a larger area (Chu, 2004; Huang et al., 2004). There is no change in the dimensions of the materials, but an improvement in hardness and elastic modulus are observed after PIII. The role of PIII in controlling surface degradation as well as in the surface design of biodegradable Mg alloys was addressed by Chu (2013) and Wu, Jamesh, and Chu (2013). Chu and his coworkers extensively studied the effect of PIII of Mg alloys using a variety of ions, either by single-ion or dual-ion implantation. The type of ions implanted in Mg alloys, along with their characteristics and corrosion behavior, are compiled in Table 2.1. Implantation of Zn, Cr, and Ti accelerate the degradation of Mg alloys because of their existence in a metallic state in the implanted layer and the formation of galvanic corrosion cells with the substrate material. Dual-ion implantation offered better resistance against corrosion for Mg alloys when compared with single-ion implantation (Figure 2.14) (Zhao, Wu et al., 2013). The *in vivo* stimulation of bone formation by Al and oxygen dual-implanted Mg implants was studied by

Table 2.1 Types of ions implanted in magnesium alloys, their characteristics and corrosion behaviour, and factors limiting their acceptance for biomedical applications

Types of ions implanted in Mg alloy	Characteristics and corrosion behavior of the implanted Mg alloy	Factors limiting their widespread acceptance for biomedical applications
Single-ion implantation Zn Cr Ti O N Y Ta Al Ag Si Ce Dual-ion implantation Cr and O Al and O Ti and O N and Ti N and Al	Formation of a compact oxide layer of the implanted metal ion(s) Formation of an intermixed layer and intermetallic compounds Increase in thickness of the oxide layer Increase in surface roughness Higher irradiation dose generates defects Grain boundaries become less and less clear following ion bombardment Improvement in corrosion resistance is observed only when these ions are implanted under optimum conditions	Galvanic corrosion (for Zn, Cr, and Ti) Presence of local defects in the film Susceptibility to localized corrosion Effective only during the early stages of corrosion, and it may be difficult to achieve a long-term resistance to corrosion

Ag, silver; Al, aluminum; Ce, cerium; Cr, chromium; N, nitrogen; O, oxygen; Si, silicon; Ta, tantalum; Ti, titanium; Y, yttrium; Zn, zinc.

Wong et al. (2013). There are many factors that limit the widespread acceptance of PIII of Mg alloys for biomedical applications. Ion bombardment is similar to cold working. Hence, the internal stress generated during PIII could cause deformation of the surface lattice structures and a large number of amorphous structure defects on the surface, which could accelerate galvanic corrosion. Implantation at lower doses failed to improve corrosion resistance, and the use of a much higher dose generated defects in the treated surface. A longer treatment time at a particular dose also caused a negative influence on the corrosion resistance. A hybrid treatment that involves solid solution aging followed by duplex PIII of nitrogen and Al on an AZ91D Mg alloy was explored by Hongxi, Qian, Damin, Bo, and Chunlei (2013). This hybrid treatment changed the surface microstructure, produced a transition layer consisting of $MgAl_2O_4$ with a small amount of Mg_3N_2 and AlN phases, improved the Mg–O and Al–O bonding states, and increased the thickness of the oxide layer as well as homogenous surface properties, resulting in enhanced resistance against corrosion. In spite of these

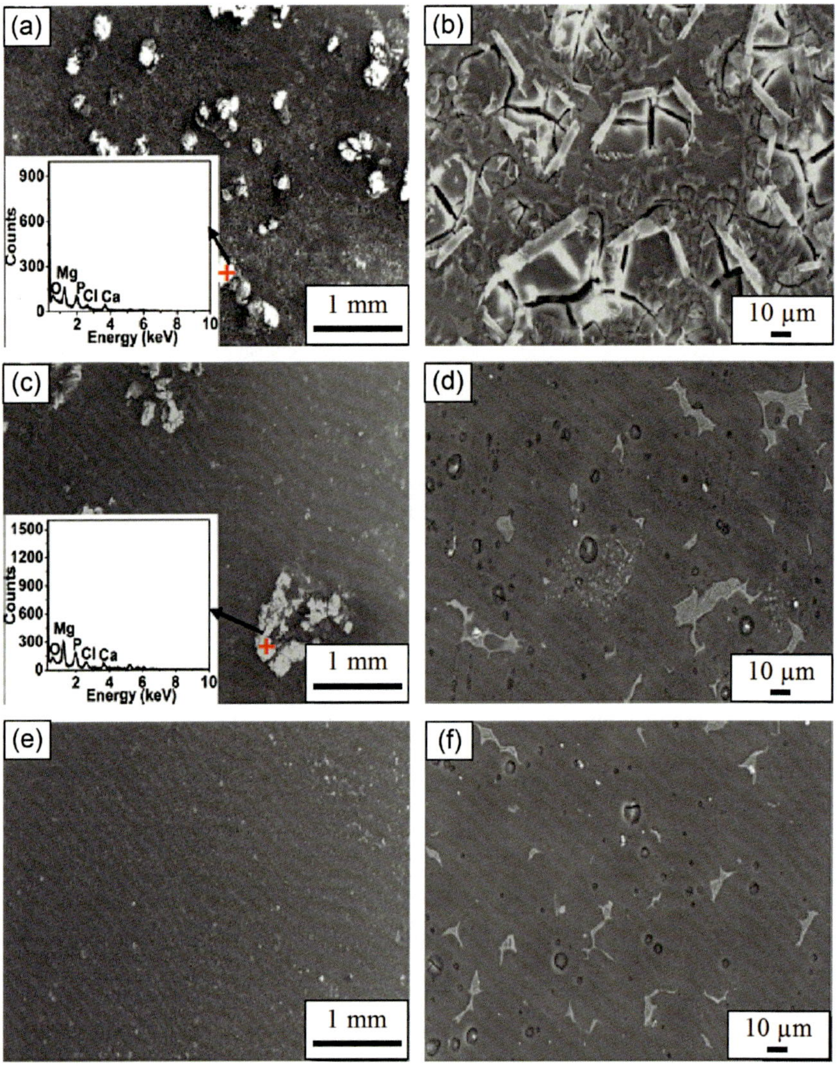

Figure 2.14 Surface morphology of untreated (a, b) and ion-implanted (titanium (c, d) and titanium–oxygen (e, f)) WE43 magnesium alloys after immersion in simulated body fluid for 3 h. Insets in (a) and (c) are the energy-dispersive X-ray spectroscopy analysis performed at the white corrosion products marked with a red +.
Adapted from Zhao, Wu et al. (2013) with permission from Elsevier.

modifications, the surface of Mg alloys treated by PIII always exhibited pitting corrosion, and their ability to provide a long-term corrosion resistance remains open. Based on the characteristic properties of the PIII-treated Mg alloys, it seems that the window of opportunity of using this method for surface engineering Mg toward the development of degradable implants is rather limited.

2.6 Laser surface modification

Depending on the kind of interactions between the laser and the material being treated, as well as the type of microstructural/compositional changes induced at the surface, laser-based surface modification can be classified as follows: laser surface hardening, laser surface melting (LSM), laser surface alloying (LSA), laser composite surfacing, laser surface cladding (LSC), and laser shock peening (LSP) (Dahotre, 1998; Singh & Harimkar, 2012). The role of laser surface engineering of Mg alloys was reviewed by Singh and Harimkar (2012). The effect of different types of laser treatments on improving the corrosion resistance of Mg alloys is addressed by Dutta Majumdar and Manna (2013, chap. 6). The importance of surface modification of biomaterials by lasers has been addressed by Dahotre, Paital, Samant, and Daniel (2010), Narayan and Goering (2011), and Bandyopadhyay, Balla, Roy, and Bose (2011). Pulsed laser deposition is another useful approach to modify the surface of Mg alloys.

2.6.1 Laser surface melting

LSM involves heating materials using a high-power laser source, namely, neodymium-doped yttrium—Al—garnet or carbon dioxide lasers, to their melting point followed by rapid solidification of the melted zone. Optimizing the conditions to avoid oxidation of the metal surface is necessary during LSM. The common aims of LSM are (1) refinement of the surface microstructure; (2) homogenization of composition; and (3) dissolution of precipitates (Dutta Majumdar & Manna, 2011). LSM has been explored as a surface modification method for many Mg alloys. Guo, Yue, and Man (2005) reported that the surface of a WE43 Mg alloy subjected to LSM was free of corrosion even after 4 h of immersion in 3.5% NaCl, whereas its untreated counterpart showed significant pitting under similar conditions. Abbas, Liu, and Skeldon (2005) reported that after LSM, the corrosion rate of AZ31, AZ61, and WE43 alloys in 5% NaCl with a pH of 10.5 at 20 °C was reduced by about 30%, 66%, and 87%, respectively. The improvement in corrosion resistance exhibited by these three alloys was due to the refinement of α-Mg grains and uniform redistribution of the β-phase after LSM. Among them, the ability of the AZ61 Mg alloy to offer a better corrosion resistance than the AZ31 Mg alloy was primarily due to the higher percentage of Al in it. The superior corrosion resistance offered by the WE43 Mg alloy was due to the presence of a rare earth element on the surface. In addition, in the case of the WE43 Mg alloy, the entire melted layer was not consumed during the test period.

Guan, Zhou, and Zheng (2009) studied the corrosion behavior of LSM of an AZ91D Mg alloy in SBF. The general and pitting corrosion resistance of the laser-treated AZ91D Mg alloy surface was significantly enhanced by the refined continuous network of β-$Mg_{17}Al_{12}$ phases and the increased Al concentration in the laser-melted zone. Banerjee, Singh Raman, Durandet, and Mcadam (2011) observed a marginal improvement in the resistance of an ZE41 Mg alloy to corrosion in 0.001 M NaCl only during the initial periods of immersion (1 h), which is nullified with a further increase in time. The main reason for the lack of improvement in corrosion resistance of the LSM ZE41 Mg alloy was because of the formation of surface cracks following the high thermal stress created by the high input energy of the laser irradiation. This was substantiated by the smaller extent of surface cracking at a laser power of 820 rather than 1856 W. Another major issue in LSM is the change in microstructural features and surface roughness in the overlapping tracks during treatment. Coy et al. (2010) showed a change in microstructure, increase in surface roughness, and formation of microcracks in the overlapping tracks, as well as the presence of trapped pores, during LSM of an AZ91D Mg alloy, which increased with an increase in number of laser pulses (Figure 2.15). Guan et al. (2009), Guan, Zhou, Zheng, and Li (2010),

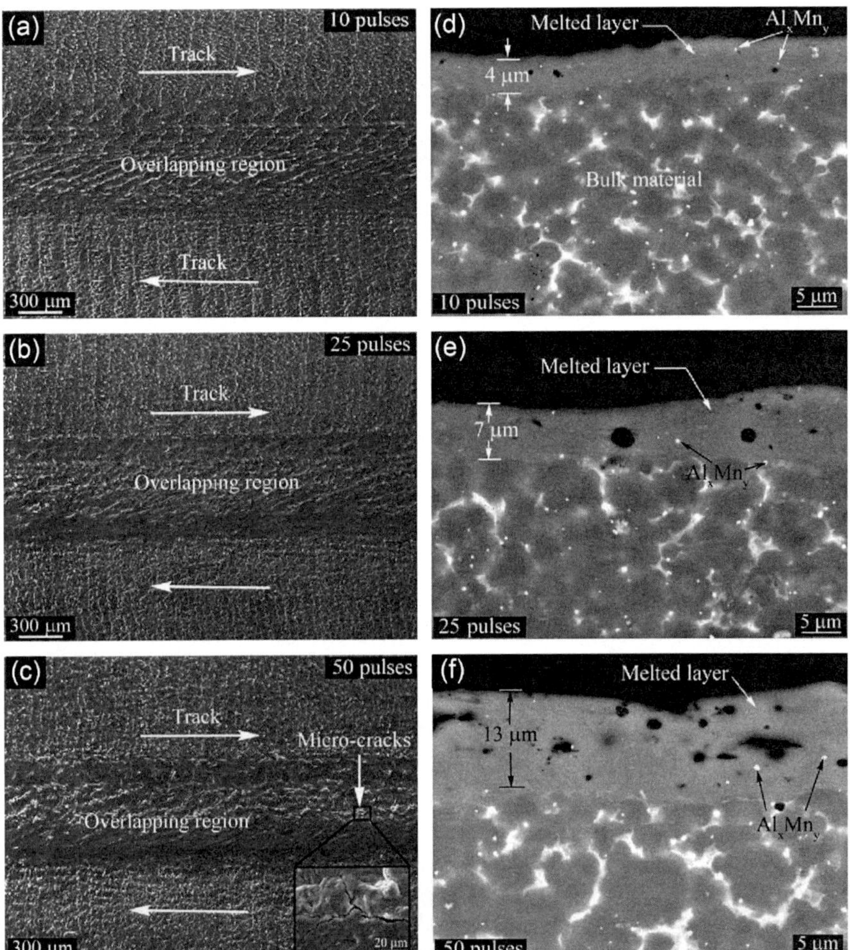

Figure 2.15 Secondary electron (a–c) and backscattered (d–f) scanning electron micrographs showing the surface appearance and cross-sectional microstructure of an AZ91D magnesium alloy after laser treatment using 10 (a, d), 25 (b, e), and 50 pulses (c, f).
Adapted from Coy et al. (2010) with permission from Elsevier.

and Guan, Zhou, Li, and Zheng (2013a) reported that the formation of a coarse structure in the overlapped area caused by scanning speed provided preferential sites for pitting corrosion in SBF. According to Guan, Zhou, Zheng, and Li (2013), an increase in laser power from 40 to 70 W led to the formation of cauliflower-like clusters with protruding particles on the surface of an AZ31B Mg alloy, which increased its average surface roughness from 0.11 to 2.69 μm. Trapping of light on the rough surface has led to a darkening effect of the treated alloy. LSM has been explored as a pretreatment for PEO of an AZ91D Mg alloy (Wang, Zhou, Liang, & Chen, 2012). The refinement of microstructure, enrichment of Al, and redistribution of the β-phase ($Mg_{17}Al_{12}$) on the AZ91D Mg alloy matrix after LSM has enabled the formation of a more compact PEO coating with fewer cracks on the surface and a higher amount of $MgAl_2O_4$ phase in the resultant coating. The corrosion resistance of the LSM−PEO-treated Mg alloy was found to be much better than those without the LSM pretreatment. Hence, it is evident that LSM could offer an improvement in corrosion resistance of Mg alloys in chloride-containing environments, including SBF, provided necessary precautions are taken during treatment to avoid the formation of a coarse microstructure in the overlapping areas and cracks due to the high thermal stress induced by the laser irradiation. In addition, LSM can be used as a pretreatment for Mg alloys for chemical conversion treatments and PEO to improve the reactivity of the alloy, to facilitate a homogeneous reaction at the surface, and to promote uniformity of the coating.

2.6.2 Laser surface alloying

LSA involves adding the preferred alloying elements to the melt pool either by direct injection or by a preplaced coating followed by laser irradiation of the surface to melt the alloying elements (injected or preplaced) and the surface of the base material, generating an enriched melt pool that subsequently solidifies to form an alloyed surface layer. Very high cooling rates (up to 1011 K/s) during resolidification enabled the formation of a refined microstructure in the alloyed surface. The depth of the alloyed layer and extent of alloying depends on laser processing parameters such as laser power, scan speed, laser spot size, and injection rate of alloying elements (or thickness of the preplaced layer) (Dahotre, 1998). The laser processing parameters need to be carefully optimized to ensure complete melting of alloying elements and the formation of strong metallurgical bonding between the alloyed layer and the substrate (Singh & Harimkar, 2012). It has been reported that an increase in laser power decreases the average alloying element content in the alloyed zone due to increased dilution. Also, an increase in scan speed decreases the average composition distribution due to the lower power input at a low interaction time (Dutta Majumdar & Manna, 2013, chap. 6). Galun and Mordike (1996) showed that LSA of Mg with Al and nickel (Ni) offers better resistance against corrosion. According to Ming, Da, and Chang (2008), LSA of an AZ91D Mg alloy with Al−Si improved its corrosion resistance in 3.5% NaCl. Paital et al. (2012) reported that LSA of an AZ31B Mg alloy with Al enabled an improvement in corrosion resistance. In spite of the ability of LSA to improve the corrosion resistance of Mg alloys in chloride-containing environments, a proper choice of alloying elements and laser processing conditions is essential to ensure a defect-free alloyed zone (Dutta Majumdar, & Manna, 2013, chap. 6). In addition, while choosing LSA of Mg alloys for use in biomedical applications, it is important to consider the toxicity of the alloying elements in the human body.

2.6.3 LSC and laser composite surfacing

LSC involves melting the material to be clad on the substrate without significant dilution from the substrate. The rapid solidification of the cladding material led to grain refinement, extension of solid solubility, and formation of nonequilibrium phases in the clad layer without significant changes in the microstructure of the substrate (Subramanian, Sircar, & Mazumder, 1991). In laser composite surfacing, hard ceramic particles are introduced in the laser-melted surface of the substrate to form a composite layer after subsequent resolidification. LSC of an Mg–Zr alloy onto Mg and its alloys showed improved resistance against corrosion (Subramanian et al., 1991). LSC of an AS41 Mg alloy with an Al–Si eutectic alloy and its composite with C-short fibers increased the corrosion resistance of the Mg alloy by 100 times (Neubert, Bakkar, & Huang, 2006). In addition, a graded multilayer coating of Ni/copper (Cu)/Al with good metallurgical bonding at the interfaces between the Mg–Al, Al–Cu, and Cu–Ni layers was successfully deposited on Mg by LSC. The LSC-treated Mg specimen exhibited a large span of passivity of over 500 mV, in contrast to the untreated Mg substrate, which showed no signs of passivation.

2.6.4 Laser shock peening (LSP)

LSP involves rapid irradiation of the component's surface with a very high power laser (power density $\sim 10^9 - 10^{12}$ W/m^2) applied at a short pulse (1–50 ns), resulting in the generation of shock waves (due to expansion of the volume of the plasma plume formed on the surface) and subsequent alteration of the microstructure/state of stress. The substrate being treated is usually covered with a thin insulating or absorbing layer (paint, metal, or plastic tape) and a transparent material (water). Upon laser irradiation, the generation of plasma induces shock waves, causing plastic deformation at the surface and introducing compressive residual stresses. In addition, the introduction of a high density of dislocations (Banas, Elsayed-Ali, Lawrence, & Rigsbee, 1990; Banas, Lawrence, Rigsbee, & Elsayed-Ali, 1990) and the formation of other phases or twins (Chu, Rigsbee, Banas, & Elsayed-Ali, 1999) during the process helped to increase the hardness and yield strength of metallic materials. Moreover, the presence of compressive residual stresses in the material led to a remarkable improvement in the fatigue strength and fatigue life of metallic materials. Because the laser spot size can be tailored suitably, the LSP can be successfully used for treating components of complex geometry, and this attribute would be beneficial in treating metallic implants such as stent materials.

The corrosion behavior of laser peened materials has been addressed by many researchers. Zhang, You et al. (2010) reported that an AZ31B Mg alloy subjected to LSP effectively reduced the initiation of stress corrosion cracking (SCC) of the alloy in 1 wt% NaOH for 500 h at room temperature. Jiang, Zhou, Fan, Huang, and Zhao

(2009) confirmed an improvement in the fatigue life of AZ31B and ZK60 Mg alloys after LSP. Sealy and Guo (2010) studied LSP of a biodegradable Mg–Ca alloy to understand the change in surface integrity and process mechanics and developed a three-dimensional semi-infinite simulation to predict the topography and residual stress fields produced by sequential peening. Guo, Sealy, and Guo (2012) reported that LSP induced unique surface topographies, high compressive residual stresses, and extended strain hardening, which helps to control the corrosion rate of Mg–Ca alloys by more than 100-fold in SBF. According to them, with LSP, the subsurface of an Mg–Ca alloy could be strain hardened up to 500 μm, and the maximum compressive residual stress of the peened surface in the planar direction might have contributed to the decrease in the corrosion rate of the Mg–Ca alloy. The ability of LSP to induce compressive residual stress, to improve fatigue life, and to reduce the susceptibility of Mg alloys to SCC will be useful in the development of Mg-based degradable load-bearing implants.

2.6.5 *Laser-assisted coloring/darkening and surface patterning*

Guan, Zhou, Li, and Zheng (2013b, 2014) studied the effect of femtosecond and krypton fluoride excimer laser irradiation on the surface evolution of an AZ31 Mg alloy, resulting in coloration and darkening of its surface following changes in surface profile and composition. According to them, by adjusting laser parameters according to the thermal properties of materials and the environmental conditions (oxygen or argon atmosphere), this type of laser irradiation offers potential applications for the development of Mg-based bio-optical implant devices. Foroozmehr, Alemohammad, Wang, Toyserkani, and Esmaeili (2011) studied patterning of Mg with silver (Ag) nanoparticles printed by a laser-assisted maskless microdeposition process. The laser-assisted maskless microdeposition process involves two steps: (1) microdeposition, in which the deposited material in the form of suspended nanoparticles is atomized to an aerosol and injected on a moving substrate in a layer-by-layer fashion; and (2) laser postprocessing, in which the laser beam is scanned over the deposited tracks of nanoparticles for sintering. The Ag film has a relatively uniform thickness with a root mean square roughness of 48 nm. It is well known that surface patterning of biocompatible materials could promote cell adhesion, infiltration, and proliferation. Being a bioactive material, Ag provides better biocompatibility and it is well known for its antibacterial effects. However, the formation of cracks on the Ag films during sintering is considered a major limitation in terms of its wide applicability for biomedical applications.

2.7 Other methods of surface modification

2.7.1 *Cold spray deposition*

Cold spray (CS) coating is a viable method to engineer the surface of Mg and its alloys. The deposition of CS coatings involves the ballistic impingement of particles, usually in the size range of 1–100 μm, that are accelerated by a high-velocity stream of gas on the substrate surface. Upon impact with the substrate, the particles undergo severe plastic deformation, resulting in flattened splats, which is also accompanied by peening of the substrate. As the temperature of the expanded gas stream that exits the spray nozzle is relatively lower than the conventional thermal spray processes, it is called a "cold spray" process. Hence, it is particularly suitable for coatings and substrate materials that are sensitive to heat or oxidation and that would transform or react under the conditions of a conventional thermal spray process (Champagne, 2007). The performance of CS coatings is largely a function of bond strength between the coating and the substrate and the residual stress, porosity, and density of the coating. To achieve a better bond strength, it is important that the mean velocity at which the particles are accelerated exceeds the minimum critical velocity so that the particles have sufficient kinetic energy to generate extensive plastic deformation of the substrate and to promote interlocking of the spray splats (Champagne, 2007). Because the particles are sprayed at relatively lower temperatures, there is little temperature-driven dimensional change, and hence the resultant coatings usually are in a state of compressive residual stress, which is beneficial to prevent delamination of the coating. The porosity of CS coatings is generally low, on the order of 1–5%, and most of them are identified on the top surface. The following factors, either individually or in combination, are considered responsible for the occurrence of porosity in CS coatings: (1) rebound of particles with a velocity lower than the critical velocity; (2) large particles that do not undergo enough plastic deformation upon impact, generating porosity at the particle–particle boundaries; (3) the absence of a significant peening effect on the particles that are already deposited, particularly at the top of the coating; and (4) the sudden incompatibility in terms of localized heating, stresses, and plastic deformation (Balani, Laha, Agarwal, Karthikeyan, & Munroe, 2005; Bu, Yandouzi, Lu, & Jodoin, 2011).

Most of the studies of CS coatings on Mg and its alloys concentrate on the deposition of Al and Al–Al_2O_3 composite coatings, particularly to understand the microstructural and interfacial characteristics, to improve the bond strength and density, to promote formation of intermetallic compounds by heat treatment after coating, and to improve the corrosion behavior (Bu et al., 2011; Spencer, Fabijanic, & Zhang, 2009; Spencer, Luzin, Matthews, & Zhang, 2012; Spencer & Zhang, 2009; Tao et al., 2010; Wang, Qiu, Xiong, Birbilis, & Zhang, 2014; Wang, Spencer, Birbilis, & Zhang, 2010). Zhang et al. address the role of CS coatings on Mg in Chapter 14 of Volume 2 of this book. Noorakma, Zuhailawati, Aishvarya, and Dhindaw (2013) recently studied the deposition of HA on an AZ51 Mg alloy by a modified CS process in which the Mg alloy was preheated to 400 °C for 1 h and the HA powder (average particle size, 4 μm) was sprayed using a high-pressure (10 bar) air nozzle spray at room temperature. According to them, this modification helped to retain the characteristics of HA, which would otherwise undergo a phase change at high temperature. In addition, the combination of a higher impact velocity of HA particles and preheating of the AZ51 Mg alloy enabled better bonding of the HA particles with the Mg alloy and the formation of a uniform coating. The average thickness and modulus of the resultant HA coating was 25 μm and 9 GPa, respectively. Immersion in SBF for up to 14 days showed that the HA-coated AZ51 Mg alloy is bioactive and facilitated apatite formation.

Being a low-temperature process, CS is particularly suitable for the deposition of bioactive coatings such as HAP on Mg and its alloys, in which it would be possible to limit the extent of oxidation of the substrate as well as the phase transformation of HAP. As long as the powder particles are accelerated at a speed that exceeds the minimum critical velocity, the bonding strength between the substrate and the coating would be good. Peening of the substrate as well as the deposited coatings is likely to induce compressive residual stress, which would be beneficial in improving the fatigue strength of the substrate as well as in preventing delamination of the coating – both attributes are critical in biomedical implant applications. Porosity of the CS coating is a matter of concern (Figure 2.16). Fortunately, none of these pores is continuous from the outermost surface of the coating down to the substrate, suggesting that the coating is impermeable to solution (Tao et al., 2010), which is an important feature for considering them for magnesium and its alloys in the development of degradable implants. Adopting after spraying a heat treatment at temperatures closer to the melting point of the substrate for a longer treatment time would offer the benefits of increasing the bond strength and relaxing the compressive residual stress (Spencer et al., 2009; Spencer & Zhang, 2009), both of which increase the adherence of the coating and prevent its delamination. This methodology has many positive attributes for surface modification of Mg and its alloys toward the development of

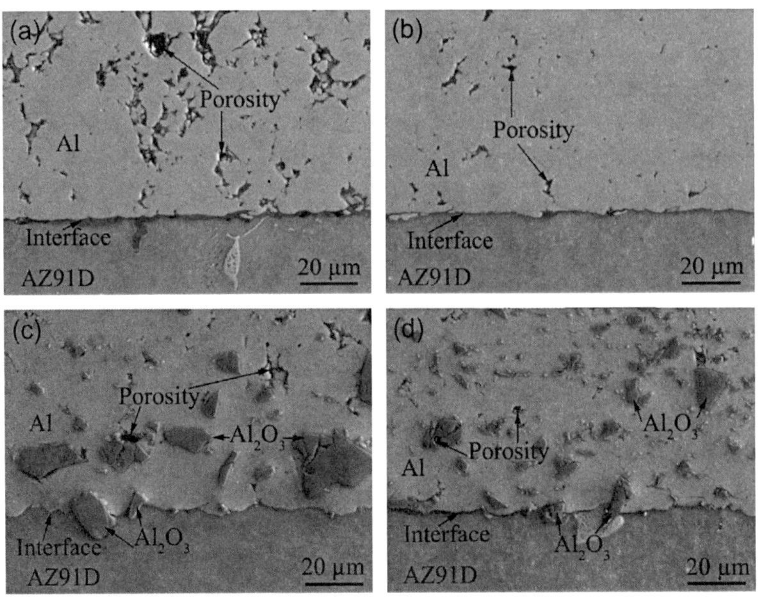

Figure 2.16 Backscattered scanning electron microscopy images of unreinforced aluminum (Al) and Al–Al_2O_3 composite coatings showing the presence of porosity in the coatings prepared by cold spray deposition: (a) Unreinforced Al coating cold sprayed at 65 °C; (b) unreinforced Al coating cold sprayed at 125 °C; (c) Al–25 vol% Al_2O_3 composite coating cold sprayed at 65 °C; and (d) Al–50 vol% Al_2O_3 composite coating cold sprayed at 65 °C. Adapted from Wang, Spencer et al. (2010) with permission from Elsevier.

degradable implants. Being a line-of-sight process, the difficulty in coating complex shapes and the internal surfaces of implants is a major limitation of the CS process (Champagne, 2007).

2.8 Summary and concluding remarks

This chapter provides a detailed outline of the various surface modification methods available to modify the surface of Mg alloys and discusses the opportunities and limitations of each in the development of Mg-based degradable biomaterials. The salient features are summarized below.

- Imparting self-passivation of Mg alloys through selective oxidation of alloying elements such as scandium and yttrium is indeed an effective method to control the rate of degradation of Mg alloys. This approach will be useful when the implant geometry restricts the use of any coatings. However, the alloying elements should not exhibit any significant toxicity in the human body.
- Microstructural modification of AZ-type Mg alloys by developing a continuous β-phase network on the alloy's surface could also be used to impart a self-passivation mechanism after the corrosion process. The formation of galvanic cells and an increase in the corrosion rate are believed to promote deposition of $Mg(OH)_2$ and $Ca_3(PO_4)_2$ (or $Ca_{10}(PO_4)_6(OH)_2$) and $Mg_3(PO_4)_2$ as the primary and secondary corrosion products, respectively. However, the porous nature of $Mg(OH)_2$ would allow permeation of the body fluid and limit the long-term ability of the Mg alloy to protect against corrosion.
- Hydrothermal treatment of Mg alloys enables the formation of an $Mg(OH)_2$ coating, which is hydrophilic and adherent and offers a higher resistance to corrosion than their untreated counterparts. However, the ability of the $Mg(OH)_2$ coating to provide long-term protection for the Mg alloys against corrosion in SBF is a matter of concern.
- AHT has the ability to reduce the rate of degradation of Mg alloys. The dense and compact nature and the amount and crystallinity of MgO are the decisive factors in controlling the rate of degradation of Mg alloys in SBF.
- Passivation in 1 M NaOH or in m-SBF has limited potential in the surface modification of Mg and its alloys in the development of degradable implants because of the insufficient thickness and compactness of the passive films formed in them.
- The difficulty imposed by the intrinsic heterogeneity of Mg alloys, slow reaction kinetics, and the defective nature of the surface conversion layer formed by the reaction between the Mg alloys and ILs suggest that there remains much to be explored in generating robust passivating films using ILs.
- The formation of well-ordered and closely packed SAMs with a high homogeneity, good order, and better chemical stability is indeed promising. However, the rapid decrease in the contact angle of some of the SAM-modified Mg alloy surfaces in aqueous solutions within a short duration of time has raised questions on their long-term chemical stability in the human body.
- The difference in reactivity imposed by the intrinsic heterogeneity of Mg alloys, the formation of a porous coating with a typical "mud crack" pattern, possibly due to the release of hydrogen gas and/or dehydration of the coating after deposition, limits the extent of protection against corrosion offered by conversion coatings. Among them, fluoride conversion coating is promising in terms of its biocompatibility. However, modification of the existing

methodologies to prepare MgF_2 coatings with desirable characteristics and with a better understanding of the mechanism of deposition and the influence of volume fraction of $Mg(OH)_{2-x}F_x$ on the resistance of the resultant coatings against corrosion needs to be explored for its potential utility.

- The susceptibility of the sol-gel-deposited BGC coatings for cracking after immersion in SBF is the major drawback that limits their acceptance as a suitable surface coating of Mg alloys in the development of degradable implants.
- Silane coating offers only a short-term protection for Mg alloys against corrosion. Alkalinization at the interface during corrosion promotes swelling and decomposition of the SiO_2 network, while accumulation of corrosion products promotes delamination and blistering of the coating, thus exposing the Mg alloy to degrade further. However, silane coatings will be useful for functionalizing compounds with biofunctional properties. Hence, increasing the bond strength and hydrolytic stability of silane coatings becomes critical in achieving reasonably good protection of Mg alloys against corrosion.
- It is possible to coat or bind bioactive compounds such as gelatin or albumin. However, the inability of the precoatings to form a completely dense layer on Mg alloys and the detachment/delamination of the precoating and/or bioactive molecule coating is a matter of concern in realizing their potential utility.
- Degradable polymer coatings deposited by both dip and spin coating methods did not offer a long-term corrosion protection. Bulk erosion/degradation leads to detachment of coatings, resulting in the formation of many micro-/macropores and cracks in the early stage of degradation. Surface-eroding polymer coatings exhibit homogeneous surface erosion from exterior to interior, and this type of coating is beneficial for the development of fully biodegradable cardiovascular stents.
- The ability to load anti-inflammatory drugs in conducting polymer coatings such as PEDOT and its subsequent release by electrical stimulation confirmed its potential application in degradable implant materials with drug delivery applications. However, the inability of PEDOT coatings to completely prevent the Mg from corrosion, its detachment after multiple electrical stimulations of drug release, and the mode of degradation of PEDOT under *in vivo* conditions are the major concerns.
- Anodizing is useful in developing a passive oxide layer on Mg alloys. However, the adverse effect on fatigue properties caused by oxidation-induced surface tensile stress, structural defects in the oxide layer, and substrate "age softening" following the heat associated with oxide film formation restrict their widespread acceptance. MAO is considered an option to reduce this risk of fatigue failure. However, because of the higher pore density on the surface of the MAO coatings of Mg alloys, the effective surface area and the tendency of the corrosive medium to adsorb and concentrate into these pores are increased, thus limiting the protection against corrosion offered by them.
- ED is a useful method for depositing HA, calcium-deficient HA, and fluoride-, strontium-, Zn-, and Si-doped HA coatings. Because Mg alloys have a lower melting point, the difficulty associated with heat treatment at high temperatures to improve the adhesion of the coatings is a major limitation in EPD. However, MAO coatings could be used as a pretreatment for deposition of a variety of inorganic materials with a better adhesion by EPD.
- The major limitation of the coatings deposited by PVD is the presence of pores and pinholes, which in most cases are difficult to eliminate completely. The formation of a transition layer to buffer the stress developed by the mismatch between the Mg alloy and the coatings, and an improvement in deposition methodology by implementing filtered cathodic arc deposition, could not completely prevent penetration of the electrolyte or sustain their protective properties after long-term exposure in SBF.

- Implantation of Zn, Cr, and Ti accelerates the degradation of Mg alloys because of their existence in a metallic state in the implanted layer and the formation of galvanic corrosion cells with the substrate material. Dual-ion implantation offered better resistance against corrosion for Mg alloys than single-ion implantation. Implantation at lower doses failed to improve the corrosion resistance, whereas the use of a much higher dose generated defects in the treated surface. The surface of Mg alloys treated by PIII always exhibited pitting corrosion, and their ability to provide a long-term corrosion resistance remains open. It seems that the window of opportunity for using PIII for surface engineering of Mg in the development of degradable implants is rather limited.
- LSM could offer an improvement in resistance of Mg alloys against corrosion in chloride-containing environments, including SBF, provided necessary precautions are taken during treatment to avoid the formation of a coarse microstructure in the overlapping area and cracks due to the high thermal stress induced by the laser irradiation. LSM can be used as a pretreatment for Mg alloys for chemical conversion treatments and PEO to improve the reactivity of the alloy, to facilitate a homogeneous reaction at the surface, and to promote the uniformity of the coating.
- A proper choice of alloying elements (nontoxic in the human body) and processing conditions are essential to ensure a defect-free alloyed zone in LSA. The ability of LSP to induce compressive residual stress, to improve fatigue life, and to reduce the susceptibility of Mg alloys to SCC will be useful in the development of Mg-based, degradable, load-bearing implants.
- CS is particularly suitable for depositing bioactive coatings such as HA on Mg alloys because it can limit the extent of oxidation of the substrate as well as the phase transformation of HA. Peening of the substrate as well as the deposited coatings is likely to induce compressive residual stress, which would be beneficial in improving the fatigue strength of the substrate as well as in preventing delamination of the coating – both attributes are critical for biomedical implant applications. The porosity of the CS coating is a matter of concern.

Hence, it is evident that there remain plenty of challenges in engineering the surface of Mg alloys toward the development of degradable biomaterials. The major limitation stems from the inability of the methods to deposit a defect-free coating. In addition, the corrosion products formed are not impervious. As Mg alloys are highly susceptible to localized corrosion in spite of surface modifications, the mechanical strength could not be retained for a longer period of time. Adopting a combination of surface engineering methods could help to achieve certain attributes. The quest to modify the surface of Mg alloys to impart both long-term resistance against corrosion and biocompatibility continues.

References

Abbas, G., Liu, Z., & Skeldon, P. (2005). Corrosion behaviour of laser-melted magnesium alloys. *Applied Surface Science, 247*(1–4), 347–353.

Alabbasi, A., Liyanaarachchi, S., & Kannan, M. B. (2012). Polylactic acid coating on a biodegradable magnesium alloy: an in vitro degradation study by electrochemical impedance spectroscopy. *Thin Solid Films, 520*(23), 6841–6844.

Amaravathy, P., Rose, C., Sathiyanarayanan, S., & Rajendran, N. (2012). Evaluation of in vitro bioactivity and MG63 oesteoblast cell response for TiO_2 coated magnesium alloys. *Journal of Sol-Gel Science and Technology, 64*(3), 694–703.

Bakhsheshi-Rad, H. R., Idris, M. H., & Abdul-kadir, M. R. (2013). Surface & coatings technology synthesis and in vitro degradation evaluation of the nano-HA/MgF_2 and DCPD/MgF_2 composite coating on biodegradable Mg–Ca–Zn alloy. *Surface and Coatings Technology, 222*, 79–89.

Bakhsheshi-Rad, H. R., Idris, M. H., Abdul-kadir, M. R., & Daroonparvar, M. (2013). Effect of fluoride treatment on corrosion behavior of Mg–Ca binary alloy for implant application. *Transactions of Nonferrous Metals Society of China, 23*(3), 699–710.

Balani, K., Laha, T., Agarwal, A., Karthikeyan, J., & Munroe, N. (2005). Effect of carrier gases on microstructural and electrochemical behavior of cold-sprayed 1100 aluminum coating. *Surface and Coatings Technology, 195*(2–3), 272–279.

Banas, G., Elsayed-Ali, H. E., Lawrence, F. V., Jr., & Rigsbee, J. M. (1990). Laser shock induced mechanical and microstructural modification of welded maraging steel. *Applied Physics, 67*, 2380–2384.

Banas, G., Lawrence, F. V., Jr., Rigsbee, J. M., & Elsayed-Ali, H. E. (1990). Laser shock hardening of welded maraging steel. *Surface Engineering, 67*, 280–290.

Bandyopadhyay, A., Balla, V. K., Roy, M., & Bose, S. (2011). Laser surface modification of metallic biomaterials. *Journal of Metals, 63*(6), 94–99.

Banerjee, P. C., Singh Raman, R. K., Durandet, Y., & Mcadam, G. (2011). Electrochemical investigation of the influence of laser surface melting on the microstructure and corrosion behaviour of ZE41 magnesium alloy – an EIS based study. *Corrosion Science, 53*(4), 1505–1514.

Ban, S., & Hasegawa, J. (2002). Morphological regulation and crystal growth of hydrothermal-electrochemically deposited apatite. *Biomaterials, 23*, 2965–2972.

Bauer, S., Schmuki, P., Mark, K. V. D., & Park, J. (2013). Engineering biocompatible implant surfaces. *Progress in Materials Science, 58*, 261–326.

Besra, L., & Liu, M. (2007). A review on fundamentals and applications of electrophoretic deposition (EPD). *Progress in Materials Science, 52*, 1–61.

Best, S. M., Porter, A. E., Thian, E. S., & Huang, J. (2008). Bioceramics: past, present and for the future. *Journal of the European Ceramic Society, 28*, 1319–1327.

Bikulčius, G., Ručinskienė, A., Bukelskienė, V., Selskis, A., & Grigucevičienė, A. (2012). Formation of conversion coatings on magnesium alloy AZ31 in solutions containing nicotinic acid. *Materials Science, 18*(2), 123–127.

Blawert, C., Dietzel, W., Ghali, E., & Song, G. (2006). Anodizing treatments for magnesium alloys and their effect on corrosion resistance in various environments. *Advanced Engineering Materials, 8*, 511–533.

Boccaccini, A. R., Cho, J., Subhani, T., Kaya, C., & Kaya, F. (2010). Electrophoretic deposition of carbon nanotube–ceramic nanocomposites. *Journal of the European Ceramic Society, 30*, 1115–1129.

Boccaccini, A. R., Van der Biest, O. O., & Talbot, J. B. (Eds.). (2002). *Electrophoretic deposition: Fundamental and applications*. Pennington: The Electrochemical Society.

Boccaccini, A. R., & Zhitomirsky, I. (2002). Application of electrophoretic and electrolytic deposition techniques in ceramics processing. *Current Opinion in Solid State and Materials Science, 6*, 251–260.

Brar, H. S., Ball, J. P., Berglund, I. S., Allen, J. B., & Manuel, M. V. (2013). A study of a biodegradable Mg-3Sc-3Y alloy and the effect of self-passivation on the in vitro degradation. *Acta Biomaterialia, 9*(2), 5331–5340.

Bu, H., Yandouzi, M., Lu, C., & Jodoin, B. (2011). Effect of heat treatment on the intermetallic layer of cold sprayed aluminum coatings on magnesium alloy. *Surface and Coatings Technology, 205*(19), 4665–4671.

Carboneras, M., García-Alonso, M. C., & Escudero, M. L. (2011). Biodegradation kinetics of modified magnesium-based materials in cell culture medium. *Corrosion Science, 53*(4), 1433–1439.

Champagne, V. K. (Ed.). (2007). *The cold spray materials deposition process – Fundamentals and applications*. Woodhead Publishing Limited and CRC Press LLC.

Chan, W. Y., Chian, K. S., & Tan, M. J. (2013). In vitro metal ion release and biocompatibility of amorphous $Mg_{67}Zn_{28}Ca_5$ alloy with/without gelatin coating. *Materials Science and Engineering C: Materials for Biological Applications, 33*(8), 5019–5027.

Chen, X.-B., Birbilis, N., & Abbott, T. B. (2011). A simple route towards a hydroxyapatite–$Mg(OH)_2$ conversion coating for magnesium. *Corrosion Science, 53*, 2263–2268.

Chen, S., Guan, S., Li, W., Wang, H., Chen, J., Wang, Y., et al. (2012). In vivo degradation and bone response of a composite coating on Mg–Zn–Ca alloy prepared by microarc oxidation and electrochemical deposition. *Journal of Biomedical Materials Research Part B, 100*, 533–543.

Chen, W., Huang, J., & Peng, J. (2013). Characterisation of TiAlN PVD coatings on AZ31 magnesium alloy. *Research on Chemical Intermediates*. http://dx.doi.org/10.1007/s11164-013-1270-5.

Chen, Y., Song, Y., Zhang, S., Li, J., Wang, H., Zhao, C., et al. (2011). Effect of fluoride coating on in vitro dynamic degradation of Mg–Zn alloy. *Materials Letters, 65*(17–18), 2568–2571.

Chen, Y., Song, Y., Zhang, S., Li, J., Zhao, C., & Zhang, X. (2011). Interaction between a high purity magnesium surface and PCL and PLA coatings during dynamic degradation. *Biomedical Materials, 6*(2), 025005. http://dx.doi.org/10.1088/1748–6041/6/2/025005 (5 pp.).

Chen, Y., Wan, G., Wang, J., Zhao, S., Zhao, Y., & Huang, N. (2013). Covalent immobilization of phytic acid on Mg by alkaline pre-treatment: corrosion and degradation behavior in phosphate buffered saline. *Corrosion Science, 75*, 280–286.

Chiu, K. Y., Wong, M. H., Cheng, F. T., & Man, H. C. (2007). Characterization and corrosion studies of fluoride conversion coating on degradable Mg implants. *Surface and Coatings Technology, 202*(3), 590–598.

Chu, P. K. (2004). Recent developments and applications of plasma immersion ion implantation. *Journal of Vacuum Science and Technology B: Microelectronics and Nanometer Structures, 22*(1), 289–296.

Chu, P. K. (2013). Control of surface degradation on biodegradable magnesium alloys by plasma-based technology. *IEEE Transactions on Plasma Science, 41*(4), 725–730.

Chu, J. P., Rigsbee, J. M., Banas, G., & Elsayed-Ali, H. E. (1999). Laser-shock processing effects on surface microstructure and mechanical properties of low carbon steel. *Materials Science and Engineering A, 260*, 260–268.

da Conceicao, T. F., Scharnagl, N., Blawert, C., Dietzel, W., & Kainer, K. U. (2010). Surface modification of magnesium alloy AZ31 by hydrofluoric acid treatment and its effect on the corrosion behaviour. *Thin Solid Films, 518*(18), 5209–5218.

da Conceição, T. F., Scharnagl, N., Dietzel, W., & Kainer, K. U. (2012). Controlled degradation of a magnesium alloy in simulated body fluid using hydrofluoric acid treatment followed by polyacrylonitrile coating. *Corrosion Science, 62*, 83–89.

Corni, I., Ryan, M. P., & Boccaccini, A. R. (2008). Electrophoretic deposition: from traditional ceramics to nanotechnology. *Journal of the European Ceramic Society, 28*(7), 1353–1367.

Coy, A. E., Viejo, F., Garcia-Garcia, F. J., Liu, Z., Skeldon, P., & Thompson, G. E. (2010). Effect of excimer laser surface melting on the microstructure and corrosion performance of the die cast AZ91D magnesium alloy. *Corrosion Science, 52*, 387–397.

Cui, X., Li, Q., Li, Y., Wang, F., Jin, G., & Ding, M. (2008). Microstructure and corrosion resistance of phytic acid conversion coatings for magnesium alloy. *Applied Surface Science, 255*, 2098–2103.

Cui, X., Yang, Y., Liu, E., Jin, G., Zhong, J., & Li, Q. (2011). Corrosion behaviors in physiological solution of cerium conversion coatings on AZ31 magnesium alloy. *Applied Surface Science, 257*(23), 9703–9709.

Dahotre, N. B. (1998). *Lasers in surface engineering*. Materials Park, OH: ASM International.

Dahotre, N. B., Paital, S. R., Samant, A. N., & Daniel, C. (2010). Wetting behaviour of laser synthetic surface microtextures on Ti-6Al-4V for bioapplication. *Philosophical Transactions Series A: Mathematical, Physical, and Engineering Sciences, 368*(1917), 1863–1889.

Degner, J., Singer, F., Cordero, L., Boccaccini, A. R., & Virtanen, S. (2013). Electrochemical investigations of magnesium in DMEM with biodegradable polycaprolactone coating as corrosion barrier. *Applied Surface Science, 282*, 264–270.

Dou, Y., Cai, S., Ye, X., Xu, G., Huang, K., Wang, X., et al. (2013). 45S5 bioactive glass–ceramic coated AZ31 magnesium alloy with improved corrosion resistance. *Surface and Coatings Technology, 228*, 154–161.

Drevet, R., & Benhayoune, H. (2013). Pulsed electrodeposition for the synthesis of strontium-substituted calcium phosphate coatings with improved dissolution properties. *Materials Science and Engineering C, 33*, 4260–4265.

Drynda, A., Hassel, T., Hoehn, R., Perz, A., Bach, F. W., & Peuster, M. (2010). Development and biocompatibility of a novel corrodible fluoride-coated magnesium-calcium alloy with improved degradation kinetics and adequate mechanical properties for cardiovascular applications. *Journal of Biomedical Materials Research Part A, 93*(2), 763–775.

Drynda, A., Seibt, J., Hassel, T., Bach, F. W., & Peuster, M. (2013). Biocompatibility of fluoride-coated magnesium-calcium alloys with optimized degradation kinetics in a subcutaneous mouse model. *Journal of Biomedical Materials Research Part A, 101*(1), 33–43.

Dutta Majumdar, J., & Manna, I. (2011). Laser material processing. *International Materials Reviews, 56*(5–6), 341–388.

Dutta Majumdar, J., & Manna, I. (2013). In G. L. Song (Ed.), *Laser treatment to improve the corrosion resistance of Mg alloys* (pp. 133–162). U.K.: Woodhead Publishing.

Efthimiadis, J., Neil, W. C., Bunter, A., Howlett, P. C., Hinton, B. R. W., MacFarlane, D. R., et al. (2010). Potentiostatic control of ionic liquid surface film formation on ZE41 magnesium alloy. *ACS Applied Materials and Interfaces, 2*(5), 1317–1323.

El Kady, A. M., Mohamed, K. R., & El-Bassyouni, G. T. (2009). Fabrication, characterization and bioactivity evaluation of calcium pyrophosphate/polymeric biocomposites. *Ceramics International, 35*, 2933–2942.

Farrokhi-Rad, M., Loghmani, S. K., Shahrabi, T., & Khanmohammadi, S. (2014). Electrophoretic deposition of hydroxyapatite nanostructured coatings with controlled porosity. *Journal of the European Ceramic Society, 34*, 97–106.

Fekry, A. M., Ghoneim, A. A., & Ameer, M. A. (2014). Electrochemical impedance spectroscopy of chitosan coated magnesium alloys in a synthetic sweat medium. *Surface and Coatings Technology, 238*, 126–132.

Feliu, S., Jr., Samaniego, A., El-Hadad, A. A., & Llorente, I. (2013). The effect of $NaHCO_3$ treatment time on the corrosion resistance of commercial magnesium alloys AZ31 and AZ61 in 0.6 M NaCl solution. *Corrosion Science, 67*, 204–216.

Feng, J., Chen, Y., Liu, X., Liu, T., Zou, L., & Wang, Y. (2013). In-situ hydrothermal crystallization $Mg(OH)_2$ films on magnesium alloy AZ91 and their corrosion resistance properties. *Materials Chemistry and Physics, 143*(1), 322–329.

Foroozmehr, E., Alemohammad, H., Wang, X., Toyserkani, E., & Esmaeili, S. (2011). Laser-assisted surface patterning of magnesium with silver nanoparticles: synthesis, characterization and modelling. *Journal of Physics D: Applied Physics, 44*(49), 495305, 13 pp.

Forsyth, M., Neil, W. C., Howlett, P. C., Macfarlane, D. R., Hinton, B. R. W., Rocher, N., et al. (2009). New insights into the fundamental chemical nature of ionic liquid film formation on magnesium alloy surfaces. *ACS Applied Materials and Interfaces, 1*(5), 1045–1052.

Galun, R. W., & Mordike, B. L. (1996). Improving the surface properties of magnesium by laser alloying. In *Proceedings of the third international magnesium conference, Manchester* (pp. 699–711).

Gao, J. H., Guan, S. K., Chen, J., Wang, L. G., Zhu, S. J., Hu, J. H., et al. (2011). Fabrication and characterization of rod-like nano-hydroxyapatite on MAO coating supported on Mg–Zn–Ca alloy. *Applied Surface Science, 257*, 2231–2237.

Gao, J.-C., Qiao, L.-Y., & Xin, R.-L. (2010). Corrosion and bone response of magnesium implants after surface modification by heat-self-assembled monolayer. *Frontiers of Materials Science in China, 4*(2), 120–125.

Ghasemi, A., Raja, V. S., Blawert, C., Dietzel, W., & Kainer, K. U. (2008). Study of the structure and corrosion behavior of PEO coatings on AM50 magnesium alloy by electrochemical impedance spectroscopy. *Surface and Coatings Technology, 202*, 3513–3518.

Grillo, C. A., Alvarez, F., & Fernández Lorenzo de Mele, F. L. (2011). Biological effects of magnesium particles degradation on UMR-106 cell line: influence of fluoride treatments. *Colloids and Surfaces B: Biointerfaces, 88*(1), 471–476.

Grubač, Z., Metikoš-Huković, M., Babić, R., Rončević, I.Š., Petravić, M., & Peter, R. (2013). Functionalization of biodegradable magnesium alloy implants with alkylphosphonate self-assembled films. *Materials Science and Engineering C: Materials for Biological Applications, 33*(4), 2152–2158.

Grubač, Z., Škugor Rončević, I., Metikoš-Huković, M., Babić, R., Petravić, M., & Peter, R. (2012). Surface modification of biodegradable magnesium alloys. *Journal of the Electrochemical Society, 159*(6), C253–C258.

Guan, Y. C., Zhou, W., Li, Z. L., & Zheng, H. Y. (2013a). Influence of overlapping tracks on microstructure evolution and corrosion behavior in laser-melt magnesium alloy. *Materials and Design, 52*, 452–458.

Guan, Y. C., Zhou, W., Li, Z. L., & Zheng, H. Y. (2013b). Femtosecond laser-induced iridescent effect on AZ31B magnesium alloy surface. *Journal of Physics D: Applied Physics, 46*, 425305 (6 pp).

Guan, Y. C., Zhou, W., Li, Z. L., & Zheng, H. Y. (2014). Effect of processing environment on laser-induced darkening evolution in magnesium alloy. *Optics and Lasers in Engineering, 52*, 35–40.

Guan, Y. C., Zhou, W., & Zheng, H. Y. (2009). Effect of laser surface melting on corrosion behaviour of AZ91D in simulated-modified body fluid. *Journal of Applied Electrochemistry, 39*, 1457–1464.

Guan, Y. C., Zhou, W., Zheng, H. Y., & Li, Z. L. (2010). Solidification microstructure of AZ91D Mg alloy after laser surface melting. *Applied Physics A, 101*, 339–344.

Guan, Y. C., Zhou, W., Zheng, H. Y., & Li, Z. L. (2013). Darkening effect on AZ31B magnesium alloy surface induced by nanosecond pulse Nd:YAG laser. *Applied Surface Science, 280*, 462–466.

Guo, Y. B., Sealy, M. P., & Guo, C. S. (2012). Significant improvement of corrosion resistance of biodegradable metallic implants processed by laser shock peening. *CIRP Annals – Manufacturing Technology, 61*, 583–586.

Guo, L. F., Yue, T. M., & Man, H. C. (2005). Excimer laser surface treatment of magnesium alloy WE43 for corrosion resistance improvement. *Journal of Materials Science, 40*, 3531–3533.

Gupta, R. K., Mensah-Darkwa, K., & Kumar, D. (2013). Effect of post heat treatment on corrosion resistance of phytic acid conversion coated magnesium. *Journal of Materials Science and Technology, 29*(2), 180–186.

Gupta, R. K., Mensah-Darkwa, K., & Kumar, D. (2014). Corrosion protective conversion coatings on magnesium disks using a hydrothermal technique. *Journal of Materials Science and Technology, 30*(1), 47–53.

Gupta, R. K., Mensah-Darkwa, K., Sankar, J., & Kumar, D. (2013). Enhanced corrosion resistance of phytic acid coated magnesium by stearic acid treatment. *Transactions of Nonferrous Metals Society of China, 23*(5), 1237–1244.

Gu, X. N., Zheng, W., Cheng, Y., & Zheng, Y. F. (2009). A study on alkaline heat treated Mg–Ca alloy for the control of the biocorrosion rate. *Acta Biomaterialia, 5*(7), 2790–2799.

Gu, X. N., Zheng, Y. F., Lan, Q. X., Cheng, Y., Zhang, Z. X., Xi, T. F., et al. (2009). Surface modification of an Mg-1Ca alloy to slow down its biocorrosion by chitosan. *Biomedical Materials (Bristol, England), 4*(4), 044109.

Heavens, S. N. (1990). *Electrophoretic deposition as a processing route for ceramics. Advanced ceramic processing and technology*. New Jersey: Noyes Publications, 255–283.

Hench, L. L. (1997). Sol–gel materials for bioceramic applications. *Current Opinion in Solid State and Materials Science, 2*, 604–610.

Hiromoto, S., & Yamamoto, A. (2010). Control of degradation rate of bioabsorbable magnesium by anodization and steam treatment. *Materials Science and Engineering C, 30*, 1085–1093.

Hoche, H., Groß, S., Foerster, R., Schmidt, J., & Adamitzki, W. (2009). Development of decorative and corrosion resistant coatings for the surface refinement of magnesium alloys by plasma-based methods. *Plasma Processes and Polymers, 6*(S1), S671–S677.

Hongxi, L., Qian, X., Damin, X., Bo, L., & Chunlei, M. (2013). Microstructure and corrosion resistance of AZ91D magnesium alloy treated by hybrid ion implantation and heat treatment. *Vacuum, 89*, 233–237.

Hornberger, H., Virtanen, S., & Boccaccini, A. R. (2012). Biomedical coatings on magnesium alloys – a review. *Acta Biomaterialia, 8*, 2442–2455.

Howlett, P. C., Efthimiadis, J., Hale, P., van Riessen, G. A., MacFarlane, D. R., & Forsyth, M. (2010). Characterization of the magnesium alloy AZ31 surface in the ionic liquid trihexyl(tetradecyl)phosphonium bis(trifluoromethanesulfonyl)amide. *Journal of the Electrochemical Society, 157*(11), C392–C398.

Hu, L., Meng, Q., Chen, S., & Wang, H. (2012). Effect of Zn content on the chemical conversion treatments of AZ91D magnesium alloy. *Applied Surface Science, 259*, 816–823.

Huang, K., Cai, S., Xu, G., Ye, X., Dou, Y., Ren, M., et al. (2013). Preparation and characterization of mesoporous 45S5 bioactive glass-ceramic coatings on magnesium alloy for corrosion protection. *Journal of Alloys and Compounds, 580*, 290–297.

Huang, Y., Ding, Q., Pang, X., Han, S., & Yan, Y. (2013). Corrosion behavior and biocompatibility of strontium and fluorine co-doped electrodeposited hydroxyapatite coatings. *Applied Surface Science, 282*, 456–462.

Huang, P., Latham, J.-A., MacFarlane, D. R., Howlett, P. C., & Forsyth, M. (2013). A review of ionic liquid surface film formation on Mg and its alloys for improved corrosion performance. *Electrochimica Acta, 110*, 501–510.

Huang, Y., Yan, Y., Pang, X., Ding, Q., & Han, S. (2013). Bioactivity and corrosion properties of gelatin-containing and strontium-doped calcium phosphate composite coating. *Applied Surface Science, 282*, 583–589.

Huang, N., Yang, P., Leng, Y. X., Wang, J., Sun, H., Chen, J. Y., et al. (2004). Surface modification of biomaterials by plasma immersion ion implantation. *Surface and Coatings Technology, 186*(1–2), 218–226.

Hussein, R. O., Zhang, P., Nie, X., Xia, Y., & Northwood, D. O. (2011). The effect of current mode and discharge type on the corrosion resistance of plasma electrolytic oxidation (PEO) coated magnesium alloy AJ62. *Surface and Coatings Technology, 206*, 1990–1997.

Ishizaki, T., Okido, M., Masuda, Y., Saito, N., & Sakamoto, M. (2011). Corrosion resistant performances of alkanoic and phosphonic acids derived self-assembled monolayers on magnesium alloy AZ31 by vapor-phase method. *Langmuir, 27*(10), 6009–6017.

Ishizaki, T., Teshima, K., Masuda, Y., & Sakamoto, M. (2011). Liquid phase formation of alkyl- and perfluoro-phosphonic acid derived monolayers on magnesium alloy AZ31 and their chemical properties. *Journal of Colloid and Interface Science, 360*(1), 280–288.

Jamesh, M., Kumar, S., & Sankara Narayanan, T. S. N. (2012). Electrodeposition of hydroxyapatite coating on magnesium for biomedical applications. *Journal of Coatings Technology and Research, 9*(4), 495–502.

Jiang, S. Q., Zhou, J. Z., Fan, Y. J., Huang, S., & Zhao, J. F. (2009). Prediction on residual stress and fatigue life of magnesium alloy treated by laser shot peening. *Materials Science Forum, 626–627*, 393–398.

Jianrui, L., Yina, G., & Weidong, H. (2006). Study on the corrosion resistance of phytic acid conversion coating for magnesium alloys. *Surface and Coatings Technology, 201*(3–4), 1536–1541.

Jo, J. H., Kang, B. G., Shin, K. S., Kim, H. E., Hahn, B. D., Park, D. S., et al. (2011). Hydroxyapatite coating on magnesium with MgF_2 interlayer for enhanced corrosion resistance and biocompatibility. *Journal of Materials Science: Materials in Medicine, 22*(11), 2437–2447.

Killian, M. S., Wagener, V., Schmuki, P., & Virtanen, S. (2010). Functionalization of metallic magnesium with protein layers via linker molecules. *Langmuir: The ACS Journal of Surfaces and Colloids, 26*(14), 12044–12048.

Kim, S.-B., Jo, J.-H., Lee, S.-M., Kim, H.-E., Shin, K.-H., & Koh, Y.-H. (2013). Use of a poly (ether imide) coating to improve corrosion resistance and biocompatibility of magnesium (Mg) implant for orthopedic applications. *Journal of Biomedical Materials Research Part A, 101*(6), 1708–1715.

Lalk, M., Reifenrath, J., Angrisani, N., Bondarenko, A., Seitz, J. M., Mueller, P. P., et al. (2013). Fluoride and calcium-phosphate coated sponges of the magnesium alloy AX30 as bone grafts: a comparative study in rabbits. *Journal of Materials Science: Materials in Medicine, 24*(2), 417–436.

Latham, J.-A., Howlett, P. C., MacFarlane, D. R., & Forsyth, M. (2011). Electrochemical reactivity of trihexyl(tetradecyl)phosphonium bis(2,4,4-trimethylpentyl)phosphinate ionic liquid on glassy carbon and AZ31 magnesium alloy. *Electrochimica Acta, 56*(15), 5328–5334.

Lei, T., Ouyang, C., Tang, W., Li, L.-F., & Zhou, L.-S. (2010). Enhanced corrosion protection of MgO coatings on magnesium alloy deposited by an anodic electrodeposition process. *Corrosion Science, 52*, 3504–3508.

Lellouche, J., Kahana, E., Elias, S., Gedanken, A., & Banin, E. (2009). Antibiofilm activity of nanosized magnesium fluoride. *Biomaterials, 30*(30), 5969–5978.

Levy, G., & Aghion, E. (2013). Effect of diffusion coating of Nd on the corrosion resistance of biodegradable Mg implants in simulated physiological electrolyte. *Acta Biomaterialia, 9*(10), 8624–8630.

Li, M., Cheng, Y., Zheng, Y. F., Zhang, X., Xi, T. F., & Wei, S. C. (2012). Surface characteristics and corrosion behaviour of WE43 magnesium alloy coated by SiC film. *Applied Surface Science, 258*(7), 3074–3081.

Li, M., Cheng, Y., Zheng, Y. F., Zhang, X., Xi, T. F., & Wei, S. C. (2013). Plasma enhanced chemical vapor deposited silicon coatings on Mg alloy for biomedical application. *Surface and Coatings Technology, 228*(Suppl. 1), S262–S265.

Li, L., Gao, J., & Wang, Y. (2004). Evaluation of cyto-toxicity and corrosion behavior of alkali-heat-treated magnesium in simulated body fluid. *Surface and Coatings Technology, 185*(1), 92–98.

Li, N., Li, Y. D., Wang, Y. B., Li, M., Cheng, Y., Wu, Y. H., et al. (2013). Corrosion resistance and cytotoxicity of a MgF_2 coating on biomedical Mg-1Ca alloy via vacuum evaporation deposition method. *Surface and Interface Analysis, 45*(8), 1217–1222.

Li, G. Y., Lian, J. S., Niu, L. Y., & Jiang, Z. H. (2006). Influence of pH of phosphating bath on the zinc phosphate coating on AZ91D magnesium alloy. *Advanced Engineering Materials, 8*(1–2), 123–127.

Li, C.-F., Wang, M.-J., Ho, W.-H., Li, H.-N., & Yen, S.-K. (2011). Effects of electrolytic MgO coating parameters on corrosion resistance of AZ91D magnesium alloy. *Journal of the Electrochemical Society, 158*(2), C11–C16.

Liu, X., Chu, P. K., & Ding, C. (2004). Surface modification of titanium, titanium alloys, and related materials for biomedical applications. *Materials Science and Engineering: R: Reports, 47*(3–4), 49–121.

Liu, P., Pan, X., Yang, W., Cai, K., & Chen, Y. (2012). Al_2O_3–ZrO_2 ceramic coatings fabricated on WE43 magnesium alloy by cathodic plasma electrolytic deposition. *Materials Letters, 70*, 16–18.

Liu, Y., Yu, Z., Zhou, S., & Wu, L. (2006). Self-assembled monolayers on magnesium alloy surfaces from carboxylate ions. *Applied Surface Science, 252*(10), 3818–3827.

Liu, X., Yue, Z., Romeo, T., Weber, J., Scheuermann, T., Moulton, S., et al. (2013). Biofunctionalized anti-corrosive silane coatings for magnesium alloys. *Acta Biomaterialia, 9*(10), 8671–8677.

Lorenz, C., Brunner, J. G., Kollmannsberger, P., Jaafar, L., Fabry, B., & Virtanen, S. (2009). Effect of surface pre-treatments on biocompatibility of magnesium. *Acta Biomaterialia, 5*, 2783–2789.

Luo, X., & Cui, X. T. (2011). Electrochemical deposition of conducting polymer coatings on magnesium surfaces in ionic liquid. *Acta Biomaterialia, 7*(1), 441–446.

Malayoglu, U., Tekin, K. C., & Shrestha, S. (2010). Influence of post-treatment on the corrosion resistance of PEO coated AM50B and AM60B Mg alloys. *Surface and Coatings Technology, 205*, 1793–1798.

Meng, E. C., Guan, S. K., Wang, H. X., Wang, L. G., Zhu, S. J., Hu, J. H., et al. (2011). Effect of electrodeposition modes on surface characteristics and corrosion properties of fluorine-doped hydroxyapatite coatings on Mg–Zn–Ca alloy. *Applied Surface Science, 257*, 4811–4816.

Ming, Q., Da, L., & Chang, J. (2008). Microstructure and corrosion characteristics of laser-alloyed magnesium alloy AZ91D with Al–Si powder. *Science and Technology of Advanced Materials, 9*, 025002 (7 pp).

Mohan, L., Durgalakshmi, D., Geetha, M., Sankara Narayanan, T. S. N., & Asokamani, R. (2012). Electrophoretic deposition of nanocomposite (HAp + TiO_2) on titanium alloy for biomedical applications. *Ceramics International, 38*, 3435–3443.

Murase, K., Sasaki, I., Kitada, A., Uchimoto, Y., Ichii, T., & Sugimura, H. (2013). Enhanced anodic dissolution of magnesium in quaternary-ammonium-based ionic liquid containing a small amount of water. *Journal of the Electrochemical Society, 160*(10), D453–D458.

Nan, M., Yangmei, C., & Bangcheng, Y. (2013). Magnesium metal – a potential biomaterial with antibone cancer properties. *Journal of Biomedical Materials Research Part A, 102*(8), 2644–2651.

Narayan, R., & Goering, P. (2011). Laser micro- and nanofabrication of biomaterials. *MRS Bulletin, 36*(12), 973–982.

Neubert, V., Bakkar, A., & Huang, C. A. (2006). Coating and surface treatment of magnesium based alloys and composites. In Karl U. Kainer (Ed.), *Proceedings of the 7th international conference on magnesium alloys and their applications* (pp. 842–848). Weinheim: Wiley-VCH.

Noorakma, A. C. W., Zuhailawati, H., Aishvarya, V., & Dhindaw, B. K. (2013). Hydroxyapatite-coated magnesium-based biodegradable alloy: cold spray deposition and simulated body fluid studies. *Journal of Materials Engineering and Performance, 22*(10), 2997–3004.

de Oliveira, M. C. L., Pereira, V. S. M., Correa, O. V., & Antunes, R. A. (2013). Corrosion performance of anodized AZ91D magnesium alloy: effect of the anodizing potential on the film structure and corrosion behavior. *Journal of Materials Engineering and Performance, 23*, 593–603.

van Ooij, W. J., Zhu, D. Q., Prasad, G., Jayaseelan, S., Fu, Y., & Teredesai, N. (2000). Silane based chromate replacements for corrosion control, paint adhesion, and rubber bonding. *Surface Engineering, 16*, 386–396.

Ostrowski, N. J., Lee, B., Roy, A., Ramanathan, M., & Kumta, P. N. (2013). Biodegradable poly(lactide-*co*-glycolide) coatings on magnesium alloys for orthopedic applications. *Journal of Materials Science: Materials in Medicine, 24*(1), 85–96.

O'Brien, B., & Carroll, W. (2009). The evolution of cardiovascular stent materials and surfaces in response to clinical drivers: a review. *Acta Biomaterialia, 5*, 945–958.

Paital, S. R., Bhattacharya, A., Moncayo, M., Ho, Y. H., Mahdak, K., Nag, S., et al. (2012). Improved corrosion and wear resistance of Mg alloys via laser surface modification of Al on AZ31B. *Surface and Coatings Technology, 206*(8–9), 2308–2315.

Park, M., Lee, J. E., Park, C. G., Lee, S. H., Seok, H. K., & Choy, Y. B. (2013). Polycaprolactone coating with varying thicknesses for controlled corrosion of magnesium. *Journal of Coatings Technology and Research, 10*(5), 695–706.

Paulmier, T., Bell, J. M., & Fredericks, P. M. (2007). Development of a novel cathodic plasma/electrolytic deposition technique part 1: production of titanium dioxide coatings. *Surface and Coatings Technology, 201*(21), 8761–8770.

Pereda, M. D., Alonso, C., Burgos-Asperilla, L., del Valle, J. A., Ruano, O. A., Perez, P., et al. (2010). Corrosion inhibition of powder metallurgy Mg by fluoride treatments. *Acta Biomaterialia, 6*(5), 1772–1782.

Pereda, M. D., Alonso, C., Gamero, M., del Valle, J. A., & Fernández Lorenzo de Mele, M. A. (2011). Comparative study of fluoride conversion coatings formed on biodegradable powder metallurgy Mg: the effect of chlorides at physiological level. *Materials Science and Engineering C, 31*(5), 858–865.

Peters, T., Jr. (1996). *All about albumin*. San Diego: Academic Press.

Pinto, R., Carmezim, M. J., Ferreira, M. G. S., & Montemor, M. F. (2010). A two-step surface treatment, combining anodisation and silanisation, for improved corrosion protection of the Mg alloy WE54. *Progress in Organic Coatings, 69*(2), 143–149.

Plueddemann, E. P. (1991). *Silane coupling agents*. Berlin: Springer.

Qiu, X., Wan, P., Tan, L. L., Fan, X. M., & Yang, K. (2014). Preliminary research on a novel bioactive silicon doped calcium phosphate coating on AZ31 magnesium alloy via electrodeposition. *Materials Science and Engineering C, 36*, 65–76.

Razavi, M., Fathi, M., Savabi, O., Beni, B. H., Mohammad Razavi, S., Vashaee, D., et al. (2014). Coating of biodegradable magnesium alloy bone implants using nanostructured diopside ($CaMgSi_2O_6$). *Applied Surface Science, 288*, 130–137.

Razavi, M., Fathi, M., Savabi, O., Beni, B. H., Vashaee, D., & Tayebi, L. (2013). Surface microstructure and in vitro analysis of nanostructured akermanite ($Ca^{2MgSi}{}_2O^7$) Coating on biodegradable magnesium alloy for biomedical applications. *Colloids and Surfaces B: 117*, 432–440.

Razavi, M., Fathi, M., Savabi, O., Mohammad Razavi, S., Beni, B. H., Vashaee, D., et al. (2013). Surface modification of magnesium alloy implants by nanostructured bredigite coating. *Materials Letters, 113*, 174–178.

Razavi, M., Fathi, M., Savabi, O., Mohammad Razavi, S., Beni, B. H., Vashaee, D., et al. (2014). Controlling the degradation rate of bioactive magnesium implants by electrophoretic deposition of akermanite coating. *Ceramics International, 40*(3), 3865–3872.

Ren, M., Cai, S., Xu, G., Ye, X., Dou, Y., Huang, K., et al. (2013). Influence of heat treatment on crystallization and corrosion behavior of calcium phosphate glass coated AZ31 magnesium alloy by sol–gel method. *Journal of Non-Crystalline Solids, 369*, 69–75.

Rojaee, R., Fathi, M., & Raeissi, K. (2013a). Electrophoretic deposition of nanostructured hydroxyapatite coating on AZ91 magnesium alloy implants with different surface treatments. *Applied Surface Science, 285*, 664–673.

Rojaee, R., Fathi, M., & Raeissi, K. (2013b). Controlling the degradation rate of AZ91 magnesium alloy via sol-gel derived nanostructured hydroxyapatite coating. *Materials Science and Engineering C: Materials for Biological Applications, 33*(7), 3817–3825.

Salman, S. A., & Okido, M. (2012). Self-assembled monolayers formed on AZ31 Mg alloy. *Journal of Physics and Chemistry of Solids, 73*(7), 863–866.

Sarkar, P., & Nicholson, P. S. (1996). Electrophoretic deposition (EPD): mechanism, kinetics and application to ceramics. *Journal of American Ceramic Society, 79*, 1987–2002.

Schwartz, D. K. (2001). Mechanisms and kinetics of self-assembled monolayer formation. *Annual Reviews in Physical Chemistry, 52*, 107–137.

Sealy, M. P., & Guo, Y. B. (2010). Surface integrity and process mechanics of laser shock peening of novel biodegradable magnesium–calcium (Mg–Ca) alloy. *Journal of the Mechanical Behaviour of Biomedical Materials, 3*, 488–496.

Sebaa, M. A., Dhillon, S., & Liu, H. (2013). Electrochemical deposition and evaluation of electrically conductive polymer coating on biodegradable magnesium implants for neural applications. *Journal of Materials Science: Materials in Medicine, 24*, 307–316.

Shi, Y., Qi, M., Chen, Y., & Shi, P. (2011). MAO-DCPD composite coating on Mg alloy for degradable implant applications. *Materials Letters, 65*, 2201–2204.

Singh, A., & Harimkar, S. P. (2012). *Laser Surface Engineering of Magnesium Alloys: A Review, 64*(6), 716–733.

Song, Y. W., Shan, D. Y., & Han, E. H. (2008). Electrodeposition of hydroxyapatite coating on AZ91D magnesium alloy for biomaterial application. *Materials Letters, 62*, 3276–3279.

Song, G., & Song, S. (2007). A possible biodegradable magnesium implant material. *Advanced Engineering Materials, 9*, 298–302.

Spencer, K., Fabijanic, D. M., & Zhang, M. X. (2009). The use of $Al–Al_2O_3$ cold spray coatings to improve the surface properties of magnesium alloys. *Surface and Coatings Technology, 204*, 336–344.

Spencer, K., Luzin, V., Matthews, N., & Zhang, M. X. (2012). Residual stresses in cold spray Al coatings: the effect of alloying and of process parameters. *Surface and Coatings Technology, 206*(19–20), 4249–4255.

Spencer, K., & Zhang, M. X. (2009). Heat treatment of cold spray coatings to form protective intermetallic layers. *Scripta Materialia, 61*, 44–47.

Srinivasan, A., Ranjani, P., & Rajendran, N. (2013). Electrochemical polymerization of pyrrole over AZ31 Mg alloy for biomedical applications. *Electrochimica Acta, 88*(2010), 310–321.

Staiger, M. P., Pietak, A. M., Huadmai, J., & Dias, G. (2006). Magnesium and its alloys as orthopedic biomaterials: a review. *Biomaterials, 27*, 1728–1734.

Subramanian, R., Sircar, S., & Mazumder, J. (1991). Laser cladding of zirconium on magnesium for improved corrosion properties. *Journal of Materials Science, 26*(4), 951–956.

Surmenev, R. A. (2011). A review of plasma-assisted methods for calcium phosphate-based coatings fabrication. *Surface and Coatings Technology, 206*(8–9), 2035–2056.

Tacikowski, M., Banaszek, M., & Smolik, J. (2014). Corrosion-resistant composite titanium nitride layers produced on the AZ91D magnesium alloy by a hybrid method. *Vacuum, 99*, 298–302.

Tamada, J. A., & Langer, R. (1993). Erosion kinetics of hydrolytically degradable polymers. *Proceedings of the National Academy of Sciences United States of America, 90*, 552–556.

Tang, H., Xin, T., & Wang, F. (2013). Calcium phosphate/titania sol-gel coatings on AZ31 magnesium alloy for biomedical applications. *International Journal of Electrochemical Science, 8*, 8115–8125.

Tao, Y., Xion, T., Sun, C., Kong, L., Cui, X., Li, T., et al. (2010). Microstructure and corrosion performance of a cold sprayed aluminium coating on AZ91D magnesium alloy. *Corrosion Science, 52*(10), 3191–3197.

Tay, B. K., Zhao, Z. W., & Chua, D. H. C. (2006). Review of metal oxide films deposited by filtered cathodic vacuum arc technique. *Materials Science and Engineering: R: Reports, 52*, 1–48.

Thomann, M., Krause, C., Angrisani, N., Bormann, D., Hassel, T., Windhagen, H., et al. (2010). Influence of a magnesium-fluoride coating of magnesium-based implants (MgCa0.8) on degradation in a rabbit model. *Journal of Biomedical Materials Research Part A, 93*(4), 1609–1619.

Turhan, M. C., Weiser, M., Jha, H., & Virtanen, S. (2011). Optimization of electrochemical polymerization parameters of polypyrrole on Mg–Al alloy (AZ91D) electrodes and corrosion performance. *Electrochimica Acta, 56*(15), 5347–5354.

Van der Biest, O. O., & Vandeperre, L. J. (1999). Electrophoretic deposition of materials. *Annual Reviews in Materials Science, 29*, 327–352.

Van Phuong, N., Lee, K. H., Chang, D., Kim, M., Lee, S., & Moon, S. (2013). Zinc phosphate conversion coatings on magnesium alloys: a Review. *Metals and Materials International, 19*(2), 273–281.

Van Phuong, N., Lee, K. H., Chang, D., & Moon, S. (2013). Effects of Zn^{2+} concentration and pH on the zinc phosphate conversion coatings on AZ31 magnesium alloy. *Corrosion Science, 74*, 314–322.

Virtanen, S. (2011). Biodegradable Mg and Mg alloys: corrosion and biocompatibility. *Materials Science and Engineering B, 176*, 1600–1608.

Wagener, V., Killian, M. S., Turhan, C. M., & Virtanen, S. (2013). Albumin coating on magnesium via linker molecules-comparing different coating mechanisms. *Colloids and Surfaces B: Biointerfaces, 103*, 586–594.

Wang, H. X., Guan, S. K., Wang, Y., Liu, H., Wang, H., Wang, L. G., et al. (2011). In vivo degradation behavior of Ca-deficient hydroxyapatite coated Mg–Zn–Ca alloy for bone implant application. *Colloids and Surfaces B: Biointerfaces, 88*, 254–259.

Wang, H. X., Guan, S. K., Wang, X., Ren, C. X., & Wang, L. G. (2010). In vitro degradation and mechanical integrity of Mg–Zn–Ca alloy coated with Ca-deficient hydroxyapatite by the pulse electrodeposition process. *Acta Biomaterialia, 6*, 1743–1748.

Wang, J., He, Y., Maitz, M. F., Collins, B., Xiong, K., Guo, L., et al. (2013). A surface-eroding poly(1,3-trimethylene carbonate) coating for fully biodegradable magnesium based stent applications: toward better biofunction, biodegradation and biocompatibility. *Acta Biomaterialia, 9*(10), 8678–8689.

Wang, B., Huang, P., Ou, C., Li, K., Yan, B., & Lu, W. (2013). In vitro corrosion and cytocompatibility of ZK60 magnesium alloy coated with hydroxyapatite by a simple chemical conversion process for orthopedic applications. *International Journal of Molecular Science, 14*, 23614–23628.

Wang, Q., Qiu, D., Xiong, Y., Birbilis, N., & Zhang, M. X. (2014). High resolution microstructure characterization of the interface between cold sprayed Al coating and Mg alloy substrate. *Applied Surface Science, 289*, 366–369.

Wang, Q., Spencer, K., Birbilis, N., & Zhang, M. X. (2010). The influence of ceramic particles on bond strength of cold spray composite coatings on AZ91 alloy substrate. *Surface and Coatings Technology, 205*(1), 50–56.

Wang, J., Tang, J., Zhang, P., Li, Y., Wang, J., Lai, Y., et al. (2012). Surface modification of magnesium alloys developed for bioabsorbable orthopedic implants: a general review. *Journal of Biomedical Materials Research Part B, 100*, 1691–1701.

Wang, L., Zhou, J., Liang, J., & Chen, J. (2012). Microstructure and corrosion behavior of plasma electrolytic oxidation coated magnesium alloy pre-treated by laser surface melting. *Surface and Coatings Technology, 206*(13), 3109–3115.

Weetall, H. H. (1993). Preparation of immobilized proteins covalently coupled through silane coupling agents to inorganic supports. *Applied Biochemistry Biotechnology, 41*, 157–188.

Witte, F. (2010). The history of biodegradable magnesium implants: a review. *Acta Biomaterialia, 6*, 1680–1692.

Witte, F., Fischer, J., Nellesen, J., Vogt, C., Vogt, J., & Donath, T. (2010). In vivo corrosion and corrosion protection of magnesium alloy LAE442. *Acta Biomaterialia, 6*(5), 1792–1799.

Witte, F., Hort, N., Vogt, C., Cohen, S., Kainer, K. U., Willumeit, R., et al. (2008). Degradable biomaterials based on magnesium corrosion. *Current Opinion in Solid State and Materials Science, 12*, 63–72.

Wong, H. M., Yeung, K. W. K., Lam, K. O., Tam, V., Chu, P. K., Luk, K. D. K., et al. (2010). A biodegradable polymer-based coating to control the performance of magnesium alloy orthopaedic implants. *Biomaterials, 31*(8), 2084–2096.

Wong, H. M., Zhao, Y., Tam, V., Wu, S., Chu, P. K., Zheng, Y., et al. (2013). In vivo stimulation of bone formation by aluminum and oxygen plasma surface-modified magnesium implants. *Biomaterials, 34*(38), 9863–9876.

Wu, G. (2007). Fabrication of Al and Al/Ti coatings on magnesium alloy by sputtering. *Materials Letters, 61*(18), 3815–3817.

Wu, G., Jamesh, M. I., & Chu, P. K. (2013). Surface design of biodegradable magnesium alloys – a review. *Surface and Coatings Technology, 233*, 2–12.

Wu, C., Wen, Z.-H., Dai, C.-S., Lu, Y., & Yang, F. (2010). Fabrication of calcium phosphate/chitosan coatings on AZ91D magnesium alloy with a novel method. *Surface and Coatings Technology, 204*(20), 3336–3347.

Wu, G., Zhao, Y., Zhang, X., Ibrahim, J. M., & Chu, P. K. (2013). Self-protection against corrosion of aged magnesium alloy in simulated physiological environment. *Corrosion Science, 68*, 279–285.

Xin, Y., Liu, C., Huo, K., Tang, G., Tian, X., & Chu, P. K. (2009). Corrosion behavior of ZrN/Zr coated biomedical AZ91 magnesium alloy. *Surface and Coatings Technology, 203*(17–18), 2554–2557.

Xin, Y., Liu, C., Zhang, W., Huo, K., Tang, G., Tian, X., et al. (2008). Corrosion resistance of ZrO_2–Zr-coated biodegradable surgical magnesium alloy. *Journal of Materials Research, 23*(2), 312–319.

Xin, Y., Liu, C., Zhang, W., Jiang, J., Tang, G., Tian, X., et al. (2008). Electrochemical behavior Al_2O_3/Al coated surgical AZ91 magnesium alloy in simulated body fluids. *Journal of the Electrochemical Society, 155*(5), C178–C182.

Xu, L., & Yamamoto, A. (2012). Characteristics and cytocompatibility of biodegradable polymer film on magnesium by spin coating. *Colloids and Surfaces B: Biointerfaces, 93*, 67–74.

Xu, L., Zhang, E., & Yang, K. (2009). Phosphating treatment and corrosion properties of Mg–Mn–Zn alloy for biomedical application. *Journal of Materials Science: Materials in Medicine, 20*, 859–867.

Xue, D., Tan, Z., Schulz, M. J., Vanooij, W. J., Sankar, J., Yun, Y., et al. (2012). Corrosion studies of modified organosilane coated magnesium–yttrium alloy in different environments. *Materials Science and Engineering C, 32*(5), 1230–1236.

Xue, D., Yun, Y., Schulz, M. J., & Shanov, V. (2011). Corrosion protection of biodegradable magnesium implants using anodization. *Materials Science and Engineering C, 31*, 215–223.

Yamazoe, H., Oyane, A., Nashima, T., & Ito, A. (2010). Reduced platelet adhesion and blood coagulation on cross-linked albumin films. *Materials Science and Engineering C, 30*(6), 812–816.

Yamazoe, H., & Tanabe, T. J. (2010). Drug-carrying albumin film for blood-contacting biomaterials. *Journal of Biomaterials Science Polymer Edition, 21*(5), 647–657.

Yang, J., Cui, F., & Lee, I. S. (2011). Surface modifications of magnesium alloys for biomedical applications. *Annals of Biomedical Engineering, 39*, 1857–1871.

Yang, J., Cui, F. Z., Lee, I. S., & Wang, X. (2010). Plasma surface modification of magnesium alloy for biomedical application. *Surface and Coatings Technology, 205*(Suppl. 1), S182–S187.

Yang, Y. C., Tsai, C. Y., Huang, Y. H., & Lin, C. S. (2012). Formation mechanism and properties of titanate conversion coating on AZ31 magnesium alloy. *Journal of the Electrochemical Society, 159*(5), C226–C232.

Yan, T., Tan, L., Xiong, D., Liu, X., Zhang, B., & Yang, K. (2010). Fluoride treatment and in vitro corrosion behavior of an AZ31B magnesium alloy. *Materials Science and Engineering C, 30*(5), 740–748.

Ye, X., Cai, S., Dou, Y., Xu, G., Huang, K., Ren, M., et al. (2012). Bioactive glass–ceramic coating for enhancing the in vitro corrosion resistance of biodegradable Mg alloy. *Applied Surface Science, 259*, 799–805.

Ye, X.-Y., Chen, M.-F., You, C., & Liu, D.-B. (2010). The influence of HF treatment on corrosion resistance and in vitro biocompatibility of Mg–Zn–Zr alloy. *Frontiers of Materials Science in China, 4*(2), 132–138.

Ye, C. H., Zheng, Y. F., Wang, S. Q., Xi, T. F., & Li, Y. D. (2012). In vitro corrosion and biocompatibility study of phytic acid modified WE43 magnesium alloy. *Applied Surface Science, 258*(8), 3420–3427.

Yerokhin, A. L., Shatrov, A., Samsonov, V., Shashkov, P., Leyland, A., & Matthews, A. (2004). Fatigue properties of Keronite® coatings on a magnesium alloy. *Surface and Coatings Technology, 182*(1), 78–84.

Zeng, R., Dietzel, W., Witte, F., Hort, N., & Blawert, C. (2008). Progress and challenge for magnesium alloys as biomaterials. *Advanced Engineering Materials, 10*, B3–B14.

Zhang, J., Dai, C.-S., Wei, J., & Wen, Z.-H. (2012). Study on the bonding strength between calcium phosphate/chitosan composite coatings and a Mg alloy substrate. *Applied Surface Science, 261*, 276–286.

Zhang, E., Xu, L., & Yang, K. (2005). Formation by ion plating of Ti-coating on pure Mg for biomedical applications. *Scripta Materialia, 53*, 523–527.

Zhang, Y., Yang, J. X., Cui, F. Z., Lee, I. S., & Lee, G. (2010). Characterization and degradation comparison of DLC film on different magnesium alloys. *Surface and Coatings Technology, 205*, S15–S20.

Zhang, Y., You, J., Lu, J., Cui, C., Jiang, Y., & Ren, X. (2010). Effects of laser shock processing on stress corrosion cracking susceptibility of AZ31B magnesium alloy. *Surface and Coatings Technology, 204*(24), 3947–3953.

Zhao, Q., Guo, X., Dang, X., Hao, J., Lai, J., & Wang, K. (2013). Preparation and properties of composite MAO/ECD coatings on magnesium alloy. *Colloids and Surfaces B, 102*, 321–326.

Zhao, Y., Wu, G., Jiang, J., Wong, H. M., Yeung, K. W. K., & Chu, P. K. (2012). Improved corrosion resistance and cytocompatibility of magnesium alloy by two-stage cooling in thermal treatment. *Corrosion Science, 59*, 360–365.

Zhao, Y., Wu, G., Lu, Q., Wu, J., Xu, R., Yeung, K. W. K., et al. (2013). Improved surface corrosion resistance of WE43 magnesium alloy by dual titanium and oxygen ion implantation. *Thin Solid Films, 529*, 407–411.

Zhitomirsky, D., Roether, J. A., Boccaccini, A. R., & Zhitomirsky, I. (2009). Electrophoretic deposition of bioactive glass/polymer composite coatings with and without HA nanoparticles inclusions for biomedical applications. *Journal of Materials Processing Technology, 209*, 1853–1860.

Zhu, Y., Wu, G., Zhang, Y.-H., & Zhao, Q. (2011). Growth and characterization of $Mg(OH)_2$ film on magnesium alloy AZ31. *Applied Surface Science, 257*, 6129–6137.

Zhu, Y., Zhao, Q., Zhang, Y.-H., & Wu, G. (2012). Hydrothermal synthesis of protective coating on magnesium alloy using de-ionized water. *Surface and Coatings Technology, 206*, 2961–2966.

Zomorodian, A., Brusciotti, F., Fernandes, A., Carmezim, M. J., Moura e Silva, T., Fernandes, J. C. S., et al. (2012). Anti-corrosion performance of a new silane coating for corrosion protection of AZ31 magnesium alloy in Hank's solution. *Surface and Coatings Technology, 206*(21), 4368–4375.

Surface design of biodegradable magnesium alloys for biomedical applications

3

P.K. Chu, G.S. Wu
City University of Hong Kong, Hong Kong, China

3.1 Introduction

Surface is crucial to biomaterials because it not only provides a platform for cell/biomaterial interactions and associated chemical reactions, but also acts as a shield to resist corrosion or wear in the physiological environment. Magnesium (Mg), as the fourth most abundant cation in the human body, is essential to human metabolism and naturally found in bone tissues (Staiger, Pietak, Huadmai, & Dias, 2006). However, compared to traditional biometals such as Ti-based alloys, biodegradable Mg alloys have a more reactive surface and normally degrade too quickly, especially in the initial stage under physiological conditions (Jamesh, Kumar, & Narayanan, 2011; Song, 2007; Witte et al., 2006; Wu, Zhao, Zhang, Jamesh, & Chu, 2013). Adverse effects can occur to the patients if excessive Mg ions and other corrosion products cannot be effectively absorbed by the surrounding tissues or excreted properly. Moreover, rapid degradation will bring about premature failure, thereby inducing Mg-based implants to lose the desirable ability of osteosynthesis and counteracting the capability of Mg alloys to mitigate the stress-shielding effect as a result of their Young's modulus ($E = 41-45$ GPa) being similar to that of bones ($E = 3-20$ GPa) (Wu, Zhao, et al., 2013; Zberg, Uggowitzer, & Löffler, 2009).

Surface modification is one of the desirable methods to overcome the drawbacks and expedite the development of new-generation biodegradable metals (Narayanan, Park, & Lee, 2014; Wu, Jamesh, & Chu, 2013). A temporary surface fabricated on Mg-based materials can be used to tailor the mechanical performance, corrosion behavior, and biological properties to meet clinical requirements. In this chapter, the common surface treatment techniques suitable for Mg alloys and the design principles in the development of degradable Mg alloys to address clinical needs are discussed. The role of the various coating techniques and ion implantation are described with examples.

3.2 Surface modification techniques

Several coating technologies are applicable to Mg and its alloys, including electrochemical plating, conversion coatings, anodizing, organic coatings, and vapor-phase processes. Each of them has special advantages and limitations, and they are briefly described in the following sections.

Surface Modification of Magnesium and its Alloys for Biomedical Applications. http://dx.doi.org/10.1016/B978-1-78242-077-4.00003-6

3.2.1 Electrochemical deposition

Electrochemical deposition or plating is an old but practical technique to deposit metallic coatings. It is usually subdivided into two types: electroplating and electroless plating. In both cases, the metal salt in a solution is reduced to the metallic form on the surface of the substrate. The distinction between electroplating and electroless plating is that the electrons for reduction in electroplating are supplied by an external source, whereas the reducing electrons in electroless plating are supplied by a chemical reducing agent in the solution or, in the case of immersion plating, the substrate itself (Gray & Luan, 2002).

3.2.1.1 Electrodeposition

In electrodeposition or electroplating as schematically illustrated in Figure 3.1, the surface of a substrate is modified in an aqueous or nonaqueous electrolytic environment by applying external power. In the electrolysis cell, the sample being plated is the cathode and immersed in a solution containing the required metal in an oxidized form. The anode is made of the metal to be deposited on the substrate. When power is turned on, the metal atoms are oxidized and dissolved in the solution according to the following reaction: $Me(s) \rightarrow Me^{n+}(aq) + ne^-$. On the cathode, the dissolved metal ions in the electrolyte are reduced at the interface between the solution and the cathode according to the reaction: $Me^{n+}(aq) + ne^- \rightarrow Me(s)$. Usually, the rate at which the anode is dissolved is equal to the rate at which the cathode is plated. Hence, the ions in the bath are continuously replenished by the anode. In addition, a nonconsumable anode such as lead or carbon can be used in electrodeposition. In this case, ions of the metal to be plated must be periodically replenished in the bath after they have been extracted from the solution. At present, electroplating is widely used in the industry to coat

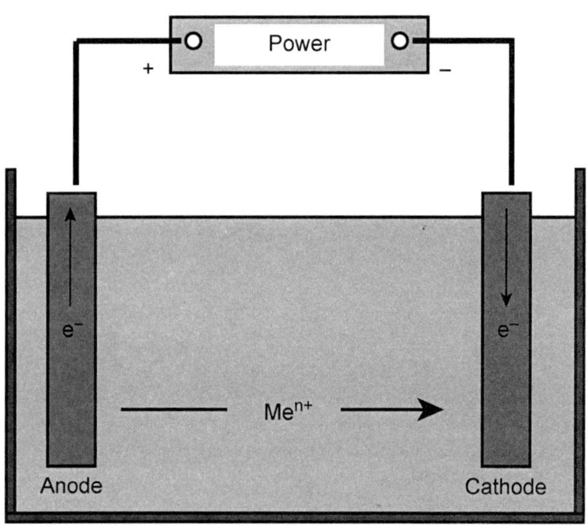

Figure 3.1 Schematic illustrating the setup in electrodeposition.
Source: Sudagar et al. (2013) and Carraro et al. (2007) with permission from Elsevier.

metallic products. The technique can provide protection against species in aggressive environments and even render the products' special surface properties such as decorative effects (Carraro, Maboudian, & Magagnin, 2007; Kanani, 2005; Schlesinger & Paunovic, 2010; Sudagar, Lian, & Sha, 2013).

3.2.1.2 Electroless deposition

In comparison to electrodeposition, electroless deposition or plating uses only one electrode and no external power source. The electroless deposition process can be divided into two types: autocatalytic deposition and galvanic displacement. In autocatalytic deposition, reduction of metallic ions in the solution and film deposition can be carried out by oxidation of the chemical compound in the solution itself, that is, a reducing agent. This reducing agent at a defined temperature that depends on the reducing agent and bath composition can spontaneously oxidize and free electrons for the reduction of metallic ions. Thus, it is named autocatalytic because oxidation of the reducing agent can start or become self-sustained only on the deposited metal surface. Figure 3.2 shows a schematic of electroless deposition with the reducing agent as the source of electrons. Galvanic displacement or immersion plating has a mechanism different from that of autocatalytic deposition. In immersion plating, reducing agents are not required because the base materials can behave as the reducing agent. Galvanic displacement takes place when the base material is displaced by a metallic ion in the solution having a lower oxidation potential than the displaced metal ion. As a sequence, the base material is dissolved in the solution and the metallic ions in

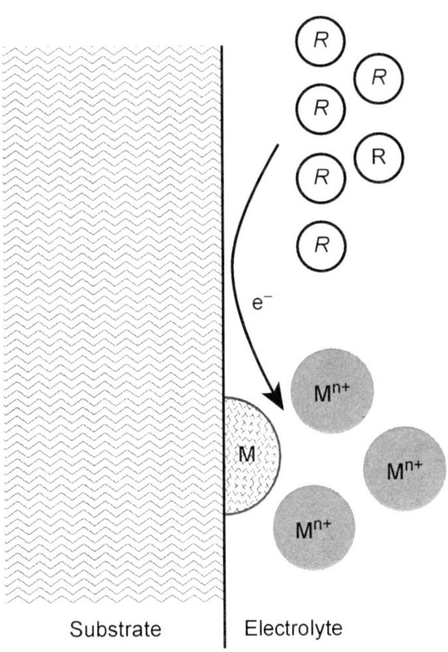

Figure 3.2 Electroless deposition with reducing agent (*R*) as the source of electrons. *Source:* Sudagar et al. (2013) and Carraro et al. (2007) with permission from Elsevier.

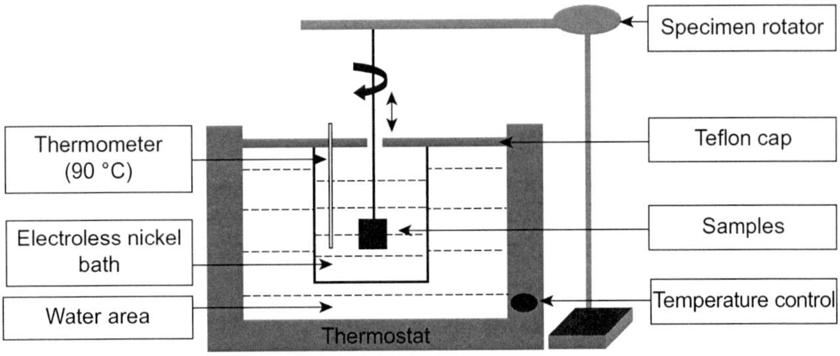

Figure 3.3 Experimental apparatus of electroless nickel plating.
Source: Li et al. (2006) with permission from Elsevier.

the solution are reduced on the surface of the base material (Carraro et al., 2007; Kanani, 2005; Schlesinger & Paunovic, 2010; Sudagar et al., 2013).

Figure 3.3 depicts a simplified schematic of the apparatus used in electroless nickel plating. This apparatus consists of a plating tank, thermostat for temperature control, and specimen rotator. The samples are placed in a glass tank covered with a Teflon cap with the thermostat (Li, An, & Wu, 2006). Electroless plating has advantages over electroplating because power sources are not needed. Autocatalytic deposition can also avoid the effects of current distribution, thereby improving the thickness uniformity. Therefore, it is more suitable for plating components with a complex shape. The typical disadvantages are that the plating process is usually slower and cannot create thick coatings. Consequently, electroless deposition is commonly used for decorative purposes under mild working conditions.

3.2.2 Chemical conversion coatings

3.2.2.1 General aspects

Chemical conversion is one of the important coating techniques to prepare coatings on metals by converting the part of the surface into the coating by means of a chemical or electrochemical process. The produced surface layer can be composed of metal oxides, chromates, phosphates, or other compounds that are chemically bonded to the surface. Because conversion coatings are formed in situ, adhesion to the substrate is generally very good, and so conversion as a pretreatment is effective in improving adhesion of the final coating. Several different types of conversion coatings have been developed by, for instance, chromate, phosphate/permanganate, and fluorozirconate treatments. One of the main disadvantages of conversion coatings is the toxicity of the treatment solutions. The conventional conversion coatings are based on chromium compounds that have been shown to be toxic and carcinogenic, and it is imperative to develop environmentally friendly processes. Besides, conversion coatings suffer from the nonuniform surface composition. If a conversion coating with uniform composition is needed, all the elements should be present in the alloy uniformly. At present, conversion coatings are mainly used for corrosion protection, hardness improvement, and color change as well as paint primers (Gray & Luan, 2002; Hornberger, Virtanen, & Boccaccini, 2012).

3.2.2.2 Anodization and microarc oxidation

Anodization is a type of classical electrochemical conversion. It encompasses electrode reactions in combination with an electric field-driven metal and oxygen ion diffusion, leading to the formation of an oxide film on the anode surface. The structural and chemical properties of the anodic oxides can be varied over a wide range by altering the process parameters, such as anode potential, electrolyte composition, temperature, and current. Anodic oxidation is a well-established method and can produce different types of protective oxide films on metals with excellent adhesion and bonding (Liu, Chu, & Ding, 2004).

Microarc oxidation (MAO), also known as plasma electrolytic oxidation (PEO), is an electrochemical surface treatment process based on anodizing. Compared to conventional anodizing, it uses much higher potentials. Figure 3.4 presents the schematic of microarc oxidation and the related electrical circuit. The sample is immersed in a bath containing the special electrolyte and forms one of the electrodes in the electrochemical cell, with the other counter-electrode being made from an inert material such as stainless steel. Potentials of more than 200 V are applied between these two electrodes. They may be in the forms of continuous or pulsed direct current (DC) or alternating pulses in which the stainless steel counter electrode may be grounded. When the potential exceeds the dielectric breakdown potential of the oxide film, discharges occur, resulting in localized plasma reactions to modify the oxide. Similar to conventional anodic oxidation coating, the coating adheres better to the substrate

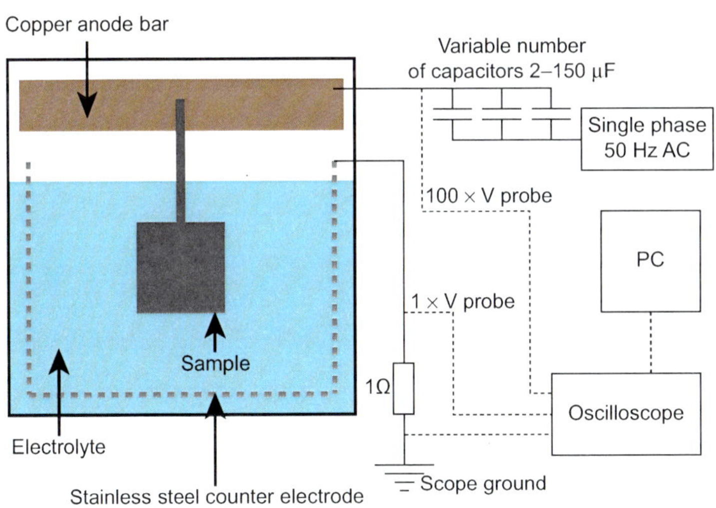

Figure 3.4 Schematic of microarc oxidation.
Source: Dunleavy et al. (2009) with permission from Elsevier.

metal than that formed by plasma spraying, because the MAO coating is a chemical conversion of the substrate metal into its oxide growing both inward and outward from the original metal surface. This process can produce thick (tens or hundreds of micrometers) and largely crystalline oxide coatings on metals such as titanium (Ti), aluminum (Al), and Mg. The coating provides high hardness, a continuous barrier against wear, corrosion, and heat, and electrical insulation. Usually the coating properties depend on the substrate as well as composition of the electrolyte and electrical regime (Curran & Clyne, 2005a, 2005b, 2006; Dunleavy, Golosnoy, Curran, & Clyne, 2009).

3.2.3 Physical vapor deposition

In physical vapor deposition (PVD), atoms or molecules are deposited onto a substrate from a vapor phase. The process can be roughly divided into three steps: generation of particles from the target materials, transportation, and film growth. To obtain the desirable films, substrate temperature, particle energy, reactive gas properties, and film density need to be tailored in the PVD process. Compared to wet coating methods such as electroplating and microarc oxidation, PVD is a dry coating method and boasts unique merits such as low deposition temperature, strong adhesion, multicomponent layers, and high coating density. Evaporation, sputter deposition, and ion plating are the three main types of PVD (Gray & Luan, 2002; Liu et al., 2004).

3.2.3.1 Evaporation

Evaporation is one of the most widely used thin film deposition techniques. A solid material is heated in a vacuum chamber to a temperature that generates some vapor from the material. In vacuum, the vaporized particles can travel directly to the target or substrate without encountering too many collisions and condense to the solid state, forming a film. Generally speaking, an evaporation system requires a main chamber, vacuum pump, and energy source that evaporates the materials. The source is normally placed on the bottom of the chamber, often in an upright crucible because it becomes liquid during heating in most cases. The substrates are held inverted by suitable fixtures on the top of the chamber with surfaces to be coated facing down toward the vapor source. A filament or electron beam source is typically used to vaporize the materials (Harsha, 2006; Mattox, 2010a; Wasa, Kanno, & Kotera, 2012).

3.2.3.2 Sputtering

Sputtering is another widely used thin film fabrication technique. Sputtering is a process in which ionized atoms are accelerated to a target surface to eject atoms from the surface. The ejected atoms are then condensed onto a sample to be plated, forming a thin film composed of sputtered materials. Sputter deposition has many advantages over other deposition methods such as evaporation, electroplating, and chemical vapor deposition (CVD). For example, sputter deposition can form smooth, dense, conformal, and continuous films more easily than evaporation because it produces a high-energy flux that leads to high surface mobility on the substrate surface. Usually, the sputtering rates of common metals vary within an order of magnitude, and thus another point distinguishing sputter deposition from evaporation or CVD is that sputtering preserves the stoichiometry of the target source because the physical bombardment mechanism of particle ejection results in a consistent stoichiometry on the sample surface (Mattox, 2010b; Wasa et al., 2012).

3.2.3.3 Ion plating

Ion plating is an atomistic vacuum coating process in which the deposited film is continuously or periodically bombarded by energetic atomic inert or reactive particles that can affect the growth and properties of the film. The depositing atoms can come from vacuum evaporation, sputtering, or arc vaporization. Bombardment prior to deposition is used to sputter clean the substrate surface, while bombardment during deposition is used to modify and control the properties of the film. It is crucial that bombardment is continuous between cleaning and deposition in the process to maintain an atomically clean interface. The bombarding species are generally ions accelerated from a plasma in the deposition chamber (ions for bombardment are extracted from the plasma and so termed plasma-based ion plating) or ions from an ion source (ion plating is performed in a vacuum environment and so termed vacuum-based ion plating). Figure 3.5 shows the two variations. The individual processes in ion plating can be separated into surface preparation, nucleation and interface formation, and film growth. Ion plating can also be considered a special process that varies from common

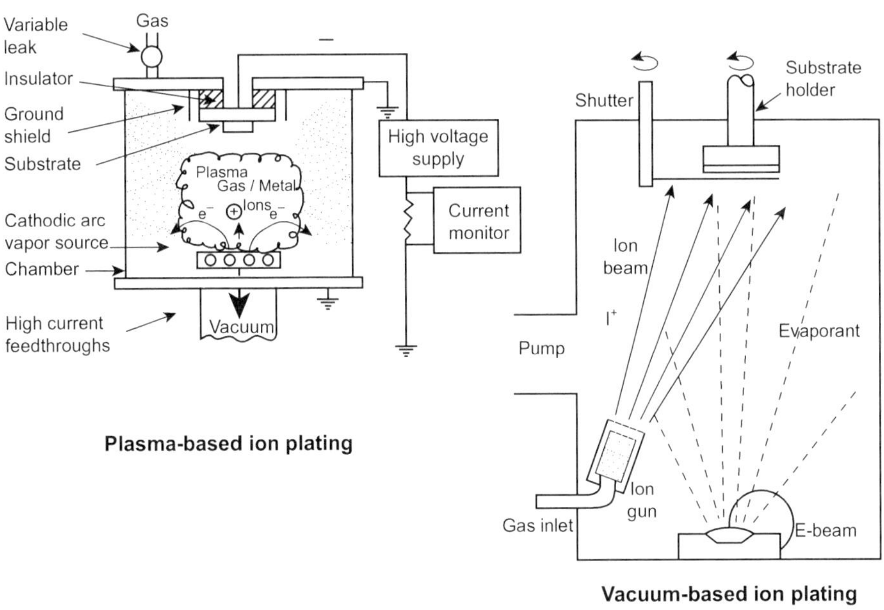

Figure 3.5 Plasma-based ion plating system equipped with a cathodic arc vaporization source using bombardment from the plasma and a vacuum-based ion plating system using thermal evaporation and an ion gun for bombardment.
Source: Mattox (2000) with permission from Elsevier.

deposition processes such as thermal evaporation, sputter deposition, and arc vapor deposition because the process name comes from the source of materials being deposited. Therefore, the ion plating process is also known by a number of other names, for example, ion vapor deposition (IVD), ion assisted deposition (IAD), bias sputtering, sputter ion plating (SIP), energy-assisted deposition, and ion beam assisted deposition (IBAD) (Mattox, 2000).

3.2.4 Ion implantation

Ion implantation is a process in which ions of a material are accelerated by an electrical field to impact a solid. If the ions differ in composition from the target, namely, the

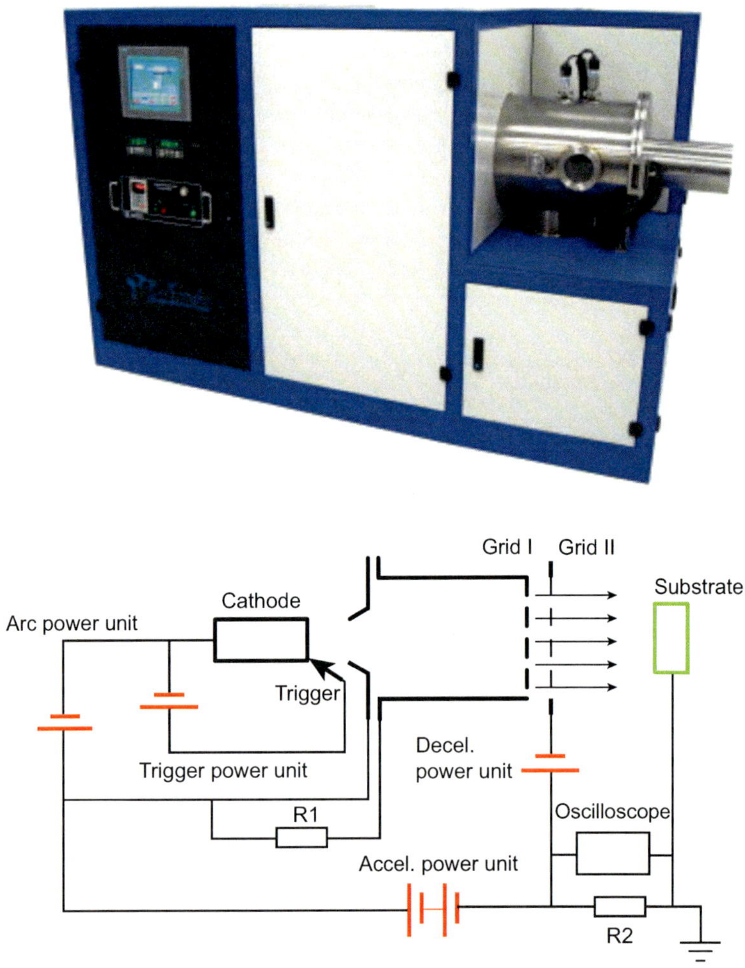

Figure 3.6 Picture and schematic diagram of HEMII-80 high energy metal ion implanter. *Source:* Feng et al. (2012) with permission from Elsevier.

specimen to be implanted, they will alter the elemental composition of the target and possibly change the physical, chemical, and/or electrical properties of the specimen. In particular, the use of energetic ions affords the possibility of introducing a wide range of atomic species independent of thermodynamic factors, thus making it possible to obtain impurity concentrations and distributions of particular interest. Ion implantation equipment consists of an ion source to produce ions of the desired element, an accelerator to accelerate the ions to a high energy, and a target chamber. Conventional beam-line ion implantation is a line-of-sight process in which ions are extracted from an ion source, accelerated to high energy, and then bombard the workpiece. Figure 3.6 shows the picture of a conventional HEMII-80 high-energy metal ion implantation machine in City University of Hong Kong and the simplified schematic diagram. The ion beam is usually quite small, and thus either the ion beam or sample is raster scanned to achieve uniform implantation over a large area. For samples with a very complicated geometry, it may be difficult to obtain conformal ion implantation by this method (Chu, Qin, Chan, Cheung, & Larson, 1996; Liu et al., 2004).

In comparison to conventional ion implantation, plasma immersion ion implantation (PIII) is an advanced technique that can more conveniently process samples with a complex shape. In PIII, the specimens are surrounded by a plasma and pulse-biased to a high negative potential relative to the chamber wall. Ions in the overlying plasma are accelerated across the plasma sheath generated around the specimens and implanted into the surface conformally. Figure 3.7 displays the schematic illustration of PIII. The plasma is produced in the vacuum chamber by various plasma sources such as electron cyclotron resonance (ECR) or radio frequency (RF). If metal plasmas are involved, plasma immersion ion implantation and deposition (PIII&D), which is a hybrid process that involves ion implantation and deposition, can be conducted to form an atomically intermixed layer between the substrate and coating in addition to ion implantation and coating. It is an ion plating technique in the broad sense (Anders, 1997; Chu, Chen, Wang, & Huang, 2002; Liu, Chu, & Ding, 2010).

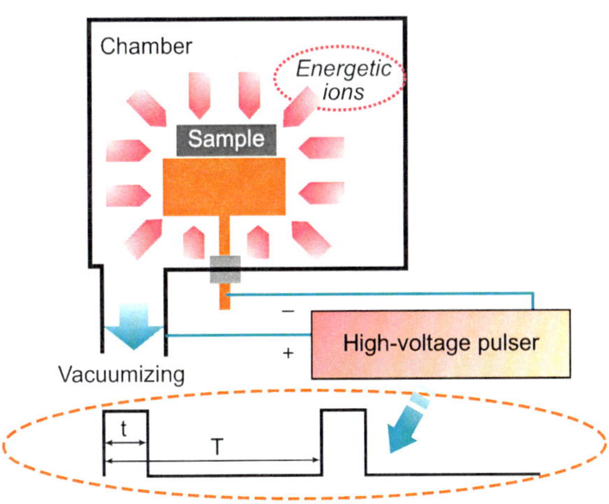

Figure 3.7 Schematic diagram of plasma immersion ion implantation (PIII).

3.3 Surface design principles for biodegradable magnesium alloys

3.3.1 Role of surface modification in traditional biometals

Metals play an essential role in biomedical devices because they are suitable for load-bearing applications. Different metals, such as titanium alloys and stainless steels, have been adopted in artificial hip joints, knee replacements, and fracture fixation devices because they possess excellent bulk properties, such as high mechanical strength, corrosion resistance, and superior fatigue properties (Cao, Liu, Meng, & Chu, 2011; Wu, Liu, et al., 2008). The biocompatibility of materials is generally related to the behavior of cells in contact, particularly cell adhesion onto the surface. Unfortunately, most artificial materials integrate poorly with host tissues, resulting in postoperation infection and other negative body responses. In this respect, surface modification plays an important role in changing the response of the biological environment on the artificial medical devices. The proper techniques not only preserve the excellent bulk attributes of the biometals, but also endow the metals with specific surface properties required by different clinical applications (Liu et al., 2004). Therefore, surface design and associated treatments are critical to biometal engineering.

3.3.2 Key issues in developing biodegradable magnesium alloys

Since biodegradable Mg-based implants dissolve as new tissues are formed, Mg alloys form a dynamic interface in the physiological environment, and it is different from those on traditional biomedical metals such as Ti alloys and stainless steels that do not degrade and thus have a static surface, as illustrated in Figure 3.8 (Wu, Jamesh,

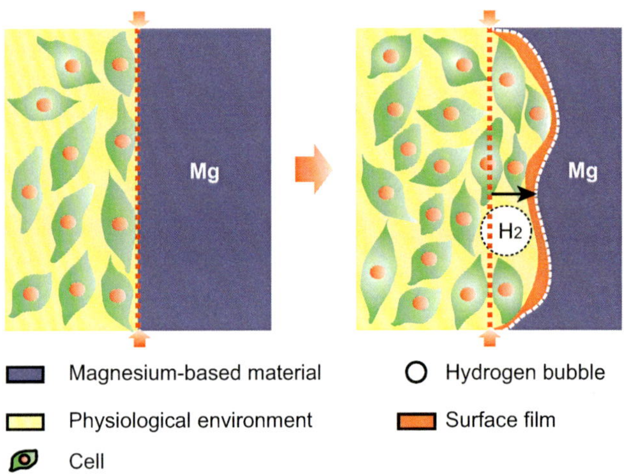

Figure 3.8 Dynamic interface between the Mg-based materials and bioenvironment during surface degradation.
Source: Wu, Jamesh, et al. (2013) with permission from Elsevier.

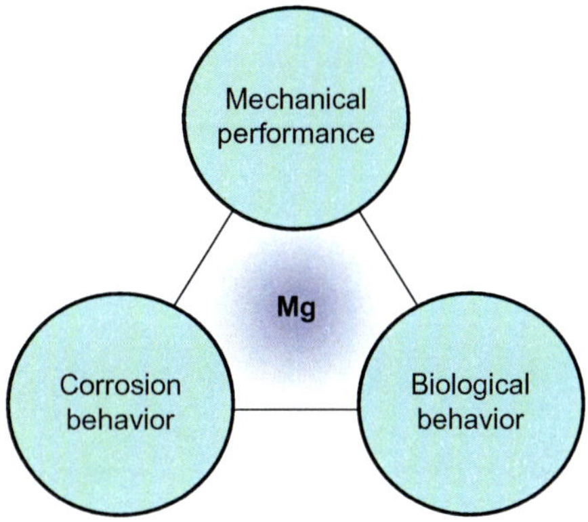

Figure 3.9 Critical factors in the design of Mg-based biomaterials.

et al., 2013). Mg corrodes in aqueous media via the following chemical reaction: $Mg + 2H_2O \rightarrow Mg^{2+} + 2OH^- + H_2\uparrow$ (Song, 2007; Zberg et al., 2009). Consequently, hydrogen evolution, localized basification, and degradation occur on the active interface to complicate the cells/biomaterials interactions. Biodegradable Mg-based devices, which have the advantage of avoiding a second surgical operation to remove the components thus minimizing patient trauma, are not only considered a revolutionary concept in bioengineering but also of scientific interest. Unfortunately, rapid degradation in the physiological environment has impeded their use as metallic implants. On the one hand, improper degradation compromises the mechanical properties before the tissues have a chance to recover or heal completely. On the other hand, the rapid degradation may lead to adverse biological response if Mg and other elements are released into the surroundings too rapidly. Gas bubbles and subcutaneous inflammation are possible causes of excessive Mg^{2+} or H_2 release *in vivo*, further harming tissue growth on the surface and loosening the bonding between the tissues and Mg-based implants (Kirkland, 2012). Owing to the dynamic interface, it is more difficult to develop new-generation biodegradable Mg alloys compared to old metallic biomaterials such as Ti alloys. In particular, the corrosion as well as biological behavior and mechanical performance are critical concerns of Mg-based biomaterials in the physiological environment (Figure 3.9).

3.3.3 Aims of surface design for biodegradable magnesium alloys

Biodegradation of Mg alloys is a dynamic process and often concomitant with continuously changing surface properties and loss of mechanical strength. In addition, the human body is a very complex environment for biodegradation, thus making degradation unpredictable in the healing stage. The aim is thus to ensure safety when using Mg-based implants *in vivo*. A proposed degradation mode is shown in Figure 3.10. It is obviously easier if the Mg-based implants have an approximately intact initial contour in the serving period to retain the designed load-bearing capacity. With enough

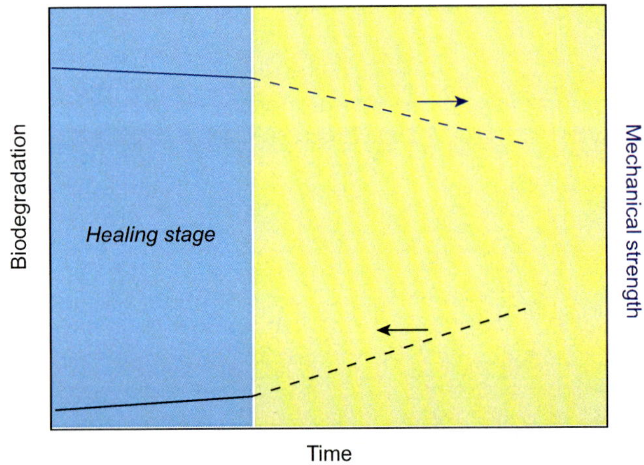

Figure 3.10 Schematic of desirable biodegradation and mechanical strength as a function of time.

corrosion resistance, slow degradation can ensure the designed mechanical strength and structure integrity in the healing stage. The dynamic surface becomes relatively stable to foster cell/tissue growth without producing unacceptable hydrogen evolution and other side effects. After the healing stage, faster degradation is allowable, and the mechanical strength of the implant is permitted to decrease gradually because the surrounding tissues can bear the load gradually. Adverse effects on human physiology are usually not expected, and thus it is still important to be concerned with degradation from the perspective of the actual clinical requirements. This aspect mainly depends on the alloy design and selection.

To achieve this designed degradation process, the surface design must aim to construct a temporary surface on Mg alloys to render the proper mechanical properties, corrosion behavior, and biocompatibility. This temporary surface should effectively impede corrosion in the healing stage, but gradually change its role as a corrosion barrier during healing. This temporary surface must satisfy some special requirements. First, compared to bulk properties, the surface needs better resistance against corrosion during the initial period after implantation and must be biocompatible. Second, the surface is biodegradable and expected to disappear eventually via biological or chemical reactions. Third, the designed surface is further required to allow controlled corrosion after the healing stage and must fully degrade afterward without introducing or leaving harmful materials.

3.3.4 Strategies in surface design

In the surface design of biodegradable Mg alloys, it is important to realize that their primary applications are in bone fixation devices and cardiovascular stents. It is also essential to fathom the bulk properties of Mg alloys, including mechanical strength and degradation behavior, before taking the next step to select the proper materials for the temporary surface as mentioned above. In this process, it is necessary to establish the surface composition, microstructure, and associated preparation methods to achieve the desirable corrosion resistance, surface mechanical performance, and biocompatibility according to clinical requirements.

Biocompatibility, which usually encompasses tissue biocompatibility and hemolytic biocompatibility, is the primary factor in designing new biomaterials, and different applications require different properties. Tissue biocompatibility refers to cytotoxicity, inflammatory reactions, and mutagenic or carcinogenic reactions, whereas hemolytic biocompatibility is associated with hemolytic response and blood coagulation. Generally speaking, materials with good biocompatibility induce a minimal level of cytotoxicity, hemolysis, blood coagulation, and inflammatory reaction while imposing no risk of inducing mutagenic or carcinogenic reactions (Wang, Cao, Qiu, & Bi, 2011). Conventional corrosion-resistant biometals can be more easily rendered biocompatible to meet clinical needs, but the issue is more complicated for biodegradable Mg alloys. Some alloying elements such as zinc (Zn), calcium (Ca), and manganese can cause deleterious effects if the release rate is too high (Kirkland, 2012). Therefore, a temporary surface is required to produce good biocompatibility in the initial stage and control the degradation process to ensure acceptable release from the bulk materials. Calcium-Phosphorus (Ca-P) coatings such as hydroxyapatite (HA) are good candidates to obtain excellent biocompatibility in osteosynthesis.

In surface design, emphasizing only the properties of the surface is often not enough, and the substrate must also be considered with the surface as one entity. This concept is not only confined to biocompatibility and can be generalized to corrosion and wear resistance. Using anticorrosion coatings as an example, pinholes, pores, and cracks are inevitably present in the coatings and the surrounding electrolyte in the bioenvironment can penetrate the coating via these cracks and pores. Since Mg is chemically active and has a smaller standard potential than many permanent metals such as Ti alloys, Al alloys, and stainless steels (Song & Atrens, 1999), the galvanic effect must be considered in materials selection. Figure 3.11 depicts the schematic diagram of the corrosion failure mechanism of the coated Mg-based materials. If the coating is conductive, a galvanic cell is formed between the coating and substrate

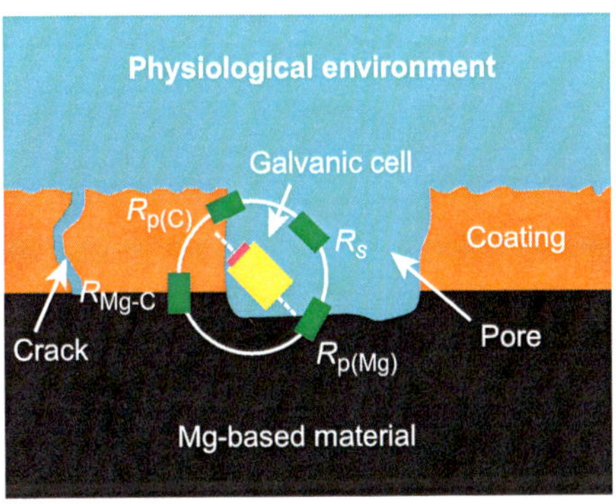

Figure 3.11 Schematic diagram illustrating the corrosion failure mechanism on coated Mg-based materials.
Source: Wu, Jamesh, et al. (2013) with permission from Elsevier.

and the galvanic current is given by the following formula (Song, Johannesson, Hapugoda, & StJohn, 2004; Wu, Jamesh, et al. (2013)):

$$I = \frac{E_C - E_{Mg}}{R_{p(Mg)} + R_{p(C)} + R_s + R_{Mg-C}} \quad (3.1)$$

where E_C and E_{Mg} are the corrosion potentials of the cathode and anode, respectively; $R_{p(Mg)}$ is the polarization resistance of the anode; $R_{p(C)}$ is the polarization resistance of the cathode; R_s is the electrical resistance of the electrolyte; and R_{Mg-C} is the electrical resistance between the anode and cathode. When a Mg-based implant is in a fixed bioenvironment, $R_{p(Mg)}$ and R_s do not change easily, and the current between the anode and cathode depends on $E_C - E_{Mg}$, $R_{p(C)}$, and R_{Mg-C}. Obviously, a small galvanic current is generated by decreasing $E_C - E_{Mg}$ and increasing R_{Mg-C} and $R_{p(C)}$, and it is better to use insulating and chemically inert coatings such as microarc oxidation (MAO) coatings.

For Mg alloys used in artificial joints, friction and wear raise some unwanted issues such as possible inflammation caused by the debris. To improve the wear resistance, a surface with a small friction coefficient and high hardness is called for. Although harder coatings can improve the surface mechanical properties, the hardest materials may not always be the best because Mg alloys are relatively soft. Only when sufficient adhesion is ensured can hard coatings deliver the desirable performance. For example, diamond-like carbon (DLC) with high hardness does not adhere to Mg well unless an interlayer such as chromium is inserted (Wu, Sun, Dai, Song, & Wang, 2010). It should also be mentioned that the conductive interlayer may be dangerous when the coating system is damaged if its electrode potential is higher than that of the Mg alloy. It is necessary to select other appropriate interlayers or alternative methods. Ion implantation or laser treatment is a moderate way to modify the surface hardness. Although its improvement is finite compared to that rendered by hard coatings, it has other advantages. For example, the interfacial bonding strength is not an issue because of the metallurgical bond.

There are many surface modification techniques that can be used when designing Mg biomaterials, for example, microarc oxidation, physical vapor deposition, electrodeposition, ion implantation, etc. By using these techniques, it is possible to construct an appropriate temporary surface on Mg-based biomaterials, and in the following sections, some recent research examples pertaining to surface modification of biomedical Mg alloys will be discussed.

3.4 Research examples

3.4.1 Microarc oxidation coatings

MAO is a simple way to fabricate thick, porous, and hard ceramic coatings on metals. The properties of MAO coatings mainly depend on the processing parameters, substrate, and chemical composition of the electrolyte. In the case of Mg, these properties should be further fine-tuned to adapt to different clinical applications, and some recent studies are presented here.

Gu et al. (2011) used MAO to prepare a coating on the Mg-Ca (1 wt%) alloy. The microarc oxidation process was conducted in a 10 kW microarc oxidation setup comprising a high-power plasma source with a frequency of 700 Hz and a duty cycle of 30% as well as a stirring and cooling system. The aqueous electrolytes were prepared from solutions of 10 g/L sodium silicate with 3.5 g/L sodium hydroxide. MAO was conducted for 10 min at a fixed applied voltage in the range of 300–400 V and the effects of the applied voltages on the surface morphology, phase constituents, hydrogen evolution, pH variation in the solution, and *in vitro* biocompatibility were studied. The thickness and pore size of the MAO coating increased with increasing applied voltages as shown in Figure 3.12. Immersion in Hank's solution for 50 days revealed that the MAO coating produced at 360 V had the best

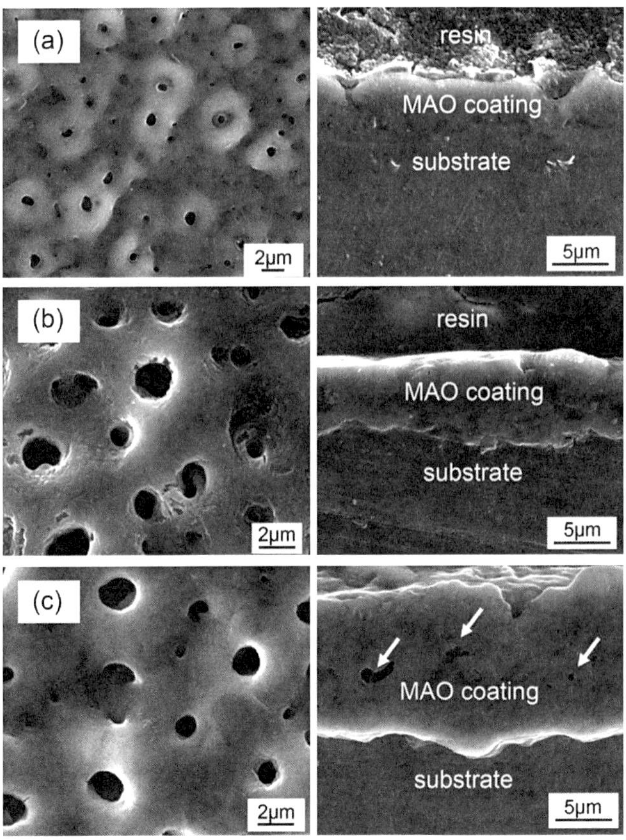

Figure 3.12 Surface and cross-sectional morphologies of MAO-treated Mg-Ca alloys at different voltages: (a) 300 V, (b) 360 V, and (c) 400 V.
Source: Gu et al. (2011) with permission from Elsevier.

long-term corrosion resistance. Adhesion, proliferation, and differentiation of MG63 cells were promoted on the MAO coatings because of reduced Mg ion release and pH value variation. Besides the voltage, the pulsing frequency was crucial to the performance of the MAO coatings. Gu et al. (2012) deposited MAO coatings on AZ31 Mg alloys using 30 g/L Na_3PO_4 aqueous solution and a constant DC voltage of 325 V was applied for 5 min in the coating preparation. Four different pulse frequencies at a constant pulse ratio of 0.3 (300 Hz, 500 Hz, 1000 Hz and 3000 Hz) were investigated, and it was found that the MAO coating produced using 3000 Hz exhibited the best corrosion resistance.

Fischerauer et al. (2013) studied the *in vivo* degradation behavior of MAO-modified ZX50 Mg alloy. They implanted the samples into the femoral legs of 20 male Sprague–Dawley rats and monitored them using microcomputed tomography over an observation period of 24 weeks. The MAO-modified samples showed almost no corrosion in the first week, but after 3 weeks the degradation rate increased and it was even higher than that of the untreated sample. Based on the fact that MAO implants degraded inhomogeneously via localized corrosion attacks, it was considered that this increase was due to an increase in the surface-area-to-volume ratio of the MAO implants. Histological analysis performed after 4, 12, and 24 weeks showed that the initially improved corrosion resistance observed from the MAO implants had a positive effect on bone and tissue response. The reduced hydrogen evolution increased osteoblast apposition from the very beginning, thus generating a stable bone–implant interface. Since MAO can delay initial degradation after implantation, it improves the fracture stabilization, minimizes the burden on the postoperatively irritated surrounding tissues, and generates good bone–implant bonding. Actually, if a good bone–implant interface is formed in the initial stage of bone healing, accelerated degradation in the later stage will be of interest to osteosynthetic applications.

Wang et al. (2011) used MAO coatings to tailor the blood compatibility on Mg alloys. They found that the MAO Mg-1.0 wt% Zn-1.0 wt% Ca alloy exhibited favorable blood compatibility. In their experiments, the MAO Mg alloy showed a decreased hemolytic ratio (2.25%) compared to the untreated one (24.58%). The MAO Mg alloy also showed significantly shorter prothrombin and thrombin time and significantly longer activated partial thromboplastin time than the untreated Mg alloy. Arachidonic acid- and adenosine diphosphate-induced platelet aggregation was significantly reduced by the untreated Mg alloy extract, but it was less affected by the extract of the MAO-treated Mg alloy.

Good antimicrobial properties are also very important to surgical operation due to the risk of bacterial infection (Wu, Liu, et al., 2011), and an antibacterial surface to inhibit bacterial colonization is of practical interest. Silver and silver-bearing compounds are well known to exhibit antimicrobial activity and have been incorporated into the surfaces of a variety of medical devices. Ryu and Hong (2010) developed an approach to fabricate MAO coatings on the AZ31 magnesium alloy using $AgNO_3$-containing electrolytes. The Ag-containing MAO coatings exhibited higher corrosion resistance than the Ag-free MAO coatings and also excellent antibacterial activity of over 99.9% against two strains of bacteria, *Staphylococcus aureus* and *Escherichia coli*.

3.4.2 Ca-P based coatings

Ca-P coatings such HA and tricalcium phosphate (TCP) are often considered as osteoconductive materials and have been widely used to construct new bones and promote osteointegration on biomedical implants because calcium and phosphorus are two major elements in bone tissues. Recently, various techniques such as electrodeposition and hydrothermal methods have been applied to deposit Ca-P coatings on Mg alloys.

Electro deposition is a promising method for biomedical Mg-based implants because it can form uniform coatings on porous substrates or implants with a complex shape at a low deposition temperature. Song, Shan, and Han (2008) used electrodeposition to coat AZ91D Mg alloy with bioactive Ca-P coatings. The electrolyte solution was prepared from 0.1 M $Ca(NO_3)_2$, 0.06 M $NH_4H_2PO_4$, and 10 mL/L H_2O_2 and the pH value was adjusted to 4.3. Electrodepositon was carried out at a stable cathodic potential of 4 V for 2 h at room temperature. The as-deposited coating consisted of dicalcium phosphate dehydrate ($CaHPO_4 \cdot 2H_2O$, DCPD) and β-tricalcium phosphate ($Ca_3(PO_4)_2$, β-TCP) (Figure 3.13(a)).

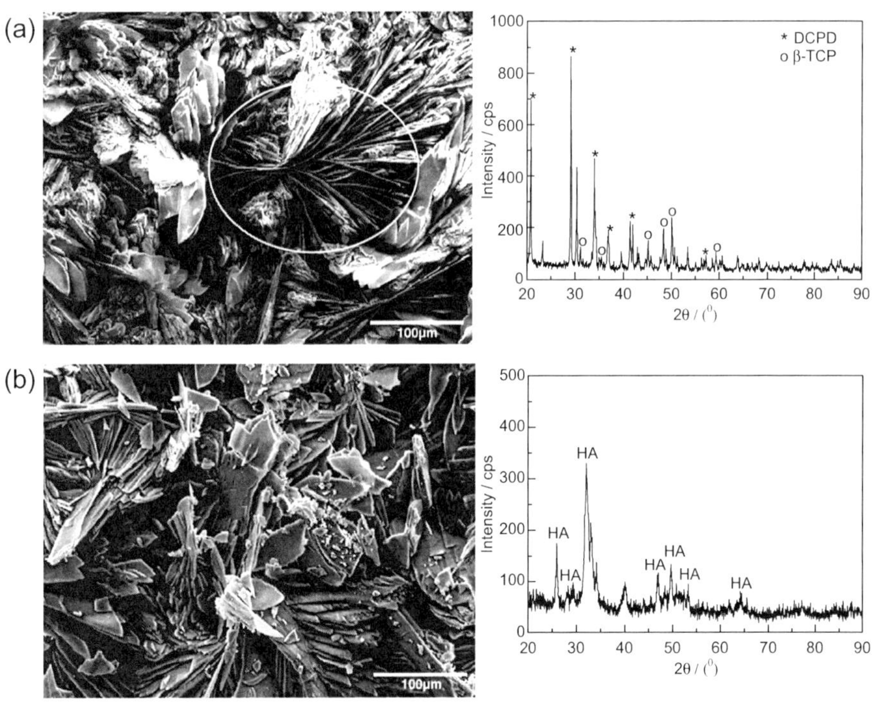

Figure 3.13 Surface morphology and XRD patterns: (a) As-deposited coating and (b) HA coating.
Source: Song et al. (2008) with permission from Elsevier.

Usually, the electrodeposition reactions on the Mg alloy surface are as follows (Kuo & Yen, 2002; Song et al., 2008):

Stage I: Reduction reaction of $H_2PO_4^-$ and HPO_4^{2-}

$$2H_2PO_4^- + 2e^- \rightarrow 2HPO_4^{2-} + H_2\uparrow \quad (3.2)$$

$$2HPO_4^{2-} + 2e^- \rightarrow 2PO_4^{3-} + H_2\uparrow \quad (3.3)$$

Stage II: Ca^{2+} reacting with HPO_4^{2-} and PO_4^{3-} to form $CaHPO_4 \cdot 2H_2O$ (DCPD) and $Ca_3(PO_4)_2$ (β-TCP), respectively.

$$Ca^{2+} + HPO_4^{2-} + 2H_2O \rightarrow CaHPO_4 \cdot 2H_2O \quad (3.4)$$

$$3Ca^{2+} + 2PO_4^{3-} \rightarrow Ca_3(PO_4)_2 \quad (3.5)$$

DCPD and β-TCP are the precursors of HA, which is the stable calcium phosphate form in an alkaline solution. The as-deposited coating was further immersed in 1 M NaOH solution for 2 h at 80 °C to obtain the HA coating. Afterward, the coating was transformed into a uniform HA ($Ca_{10}(PO_4)_6(OH)_2$) one, as shown in Figure 3.13(b), and the corrosion resistance of AZ91D Mg alloy in the simulated body fluid (SBF) was increased.

Coatings fabricated by the traditional cathodic electrodeposition process using a static potential tend to be loose and porous and have low adhesion. One reason is that a concentration polarization is easily formed because ion diffusion from the solution to the substrate surface is too slow. The other is that H_2 is produced on the cathode as a result of reduction of H_2O. To overcome these hurdles, Wang, Guan, Wang, Ren, and Wang (2010) used pulsed power. The electrolyte was prepared from 0.042 mol/L $Ca(NO_3)_2$, 0.025 mol/L $NH_4H_2PO_4$, and 0.1 mol/L $NaNO_3$ and the pH was adjusted to 5.0 by diluted HNO_3 and $(CH_2OH)_3CNH_2$. They obtained a soluble Ca-deficient hydroxyapatite (Ca-def HA) coating on the Mg-Zn-Ca alloy substrate by pulsed eletrodeposition. The Ca/P atomic ratio of the as-deposited coating was about 1.33 (within the range between 1.33 and 1.65). By regulating the pulse amplitude and width, the Ca-def HA coating had better adhesion to the Mg-Zn-Ca alloy and the lap shear strength increased to 41.8 ± 2.7 MPa. The ultimate tensile strength and time of fracture measured from the coated Mg-Zn-Ca alloy were larger than those from the uncoated one, thus offering benefits in supporting fractured bone healing for a longer time. In addition, the Ca-def HA coating also improved the corrosion resistance appreciably in SBF.

Chemical solution deposition was developed by Hiromoto and Tomozawa (2011) to prepare Ca-P coatings on Mg alloys. In their experiment, the solution was prepared with ethylenediaminetetraacetic acid calcium disodium salt hydrate ($C_{10}H_{12}N_2O_8$-Na_2Ca, Ca-EDTA), potassium dihydrogenphosphate (KH_2PO_4), and sodium hydroxide (NaOH). They first mixed 250 mmol/L Ca-EDTA and 250 mmol/L KH_2PO_4 and then used a NaOH solution to adjust the pH to 8.9. HA coatings were obtained on

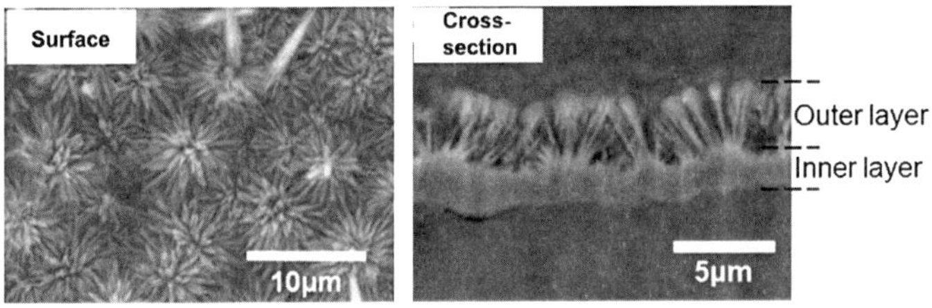

Figure 3.14 Surface and cross-sectional morphology of the hydroxyapatite (HA) coated AZ31 magnesium alloy.
Source: Hiromoto and Tomozawa (2011) with permission from Elsevier.

AZ31 Mg alloys after treatment at 363 K for 6 h. Those HA coatings possessed a novel microstructure consisting of an inner dense layer and outer coarse layer as shown in Figure 3.14. The inner layer on the AZ31 was composed of dome-shaped and densely packed precipitates, while the outer layer comprised rod-like crystals growing from each dome in the radial direction. It was also found from the corrosion tests that the Mg ion release and corrosion current density were remarkably reduced. Hiromoto, Tomozawa, and Maruyama (2013) studied the fatigue properties of HA-coated AZ31 Mg alloys. The HA coating prepared by single-step chemical solution deposition consisted of an outer porous HA layer, an inner continuous HA layer, and a thin intermediate MgO layer. In the tensile test, the HA coating microscopically showed neither cracks nor detachment at 5% static elongation (1.5% residual strain). With further elongation under tensile stress, cracks were formed perpendicularly to the tensile direction and fragments of the coating detached from the fracture inside the inner continuous HA layer. The fatigue strength at 10^7 cycles (fatigue limit) of the HA-coated and mechanically polished AZ31 was about 80 MPa and 90 MPa, respectively. The slight decrease in the fatigue limit observed from the HA coating was attributed to small pits with a depth of about 10 μm. The HA coating remained on the specimen without cracks after 10^7 cycles at the fatigue limit providing about 3% cyclic elongation.

3.4.3 Polymer-based coatings

Biodegradable polymers such as poly L-lactic acid (PLLA), poly ε-caprolactone (PCL), and poly glycolic acid (PGA) have been approved for human clinical applications and become a promising option to improve the initial corrosion resistance and cell compatibility on Mg alloys to meet healing requirements. Dipping, spraying, and spinning are common preparation methods and some examples are described below.

Wong et al. (2010) prepared porous polymeric membranes on AZ91 Mg alloy. They first mixed polycaprolactone (PCL) with the average molecular weight (M_n) of about 80,000 g/mol and dichloromethane (DCM) and then deposited the polymer-based membrane layer by layer on the sample surface using a custom-designed spraying device. In the spraying process, the device was equipped with air-flow and temperature control, thereby standardizing the thickness, homogeneity, and adhesiveness of the polymer-based membrane. The air-flow pressure and spraying temperature were 276 kPa and 37 °C, respectively. The spraying process was conducted at 50% humidity, 22 °C, and atmospheric pressure. The polymeric membranes reduced the degradation rate while preserving the bulk mechanical properties during degradation. The polymer-coated samples showed better cytocompatibility with eGFP and SaOS-2 osteoblasts than the uncoated samples and higher volumes of new bone were observed on the coated samples by microcomputed tomography. Histological analysis indicated no inflammation, necrosis, and hydrogen gas accumulation during degradation.

Xu and Yamamoto (2012) used spin coating to prepare uniform, nonporous, and amorphous poly L-lactic acid (PLLA) and semi-crystalline poly ε-caprolactone (PCL) films on extruded Mg substrates. Spinning is a typical coating process that involves depositing a polymer solution onto a solid substrate as the substrate was rotated at a high speed. When the solvent evaporates and the dissolved polymer covers the substrate, a layer of homogeneous thin polymer film is produced. The experimental details are shown as follows. PLLA with two different molecular weights of 50,000 (low molecular weight, LMW) and 80,000−100,000 (high molecular weight, HMW) and PCL with two different molecular weights of 40,000 (LMW) and 70,000−100,000 (HMW) were used. The polymers were weighed to the desired proportions and dissolved in chloroform to obtain 5% (w/v) solutions. Each polymer solution was dropped onto the Mg sample placed on a spin coater by a micropipette, and the polymer film was prepared by spin-coating. They investigated the effects of the coating thickness, adhesion strength between the coatings and substrates, polymer molecular weight, and different polymers on the corrosion resistance and differentiated cell functions on the coated Mg-based materials. The PLLA film showed better adhesion strength to the Mg substrate than the PCL one. For both PLLA and PCL, the LMW films were thinner and exhibited better adhesion strength than the HMW ones. According to the pH measurements of the cell culture medium and quantification of released Mg^{2+} during the cell culture, the corrosion resistance of the Mg substrate was improved by the polymer films to a different degree. In addition, all the polymeric films enhanced the cytocompatibility during incubation for 7 days.

3.4.4 Physical vapor deposition coatings

PVD is a modern coating technique commonly used by the industry. Compared to other methods such as CVD and thermal spraying, the required deposition temperature in PVD is often lower and so it is very suitable for Mg alloys. Moreover, PVD is friendly to the environment and is regarded as a green technique that can substitute for industrial electrochemical plating and anodic oxidation. Last but not least, many coating species from metals to ceramics can be used. Researchers have used it to improve the corrosion resistance of Mg and Mg alloys in saline solutions. Hoche et al. (2005) used sputtering to prepare CrN and Al_2O_3 coatings on AZ91D Mg die cast alloy, and Altun and Sen (2005) conducted DC sputtering to deposit AlN coatings on AZ31, AZ61, AZ63 and AZ91 Mg alloys. Wu (2007), Wu, Zeng, and Yuan (2008) and Wu, Ding, Zeng, Wang, and Yao (2009) also used sputtering to coat AZ31 Mg alloys with Al and Ti metallic coatings. Most of the samples showed improved corrosion resistance in saline solutions.

Xin, Liu, Zhang, Huo, et al. (2008) used cathodic arc-deposition to deposit ZrO_2 coatings on Mg alloys for biomedical applications because zirconia has good chemical stability and favorable biocompatibility. Commercial extruded AZ91 Mg alloys were used in their experiment. Argon sputter cleaning was conducted at a bias of 1000 V for about 30 min before deposition. A zirconium transition layer was first deposited for about half an hour, followed by deposition of the zirconia coating for about 3 h at a bias of 100 V. Finally, a 1.5 μm-thick ZrO_2/Zr bilayered structure with good adhesion was obtained. EIS measurements disclosed that the corrosion resistance of the coated alloy was significantly improved. Electrolyte penetration eventually deteriorated the protection of the coating after long exposure in the SBF. In addition, they also deposited Al_2O_3 (Xin, Liu, Zhang, Jiang, et al., 2008) and ZrN (Xin et al., 2009) on Mg alloys. Most of these PVD coatings are nonbiodegradable, and so inflammation may occur if they are broken and remain in the human body for a prolonged period. Synthetic apatites exhibit excellent biological properties, such as biocompatibility, bioactivity, lack of toxicity, or inflammatory and immunitary responses, and also have relatively high bioresorbability (Jaime, Michele, José, Stéphanie, & Christophe, 2013). They can also be prepared by physical vapor deposition such as sputtering (Boyd, Duffy, McCann, & Meenan, 2008; Yamashita, Matsuda, Arashi, & Umegaki, 1998). Thus, it is a good candidate using PVD on Mg alloy in the future.

3.4.5 Ion implantation

Ion implantation is different from the aforementioned coating techniques. It can provide the possibility of introducing different species into a substrate independent of thermodynamic limitations such as solubility. Besides, an ion implanted layer does not have an abrupt interface, thereby avoiding layer delamination that plagues coatings. Conventional line-beam metal ion implantation has been attempted to modify the properties of magnesium substrates. For example, cerium (Ce) ion implantation improves the corrosion resistance of AZ31 Mg alloy (Wang, Zeng, Yao, Wu, & Lai, 2008) and yttrium (Y) ion implantation enhances the oxidation resistance of AZ31 magnesium alloy (Wang, Zeng, Wu, Yao, & Lai, 2007).

Zn as one of the vital elements in the human body has been considered and implanted into pure Mg substrate by conventional beam-line metal ion implantation. However, after implanting 2.5×10^{17} ions·cm^{-2} of Zn using a cathodic arc source into pure Mg at 35 kV, the degradation rate increased significantly in SBF. It was believed to be due to galvanic effects between the metallic Zn-rich surface and Mg matrix beneath (Wu, Gong, et al., 2011). Al has an electrode potential close to that of Mg in aqueous solutions, and Al ion implantation can tailor the surface corrosion resistance of pure Mg. The corrosion resistance in SBF improves significantly and this enhancement is attributed to the formation of a gradient surface structure involving a gradual transition from an Al-rich oxide layer to Al-rich metal layer. However, when the biological properties and toxicity of the alloying elements are considered, the use of Al is suspected because Al is suspected to be involved with Alzheimer's disease and may also cause muscle fiber damage (Wu, Xu, et al., 2012).

Mg corrosion is affected by the thin surface oxide film and dissolution typically occurs in the oxide-free areas (Galicia, Pébère, Tribollet, & Vivier, 2009). The native surface oxide film formed on exposure to air consists of mainly MgO, but MgO is not stable in an aqueous solution according to thermodynamics and converted to magnesium hydroxide. Cl^- in the aqueous solutions can substitute OH^- forming chloride, which expedites dissolution of the surface structure (Wu, Feng, et al., 2012). To improve the chemical stability of the temporary surface on the biodegradable Mg alloy, some chemically stable phases such as Cr_2O_3 are needed. Xu et al. (2011) performed Cr ion implantation into pure magnesium, but it induced rapid degradation in SBF similar to Zn ion implantation due to galvanic corrosion. Ensuing oxygen ion implantation produced a thicker oxidized layer composed of chromium oxide that successfully retarded surface degradation. Wu, Feng, et al. (2012) applied oxygen ion implantation to modify the Mg-Nd-Zn-Zr alloy. But unfortunately, no significant improvement was observed. They performed Cr ion implantation prior to oxygen ion implantation and attained improved corrosion resistance in SBF due to the formation of Cr-rich oxide in the surface layer.

Ti and Zr are biologically friendly to the human body and have also been implanted. Zhao et al. (2013) showed that the surface corrosion resistance on WE43 alloy in SBF was significantly improved after Ti ion implantation in conjunction with oxygen PIII. They also conducted Zr and O dual implantation to modify Mg-Ca and Mg-Sr alloys (Zhao et al., 2014). Besides the improved corrosion resistance in simulated physiological environments, the amounts of adherent bacteria on the Zr-O-implanted and Zr-implanted samples diminished remarkably compared to the unimplanted alloy (Figure 3.15) and significantly enhanced cell adhesion and proliferation were observed from the Zr-O-implanted sample. The results suggest that dual zirconium and oxygen ion implantation is a possible means to avoid inflammation in clinical applications of biodegradable magnesium alloys.

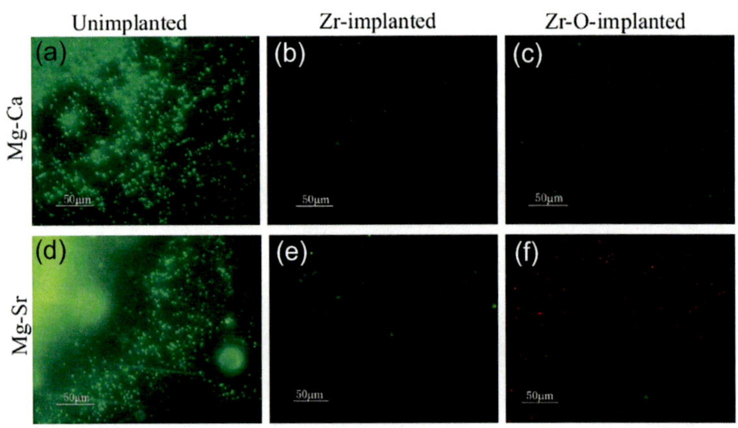

Figure 3.15 Fluorescent microscopic views of magnesium samples after bacteria culturing for 30 min: (a) Unimplanted Mg-Ca, (b) Zr-implanted Mg-Ca, (c) Zr-O-implanted Mg-Ca, (d) Unimplanted Mg-Sr, (e) Zr-implanted Mg-Sr, and (f) Zr-O-implanted Mg-Sr.
Source: Zhao et al. (2014) with permission from Elsevier.

Plasma immersion ion implantation (PIII) can process samples with a complex shape and is a viable technique to process biomedical artificial joints. Wu et al. (2014) introduced C_2H_2 gas into the PIII process and conducted plasma immersion ion implantation and deposition (PIII&D) on Mg-Nd-Zn-Zr alloys. PIII&D combines energetic ion implantation and low-energy plasma deposition, resulting in a thin diamond-like carbon film formation on the Mg-Nd-Zn-Zr alloy (Figure 3.16). Both electrochemical and immersion tests reveal enhanced corrosion resistance in the 0.9 wt% NaCl solution. Although the diamond-like carbon film has an excellent barrier effect against corrosion, defects generated in the deposition process induce corrosion failure of the plasma-modified Mg-Nd-Zn-Zr alloy in aqueous solutions eventually.

3.4.6 Composite coatings

Composite coatings are attractive as advanced coatings. When two or more constituents are combined to form a layered or mixed structure, the properties of a traditional coating can be refined to address specific requirements. Various combinations such as ceramics/ceramics and ceramics/polymers have been proposed.

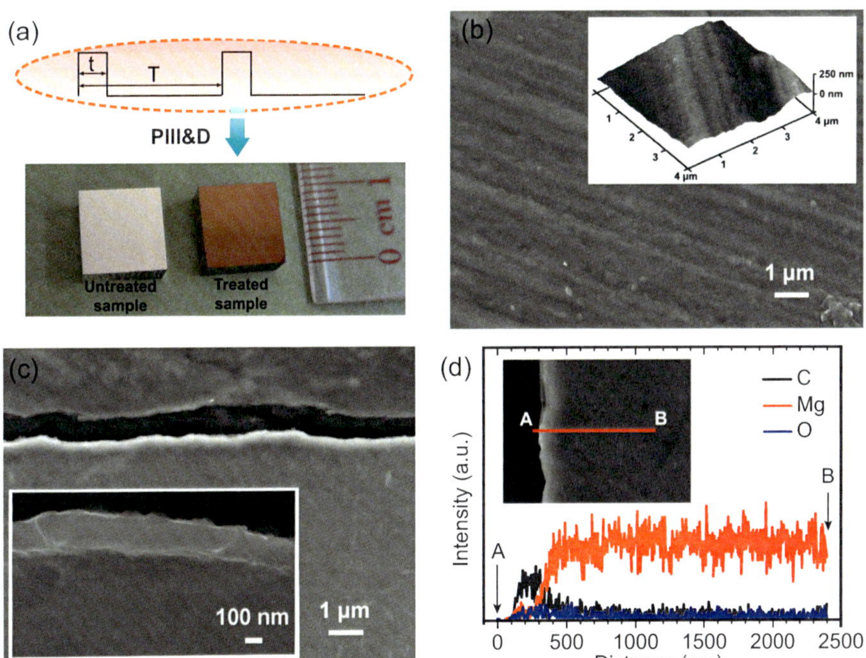

Figure 3.16 (a) Surface appearance of the untreated and treated samples. (b) SEM picture of surface morphology of the film with the inset showing the magnified surface obtained by AFM. (c) SEM view of the cross-section of the plasma-modified sample, with the inset showing a magnified picture of the film. (d) EDS line scan of the film.
Source: Wu et al. (2014) with permission from Elsevier.

Ca-P coatings not only retard degradation of Mg alloys under physiological conditions, but also have good biocompatibility. However, their fragile nature and structural heterogeneity induce the loss of integrity possibly reducing the corrosion resistance of the coated Mg alloys. Wang, Zhao, Chen, Li, and Zhang (2012) used polycaprolactone (PCL) to preserve the integrity of dicalcium phosphate dihydrate ($CaHPO_4 \cdot 2H_2O$, DCPD) coatings for a longer time because it has good plasticity and uniform structure. Ca-P coating can avoid direct contact between the PCL and Mg matrix because there is a negative effect in the interaction in the later degradation stage. Therefore, they fabricated a layered composite coating composed of dicalcium phosphate dihydrate (DCPD) and polycaprolactone (PCL) on the Mg-Zn alloy as shown in Figure 3.17. The DCPD coating was synthesized in a 0.042 mol/L $Ca(NO_3)_2 \cdot 4H_2O$ and 0.025 mol/L $NH_4H_2PO_4$ solution by electrodeposition and the DCPD-coated sample was immersed in a 2 wt% PCL chloroform solution before drying in air. Compared to the DCPD-coated alloy, the DCPD/PCL-coated alloy had higher corrosion resistance as manifested by the elevated corrosion potential, reduced corrosion current density, and smaller amount of released hydrogen.

Chitosan is a natural biopolymer that exhibits various biological activity including excellent biocompatibility, biodegradability, osteoconductivity, and antimicrobial properties. Hahn et al. (2011) prepared a dense and well-adherent HA-chitosan composite coating on AZ31 Mg alloy. They prepared HA-chitosan powder mixtures containing up to 20 wt% of chitosan and commercial HA nanocrystalline powders having a volumetric mean diameter of 15 nm and chitosan powders with a degree of deacetylation of about 85% were used as the starting materials. To obtain powders with an appropriate particle size, the as-received HA powders were heated at 1050 °C for 2 h, and the chitosan powders were dry ball-milled for 12 h in a planetary mill using ZrO_2 balls in a jar. The heat-treated HA powders were mechanically mixed with the ball-milled chitosan powders by dry ball milling and the HA-chitosan composite coating was deposited on AZ31 Mg alloy by aerosol deposition (AD), because AD offered the advantage of room temperature deposition. Finally, various 5 μm-thick

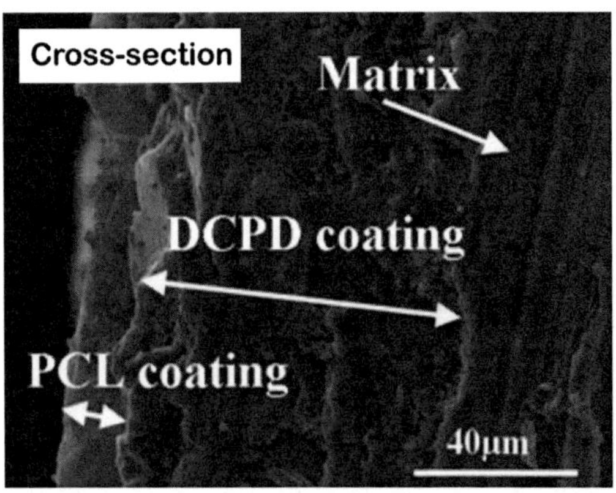

Figure 3.17 Cross-section of the DCPD-PCL coated Mg-Zn alloy. *Source:* Wang et al. (2012) with permission from Elsevier.

HA-chitosan composite coatings were deposited on the Mg alloys. The composition of the coatings was tailored by adjusting the HA and chitosan concentrations in the powder mixtures. All the coatings exhibited high adhesion strength ranging from 24.6 to 27.7 MPa and better corrosion resistance than the bare Mg alloy. Moreover, the biocompatibility of the coated alloy such as cell adhesion was improved appreciably as demonstrated by Figure 3.18.

Although Ca-P ceramics have favorable biocompatibility and osteoconductive properties, they usually induce slow bone formation *in vivo* (Arinzeh, Tran, Mcalary, & Daculsi, 2005). Compared to Ca-P ceramics, $CaSiO_3$ ceramics can promote proliferation and differentiation of osteoblast-like cells and accelerate the formation of HA in SBF, but, unfortunately, $CaSiO_3$ degrades rapidly in the physiological environment (Siriphannon, Kameshima, Yasumori, Okada, & Hayashi, 2000; Ni, Chang, Chou, & Zhai, 2007; Ni, Lin, Chang, & Chou, 2008). To improve the corrosion properties and cell compatibility, Du et al. (2011) produced a microporous calcium silicate and calcium phosphate ($CaSiO_3$-$CaHPO_4 \cdot 2H_2O$) composite coating on Mg-Zn-Mn-Ca alloy by a chemical reaction. The layer was mainly composed of $CaHPO_4 \cdot 2H_2O$ with a small amount of $CaSiO_3$. In vitro cell experiments indicated that the surface cytocompatibility of the coated Mg alloy was significantly improved as manifested by more cell adhesion, growth, and proliferation.

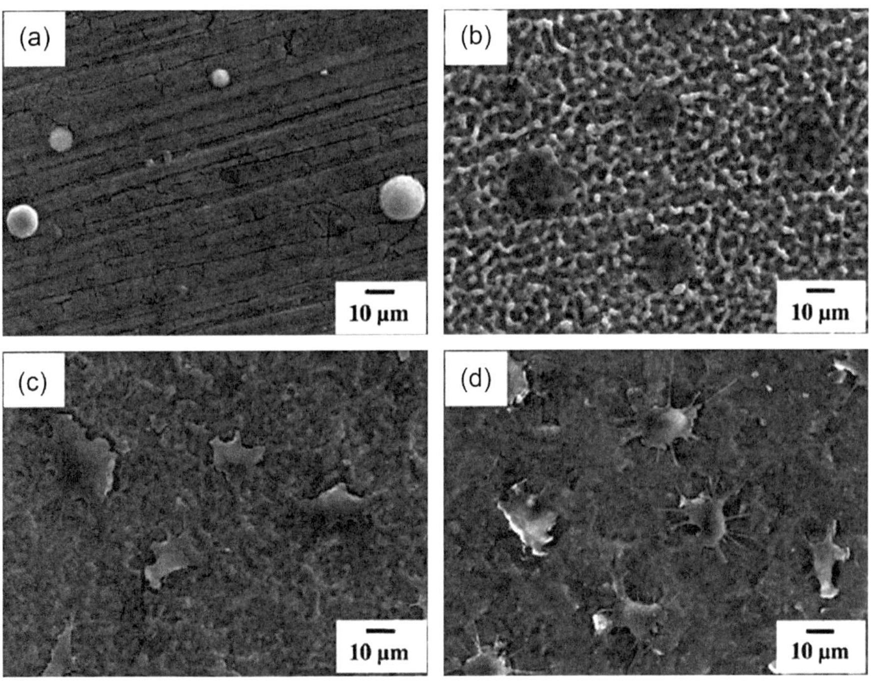

Figure 3.18 SEM micrographs of the MC3T3-E1 cells attached to the samples: (a) Uncoated AZ31 substrate, (b) HA coating, and HA-chitosan composite coatings with (c) 5 wt% and (d) 20 wt% chitosan.
Source: Hahn et al. (2011) with permission from Elsevier.

3.5 Summary and future trends

Biodegradability is a prominent advantage of Mg alloys in biomedical components such as cardiovascular stents and bone fixation. Although this is a revolutionary concept in biomaterials science contrary to traditional corrosion-resistant permanent biometals, there are several practical difficulties. The dynamic interface between the Mg alloys and biological environment is quite complicated and requires serious consideration. Rapid corrosion is a big issue especially in the initial healing stage, and proper control is crucial. Therefore, it is imperative to construct a temporary surface on Mg alloys to control the corrosion, improve the biocompatibility, and preserve the mechanical performance during the healing stage. Surface modification techniques including coating and ion implantation can be conveniently used to alter selected features to address different clinical requirements. This chapter discusses some recent research activities on Ca-P based coatings, polymer-based coatings, MAO coatings, PVD coatings, and ion implantation performed on Mg alloys. Most research activities have hitherto focused on *in vitro* investigations, and many aspects of *in vivo* degradation have not been completely understood.

Developing new biomedical Mg alloys is a main trend to substitute for traditional alloys in some applications, and there are inevitably questions and problems. New types of coatings or modified surface layers must be designed to meet practical requirements. In addition, hybrid surface treatment techniques will also be emphasized because of the flexibility. With the aid of hybrid techniques, the temporary surface can be endowed with more functions including the desirable drug controlled delivery capability. More research is needed to better comprehend biodegradation *in vitro* and *in vivo* in order to expedite clinical acceptance of the materials.

Acknowledgements

The work was supported by Hong Kong Research Grants Council (RGC) General Research Funds (GRF) Nos. 112510 and 112210 and City University of Hong Kong Applied Research Grants (ARG) Nos. 9667066 and 9667069.

References

Altun, H., & Sen, S. (2005). The effect of DC magnetron sputtering AlN coatings on the corrosion behaviour of magnesium alloys. *Surface and Coatings Technology, 197*, 193–200.

Anders, A. (1997). Metal plasma immersion ion implantation and deposition: a review. *Surface and Coatings Technology, 93*, 158–167.

Arinzeh, T. L., Tran, T., Mcalary, J., & Daculsi, G. (2005). A comparative study of biphasic calcium phosphate ceramics for human mesenchymal stem-cell-induced bone formation. *Biomaterials, 26*, 3631–3638.

Boyd, A. R., Duffy, H., McCann, R., & Meenan, B. J. (2008). Sputter deposition of calcium phosphate/titanium dioxide hybrid thin films. *Materials Science and Engineering: C, 28*, 228–236.

Cao, H., Liu, X., Meng, F., & Chu, P. K. (2011). Biological actions of silver nanoparticles embedded in titanium controlled by micro-galvanic effects. *Biomaterials, 32*, 693–705.

Carraro, C., Maboudian, R., & Magagnin, L. (2007). Metallization and nanostructuring of semiconductor surfaces by galvanic displacement processes. *Surface Science Reports, 62*, 499–525.

Chu, P. K., Qin, S., Chan, C., Cheung, N. W., & Larson, L. A. (1996). Plasma immersion ion implantation—a fledgling technique for semiconductor processing. *Mater Sci Eng R, 17*, 207–280.

Chu, P. K., Chen, J. Y., Wang, L. P., & Huang, N. (2002). Plasma-surface modification of biomaterials. *Materials Science and Engineering: R: Report, 36*, 143–206.

Curran, J. A., & Clyne, T. W. (2005a). The thermal conductivity of plasma electrolytic oxide coatings on aluminium and magnesium. *Surface and Coatings Technology, 199*, 177–183.

Curran, J. A., & Clyne, T. W. (2005b). Thermo-physical properties of plasma electrolytic oxide coatings on aluminium. *Surface and Coatings Technology, 199*, 168–176.

Curran, J. A., & Clyne, T. W. (2006). Porosity in plasma electrolytic oxide coatings. *Acta Materialia, 54*, 1985–1993.

Dunleavy, C. S., Golosnoy, I. O., Curran, J. A., & Clyne, T. W. (2009). Characterisation of discharge events during plasma electrolytic oxidation. *Surface and Coatings Technology, 203*, 3410–3419.

Du, H., Wei, Z., Wang, H., Zhang, E., Zuo, L., & Du, L. (2011). Surface microstructure and cell compatibility of calcium silicate and calcium phosphate composite coatings on Mg-Zn-Mn-Ca alloys for biomedical application. *Colloids and Surfaces B: Biointerfaces, 83*, 96–102.

Feng, K., Wu, G., Hu, T., Li, Z., Cai, X., & Chu, P. K. (2012). Dual Ti and C ion-implanted stainless steel bipolar plates in polymer electrolyte membrane fuel cells. *Surface and Coatings Technology, 206*, 2914–2921.

Fischerauer, S. F., Kraus, T., Wu, X., Tangl, S., Sorantin, E., Hanzi, A. C., et al. (2013). In vivo degradation performance of micro-arc-oxidized magnesium implants: a micro-CT study in rats. *Acta Biomaterialia, 9*, 5411–5420.

Galicia, G., Pébère, N., Tribollet, B., & Vivier, V. (2009). Local and global electrochemical impedances applied to the corrosion behaviour of an AZ91 magnesium alloy. *Corrosion Science, 51*, 1789–1794.

Gray, J. E., & Luan, B. (2002). Protective coatings on magnesium and its alloys - a critical review. *Journal of Alloys and Compounds, 336*, 88–113.

Gu, Y., Chen, C. F., Bandopadhyay, S., Ning, C., Zhang, Y., & Guo, Y. (2012). Corrosion mechanism and model of pulsed DC microarc oxidation treated AZ31 alloy in simulated body fluid. *Applied Surface Science, 258*, 6116–6126.

Gu, X. N., Li, N., Zhou, W. R., Zheng, Y. F., Zhao, X., Cai, Q. Z., et al. (2011). Corrosion resistance and surface biocompatibility of a microarc oxidation coating on a Mg-Ca alloy. *Acta Biomaterialia, 7*, 1880–1889.

Hahn, B. D., Park, D. S., Choi, J. J., Ryu, J., Yoon, W. H., Choi, J. H., et al. (2011). Aerosol deposition of hydroxyapatite-chitosan composite coatings on biodegradable magnesium alloy. *Surface and Coatings Technology, 205*, 3112–3118.

Harsha, K. S. S. (2006). *Principles of physical vapor deposition of thin films*. Amsterdam: Elsevier.

Hiromoto, S., & Tomozawa, M. (2011). Hydroxyapatite coating of AZ31 magnesium alloy by a solution treatment and its corrosion behavior in NaCl solution. *Surface and Coatings Technology, 205*, 4711–4719.

Hiromoto, S., Tomozawa, M., & Maruyama, N. (2013). Fatigue property of a bioabsorbable magnesium alloy with a hydroxyapatite coating formed by a chemical solution deposition. *Journal of Mechanical Behavior of Biomedical Materials, 25*, 1–10.

Hoche, H., Rosenkranz, C., Delp, A., Lohrengel, M. M., Broszeit, E., & Berger, C. (2005). Investigation of the macroscopic and microscopic electrochemical corrosion behaviour of PVD-coated magnesium die cast alloy AZ91. *Surface and Coatings Technology, 193*, 178–184.

Hornberger, H., Virtanen, S., & Boccaccini, A. R. (2012). Biomedical coatings on magnesium alloys—a review. *Acta Biomaterialia, 8*, 2442–2455.

Jaime, G. M., Michele, I., José, M. D. L., Stéphanie, S., & Christophe, D. (2013). Progress on the preparation of nanocrystalline apatites and surface characterization: overview of fundamental and applied aspects. *Progress in Crystal Growth and Characterization of Materials, 59*, 1–46.

Jamesh, M., Kumar, S., & Narayanan, T. S. N. S. (2011). Corrosion behavior of commercially pure Mg and ZM21 Mg alloy in Ringer's solution – long term evaluation by EIS. *Corrosion Science, 53*, 645–654.

Kanani, N. (2005). *Electroplating-basic principles, processes and practice*. Amsterdam: Elsevier.

Kirkland, N. T. (2012). Magnesium biomaterials: past, present and future. *Corrosion Engineering Science and Technology, 47*, 322–328.

Kuo, M. C., & Yen, S. K. (2002). The process of electrochemical deposited hydroxyapatite coatings on biomedical titanium at room temperature. *Materials Science and Engineering: C, 20*, 153–160.

Li, L., An, M., & Wu, G. (2006). A new electroless nickel deposition technique to metallise SiCp/Al composites. *Surface and Coatings Technology, 200*, 5102–5112.

Liu, X., Chu, P. K., & Ding, C. (2004). Surface modification of titanium, titanium alloys, and related materials for biomedical applications. *Materials Science and Engineering: R: Reports, 47*, 49–121.

Liu, X., Chu, P. K., & Ding, C. (2010). Surface nano-functionalization of biomaterials. *Materials Science and Engineering: R: Report, 70*, 275–302.

Mattox, D. M. (2000). Ion plating – past, present and future. *Surface and Coatings Technology, 133–134*, 517–521.

Mattox, D. M. (2010a). Physical sputtering and sputter deposition (sputtering). In D. M. Mattox (Ed.), *Handbook of physical vapor deposition (PVD) processing* (pp. 237–286). Elsevier Inc.

Mattox, D. M. (2010b). Vacuum evaporation and vacuum deposition. In D. M. Mattox (Ed.), *Handbook of physical vapor deposition (PVD) processing* (pp. 195–235). Elsevier Inc.

Narayanan, T. S. N. S., Park, I. S., & Lee, M. H. (2014). Strategies to improve the corrosion resistance of microarc oxidation (MAO) coated magnesium alloys for degradable implants: prospects and challenges. *Progress in Materials Science, 60*, 1–71.

Ni, S., Chang, J., Chou, L., & Zhai, W. (2007). Comparison of osteoblast-like cell responses to calcium silicate and tricalcium phosphate ceramics in vitro. *Journal of Biomedical Materials Research Part B: Applied Biomaterials, 80*, 174–183.

Ni, S., Lin, K., Chang, J., & Chou, L. (2008). β-$CaSiO_3$/β-$Ca_3(PO_4)_2$ composite materials for hard tissue repair: In vitro studies. *Journal of Biomedical Materials Research Part A, 85*, 72–82.

Ryu, H. S., & Hong, S. H. (2010). Corrosion resistance and antibacterial properties of Ag-containing MAO coatings on AZ31 magnesium alloy formed by microarc oxidation. *Journal of the Electrochemical Society, 157*, C131–C136.

Schlesinger, M., & Paunovic, M. (2010). *Modern electroplating*. Hoboken: John Wiley & Sons.

Siriphannon, P., Kameshima, Y., Yasumori, A., Okada, K., & Hayashi, S. (2000). Influence of preparation conditions on the microstructure and bioactivity of α-$CaSiO_3$ ceramics: formation of hydroxyapatite in simulated body fluid. *Journal of Biomedical Materials Research, 52*, 30–39.

Song, G. (2007). Control of biodegradation of biocompatable magnesium alloys. *Corrosion Science, 49*, 1696–1701.

Song, G. L., & Atrens, A. (1999). Corrosion mechanisms of magnesium alloys. *Advanced Engineering Materials, 1*, 11–33.

Song, G., Johannesson, B., Hapugoda, S., & StJohn, D. (2004). Galvanic corrosion of magnesium alloy AZ91D in contact with an aluminium alloy, steel and zinc. *Corrosion Science, 46*, 955–977.

Song, Y. W., Shan, D. Y., & Han, E. H. (2008). Electrodeposition of hydroxyapatite coating on AZ91D magnesium alloy for biomaterial application. *Materials Letters, 62*, 3276–3279.

Staiger, M. P., Pietak, A. M., Huadmai, J., & Dias, G. (2006). Magnesium and its alloys as orthopedic biomaterials: a review. *Biomaterials, 27*, 1728–1734.

Sudagar, J., Lian, J., & Sha, W. (2013). Electroless nickel, alloy, composite and nano coatings—a critical review. *Journal of Alloys and Compounds, 571*, 183–204.

Wang, D. W., Cao, Y., Qiu, H., & Bi, Z. G. (2011). Improved blood compatibility of Mg-1.0Zn-1.0Ca alloy by micro-arc oxidation. *Journal of Biomedical Materials Research Part A, 99*, 166–172.

Wang, H. X., Guan, S. K., Wang, X., Ren, C. X., & Wang, L. G. (2010). In vitro degradation and mechanical integrity of Mg-Zn-Ca alloy coated with Ca-deficient hydroxyapatite by the pulse electrodeposition process. *Acta Biomaterialia, 6*, 1743–1748.

Wang, X., Zeng, X., Wu, G., Yao, S., & Lai, Y. (2007). Surface analysis and oxidation behavior of Y-ion implanted AZ31 magnesium alloys. *Applied Surface Science, 253*, 3574–3580.

Wang, X., Zeng, X., Yao, S., Wu, G., & Lai, Y. (2008). The corrosion behavior of Ce-implanted magnesium alloys. *Material Characterization, 59*, 618–623.

Wang, H., Zhao, C., Chen, Y., Li, J., & Zhang, X. (2012). Electrochemical property and in vitro degradation of DCPD-PCL composite coating on the biodegradable Mg-Zn alloy. *Materials Letters, 68*, 435–438.

Wasa, K., Kanno, I., & Kotera, H. (2012). *Handbook of sputter deposition technology: Fundamentals and applications for functional thin films, nanomaterials, and MEMS*. Amsterdam: Elsevier.

Witte, F., Fischer, J., Nellesen, J., Crostack, H. A., Kaese, V., Pisch, A., et al. (2006). In vitro and in vivo corrosion measurements of magnesium alloys. *Biomaterials, 27*, 1013–1018.

Wong, H. M., Yeung, K. W. K., Lam, K. O., Tam, V., Chu, P. K., Luk, K. D. K., et al. (2010). A biodegradable polymer-based coating to control the performance of magnesium alloy orthopaedic implants. *Biomaterials, 31*, 2084–2096.

Wu, G. (2007). Fabrication of Al and Al/Ti coatings on magnesium alloy by sputtering. *Materials Letters, 61*, 3815–3817.

Wu, G., Ding, K., Zeng, X., Wang, X., & Yao, S. (2009). Improving corrosion resistance of titanium-coated magnesium alloy by modifying surface characteristics of magnesium alloy prior to titanium coating deposition. *Scripta Materialia, 61*, 269–272.

Wu, G., Feng, K., Shanaghi, A., Zhao, Y., Xu, R., Yuan, G., et al. (2012). Effects of surface alloying on electrochemical behavior of oxygen-plasma-modified biomedical magnesium alloy. *Surface and Coatings Technology, 206*, 3186–3195.

Wu, G., Gong, L., Feng, K., Wu, S., Zhao, Y., & Chu, P. K. (2011). Rapid degradation of biomedical magnesium induced by zinc ion implantation. *Materials Letters, 65*, 661–663.

Wu, G., Jamesh, M. I., & Chu, P. K. (2013). Surface design of biodegradable magnesium alloys—a review. *Surface and Coatings Technology, 233*, 2–12.

Wu, S., Liu, X., Hu, T., Chu, P. K., Ho, J. P. Y., Chan, Y. L., et al. (2008). A biomimetic hierarchical scaffold: natural growth of nanotitanates on three-dimensional microporous Ti-based metals. *Nano Letters, 8*, 3803–3808.

Wu, S., Liu, X., Yeung, A., Yeung, K. W. K., Kao, R. Y. T., Wu, G., et al. (2011). Plasma-modified biomaterials for self-antimicrobial applications. *ACS Applied Materials and Interfaces, 3*, 2851–2860.

Wu, G., Sun, L., Dai, W., Song, L., & Wang, A. (2010). Influence of interlayers on corrosion resistance of diamond-like carbon coating on magnesium alloy. *Surface and Coatings Technology, 204*, 2193–2196.

Wu, G., Xu, R., Feng, K., Wu, S., Wu, Z., Sun, G., et al. (2012). Retardation of surface corrosion of biodegradable magnesium-based materials by aluminum ion implantation. *Applied Surface Science, 258*, 7651–7657.

Wu, G., Zeng, X., & Yuan, G. (2008). Growth and corrosion of aluminum PVD-coating on AZ31 magnesium alloy. *Materials Letters, 62*, 4325–4327.

Wu, G., Zhang, X., Zhao, Y., Jamesh, M. I., Yuan, G., & Chu, P. K. (2014). Plasma modified Mg-Nd-Zn-Zr alloy with enhanced surface corrosion resistance. *Corrosion Science, 78*, 121–129.

Wu, G., Zhao, Y., Zhang, X., Jamesh, M. I., & Chu, P. K. (2013). Self-protection against corrosion of aged magnesium alloy in simulated physiological environment. *Corrosion Science, 68*, 279–285.

Xin, Y., Liu, C., Huo, K., Tang, G., Tian, X., & Chu, P. K. (2009). Corrosion behavior of ZrN/Zr coated biomedical AZ91 magnesium alloy. *Surface and Coatings Technology, 203*, 2554–2557.

Xin, Y., Liu, C., Zhang, W., Huo, K., Tang, G., Tian, X., et al. (2008). Corrosion resistance of ZrO_2-Zr-coated biodegradable surgical magnesium alloy. *Journal of Materials Research, 23*, 312–319.

Xin, Y., Liu, C., Zhang, W., Jiang, J., Tang, G., Tian, X., et al. (2008). Electrochemical behavior Al_2O_3/Al coated surgical AZ91 magnesium alloy in simulated body fluids. *Journal of the Electrochemical Society, 155*, C178–C182.

Xu, L., & Yamamoto, A. (2012). Characteristics and cytocompatibility of biodegradable polymer film on magnesium by spin coating. *Colloids and Surfaces B: Biointerfaces, 93*, 67–74.

Xu, R., Wu, G., Yang, X., Hu, T., Lu, Q., & Chu, P. K. (2011). Controllable degradation of biomedical magnesium by chromium and oxygen dual ion implantation. *Materials Letters, 65*, 2171–2173.

Yamashita, K., Matsuda, M., Arashi, T., & Umegaki, T. (1998). Crystallization, fluoridation and some properties of apatite thin films prepared through rf-sputtering from CaO-P_2O_5 glasses. *Biomaterials, 19*, 1239–1244.

Zberg, B., Uggowitzer, P. J., & Löffler, J. F. (2009). MgZnCa glasses without clinically observable hydrogen evolution for biodegradable implants. *Nature Materials, 8*, 887–891.

Zhao, Y., Jamesh, M. I., Li, W. K., Wu, G., Wang, C., Zheng, Y., et al. (2014). Enhanced antimicrobial properties, cytocompatibility, and corrosion resistance of plasma-modified biodegradable magnesium alloys. *Acta Biomaterialia, 10*, 544–556.

Zhao, Y., Wu, G., Lu, Q., Wu, J., Xu, R., Yeung, K. W. K., et al. (2013). Improved surface corrosion resistance of WE43 magnesium alloy by dual titanium and oxygen ion implantation. *Thin Solid Films, 529*, 407–411.

Part Two

Biological and mechanical properties of surface modified magnesium and magnesium alloys

Bioabsorbable behaviour of magnesium alloys – an *in vivo* approach

4

Martin Durisin
Medical University of Hannover, Hannover, Germany

4.1 Introduction

This chapter describes the important aspects of the *in vivo* biodegradation and biocompatibility of magnesium and its alloys.

Section 4.2 sets out the general requirements for magnesium implants in relation to mechanical properties, biocompatibility, and biodegradation. Following an outline of magnesium alloy categorisation, these alloys are discussed in relation to the organ intended for transplantation.

In Section 4.3, the applicability of *in vitro* findings to *in vivo* experiments is discussed. Both the problems and potential solutions are presented.

Section 4.4 explains the processes involved in magnesium degradation. The physiological mechanisms are outlined, as are the individual degradation products. Particular attention is devoted to the specific factors that influence degradation kinetics, as well as means of altering magnesium's corrosion properties. The various *in vivo* methods are discussed separately, looking at their advantages and disadvantages.

Section 4.5 begins by outlining the issue of biocompatibility before dealing with the pharmacophysiology of magnesium and its alloys. It then outlines the techniques that enable the analysis of specimens following or during the *in vivo* experiments. The advantages and disadvantages of these methods are critically discussed.

In Section 4.6, specific aspects of the use of magnesium and its alloys, both in and on bone, are explored. To gain an overview of the current status of research, the metallic nonresorbable implants and polymers are also briefly outlined, including characteristic benefits and drawbacks. Magnesium implants are then looked at in relation to the implant's location in the bone and the type of alloy. Finally, the scope for functional testing of the implant–bone compound is discussed.

In Section 4.7, coronary stents made of magnesium are described. As in Chapter 1, the stents currently in use are presented with their advantages and disadvantages. The current status of knowledge on magnesium stents is outlined, and three specific examinations – Intravascular ultrasound (IVUS), angiography and optical coherence tomography (OCT) – are presented.

Section 4.8 discusses the particular challenges in the development of biodegradable magnesium implants. Possible approaches are addressed, as are promising new therapeutic applications.

Surface Modification of Magnesium and its Alloys for Biomedical Applications. http://dx.doi.org/10.1016/B978-1-78242-077-4.00004-8

In Section 4.9, the reader is provided with useful information and recommendations.

4.2 Requirements with regard to magnesium alloys for *in vivo* use

4.2.1 Magnesium and magnesium alloys in biomedicine – general considerations

The majority of implants currently in use are manufactured from nonresorbable material, especially titanium or steel (Hofmann, 1995; Moses et al., 2002; Rehm, Helling, & Gatzka, 1997). A major disadvantage of these implants is that they either have to remain permanently within the body to replace the missing function or need to be explanted, entailing a second surgical procedure. The high rigidity of the implants leads to stress shielding in bone tissue (Exner, Müller, & Schmidt, 2004), which prevents remodelling. Biodegradable implants may be based on ceramics, polymers, or metals. Of the ceramics, those most used in medicine are calcium phosphates such as hydroxyapatite (Hap) and tricalcium phosphate (TCP) (Heimann, Itiravivong, & Promasa, 2004; Wiltfang et al., 2002; Zöllner, Büsing, & Strutz, 1984). Of the polymers, those showing particular clinical relevance are the polyglycolides (PGAs), polyactides (PLAs), and poly-β-hydroxybutyrates (PHBA) (Törmala, Pohjonen, & Rokkanen, 1998). Clinically relevant studies have been performed on degradable metallic implants using only magnesium.

As a biodegradable material (Wintermantel et al., 2002; Wintermantel et al., 2009), magnesium also exhibits very good biocompatibility in comparison with the established polymers (Gu, Zheng, Cheng, Zhong, & Xi, 2009; Witte et al., 2005; Xu, Yu, Zhang, Pan, & Yang, 2007). It has higher rigidity than these polymers, making it promising in terms of clinical use (Huang et al., 2007). With magnesium implants, the particular challenge is achieving a controlled degradation process and the associated decomposition of degradation products. This challenge has been taken up in recent years, especially in the form of new production techniques and the development of new magnesium alloys (StJohn et al., 2005; Heublein, 2003; Hänzi, Gunde, Schinhammer, & Uggowitzer, 2009; Feyerabend et al., 2010; Rettig & Virtanen, 2009). The ductility of magnesium materials can be optimised by means of simple and multiple deformations in forming processes by altering particle size (Chen, Lin, Jin, Zeng, & Lu, 2008; Ma et al., 2009; Westengen & Aune, 2006). This process also has a positive influence on the textures in the material, which further enhances the mechanical properties (Del Valle, Carreño, & Ruano, 2006).

The mechanical properties of magnesium can be improved by alloying different elements. It must be borne in mind that, essentially, only primary magnesium (known as pure magnesium) should be used for the production of alloys, because even concentrations of the contaminated elements of less than 0.2% by weight can lead to a significant increase in corrosion rate (Song & Atrens, 1999). The alloys of current clinical relevance can be classified into three groups: alloys with a low proportion of other elements, alloys containing aluminium, and alloys containing no aluminium (Witte, Hort et al., 2008). AZ31, AZ91, AE21, and LAE442 are among the well-characterised alloys in the field of biomedical engineering. Key attributes of the AZ alloys are their good biocompatibility but limited ductility (Kainer & von Buch, 2006). The AE21 and LAE442 alloys belong to the group of typical casting alloys with improved ductility and corrosion properties (Bach, Schaper, & Jaschik, 2003; Kainer & von Buch, 2006; Mordike & Ebert, 2001). The group containing no aluminium chiefly consists of WE and alloys containing calcium, zinc and zirconium (Kainer & von Buch, 2006; Mordike & Ebert, 2001; Witte, Hort et al., 2008). For alloys with rare-earth metals, an element-dependent increase in rigidity is observed (Nd > Pr > Ce > La, containing up to 7% rare earths by weight, as-cast condition, room temperature) (Rokhlin, 2003). It must, however, be borne in mind that the rare earths differ in terms of biocompatibility. Feyerabend et al. (2010) have described neodymium and praseodymium as suitable elements with regard to biocompatibility for biomedical application. Reproducibility and impurities can also bring about considerable changes in the alloys' properties (Feyerabend et al., 2010; Ghali, Dietzel, & Kainer, 2004; Song & Atrens, 1999). Neodymium, in particular, is a component of numerous magnesium alloys that are currently commercially available, including QE22 (Mg, 2% Ag by wt., 2% Nd by wt., 0.5% Zr by wt.), WE54, and WE43 (Kopp, Lefebvre, & Pareige, 2011). The *in vivo* biocompatibility of magnesium alloys has been also demonstrated in clinical studies (Erbel, di Mario et al., 2007). As a possible replacement for LAE422, LANd442 exhibited more rapid corrosion and noninflammatory formation of new bone (Hampp et al., 2012; Ullmann et al., 2011) (see Table 4.1).

4.2.2 Target human organ define the necessary mechanical properties of magnesium alloys

Implants made of magnesium alloys are currently used, in particular, on the cardiovascular system (Schilling et al., 2010; di Mario et al., 2004; Erbel, di Mario et al., 2007) and as osteosynthetic systems (Ullmann et al., 2011). In their studies, Seelig (1924) showed that tissue type had a crucial effect on the corrosion rate of magnesium. Subcutaneously implanted magnesium corroded more slowly than magnesium in regions well supplied with blood, such as muscle tissue (Seelig, 1924). The tissue-dependent degradation of magnesium was also demonstrated by McBride in his experiments (McBride, 1938a). Degradation of screws inserted into bone proceeded far more slowly than the screw heads at the soft-tissue boundary (McBride, 1938b). The degradation behaviour of the AZ31, AZ91, WE43, and LAE442 alloys differed in studies performed *in vivo*. In intramedullary application, WE43 and LAE442 show slow and uniform corrosion in comparison with AZ31 and AZ91. Additionally, less gas was formed with the WE43 and LAE422 alloys (Witte et al., 2005, 2006) (see Figure 4.1). In an intracutaneous test, none of the four alloys induced allergy (Witte, Abeln, et al., 2008; Witte, Ulrich, Rudert, & Willbold, 2007). No gas formation was evident in investigations of *in vivo* degradation of MgZn1Mn1.2 in bone, although degradation proceeded more rapidly in the vicinity of bone marrow (Xu et al.,

Table 4.1 **Mechanical and corrosion properties of eight magnesium-based alloys**

Alloy	AZ31[a]	AZ91[a]	AE21[b]	LAE442[a]	WE43[a]	MgCa0.8[a]	ZEK100[b]	Mg[a]
Elastic limit (Mpa)	161	244	–	148	198	125	203	102
Tensile strength (Mpa)	254	341	240	247	277	215	234	126
Breaking elongation (%)	14,2	13	–	18	17	15	23,7	12,6
Corrosion rate (medium)	0.25[c,e]	0.003[c,e]	–	5.54[d,f]	0.085[c,f]	0.04[c,f]	1.28[d,g]	0.085[c,f]

[a]Extruded.
[b]Rolled.
[c]$mg/cm^2/h$.
[d]mm/year.
[e]NaCl 3.5%.
[f]Simulated body fluid.
[g]In vivo.
Bohlen, Nürnberg, Senn, Letzig, and Agnew (2007), Chang, Wang, O, and Lee (2003), Dziuba et al. (2013), Gu & Zheng (2010), Kubota, Mabuchi, and Higashi (1999), Pardo et al. (2008), Seitz et al. (2011), Somekawa & Mukai (2005), Witte et al. (2008).

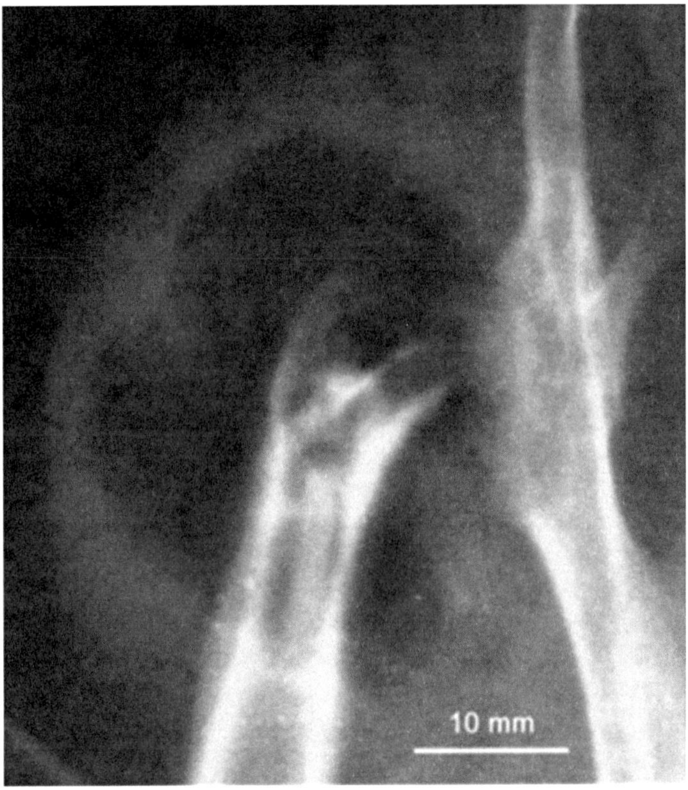

Figure 4.1 Subcutaneous gas bubbles observed on postoperative radiographs for 4 weeks during magnesium implant degradation (Witte et al., 2005).

2007). In their *in vivo* experiments with bone, Li et al. demonstrated that, using MgCa1.0, a high level of activity of osteoblasts and osteocytes – in conjunction with appreciable formation of new bone – occurred around the implant. No changes were observed in serum magnesium levels (Li, Gu, Lou, & Zheng, 2008).

4.3 Transferability of *in vitro* findings to *in vivo* trials: a suitable indicator for *in vivo* studies?

The purpose of the *in vitro* tests is to predict the biocompatibility and degradation behaviour of magnesium and its alloys for *in vivo* testing and thus to achieve the ideal properties for a biodegradable implant.

In the literature, *in vitro* tests are frequently based on standardised protocols (ISO10993–5:2009), so that this testing indirectly involves contact with a mix of degradation products. Additionally, many of these tests were performed under static conditions, which, in particular, do not correspond to the buffer capacity of the *in vivo* milieu (Levesque, Hermawan, Dube, & Mantovani, 2008; Xin et al., 2011) (see Tables 4.2 and 4.3). These mixtures do not contain all the degradation products of magnesium or one of its alloys. During the production of such mixes, the hydrogen evaporates and particles are separated out in the centrifuging or precipitation process (Gu et al., 2009; Witte, Hort, et al., 2008). A further problem (as described by Lorenz et al., 2009) associated with *in vitro* testing is mismatch between the small quantities of the medium and the surface of the test material. This imbalance may lead investigators to think corrosion is far greater than it really is. Purnama, Hermawan, Couet, and Mantovani (2010) recommend that biocompatibility testing for degradable materials should not be derived directly from tests for nonbiodegradable materials.

Magnesium corrosion involves the formation of a degradation layer on the surface, which exhibits a corrosion-inhibiting effect (Shaw, 2003). However, this layer is stable only in an alkaline environment, so that corrosion takes place a good deal more rapidly at a stable pH value (Witte et al., 2005). If it is presumed that, in the human body, diverse buffer systems keep the pH level constant, then corrosion would have to proceed faster *in vivo* than *in vitro*. However, as early as 1910, Lespinase assumed that the corrosion rate *in vitro* does not correspond to that in *in vivo* experiments (Lespinase, 1910). Witte et al. (2006) showed in their study that the degradation behaviour of AZ91 and LAE442 differs between *in vitro* and *in vivo* tests. Corrosion *in vivo* was considerably slower than that *in vitro*, with *in*

Table 4.2 Ion concentrations in five common solutions

	0.9% NaCl	PBS	Hanks'	DMEM	c-SBF
Na^+ (mmol/L)	153	157	142	127.3	142
K^+ (mmol/L)	–	4.1	5.9	5.3	5.0
Ca^{2+} (mmol/L)	–	–	1.3	1.8	2.5
Mg^{2+} (mmol/L)	–	–	0.8	0.8	1.5
HCO_3^- (mmol/L)	–	–	4.2	44.1	4.2
Cl^- (mmol/L)	153	140	145	90.8	147
HPO_4^{2-} (mmol/L)	–	11.5	0.8	0.9	1
SO_4^{2-} (mmol/L)	–	–	0.8	0.8	0.5
Tris (g/L)	–	–	–	–	6.069
Protein (g/L)	–	–	–	–	–
Amino acids (g/L)	–	–	–	1.6	–
Glucose (mmol/L)	–	–	1	4.5	–
Hepes (g/L)	–	–	–	5.96	–

Xin, Hu, and Chu (2011).

Table 4.3 Constituents and concentrations of buffering agents in several test solutions

Concentration (mmol/L)	HCO_3^-	HPO_4^{2-}	HPr	Hepes	Tris–HCl	Total
Plasma	27	0.5	16–18	–	–	43.5–45.5
0.9% NaCl	–	–	–	–	–	0
PBS	–	11.5	–	–	–	11.5
Hanks'	4.2	0.8	–	–	–	5
DMEM	44.1	0.9	–	–	25	70
c-SBF	4.2	0.5	–	–	40	44.7

Xin et al. (2011).

vivo experiments revealing that the corrosion rate of AZ91 and LAE442 exhibited a mutually opposing trend in favour of the latter (Witte et al., 2006). A number of studies indicate that changes over time in volume, form, and density are of importance in this regard (Jo et al., 2011; Wang et al., 2011). Difficulties with comparing *in vivo* and *in vitro* experiments are compounded by pitting corrosion (Mueller, De Mele, Nascimento, & Zeddies, 2009). Moreover, it was demonstrated that proteins in particular are instrumental in slowing the rate of corrosion (Kirkland et al., 2010; Yamamoto & Hiromoto, 2009). These findings show that it is not yet possible to broadly transfer *in vitro* findings to *in vivo* tests. Several authors posit that new testing systems that more closely imitate the *in vivo* conditions would be the most suitable (Gu et al., 2010; Levesque et al., 2008; Yamamoto & Hiromoto, 2009). These findings illustrate the problem of establishing such a system that reflects the biological, chemical and physical attributes of the human body.

4.3.1 Ex vivo test on bovine udder via microdialysis

Microdialysis is a well-established technique for analysing metabolic processes in extracellular space. The most important aspects of microdialysis catheters to consider are components such as length, pore size, and membrane material (Horal, Ungerstedt, Persson, Westgren, & Marcus, 1995); the driving force involved is diffusion. Ultrafiltration and osmosis, which may occur under physiological conditions, have an adverse effect on recovery (Kehr, 1993). To keep ultrafiltration and osmosis to a minimal level, the membrane pores need to be as small as possible, with flow rate and catheter length also reduced to a minimum. The concentration gradient can be influenced by varying flow rate (Benveniste & Huttemeier, 1990). One criticism of this method is that, in most cases in which it is applied *in vivo*, metabolic exchange between the extracellular space and the perfusion medium does not fully occur, preventing equilibrium from being achieved (Pasnik, Moll, Cywinska-Bernas, Sysa, & Zeman, 2007). The efficiency

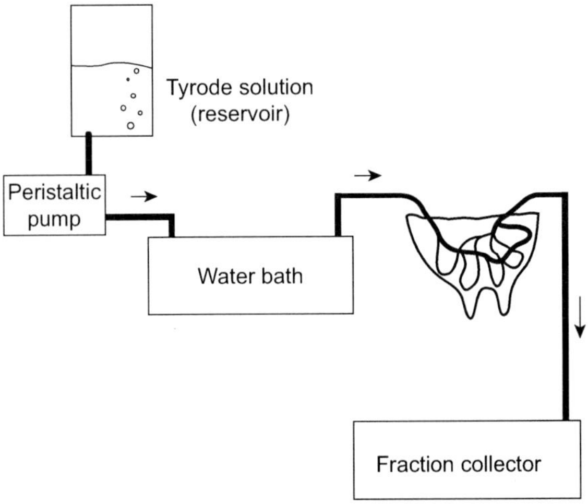

Figure 4.2 Schematic diagram of the isolated perfused bovine udder (Kietzmann et al., 1993).

of microdialysis is described in terms of relative and absolute recovery, with recovery determined by means of calibration (Jacobson, Sandberg, & Hamberger, 1985; Kehr, 1993; Lonnroth, Jansson, & Smith, 1987). Use of relative-recovery rates obtained by *in vitro* microdialysis for assessing extracellular concentration *in vivo* frequently results in underestimation of the actual concentration, because the extent of *in vivo* relative recoveries tends to be far lower than those *in vitro* (Grubb, Chadburn, & Boucher, 2002; Kovar, Nolting, & Derendorf, 1997).

The bovine udder model is an example of an *ex vivo* model. Both efferent and afferent vessels serving this organ had to be cannulated as soon as the animal was euthanised, and supplied with oxygenated Tyrode's solution. Controlled perfusion of the organ allows the tissue to remain vital for up to eighth (Kietzmann, Löscher, Arens, Maass, & Lubach, 1993) (see Figure 4.2). Microanalysis, enabled by the implantation of magnesium material in the bovine udder, makes it possible to analyse metabolic processes in close proximity to the material–tissue interface. This method also allows the concentration of the individual components to be determined (de Lange, de Boer, & Breimer, 2000). In their study, Schumacher et al. (2011) demonstrated that pure magnesium exhibits excellent biocompatibility and does not lead to an increase in proinflammatory cells. These findings are in line with those of the histological examinations.

4.4 *In vivo* biodegradation of magnesium alloys

4.4.1 *Biodegradation – general considerations*

To understand biodegradation of magnesium and magnesium alloys, it is vital to appreciate the biological importance of magnesium in humans. Magnesium is one of the essential elements and is involved in numerous metabolic enzyme processes (Kretz & Schäffer, 2008; Müller, Westenhöfer, Bosy-Westphal, Loser, & Selberg, 2007; Saris, Mervaala, Kappanan, Khawaja, & Lewenstam, 2000). Approximately half of all the magnesium in the body is concentrated in bone (Topf & Murray, 2003; Wacker, 1980), with the remainder stored predominantly in the muscles and organs, mainly the liver. Blood plasma accounts for only around one percent of the total (Niestroj, 2000; Vormann, 2003; Wacker, 1980). Magnesium occurs in the body in three different forms: ionised (60%), protein-bound (30%), and bound to serum anions (10%) (Topf & Murray, 2003). Uptake of magnesium in the gastrointestinal tract and its elimination via the kidney plays a crucial part in the regulation of magnesium levels, with the gall bladder and sweat having a more marginal role (Beyenbach, 1990; Niestroj, 2000; Wacker, 1980) (see Figure 4.3)

Hypermagnesaemia and hypomagnesaemia are generally rare and lead to impaired muscle excitability (Iannello & Belfiore, 2001). A number of cellular transport mechanisms are involved in magnesium homeostasis, primarily the sodium/magnesium exchanger (Gunther, Vormann, & Forster, 1984). The increase in intracellular magnesium has a positive effect on protein synthesis, although this process is inhibited at excessive concentrations (Rubin et al. 1997).

Other degradation and resorption processes are to be expected when magnesium is used in implants. Magnesium is a light metal with a density of 1.74 g/cm^3 and a higher

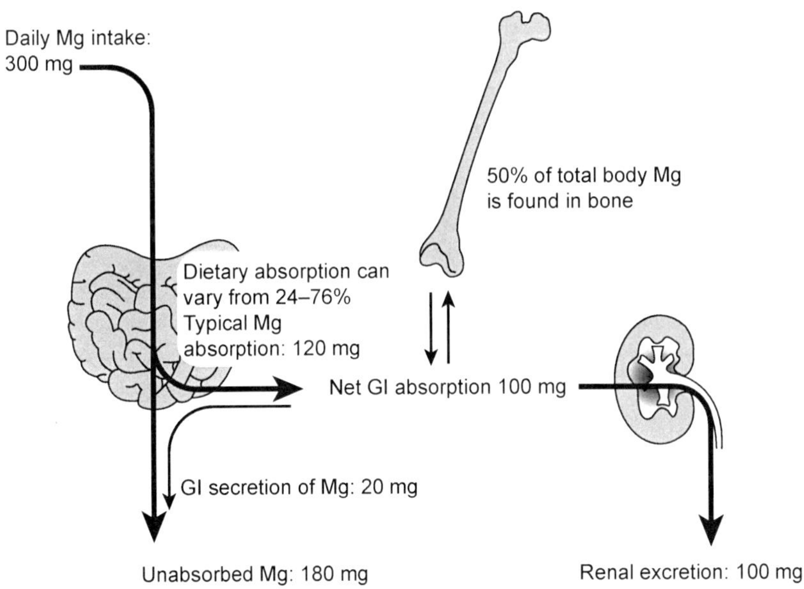

Figure 4.3 Magnesium metabolism. Of the 300 mg of magnesium ingested, approximately 120 mg are absorbed from the gut. A total of 20 mg are lost in gastrointestinal secretions, leaving a net absorption of 100 mg. Patients in magnesium balance excrete all of this absorbed magnesium in the urine. Bones provide a large magnesium buffer (Topf & Murray, 2003).

fracture toughness than that of ceramic biomaterials such as hydroxyapatite (Staiger, Pietak, Huadmai, & Dias, 2006) (see Table 4.4). In human bodily fluids, magnesium and its alloys exhibit relatively rapid corrosion (Song & Atrens, 1999, 2003). Solid magnesium hydroxide, magnesium chloride, and gaseous hydrogen are formed during the corrosion process (Li et al., 2008; Makar & Kruger, 1993; Song & Atrens, 1999; Staiger et al., 2006; Wang, Wei, Gao, Hu, & Zhang, 2008) (see Figure 4.4). As degradation progresses, a thin but stable layer of magnesium hydroxide is subsequently formed beneath the magnesium hydroxide coating, leading to a local increase in pH value (Barnett, 2007; Song, 2007). The magnesium oxide coating, i.e., the degradation layer, is a white, crystalline, nonclosed outer layer that forms directly on the implant. *In vivo* tests on animals show that the degradation products include magnesium hydroxide ($Mg[OH]_2$), magnesium oxide (MgO) and magnesium chloride ($MgCl_2$), as well as magnesium calcium apatite in the form $(Ca_{1-x}Mg_x)_{10}(PO_4)_6OH_2$ (Erdmann et al., 2011; Kuwahara, Al-Abdullat, Mazaki, Tsutsumi, & Aizawa, 2001; Li et al., 2008; Staiger et al., 2006; Thomann et al., 2010; Wang et al., 2008; Witte et al., 2005; Xu et al., 2007) (see Figure 4.5). Part of the degradation layer (magnesium and alloy elements) is derived from the implant itself and other parts (calcium and phosphate), i.e., constituents of the body, are stored there (Witte, Nellesen, Crostack, & Beckmann, 2002; Xu et al., 2007). Its thickness, which tends to increase, depends on the alloy's composition (Krause et al., 2010). This degradation layer delays the initially rapid corrosion process and also provides a certain degree of corrosion resistance (Makar & Kruger, 1993). The resulting pH values close to the surface, which (temporarily) are relatively high as corrosion progresses, may lead to the formation of magnesium-containing apatite in the form of $(Ca_{1-x}Mg_x)_{10}(PO_4)_6OH_2$, which is deposited at the implant surface (Liu, Huang, Shen, & Cui, 2001; Qi et al., 2008). In addition, gas production around the implant, the quantity of which corresponds to the degradation speed, has often been observed during *in vivo* experiments (Li et al., 2008; von der Höh et al., 2006; Witte et al., 2005; Staiger et al., 2006). Nonbound Mg^{2+} ions can be involved in normal physiological processes of metabolism. Hydrogen is not formed in the human body, but may have a positive antioxidative effect on the cells. Alkalosis and acidosis are compensated in the body – primarily via the kidneys and lungs – by both bicarbonate and haemoglobin buffering systems (van den Berg, 2005).

The corrosion pattern, which is a function of the alloy composition and environmental conditions, is usually initiated as a pitting-corrosion process (Song & Atrens, 1999).

The corrosion speed depends to a large extent on the purity of magnesium, alloy components, and the fabrication process (Li et al., 2008; Pardo et al., 2008; Song, 2007). Even low levels of impurity in magnesium result, owing to galvanic effects, in higher rates of degradation (Ren et al., 2007; Song, 2007; Song & Atrens, 1999; Witte, Hort, et al., 2008). Less than a tenth of one percent of calcium can substantially increase the corrosion resistance of magnesium (Kaese, 2002).

Increasing aluminium content has a corrosion-protective effect (Huang, Ren, Jiang, Zhang, & Yang, 2007; Kaese, 2002). Recent studies have shown that alloying with zinc (Huang et al., 2007; Pardo et al., 2008; Song, 2007) or rare-earth elements (Kannan & Raman, 2008; Witte et al., 2006; Wu, Fan, Zhai, & Zhou, 2005)

Table 4.4 Summary of the physical and mechanical properties of various implant materials in comparison to natural bone

Properties	Natural bone	Magnesium	Ti alloy	Co-Cr alloy	Stainless steel	Synthetic hydroxyapatite
Density (g/cm^3)	1.8–2.1	1.74–2.0	4.4–4.5	8.3–9.2	7.9–8.1	3.1
Elastic modulus (Gpa)	3–20	41–45	110–117	230	189–205	73–117
Compressive yield strength (Mpa)	130–180	65–100	758–1117	450–1000	170–310	600
Fracture toughness ($MPam^{1/2}$)	3–6	15–40	55–115	N/A	50–200	0.7

Staiger et al. (2006).

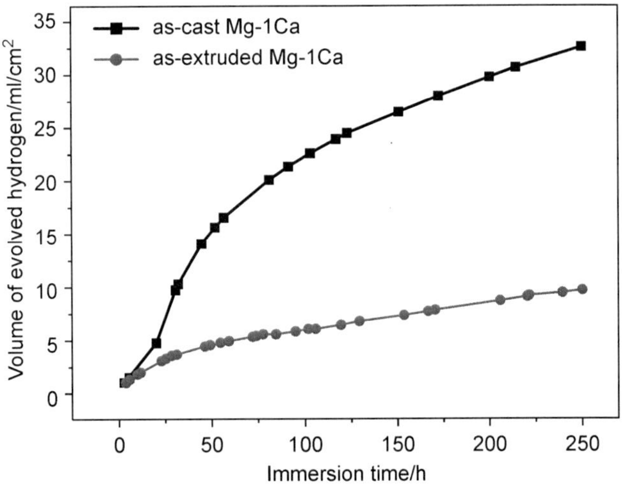

Figure 4.4 The hydrogen evolution volumes of as-cast and as-extruded Mge1Ca alloy samples as a function of the immersion time in simulated body fluid (Li et al., 2008).

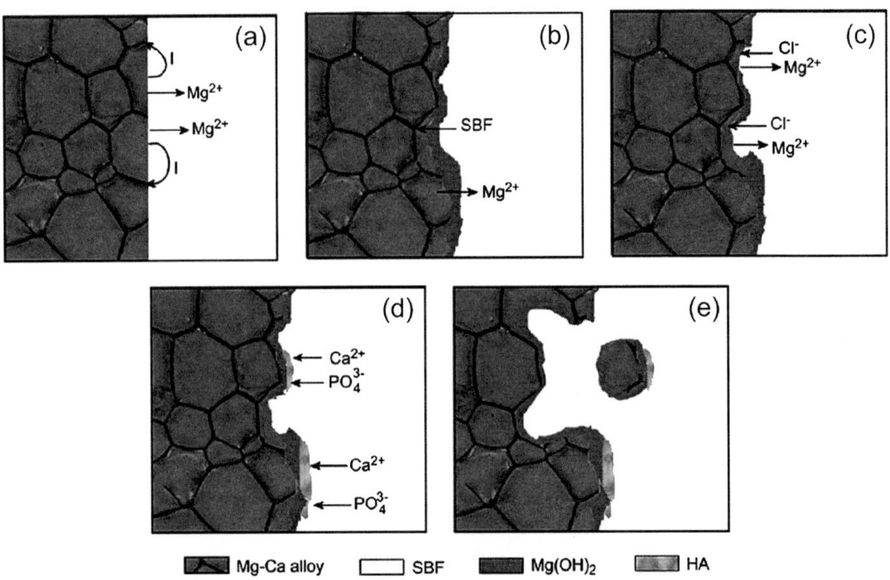

Figure 4.5 Schematic diagram of the alloy/solution biocorrosion interface: (a) the galvanic corrosion between Mg and Mg_2Ca phase, (b) the partially protective film covering the surface of Mg-Ca alloys, (c) the adsorption of chloride ions to transform $Mg(OH)_2$ into $MgCl_2$, (d) the hydroxyapatite formation by consuming Ca^{2+} and PO_4^{3-}, and (e) the disintegrated particle-shape residues falling out of the bulk substrate (Li et al., 2008).

results in a considerably slowed degradation process as compared with pure magnesium.

Apart from the alloy components, the fabrication process itself has a significant influence on the mechanical properties of magnesium (Li et al., 2008; Ma et al., 2009). By reducing the grain size, it is possible to achieve both improved elasticity and higher ductility (Koike, Ohyama, Kobayashi, Suzuki, & Maruyama, 2003; Zelin, Yang, Valiev, & Mukherjee, 1992). Extrusion provides the materials with added corrosion protection and enhanced mechanical properties (Kaese, 2002; Li et al., 2008). Extruded magnesium-calcium alloys consisting of one percent calcium by weight have excellent properties (Li et al., 2008).

Surface treatment and surface coating are further approaches to influencing degradation behaviour (Gray & Luan, 2002; Zhang, Xu, & Yang, 2005). von der Höh et al. (2006) were able to demonstrate that the smooth implant surface of magnesium-calcium alloys has a negative effect on corrosion, unlike the case with blasted surfaces (see Figures 4.6, 4.7, and 4.8). The findings of Gogolewski (2000) were similar. Increased corrosion resistance was achieved by using gas displacement to apply magnesium onto magnesium alloys (Yamamoto, Watanabe, Sugahara, Tsubakino, & Fukumoto, 2001). von Staesche developed an inexpensive procedure to coat implants with magnesium fluoride and was able to show a reduced rate of degradation in the specimens (Staesche, 1948). When an implant made of magnesium alloy has a fluoride coating, fluoride is deposited in the natural magnesium hydroxide layer, which compresses and stabilises the natural layer (Gnesca et al., 1996) and thus increases corrosion resistance (Chiu, Wong, Cheng, & Man, 2007; Staesche, 1948). A coating of bioactive hydroxyapatite (Song, Shan, & Han, 2008), the technique of plasma immersion (Liu, Xin, Tian, and Chu (2007), anodizing of the specimen surface (Song, 2007) or alkali-heat treatment (Li, Gao, & Wang, 2004; Lorenz et al., 2009) followed by slow degeneration has also been described in the literature.

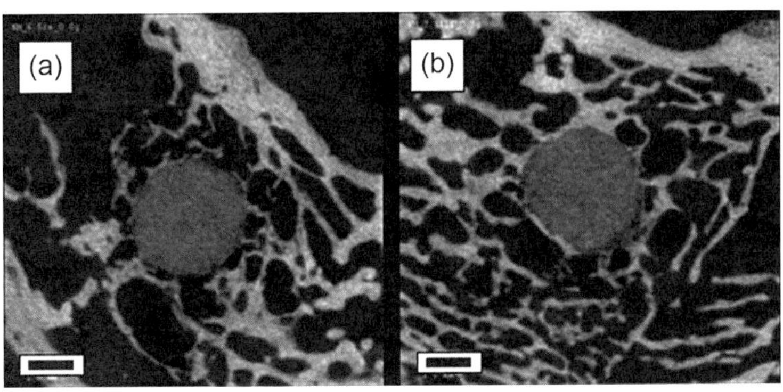

Figure 4.6 μ-Computed tomography of a smooth implant after 3 months (a) and 6 months (b) implantation duration: a small homogeneous resorption layer and a very close bone to implant contact layer is recognisable (scale, 1 mm) (Von der Höh et al., 2006).

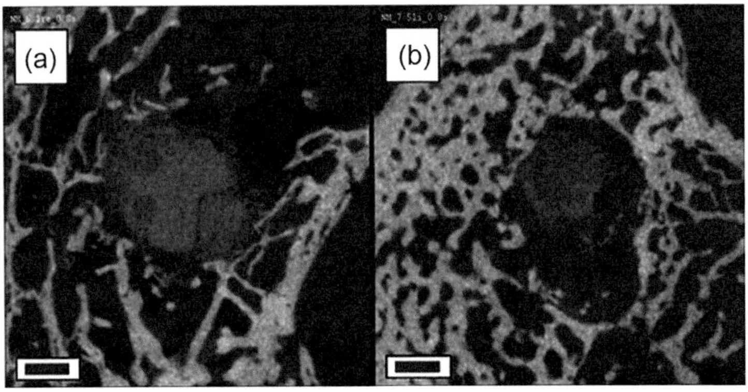

Figure 4.7 μ-Computed tomography of a sand-blasted implant after 3 months (a) and 6 months (b) implantation duration: the degradation process affected the whole implant, after 6 months the core body of the cylinder no longer existed. The bone to implant contact is less than in smooth implants and threaded cylinders (scale, 1 mm) (Von der Höh et al., 2006).

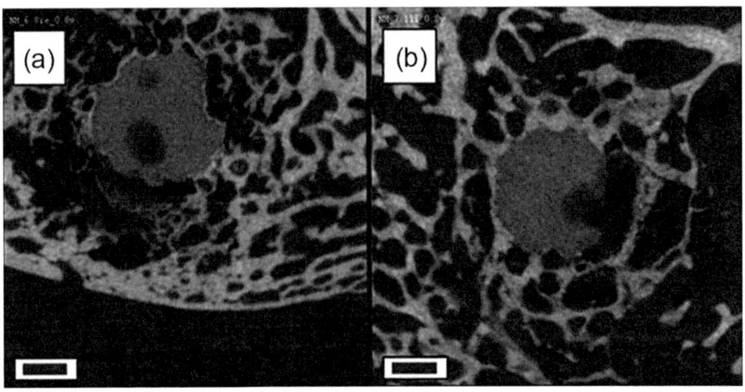

Figure 4.8 μ-Computed tomography of a threaded implant after 3 months (a) and 6 months (b) implantation duration: the hole shaped degradation at the edges and close bone contact around the implant is shown (scale, 1 mm) (Von der Höh et al., 2006).

The earliest investigations into pure magnesium's suitability as an implant material date back to the beginning of the 20th century (Lambotte, 1932). The magnesium degraded very rapidly during these experiments, and gas production was clearly evident. As early as 1932 (Lambotte, 1932), either the use of implants made of pure magnesium was viewed critically or pure magnesium was deemed unsuitable for these purposes. However, no systemic adverse effects were discovered (Lambotte, 1932; McBride,B 1938; Verbrugge, 1934). The first magnesium alloys were produced in 1932. *In vivo* experiments with magnesium–aluminium alloys, however, indicated rapid degradation and gas production, although the latter had no obvious harmful effects on the organism, and the gas was completely resorbed (Verbrugge, 1934).

Magnesium saw something of a renaissance at the beginning of the 21st century. Various alloys were developed with the aim of controlling the pattern and speed of the corrosion. Different alloy elements have an effect on degradation behaviour and/or mechanical properties (Song, 2007; Pardo et al, 2008).

4.4.2 Volumetry by means of μ-computed tomography

Several methods – with different approaches, and various advantages and disadvantages – are suitable for investigating the *in vivo* degradation process of magnesium-containing implants. These techniques are presented here, and their benefits and drawbacks outlined.

This method enables volumetric analysis of the implant during the course of animal experiments (see Figures 4.9 and 4.10). In this way, volume loss and the extent of corrosion on the alloy can thus be determined in a noninvasive procedure, and the duration of the experiment can be optimised (Holdsworth & Thornton, 2002; Paulus, Gleason, Easterly, & Foltz, 2001). The limiting factors in the performance of the investigation are twofold: the spatial distribution of the gantry in clinical μ-computed tomography (CT) scanners of up to around 70 cm and the size of the test animal. This procedure takes considerably longer than with conventional CT scanning, and the animals must be anaesthetised. Artefacts often occur when using metallic implants (Jäkel & Reiss, 2007; Shalabi, Wolke, Cuijpers, & Jansen, 2007; Stoppie, Wevers, & Naert, 2007), but are less of a problem with magnesium implants (Witte et al. 2005, 2007a).

Micro-CT provides a means of presenting two-dimensional images of the implant using three-dimensional geometry (Kiba et al., 2003; Stoppie et al., 2007). The latest devices enable *in vivo* measurements in animals with a local resolution of 10–20 μm (Brouwers, van Rietbergen, & Huiskes, 2007). Implants should be measured prior to implantation to obtain comparable outcomes. The results serve as reference values for intraoperatively performed measurements. Depending on the contrast ratios, the implant can be scanned either automatically using software or manually. It should be noted that specific thresholds must be determined and kept at a constant level during the series of tests (Erdmann et al., 2011). Volume loss can be established using a volume subtraction technique.

4.4.3 Determination of weight loss

The determination of weight loss is another method for evaluating the degenerative behaviour of magnesium implants (Li et al., 2008; Song, Bowles, & Stjohn, 2004; Xu, Zhang, Yin, Zeng, & Yang, 2008). For this purpose, specimens are weighed prior to implantation and, generally, also following euthanasia of the animal. Li et al. (2008) investigated the weight loss of magnesium–calcium implants *in vivo* and were able to show that the weight of the magnesium–calcium implants was increasingly reduced over time subsequent to implantation.

Prior to implantation, the initial weight of each individual specimen must be verified using precision analytical scales. The specimen is reweighed after explantation (although it must be borne in mind that the weight obtained includes all adherent corrosion products in the form of oxides and hydroxides). These corrosion products are not

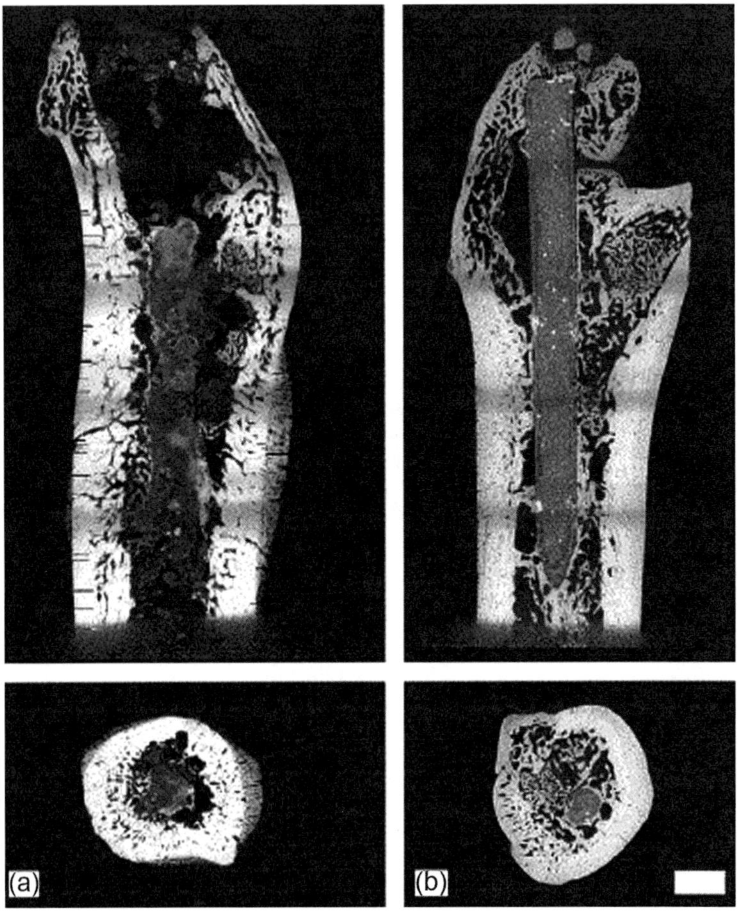

Figure 4.9 Two-dimensional reconstruction of explanted femura containing an AZ91D rod (a) and a LAE442 rod (b) after 18 weeks of implantation. Corrosion morphology and direct contact with newly formed bone can be observed for both magnesium alloys (a, b). Bar 1/4 1.5 mm (Witte et al., 2006).

part of the compact portion of the specimen; rather, they reflect the corrosion progress. An acid-cleaning process using chromic acid or hydrofluoric acid is a suitable means of removing corrosion products without causing damage to the magnesium component (Song & Atrens, 1999). Following this treatment, the specimen should be carefully rinsed in ethanol and left to air-dry.

4.4.4 Scanning electron microscopy and energy-dispersive X-ray analysis

Scanning electron microscopy (SEM) is a well-established procedure for evaluating the surface structure and degradation behaviour of magnesium implants (Duygulu, Kaya, Oktay, & Kaya, 2007; Li et al., 2008; von der Höh, von Rechenberg, Bormann,

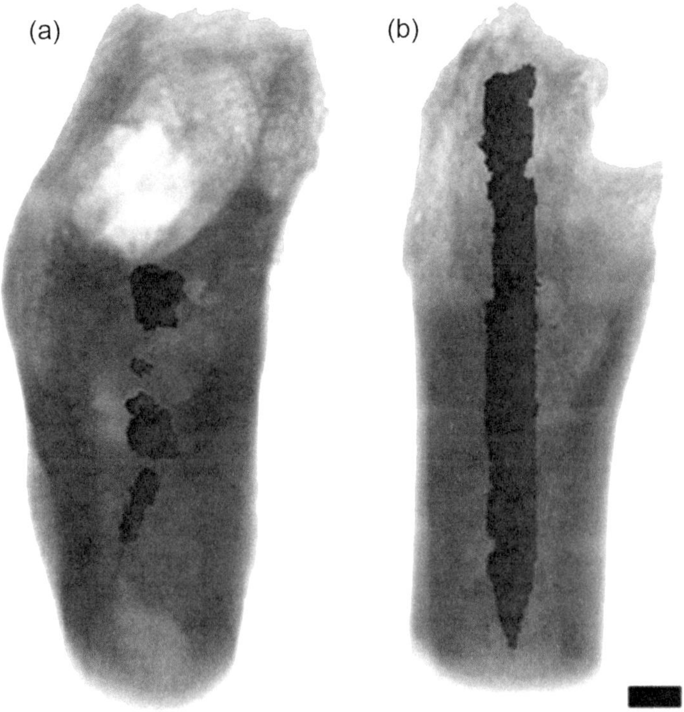

Figure 4.10 Three-dimensional reconstruction of remaining magnesium alloy (red) segmented from the bone matrix (grey) by voxel growing method. (a) AZ91D, (b) LAE442; Bar 1/4 1.5 mm (Witte et al., 2006).

Lucas, & Meyer-Linderberg, 2009; Zhang, Xu, Yu, Pan, & Yang, 2009). SEM, which can be performed before and after the acid-cleaning process, allows examination of the corrosion layer, i.e., the implant surface. Degradation in magnesium implants typically occurs in the form of pitting corrosion (Song & Atrens, 1999; von der Höh et al., 2009), which is identifiable under a scanning electron microscope as a hole-like indentation on the implant surface. However, the morphology of the magnesium alloy surface is inhomogeneous. The production process, alloy qualities, and coating properties may play a role in this.

Energy-dispersive X-ray (EDX) analysis makes use of electromagnetic radiation to investigate the implant and its interfaces with regard to the composition of the individual layers (Acarturk et al., 2008). This method may additionally indicate that the implant has an osteoconductive effect (Witte et al., 2005). In terms of evaluating the outcome, however, one disadvantage is the fact that merely the elements and their concentration can be displayed, and not any compounds of these elements.

The literature contains multiple descriptions of the formation of a corrosion layer on the surface of magnesium implants during the degradation process (Witte et al., 2005; Li et al., 2008; von der Höh, 2009; Zhang et al., 2009; Krause et al., 2010; Thomann et al., 2010) (see Figure 4.11). Krause et al. (2010) and Thomann et al. (2010)

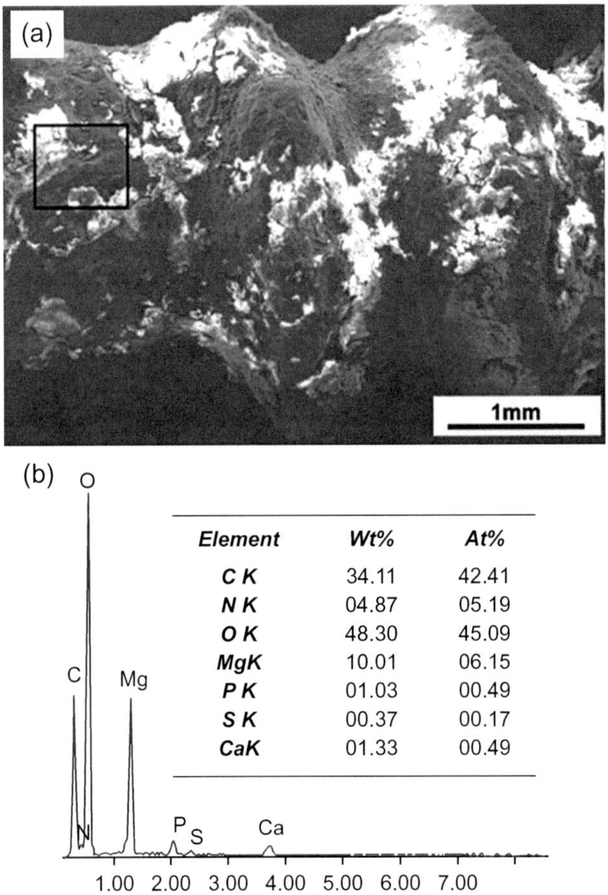

Element	Wt%	At%
C K	34.11	42.41
N K	04.87	05.19
O K	48.30	45.09
MgK	10.01	06.15
P K	01.03	00.49
S K	00.37	00.17
CaK	01.33	00.49

Figure 4.11 (a) SEM image of the screw thread part of the retrieved Mge1Ca alloy pin after 1 month implantation; (b) EDS spectra corresponding to the rectangular area in (a) (Li et al., 2008).

investigated the composition of the degradation layer of intramedullarily implanted magnesium alloys and were able to detect elements such as magnesium, oxygen, calcium, and phosphorous. Other studies found magnesium oxide, magnesium hydroxide, hydroxyapatite, and complex magnesium-calcium-phosphate compounds on the surface of corroded magnesium implants *in vivo* and *in vitro* (Li et al., 2008; Witte et al., 2005; Zhang et al., 2009).

4.5 *In vivo* biocompatibility of magnesium alloys

4.5.1 Biocompatibility – general considerations (acute, chronic, and foreign-body reaction)

Implants are generally to be regarded as foreign bodies. To test their biocompatibility, the methods to be used must first be defined. Biodegradable implants need to satisfy particular requirements for biocompatibility, because all implant components remain within the body or are progressively expelled. Pizzoferrato, Vespucci, Ciapetti, and Stea (1985) coined the terms "biocompatibility" and "biofunctionality" in the context of cell culture. The former term has been defined by various groups (Gradinger & Gollwitzer, 2006; Wintermantel & Ha, 1996). According to Gradinger and Gollwitzer (2006), the quantity, form, and nature of the substances released from the implant are crucial determinants of biocompatibility, a distinction being made between biocompatibility and bioactive implants. A material's biocompatibility is a function of its composition, form, and surface properties, the implantation site, the condition of the implant bed, the interface between implant and tissue, and the degradation of the material, as well as the surgical technique and the mechanical loading to which the implant is subjected (Bergsma, Rozema, Bos, & De, 1993; Epple, 2003; Rosengren, Bjursten, Danielsen, Persson, & Kober, 1999; Thull, 2003). Epple (2003) describes in his work that the properties of the interface between implant and tissue – such as chemical processes involving release of ions, pH value, adsorption of proteins and cells, and surface morphology – play a vital role. One basis for *in vivo* testing is that of successful *in vitro* outcomes. Various means of verifying biocompatibility are described in the literature, largely based on the recommendations of International Organization for Standardization (ISO) 10,993-6:2009 (ISO 10993-6:2009). This standard describes three test methods: epicutaneous testing, bone testing, and implantation in muscle. A prerequisite for testing biocompatibility is that the implant is not subjected to any mechanical and functional stress (ISO 10993-6:2007). Histological sections may be analysed semiquantitatively, quantitatively, or morphometrically (Bethmann & Knofler, 1987). Several authors discuss the possibility of using micro-CT images of tissue morphology to assess biocompatibility in bone (Wachter et al., 2001). The ISO 10,993-6 standard requires that the histological preparation of specimens take account of the following parameters: formation of a connective-tissue capsule, occurrence of inflammatory cells, signs of degeneration, presence of necrosis, and particulate material. At the cellular level, implanted foreign material always leads to an immune response. In the first instance, the innate immune defence system delivers the body's defence response, in particular involving the formation of macrophages, neutrophil granulocytes, and natural killer cells. The macrophages respond chemotactically and also act as a link between the innate and adaptive immune responses. They also have the ability to form foreign-body cells (Anderson, Rodriguez, & Chang, 2007). The adaptive immune response is mediated primarily by B and T lymphocytes. Biodegradable magnesium implants give rise to degradation products that need to be eliminated by the body, so that a greater incidence of macrophages should not necessarily be regarded as indicating a rejection reaction (Doernberg et al., 2006). Anderson et al. (2007) describe the body's response to implanted biomaterial over time. During the initial phase, a blood-based matrix forms at the implant site, which is the basis not only for wound healing, but also for foreign-body reactions. The next phase, i.e., that of acute inflammation, is characterised in particular by neutrophil granulocytes and is not expected to exceed 7 days. Mast cells, which secrete histamine and interleukins, perform an important regulatory function during this phase. The chronic inflammatory phase is distinguished by monocytes, lymphocytes, and plasma cells and tends to be of short duration. The foreign-body reaction involves a mix of different constituent cells, and, while similar to the chronic reaction, it is the occurrence of macrophages and foreign-body giant cells that is characteristic here. Once the inflammatory phase is over, granulation tissue is formed in conjunction with macrophages, fibroblast infiltration, and new vessel formation. Granulation tissue is regarded as the precursor stage for the formation of a fibrosis layer. It is, primarily, the quantity and time of occurrence of immune cells that are crucial in assessing the body's immunological reaction (Anderson et al., 2007) (see Figure 4.12; 4.13 and 4.14). A strong layer of connective tissue with numerous immune cells indicates the encapsulation of the implant and, indirectly, insufficient biocompatibility (Freeman & Brook, 2006; Tsai, Ruey-Mo, Chien-Ping, & Jiin-Huey Chern, 2008; Witte et al., 2007b).

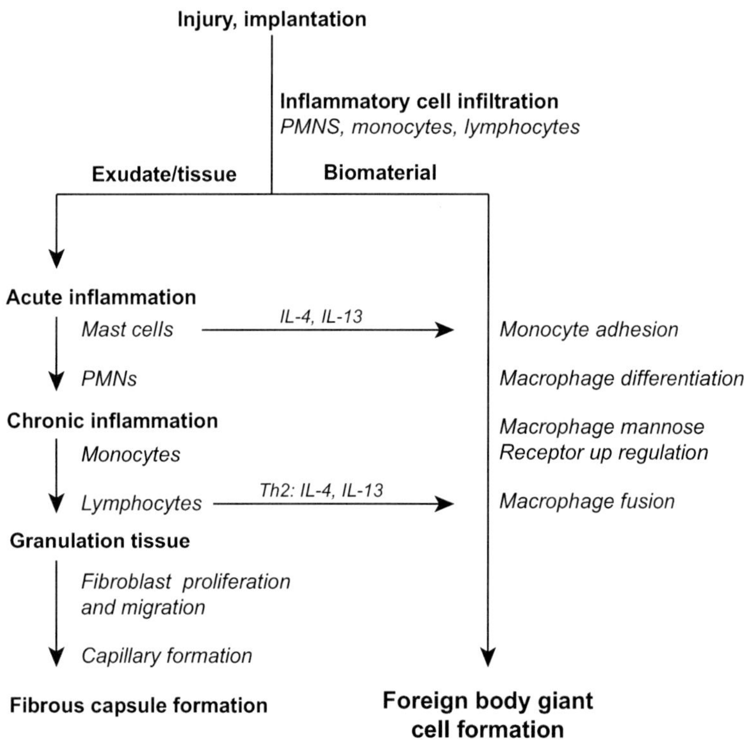

Figure 4.12 Sequence of events involved in inflammatory and wound healing responses leading to foreign body giant cell formation. This shows the potential importance of mast cells in the acute inflammatory phase and Th2 lymphocytes in the transient chronic inflammatory phase with the production of IL-4 and IL- 13, which can induce monocyte/macrophage fusion to form foreign body giant cells (Anderson et al. 2007).

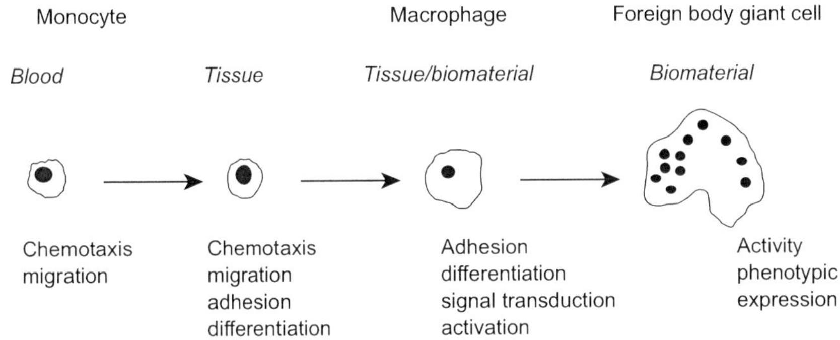

Figure 4.13 *In vivo* transition from blood-borne monocyte to biomaterial adherent monocyte/macrophage to foreign body giant cell at the tissue/biomaterial interface. There is ongoing research to elucidate the biological mechanisms that are considered to play important roles in the transition to foreign body giant cell development (Anderson et al. 2007).

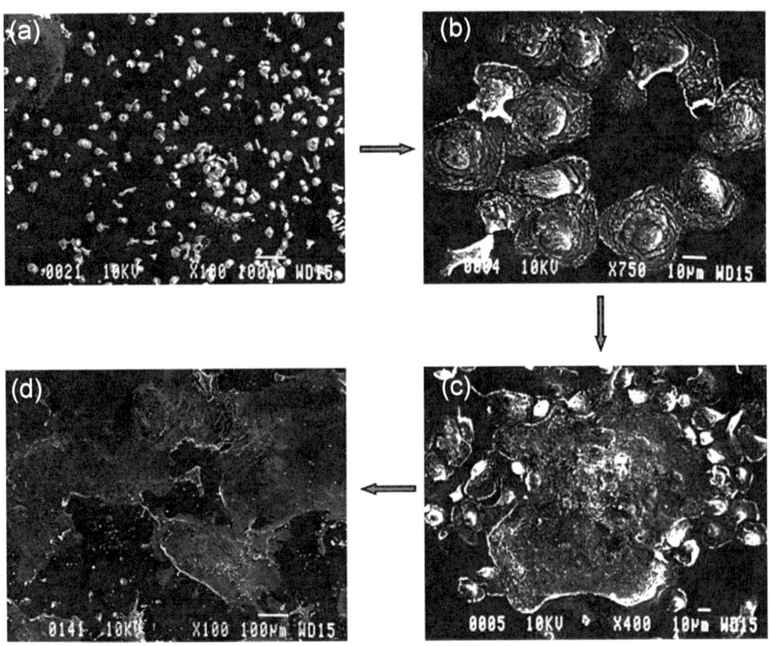

Figure 4.14 Scanning electron microscopy images of an Elasthane 80A Polyurethane surface from an *in vivo* cage study showing the morphological progression of the foreign body reaction. The sequence of events at the Polyurethane surface includes (a) monocyte adhesion (0 days), (b) monocyte-to-macrophage development (3 days), (c) ongoing macrophage–macrophage fusion (7 days), and (d) foreign body giant cells (14 days) (Anderson et al., 2007).

4.5.2 Pharmacophysiology of magnesium and selected alloy components

Magnesium is among the elements that are essential to the body and has, using the 29 Mg isotope, been detected in small mammals (Rogers & Parker, 1959; Rogers, Haven, & Mahan, 1960). Magnesium implants are generally characterised in clinical experiments by good biocompatibility (Lambotte, 1932; Verbrugge, 1934; McBrideB, 1938; Nicole, 1947; Witte et al., 2005; von der Höh et al., 2006; Witte et al., 2007; Xu et al., 2007; Li et al., 2008). Most *in vivo* studies also describe the formation (to differing extents) of hydrogen gas (Lambotte, 1932; Verbrugge, 1934; McBrideB, 1938; Nicole, 1947; Witte et al., 2005; von der Höh et al., 2006; Witte et al., 2007b; Li et al., 2008). However, the degradation products of corrosion exhibit different pharmacophysiological properties.

Because of unsuitable enzymatic conditions, hydrogen is not formed in the cells (Staiger et al., 2006). It is known, however, that hydrogen – an antioxidant – has a protective effect on cells. It also shows antiinflammatory properties in *in vivo* investigations (Atsunori, 2011; Buchholz et al., 2008; Haung et al., 2010). This formation process is described as being an important co-factor in gas production (Li et al., 2008). *In vivo*, the gas formed is resorbed by the surrounding tissue (Witte, Hort, et al., 2008).

The OH- ions that also form during the degradation process would, without regulatory mechanisms, lead to strong alkalinisation of the tissue around the implant (Heublein, 2003; Song, 2007). Magnesium chloride ($MgCl_2$) is a rapidly resorbable salt, which may lead to slight acidification of the liver and is toxic at high concentrations (Franke, 1934). Magnesium ions (Mg^{2+}) that do not contribute to the formation of salts and apatites may in principle be involved in physiological metabolism. It has been suggested that magnesium ions promote bone-remodelling processes (Janning et al., 2010; Rude et al., 2006). Apatites containing magnesium are similar to the mineral phase of bone, namely, hydroxylapatite $Ca_5(OH)(PO_4)_3$. It is therefore likely that apatite is involved in normal bone-remodelling processes. Several authors postulate that calcium phosphates exhibit osteoconductive properties (Li et al., 2004, 2008; Witte et al., 2005; Xu et al., 2007).

Alloy components also play a significant role in biocompatibility; they have, in recent years, assumed an important function in the fabrication of magnesium implants in relation to their mechanical and corrosive properties. The influence of alloy components on tissue is not fully known (Yuen & Ip, 2010) (see Table 4.5).

4.5.2.1 Aluminium

The total quantity of aluminium in the human body is (depending on weight) 0.295 g, with most of it stored in bone (Skalsky & Carchman, 1983). Aluminium is regarded as a strong neurotoxin and is thought to be one of the causative factors in Alzheimer's disease (El-Rahman, 2003; Mjöberg, Hellquist, Mallmin, & Lindh, 1997).

4.5.2.2 Calcium

Calcium is the most abundant element in the human body (which contains around 1 kg) and is hormonally regulated (Civitelli & Ziambaras, 2011). Disruption of the calcium

Table 4.5 Toxicological critical values and derived toxicological critical values for common alloying elements in magnesium alloys

	Al	Mn	Zn	Cu	Ni	Fe	Sr	Zr	Ce[c]
Source	ATSDR [11]	IRIS [17]	ATSDR [10]	ATSDR [24]	ATSDR [25] and IRIS [25]	UK FSA [28]	ATSDR [23]	Not found in FSA/ ATSDRI/ IRIS	IRIS [41]
Type of exposure limit[a]	NOAEL-a	N0AEL-h/ RfD	NOAEL-h	NOAEL-h	EPA RfD	Guidance level	NOAEL-a	Insufficient data	Insufficient data
Potential adverse systemic effects at initial overdose[b]	Neuro-toxicity	CNS effccts	Reduced erythrocyte superoxide dismutase level	Changes in blood protein and enzyme levels	Reduced body and organ mass	Reduction in serum zinc; possible increased risks of cardio-vascular disease and cancer	Abnormal bone min-eriali-sation	Allergic hyper sensitivity. Dialysis osteomala-cia accu-mulates in the brain similar to Al [42,43]	Cardiac toxicity and reduction of haemoglobin oxygen affinity
Exposure limit (mg^{-1} kg bw^{-1} day)	26	0.14	0.83	0.042	0.02	0.28	140	n/a	n/a
UF for interspecies variation	10	1	1	1	1	1	10	n/a	n/a
UF for inter-individual variation	10	1|	3	3	1	1	3	n/a	n/a

Continued

Table 4.5 Toxicological critical values and derived toxicological critical values for common alloying elements in magnesium alloys − cont'd

	Al	Mn	Zn	Cu	Ni	Fe	Sr	Zr	Ce[c]
Oral absorption efficiency (%)	0.63	5	20	36	27	15	20	n/a	n/a
Modifying factor for 100% absorption (=l/oral bio-availability)	158.7	20	5	2.78	3.703	6.6	5	n/a	n/a
UF/absorption adjusted exposure limit (mg kg bw^{-1} day)	1.64E−03	7.00E−03	553E−02	5.04E−03	5.40E−03	4.24E−02	9.33E−01	n/a	n/a
Daily exposure limit for a 60-kg adult (mg/day)	9.83E−02	4.20E−01	3.32E+00	3.02E−01	3.24E−01	2.55E+00	5.60E+01	n/a	n/a
Annual exposure limit for a 60-kg adult (mg/year)	35.88	153.30	1211.80	110.29	118.28	929.09	20.440	n/a	n/a

[a]h, human data; a, animal data NOAEL (no observed adverse effect level). RfD, reference dose; n/a, not available.

level in the blood leads to various functional disorders of certain organs (the heart, gastrointestinal tract, and kidneys) as well as the nervous system and muscles.

4.5.2.3 Lithium

Lithium, which is present in the body only in negligible amounts, may enter the cell via sodium channels. Even a mild overdose leads to muscle weakness and vegetative symptoms. A severe overdose results in damage to the lungs and kidneys (Bichet, 2006; Giles & Bannigan, 2006; Timmer, 1999).

4.5.2.4 Zinc

Zinc is among the important trace elements and is also involved in immune system function (Lastra, Pastelin, Camacho, Monroy, & Aguilar, 2001). It is regulated in the body chiefly via the gastrointestinal tract. At high concentrations, it has a neurotoxic effect, and it is believed to be involved in development of amyotrophic lateral sclerosis (ALS) (Post, Eibl, & Ross, 2008).

4.5.2.5 Rare-earth elements

Rare-earth elements are present in the body only in very small quantities, with cerium apparently the most abundant in terms of weight (Inagaki & Haraguchi, 2000). Their half-life period within the body shows considerable fluctuation and may be as much as 10 years (Hirano & Suzuki, 1996). These metals' possible antiproliferative effect on cancer cells is also discussed in the literature (Kostova, Momekov, & Stancheva, 2007). They also have a similar ionic radius to that of calcium and may act as an antagonist to calcium in the body (Feyerabend et al., 2010; Nakamura, Tsumura, Tonogai, Shibata, & Ito, 1997). Gu et al. (2009) report that the rare-earth elements may also influence haemolysis, chromosomal aberrations, and liver function.

A number of techniques are available for verifying *in vivo* biocompatibility. As these differ in the approach involved, they work together synergistically in clarifying the issue of biocompatibility.

4.5.3 Histology

To provide information about the tissue–implant compound and, therefore, to assess biocompatibility, histological examinations are essential (An, 2003). The specimens obtained must initially be fixed. Depending on the types of tissue, decalcification may be carried out prior to the further preparation of osseous and cartilaginous specimens, so sections of appropriate thickness can be created. After the specimens are embedded, the sections are produced and then stained. However, owing to the decalcification of the bone, the fine morphology and the dynamic processes in the bone are assessable only to a limited extent. A suitable method of processing decalcified samples is embedding in plastic (Wolf, Röser, Hahn, Welkerling, & Delling, 1992). The cutting-grinding technique is one of many methods that make preparation of noncalcified bone possible (Donath, 1998).

The aim of histological preparation is to visualise reactions in tissue using an appropriate stain. Use of different staining agents and suitable means of staining creates high-contrast images that allow structural analysis. A basic distinction can be made between progressive and regressive techniques. It is assumed that the different structures retain the stain to differing extents (Lang, 2006). Other options are those of indirect and direct staining and single or multiple staining (Lang, 2006). In histological examinations it is crucial that, following explantation of the sample, the materials, methods, and analysis used are comparable with other test materials, sites, and research teams. A selection of staining agents and techniques are described below.

4.5.3.1 *Haematoxylin staining*

The direct staining agent is haematein or oxidised haematoxylin (Romeis, 1989). The haematoxylin stain is yellow brown in colour and is suited for progressive or regressive staining of cell nuclei (Lang, 2006). Subsequent rinsing in water gives it its typical purple colour. Addition of eosin stains alkaline structures, such as cytoplasmic proteins, red.

4.5.3.2 *Toluidine blue*

This stain dyes cell nuclei a distinct blue colour and is suitable for making metachromasy visible (Schauer & Scheibe, 1959), a property that allows structures such as mast cell granules and cartilage matrix to be stained purple. Toluidine blue provides a good overall view and is particularly well suited for visualising bone tissue. Cells, cell nuclei, osteoid seam, osteoclasts, and osteoblasts are stained different shades of blue. Mineralised tissue appears pale blue.

4.5.3.3 *Van Gieson's stain*

This is a triple stain suitable for viewing connective tissue. Cell nuclei appear blue-black; collagen, bright red; calcified bone, red; osteoid, muscle tissue, and cytoplasm, yellow; and mast cell granules, red-brown. Amyloid, hyaline, and mucus are visible in various tones between yellow and red (Romeis, 1989).

4.5.3.4 *Masson-Trichrome-Goldner stain*

Using this multiple stain, cells of mineralised and nonmineralised bone matrix show up well. This property makes it the stain of choice for examining the morphology of non-calcified bone. The cell nuclei appear brownish black, and cytoplasmic staining makes it possible to distinguish between osteoclasts and osteoblasts (Lang, 2006; Romeis, 1989; Schwarz et al., 2007).

4.5.3.5 *Tartrate-resistant acid phosphatase*

One of the group of enzymatic histological stains, this allows reliable detection of osteoclasts,which appear a reddish-pink colour, and osteoclast progenitor cells (Ballanti et al., 1997). It has been observed that, under certain conditions, other cells, such as osteoblasts and osteocytes, may also be stained (Bianco et al., 1988; Nakano, Toyosawa, & Takano, 2004).

4.3.5.6 Movat's pentachrome stain

This is a multiple stain suitable for visualising different tissue types. Connective tissue appears red; mineralised bone tissue, yellow; mineralised cartilaginous tissue, blue-green; and nonmineralised cartilaginous tissue, reddish-yellow. Collagen fibres stain yellow; osteoid, dark red; cell nuclei, blue-black; and cytoplasm, a reddish colour.

4.5.4 Fluorescent microscopy

Another method of examining histological sections is fluorescence microscopy. Monochromeor polychrome *in vivo* fluorescence labelling with flurochromes allows the assessment of remodelling processes and the quantification of bone growth over time (Rahn, Bacellar, Trapp, & Perren, 1980; Witte et al., 2005). Rahn et al. (1980) developed dosage regimens for animals and polychrome labelling using five different-coloured staining agents: xylenol orange, calcein green, tetracycline, alizarin complex, and calcein blue. These substances are generally applied subcutaneously or intravenously in an aqueous medium. The intravital staining dyes have frequently been used *in vivo* in recent years; the only such stain to be used in humans was tetracycline (An, 2003; Iwamoto, Takeda, Sato, & Yeh, 2004; Xu et al. 2009).

4.5.5 Scanning electron microscopy and energy-dispersive X-ray spectroscopy

Scanning electron microscopy (SEM) is suitable for ultrastructural assessment between implant and tissue (An, 2003). (See Section 4.4.4 on biodegeneration.)

4.5.6 Micro-computed tomography and histomorphometry

Micro-CT allows both two- and three-dimensional visualisation of bone and provides clear views of bone structure (Bernhardt et al., 2004). The two-dimensional images can also be used to obtain morphometric parameters (Parfitt, 1988). Histomorphometry makes it possible to assess structural changes in the surrounding tissue and the implant–tissue interface (An, 2003). As well as providing options for quantitative analysis, this method has a further advantage, namely that the specimen is available for further mechanical and histological investigations and large quantities can be examined (Ruegsegger, Koller, & Mueller, 1996; Wachter et al., 2001). In recent years, this time-consuming evaluation has been simplified by computer programs (Huffer, Ruegg, Zhu, & Lepoff, 1994; Martin et al., 2002). To correctly evaluate the histomorphometric findings, it is necessary to apply the same scanning parameters, but not the standardised sectional planes with the same threshold level

(Ruegsegger et al., 1996). Analysis is carried out using quantitative and semiquantitative point systems. Parfitt (1988) standardised the nomenclature and unified the terminology (see Table 4.6). Analysis includes the measurement of bone mass (expressed as a percentage), bone volume/total volume (Smet et al., 2006), trabecular thickness, and trabecular number (Gabet et al., 2006). Wachter et al. (2001) postulate that histomorphometry is, in terms of bone assessment, superior to histological examination. Admittedly, one disadvantage is the lack of information on the biological characteristics of bone, particularly the evaluation of periosteal and endosteal remodelling. However, studies by a number of research teams show that the outcome of micro-CT analysis closely matches that of histological findings (Butz, Ogawa, Chang, & Nishimura, 2006; Stopie et al., 2007; van Oosterwyck et al., 2000) (see Figure 4.15).

4.5.7 Chemical analysis (blood, other specimens)

Especially with biodegradable materials, the question arises as to whether degradation products influence not only the implant–tissue interface but also directly or indirectly affect (or are stored in) other organs. One current means of evaluating this is to test the blood of animal subjects for magnesium. Another option is to examine the regional lymph nodes or the kidneys, the organs of elimination. Several authors report unchanged magnesium serum concentration (Li et al., 2008; Wong et al., 2010; Xu et al., 2007) or changes in organ function (Witte et al., 2010; Zhang et al., 2010, 2009) (see Table 4.7).

4.6 Testing of magnesium alloy in or on bone – special considerations

This section begins with a discussion of the advantages and disadvantages of implant materials currently used in the vicinity of bone marrow, intended to highlight the significance of magnesium for the manufacture of implants to be used in bone.

4.6.1 Metallic nonresorbable implants

Osteosynthesis materials generally used at present include high-alloyed stainless steel, pure titanium, and titanium alloys (Disegi, 2000; Jain, Podworny, Hearn, Anderson, & Schemitsch, 1997; Pohler, 2000; Singh & Dahotre, 2007). Less expensive than titanium, high-alloyed stainless steel is characterised not only by a high degree of mechanical strength and hardness, but also by good corrosion resistance (Disegi & Eschbach, 2000; Singh & Dahotre, 2007). Owing to their high density (almost double that of titanium), steel implants are substantially heavier than—and have considerably greater elasticity than—titanium, which in turn is more elastic than the cortex of the bone (Disegi & Eschbach, 2000; Pohler, 2000). Biocompatibility is affected by metal ions such as nickel, chromium, and cobalt, which are released during the corrosion process

Table 4.6 Comparison of new with old terminology for selected primary measurements (upper list) and derived indices (lower list) on cancellous bone tissue; methods of calculating the latter are given in original

Present terminology[a]	Proposed terminology[b,c]	Abbreviation	Units
Trabecular[d] bone volume (TBV)	Bone volume[e]	BV/TV[e]	%
(Relative) osteoid volume (ROV)	Osteoid volume	OV/BV	%
(Absolute) osteoid volume[f] (AOV)	Osteoid volume	OV/TV	%
(Relative) osteoid surface (ROS)	Osteoid surface	OS/BS	%
(Active[g]) osteoblast surface (AOS)	Osteoblast surface	Ob.S/BS[h]	%
(Mean[i]) osteoid seam width (MOSW)	Osteoid thickness	O.Th	mcm
(Total) resorption surface[j] (TRS)	Eroded surface	ES/BS	%
(Active[k]) resorption surface (ARS)	Osteoclast surface	Oc.S/BS[l]	%
Osteoclast index (OI)	Osteoclast number	N.Oc/T.Ar[m]	$/mm^2$
(Trabecular) specific surface[n] (tS_{sp})	Bone surface	BS/TV	mm^2/mm^3
(Mean[i]) wall thickness (MWT)	Wall thickness	W.Th.	mcm
Active[o] forming surface[p] (AFS)	Mineralising surface	MS/BS	%
Mineralisation[q] front (MF)	Mineralising surface	MS/OS	%
Calcification[r] rate (CR)	Mineral apposition rate	MAR	mcm/d
Mean[i] trabecular[s] plate thickness (MTPT)	Trabecular thickness	Tb.Th	mcm

Continued

Table 4.6 **Comparison of new with old terminology for selected primary measurements (upper list) and derived indices (lower list) on cancellous bone tissue; methods of calculating the latter are given in original – cont'd**

Present terminology[a]	Proposed terminology[b,c]	Abbreviation	Units
Mean[i] trabecular[s] plate density (MTPD)	Trabecular number[t]	Tb.N	/mm
Mean[i] trabecular[s] plate separation (MTPS)	Trabecular separation[t]	Tb.Sp	mcm
Bone formation rate, BMU level[u] ($sV_{(BMU)}$)	Adjusted apposition rate	Aj.AR	mcm/d
Bone formation rate, tissue level ($^{S}V_f$)	Bone formation rate	BFR/BS	mcm^3/mcm^2/year
Bone formation rate, volume referent[v] ($^{V}V_f$)	Bone formation rate	BFR/BV	%/year
Mineralisation lag time (MLT)	Mineralisation lag time	Mlt	day
Sigma (duration of formation) (of)	Formation period	FP	Day or year

[a]These are representative of current practice in different laboratories; it is not implied that all are used by any laboratory or that any are used by most laboratories. Qualifying terms are in parentheses if their use is inconsistent between laboratories.
[b]Measurement name only; need for inclusion of source and/or referent in name varies with context, as discussed in original.
[c]Three-dimensional expression except where otherwise stated.
[d]Source almost always included in name for this quantity, often omitted for others.
[e]The full name and abbreviation would be cancellous bone volume/tissue volume (Cn-BV/TV); see notes b and d.
[f]Also called osteoid volume density.
[g]Designation usually based on morphology.
[h]OS is another frequently used referent.
[i]Including "mean" as part of the name should imply direct rather than indirect measurement and may lead to confusion with the mean value in a group of subjects.
[j]Also termed crenated or Howship's lacunar surface.
[k]Designation usually based on presence of osteoclasts.
[l]ES sometimes used as an additional referent.
[m]Bone perimeter is an alternative referent; note that expression must be 2D, not 3D.
[n]Also called surface density.
[o]Note wide variety of meanings presently given to the term "active."
[p]Often called "labeled surface" or "tetracycline surface" (double, single, or both).
[q]Or calcification.
[r]Or mineralisation.
[s]Note ambiguity between "trabecular" as a type of bone tissue and as a type of individual structural element.
[t]Must specify whether calculated according to parallel plate or rod model or measured directly.
[u]Many other synonyms given in original.
[v]Equivalent to rate of bone turnover.
Parfitt (1988).

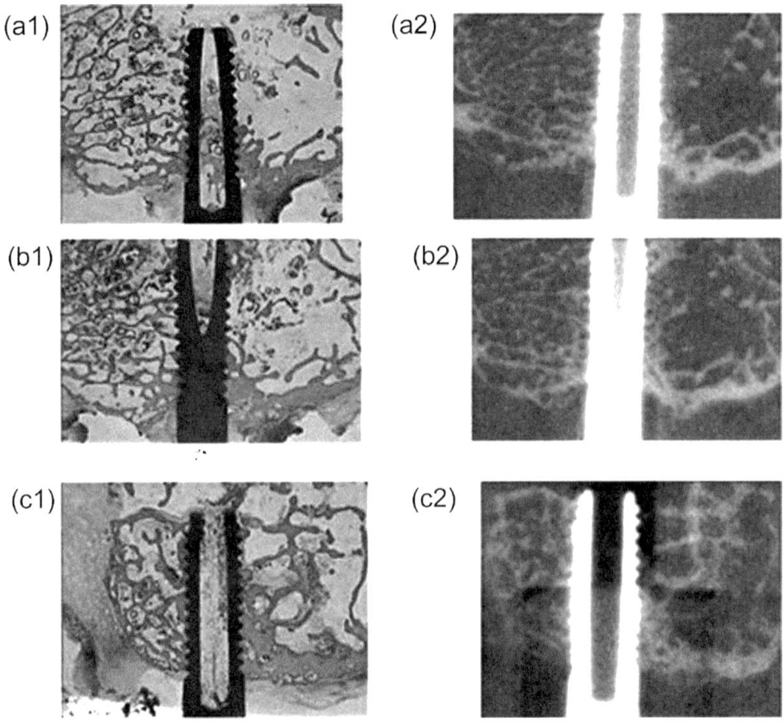

Figure 4.15 The histological sections (1) and their corresponding micro-CT images (2) for a titanium implant placed in the trabecular bone of the condyle of the left tibia of the rabbit. The left side of the implant corresponds to the ventral side of the tibia and the right side of the implant to the dorsal side of thetibia (Stopie et al., 2007).

Table 4.7 Blood biochemical examination of rat before implantation and 15 weeks post-implantation of Mg–Mn–Zn alloy in a bone

Items	Before implantation (n = 2)	15 Weeks post-implantation (n = 2)	Recommended level[13]
BUN (mmol/L)	7.07 ± 0.32	8.47 ± 0.51	5.99–14.99[14]
CREA (μmol/L)	35.6 ± 3.1	39.0 ± 3.1	29.2–53.9[14]
UA (μmol/L)	35 ± 9	50 ± 10	71.3–445.5[14]
K^+ (mmol/L)	6.20 ± 0.99	6.40 ± 1.0	3.8–5.4
Na^+ (mmol/L)	141.5 ± 0.7	145.0 ± 1.4	126–155
Cl^- (mmol/L)	100.0 ± 4.2	101.5 ± 2.1	103.0–115.1
Ca^{2+} (mmol/L)	2.95 ± 0.04	2.93 ± 0.07	3.1–5.2
P (mmol/L)	2.93 ± 0.08	2.99 ± 0.42	1.0–3.55
Mg^{2+} (mmol/L)	1.18 ± 0.05	1.28 ± 0.13	1.32 ± 0.03

Xu et al. (2007).

(Hallab, Jacobs, & Black, 2000; Singh & Dahotre, 2007; Ungethuem & Winkler-Gniewek, 1984), potentially resulting in allergies and in septic and aseptic reactions (Singh & Dahotre, 2007). Titanium and its alloys exhibit excellent biocompatibility and corrosion resistance, as well as favourable mechanical properties (Singh & Dahotre, 2007) (see Table 4.8). Their only notable disadvantages are that their elasticity is higher than that of bone and both their wear resistance and their resistance to shearing forces is lower (Singh & Dahotre, 2007). Due to their high-elasticity module, the above-mentioned metallic implants show substantially higher rigidity than bone, leading to stress shielding and, in turn, to a delay in the healing process, to

Table 4.8 Characteristics of strategic orthopaedic metallic materials

Characteristics	Stainless steels	Cobalt-base alloys	Ti and Ti-base alloys
Designation	ASTM F-138 (316 LDVMO)	ASTM F-75 ASTM F-799 ASTM F-1537 (cast and wrought)	ASTM F-67 (ISO 5832/II) ASTM F-136 (ISO 5832/II) ASTM F-1295 (cast and wrought)
Principal alloying elements (wt. %)	Fe (balance) Cr (17–20) Ni (10–14) Mo (2–4)	Co (balance) Cr (19–30) Mo (0–10) Ni (0–37)	Ti (balance) A1 (6) V (4) Nb (7)
Advantages	Cost, availability, processing	Wear resistance, corrosion resistance, fatigue strength	Biocompatibility corrosion resistance minimum modulus fatigue strength
Disadvantages	Long-term behaviour, high modulus	High modulus	Low wear resistance, low shear resistance
Application	Temporary devices (fracture plates, screws, hip nails) used for THRs stems	Dentistry casting, prostheses stems load-bearing components in TJR (wrought alloys)	In THRs (with modular Co-Cr-Mo or ceramic) femoral heads, long-term permanent devices (nails, pacemakers)

Singh and Dahotre (2007).

development of pseudoarthrosis and to pathological fractures following implant removal (Gogolewski, 2000; Hoffmann, 1995).

Tissue metallosis around the implant was found to be another drawback of nonresorbable metallic implants, potentially leading to hypersensitivity, toxicity, and cancerogenicity (Agins et al., 1988; Mcdonald, Enneking, & Sundaram, 2002; Radhi, Ibrahiem, & Al-Tweigeri, 1998; Ward, Thornbury, Lemons, & Dunham, 1990). The generation of artefacts in CT (Link et al., 2000; Mahnken et al., 2003) and magnetic resonance imaging (MRI) (Disegi & Eschbach, 2000; Pohler, 2000) is also disadvantageous.

4.6.2 Resorbable polymer-based implants

PGAs, PLAs, and their copolymers are among the main substances used for manufacturing osteosynthesis materials (Claes & Ignatius, 1998). Their elasticity model resembles bone properties and thus prevents stress protection (Hofmann, 1995). However, rapid loss of strength and rigidity irrespective of the degradation process presents a problem (Hofmann, 1995). This means that the implants are suitable only for the treatment of non-load-bearing bones (Hofmann, 1995; Rehm et al., 1997; von der Elst et al., 2000). Chemical composition, crystallinity, release of degradation products, implant design, and surface properties are major determinants of biocompatibility (Gogolewski, 2000; Hoffmann, Weller, Helling, Krettek, & Rehm, 1997; Wintermantel, 2002) (see Table 4.9). Several research teams investigating biocompatibility regard foreign-body reactions as having negative effects ranging from silent osteolysis to severe inflammation (Böstman, 1991, 1992; Claes & Ignatius, 1998; Hoffmann et al., 1997; Suganuma and Alexander, 1993; Rehm, Helling, & Claes, 1994). Another disadvantage is that the implants cannot be imaged using CT and MRI (Hofmann, 1995; Rehm et al., 1997).

4.6.3 Suitability of magnesium and its alloys for metallic implants

Magnesium and its alloys have an elasticity module that resembles that of the cortex of the bone and generally possess similar mechanical properties (Kaese, 2002; Staiger et al., 2006; Zhang et al., 2009). A large number of *in vitro* and *in vivo* studies postulate that magnesium ions, magnesium, and degradation products (i.e., magnesium hydroxide) have a positive influence on the bone-remodelling processes and an osteoconductiveeffect (Castellani et al., 2011; Janning et al., 2010; Pietak, Mahoney, Dias, & Staiger, 2008; Revell, Damien, Zhang, Evans, & Howlett, 2004; Rude et al., 2006; Witte et al., 2005; Witte et al., 2007; Yamasaki et al., 2002; Yamasaki et al., 2003; Zreiqat et al., 2002) (see Table 4.10 and Figure 4.16). It has been shown that calcium phosphate coating of implants has a positive effect on bone healing and thus ensures better implant integration (Hayakawa, Yoshinari, Nemoto, Wolke, & Jansen, 2000). The increase in pH value during degradation may also lead to osteoblast stimulation (Kaese, 2002). As soon as the capacity for resorption of the surrounding tissue has been exceeded, gas bubbles form (Witte, Hort, et al., 2008). These are resorbed

Table 4.9 **Degradation rates of various resorbable polymeric implants**

Polymer (implant form)	Retained strength (%/week)	Total strength loss (months)	Complete resorption time (months)
Polydioxanone (sutures)	60/4 (40/6)	−2	6
Poly(glycolide-*co*-trimethylene carbonate) (sutures)	55/4 (14/7)	2.5	6
Polyglycolide (sutures)	30/2	1	4
Poly(glycolide-*co*-lactide)(sutures)	30/3	1	2
Poly(L-lactide) (solid, non-oriented)	40/8	3	1−72
Poly(L-lactide) (solid, oriented)	80/12 (65/25)	1−7	36−72[a]
Poly(L/DL-lactide) 70/30% (solid, nonoriented)	40/12[a]	3	24−36[a]
Poly(L/DL-lactide) 80/20% (solid, nonoriented)	50/12[a]	4	24−36[a]
Poly (L/DL-lactide) 80/20% (porous membranes)	20/12[a]	4	12−18

[a]Values to be proven by further experiments.
Gogolewski (2000).

over a period of several weeks; no negative effect on the surrounding tissue could be demonstrated (Erdmann et al., 2011; Hampp et al., 2012; Kraus et al., 2012; Li et al., 2008; Witte et al., 2005; Zhang, Xu, et al., 2009). Several research teams suggest that compact implantation material may hinder the closure of a borehole in the cortex by preventing an influx of osteoprogenitor cells (Henslee et al., 2011). The rabbit animal model is favoured and recommended for *in vivo* studies (Meyer-Lindenberg et al., 2007; Tsai et al., 2008; Carranza-Bencano et al., 1999; Rudert, 2002; ISO 10993-6: 2007).

It is essential that bone implants are sufficiently stable, especially during the first few weeks. Based on the assumption that fractures of the thigh take around

Table 4.10 Maximum push-out force (F_{max}), ultimate shear strength (τ_u) and energy absorption to failure (EA) for each implantation period and implant type (median [first to third quartiles], Mann–Whitney *U*-test)

Implantation period		Mg-alloy	Ti-alloy	*P*
4 weeks	n	12	13	
	F_{max} (N)	49.35 (37.63–55.53)	23.58 (11.55–30.99)	0.001
	τ_u (N/mm^2)	2.43 (1.80–2.81)	1.12 (0.57–1.50)	0.002
	EA (mJ)	1.07 (0.92–1.40)	0.39 (0.17–0.79)	0.004
12 weeks	n	12	7	
	f_{max} (n)	151.87 (133.30–185.33)	100.83 (80.10–109.59)	0.003
	τ_u (N/mm^2)	6.17 (5.29–7.23)	4.14 (3.25–4.55)	0.002
	EA (mJ)	12.45 (9.08–18.70)	3.24 (2.23–7.41)	0.002
24 weeks	n	8	9	
	F_{max} (N)	185.16 (157.73–221.9S)	44.78 (30.57–90.10)	0.001
	τ_u (N/mm^2)	7.65 (6.61–8.71)	2.14 (1.26–3.53)	0.001
	EA (mJ)	22.64 (11.05–36.86)	0.70 (0.36–1.46)	0.004

Castellani et al. (2011).

12 weeks to heal (Gu et al., 2011), Staiger et al. (2006) advocate that biodegradable implants should have sufficient mechanical properties for at least this length of time. Hutmacher (2000) recommends that osteosynthesis materials used in bone should have biomechanical stability of around 2 months prior to the expected onset of controlled degradation. The degeneration rate of magnesium implants differs depending on their location in the bone; it is greater in the medullary cavity than in the cortical bone (Erdmann et al., 2011; Xu et al., 2007; Zhang, Xu, et al., 2009) (see also Section 4.2). In a study by Xu et al. (2007), the magnesium alloy in rat femurs degraded by 10–17% over the first 9 weeks, with only 50% of the original amount remaining after a total of 18 weeks. However, in the treatment of iatrogenic bone/cartilage defects with magnesium sponges made of the AZ91 alloy, the formation of degradation products prevented the onset of the hoped-for

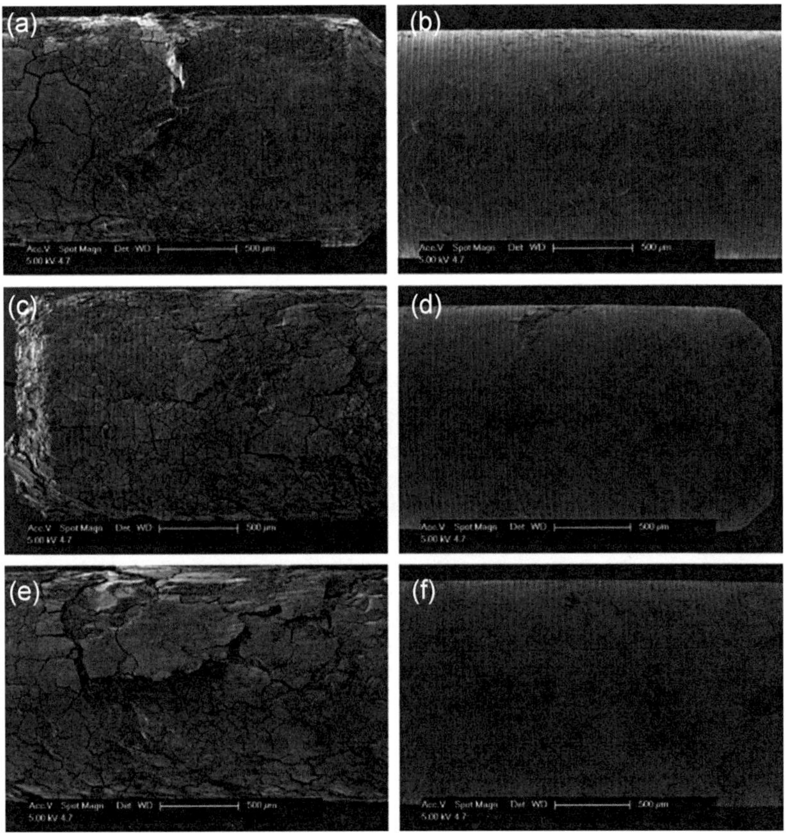

Figure 4.16 Representative scanning electron microscopy images of tested implants. Investigated biodegradable magnesium alloy rods (a, c, and e) and Ti6Al7Nb controls (b, d, and f); 4 (a and b), 12 (c and d) and 24 (e and f) weeks after implantation (Castellani et al., 2011).

osseous and cartilaginous regeneration, and the excessive rate of degradation led to insufficient mechanical stability (Reifenrath et al., 2007; Witte et al., 2007; Witte, Ulrich, Palm, & Willbold, 2007). Numerous studies report that complete degradation of the implant could not be achieved even after implant retention had lasted for 6 months (Li et al., 2008; Witte, Ulrich, Palm, 2007; Witte, Ulrich, Rudert, et al., 2007; Xu et al., 2007a).

4.6.4 Three- and four-point flexural tests

These flexural tests are among the mechanical tests that assess the relative strength of the bone and the implant attached to it (An, Kang, & Friedmann, 1996; Bramer et al., 1998; Buijs, Van, Stegenga, Bos, & Verkerke, 2007; Li, Forberg, & Hermansson, 1991). The tests, which do not require the specimens to be specially prepared, are a measure of a material's ductility. They involve the specimen being placed on the lateral support and a third, central, opposing force is applied until fracture criteria are observed (An & Draughn, 2000). The fact that, in contrast to the four-point flexural test (in which the entire length lies between supports), only one point is tested (An & Draughn, 2000), proves disadvantageous.

4.7 Testing of magnesium alloy in blood vessels – special considerations

4.7.1 Coronary stents

Various materials are currently used in vascular stenosis therapy. The most recent classification is based on the distinction between drug-eluting stents (DES) and bare-metal stents (Bennett, 2003; Costa & Simon, 2005). With a frequency of 10–30%, restenosis is one of the most common complications following stent implantation (Hoffmann & Mintz, 2000; Meads et al., 2000; van Domburg et al., 1999). DES release medication intended to prevent restenosis. Randomised double-blind studies have shown that restenosis has a lower incidence rate with DES than with bare-metal stents (Morice et al., 2002; Muni et al., 2005). A meta-analysis of several DES studies has, however, demonstrated that thrombosis occurred more often as a late complication in cases with DES than with bare-metal stents (Schomig et al., 2007). The fundamental problem with both types of stent is the fact that, in many cases, the support function is lost after a period of several weeks to months when the stent no longer serves any physiological purpose. Biodegradable metal stents or bioresorbable polymers may offer a means of preventing late complications after implantation (Erne, Schier, & Resink, 2006). Another disadvantage of nonresorbable stents is the lack of adaptability in terms of size, especially in children (Hehrlein, 2007; Peuster, Beerbaum, Bach, & Hauser, 2006). Lifelong anticoagulation therapy is generally recommended because of the risk of thrombus formation (Ong et al., 2005; Waksman, 2006).

4.7.2 Magnesium stents

Experiments have established that magnesium stents facilitate positive vessel remodelling (Hermawan, Dube, & Mantovani, 2010) and degrade completely, thus enabling the vessel to provide its original vasomotor function (Ghimire et al., 2009). Further antithrombotic and antiarrhythmic properties are described in the literature (Adams and Mitchell, 1979; Di Mario et al., 2004; Pseuter et al., 2006). One of the first studies on stents made of magnesium alloy (Mg 97%, aluminium 25%, rare-earth metals 1%) was carried out on coronary arteries of domestic pigs by Heublein et al. (2003), who demonstrated positive remodelling with complete stent degradation after 89 days. Biodegradable magnesium stents are currently being tested *in vivo* in the treatment of vascular diseases. A coronary stent made of the magnesium alloy WE43, manufactured by BIOTRONIK SE & Co. KG, is currently available on the market. Di Mario et al. (2004) reported on a clinical study in which 18 of 20 patients exhibited normal blood flow, but two showed 30–40% stenosis. No allergic or toxic reactions could be observed. However, the authors stated that the highly limited X-ray density was problematic, making it very difficult to image the stent.

Other research teams have been able to indicate that the vessel lumen with the stent could be displayed without artefacts on angiography, CT, and MRI (Eggebrecht et al., 2005; Lind, Eggebrecht, & Erbel, 2005). However, stent imaging is easily possible using ultrasound and OCT (Erbel, di Mario, et al., 2007; Pinto Slottow, Pakala, & Waksman, 2008) (see Figure 4.17). There are several case studies in which various groups of researchers describe the implantation of an AMS35 magnesium stent manufactured by BIOTRONIK SE & Co. KG into the pulmonary artery or aorta of newborn children (Schranz, Zartner, Michel-Behnke, & Akintürk, 2006; Zartner, Cesnjevar, Singer, & Weyand, 2005). According to Schranz et al. (2006), a resorbable stent is especially advantageous for growing children. Although no complications were found in a multicenter PROGRESS-AMS study on resorbable magnesium stents, restenosis occurred in 17% of the cases after 4 months due to neointima formation and negative remodelling (Erbel, Bose, et al., 2007).

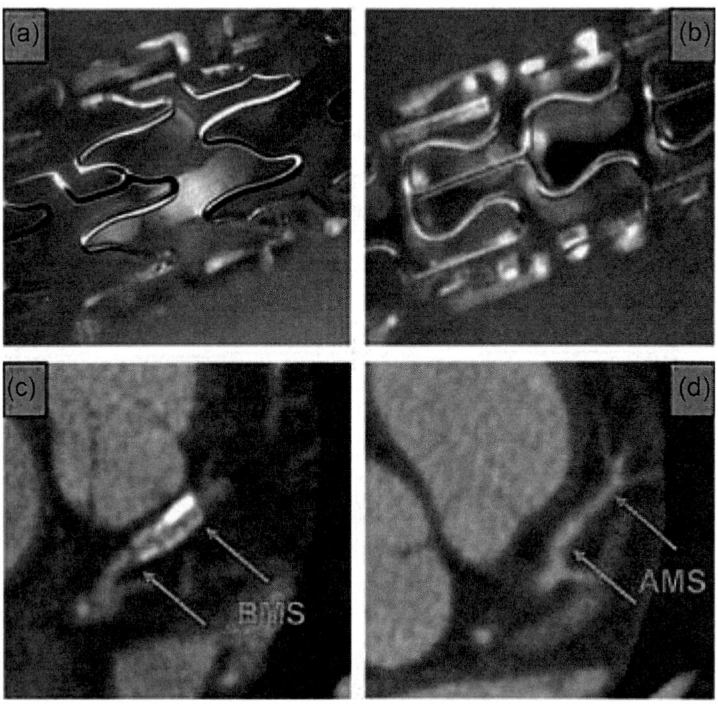

Figure 4.17 Comparison of 16-slice computed tomography of a bare-metal stent (a, c) and a magnesium stent (b, d) in segment six of the left coronary artery. The magnesium stent (Biotronik, Berlin, Germany) is not visible allowing a free imaging of the artery lumen, whereas the visualisation of the coronary artery lumen is impaired by the bare-metal stent.
Modified according to Erbel et al. (2007).

4.7.3 Intravascular ultrasound

IVUS does not differ technically from conventional ultrasound, making use of the latter over a frequency range of 20–50 MHz to generate a grey-scale image (Schoenhagen & Nissen, 2002). A detailed image of the wall structure and vessel geometry can be produced in this way (Regar et al., 1999, 2000). Magnesium stents can be directly visualised and quantified by means of IVUS (Di Mario et al., 2004; Erbel, Bose, et al., 2007; Erne et al., 2006; Heublein et al., 2003).

4.7.4 Coronary angiography

Coronary angiography, although a standard imaging procedure for the diagnostics of coronary vessels (Nemirovski, 2003), only allows the evaluation of the vessel lumen and does not provide unambiguous information on the composition of the vessel walls (Levin & Fallon, 1982). Quantitative coronary angiography allows quantification of the lumen diameter and can be used to measure the wall, but direct imaging of the magnesium stent is not possible (Erbel, Bose, et al., 2007).

4.7.5 Optical coherence tomography

OCT is a cross-sectional imaging technique similar to ultrasound, the difference being that, instead of sound waves, light is used to create the image (Brezinski et al., 1996; Hee et al., 1995; Huang et al., 1991). This method allows a resolution of 2–30 μm (Pasterkamp, Falk, Woutman, & Borst, 2000). For technical reasons, vessel imaging is not possible through blood (Brezinski, 2001/7), so that an occlusion test must be performed prior to the examination, the latter being limited to 30 s. OCT can reliably distinguish between the magnesium stent, different grades of plaque, and both necrotic and lipid-rich tissue (Yabushita et al., 2002).

4.8 Future trends

Irrespective of the specific organ, the need for a biodegradable implant is not yet fully met. Magnesium is a material that shows excellent biocompatibility in various *in vitro* and *in vivo* studies, and hence basic suitability for the development of biodegradable implants. It should, however, be mentioned that several research teams have reported on the possible cytotoxicity of magnesium and its alloys (Geng, Tan, Jin, Yang, & Yang, 2009; Gu et al., 2011; Serre, Papillard, Chavassieux, Voegel, & Boivin, 1998; Wong et al., 2010). With a magnesium implant, the particular challenge, provided the mechanical properties are satisfied, is the controlled degradation and resorption/metabolisation of degradation products of corrosion. To optimise these processes, more must first be discovered about what takes place at the interface between the implant and the tissue *in vivo*, these processes being largely unknown at present. The lack of applicability of *in vitro* findings to *in vivo* tests is to be regarded as a major disadvantage and makes apparent the need to establish a test model outside the organism that more fully reflects the situation within the living organism.

The development of new alloys offers scope for optimising mechanical properties and degradation kinetics. It should be noted, however, that magnesium alloys with combinations of rare-earth metals, which are used increasingly frequently – with the precise composition of these mixtures not defined – may lead to the production of variable batches (Freyerabend et al., 2010; Witte et al., 2008). The differing properties of the rare-earth metals may ultimately influence the biocompatibility, biodegradation, and mechanical properties of the alloy as a whole. The long-term effects on health of the individual alloy components remain to be tested.

With a view to optimising the properties of magnesium alloys, a number of research teams are investigating functionalised surface coatings such as hydroxyapatite, calcium phosphate, and polymers. What is being evaluated are the corrosion-delaying properties of these materials in contact with magnesium and the possibility of establishing local drug delivery systems (Chen et al., 2011; Shadanbaz & Dias, 2012; Zhang, Zhang, & Wei, 2009). Coating, especially conversion coating, is a further promising option for delaying degradation (Drynda et al., 2009; Thomann et al., 2010; Zhang et al., 2010).

An important point involves the differences between the way magnesium and its alloys behave in different organs. The development of a universal magnesium alloy covering all applications, therefore, appears improbable. A significant aspect with regard to osteosynthesis material made of magnesium is cavity formation in the cortex, which may in principle lead to a reduction in mechanical bone strength (Kaartinen, Paavolainen, Holmstroem, & Slaetis, 1985) and which a comparative study with titanium and polyactide revealed to be more pronounced in conjunction with magnesium (Danckwardt-Lillienstroem, 1969; Husby, Gjerdet, Erichsen, Rykkje, & Molster, 1989; Kaartinen et al., 1985). Future studies are required to clarify this parameter.

As present, one specific focus in the development of biodegradable magnesium implants is on coronary stents and osteosynthesis materials. Another potential application is that of hollow structures such as the Eustachian tube or the system of paranasal sinuses. As with coronary stents, the embedded magnesium is not surrounded by tissue on all sides, its inner surface being in contact with air or fluid. In cases of chronic sinusitis and recurrent scarring of the ventilation passages (especially those leading into the frontal sinus), silicone stents have been applied both with (Beule et al., 2008, 2009; Herrmann et al., 2004) or without drug delivery systems (Freeman & Blom, 2000), although their success is disputed (Hosemann, Schindler, Wiegrebe, & Gopferich, 2003). In the literature, use of permanent stents has been associated with complications such as dislocation, unpleasant odour, and new trauma following explantation (Perloff & Palmer, 2004). Temporary stents, which degrade over time and progressively enable the normal mucosa to regenerate, may constitute a long-term solution to this problem.

It appears desirable that the degradation of magnesium implants be deliberately timed. Rapid degradation of magnesium results in considerable formation of gas, which may in turn lead to complications.

4.9 Further information and advice

The websites selected provide an overview of institutes and research associations involved with the development of bioresorbable magnesium-based implants.

1. EU Project; "Tailored Biodegradable Magnesium Implant Materials"(MagnIM) coordinated by the Helmholtz-Zentrum Geesthacht (HZG). URL: http://www.hzg.de/public_relations/press_releases/012694/index_0012694.html.en?chunk=7.
2. Collaborative Research Centre 599 "Sustainable bioresorbable and permanent implants of metallic and ceramic materials" funded by the German Research Foundation (DFG). URL: http://www.sfb599.de/.

Acknowledgements

My special thanks apply to Thomas Lenarz, M.D., Ph.D., for his excellent support.

References

Acarturk, O., Lehmicke, M., Aberman, H., Toms, D., Hollinger, J. O., & Fulmer, M. (2008). Bone healing response to an injectable calcium phosphate cement with enhanced radiopacity. *J Biomed Mater Res B Appl Biomater, 86*, 56–62.

Adams, J. H., & Mitchell, J. R. (1979). The effect of agents which modify platelet behaviour and of magnesium ions on thrombus formation in vivo. *Thrombosis and Haemostasis, 42*, 603–610.

Agins, H. J., Alcock, N. W., Bansal, M., Salvati, E. A., Wilson, P. R., Pellicci, P. M., et al. (1988). Metallic wear in failed titanium-alloy total hip replacements, a histological and quantitative analysis. *Journal of Bone and Joint Surgery, 70*, 347–356.

An, Y. H., & Draughn, R. A. (2000). *Mechanical testing of bone and the bone-implant interface*. Boca Raton USA: CRC Press.

An, Y. H. (2003). *Handbook of histology methods for bone and cartilage* (1. Aufl). Totowa, NJ: Verlag Humana Press.

An, Y. H.., Kang, Q., & Friedmann, R. J. (1996). Mechanical symmetry of rabbit bones studied by bending and indentation testing. *American Journal of Veterinary Research, 12*, 1786–1789.

Anderson, J. M., Rodriguez, A., & Chang, D. T. (2007). Foreign body reaction to biomaterials. *Seminars in Immunology, 20*, 86–100.

Atsunori, N. (2011). Hydrogen and medical application. In T. Yoshikawa, & Y. Naito (Eds.), *Gas biology research in clinical practice* (pp. 91–99). Basel: Karger.

Bach, F., Schaper, M., & Jaschik, C. (2003). Influence of lithium on hcp magnesium alloys. *Materials Science Forum, 419–422*, 1037–1040.

Ballanti, P., Minisola, S., Pacitti, M. T., Scarnecchia, L., Rosso, R., Mazzuoli, G. F., et al. (1997). Tartrate-resistant acid phosphate activity as osteoclastic marker: sensitivity of cytochemical assessment and serum assay in comparison with standardised osteoclast histomorphometry. *Osteoporosis International, 7*, 39–43.

Barnett, M. R. (2007). Twinning and the ductility of magnesium alloys: part II. "Contraction" twins. *Materials Science and Engineering: A, 464*, 8–16.

Bennett, M. R. (2003). In-stent stenosis: pathology and implications for the development of drug eluting stents. *Heart, 89*, 218–224.

Benveniste, H., & Huttemeier, P. C. (1990). Microdialysis—theory and application. *Progress in Neurobiology, 35*, 195–215.

van den Berg, F. (2005). *Angewandte Physiologie* (1st ed.). Stuttgart: Thieme.

Bergsma, E. J., Rozema, F. R., Bos, R. R., & De, B. C. (1993). Foreign body reactions to resorbable poly(L-lactide) bone plates and screws used for the fixation of unstable zygomatic fractures. *Journal of Oral and Maxillofacial Surgery, 51*, 666–670.

Bernhardt, R., Scharnweber, D., Müller, B., Thurner, P., Schliephake, H., Wyss, P., et al. (2004). Comparison of microfocus- and synchrotron X-ray tomography for the analysis of osteointegration around Ti6Al4V implants. *European Cells & Materials, 7*, 42–51.

Bethmann, W., & Knofler, W. (1987). Tissue reactions to implant materials. *Zahnärztl Prax, 38*, 162–168.

Beule, A. G., Scharf, C., Biebler, K. E., Gopferich, A., Steinmeier, E., Wolf, E., et al. (2008). Effects of topically applied dexamethasone on mucosal wound healing using a drug-releasing stent. *Laryngoscope, 118*, 2073–2077.

Beule, A. G., Steinmeier, E., Kaftan, H., Biebler, K. E., Gopferich, A., Wolf, E., et al. (2009). Effects of a dexamethasone-releasing stent on osteoneogenesis in a rabbit model. *American Journal of Rhinology & Allergy, 23*, 433–436.

Beyenbach, K. W. (1990). Transport of magnesium across biological membranes. *Magnesium and Trace Elements, 9*, 233–254.

Bianco, P., Ballanti, P., & Bonucci, E. (1988). Tartrate-resistant acid phosphatase activity in rat osteoblasts and osteocytes. *Calcified Tissue International, 43*, 167–171.

Bichet, D. G. (2006). Lithium, cyclic AMP signaling, a-kinase anchoring proteins, and aquaporin-2. *Journal of the American Society of Nephrology, 17*, 920–922.

Bohlen, J., Nürnberg, M. R., Senn, J. W., Letzig, D., & Agnew, S. R. (2007). The texture and anisotropy of magnesium–zinc–rare earth alloy sheets. *Acta Materialia, 55*, 2101–2112.

Böstman, O. M. (1991). Osteolytic changes accompanying degradation of absorbable fracture fixation implants. *Journal of Oral and Maxillofacial Surgery, 4*, 679–682.

Böstman, O. M. (1992). Intense granulomatous inflammatory lesions associated with absorbable internal fixation devices made of polyglycolide in ankle fractures. *Clinical Orthopaedics and Related Research, 278*, 193–199.

Bramer, J. A., Barentsen, R. H., Vd, E. M., De, L. S., Patka, P., & Haarman, H. J. (1998). Representative assessment of long bone shaft biomechanical properties: an optimized testing method. *Journal of Biomechanics, 31*, 741–745.

Brezinski, M., Saunders, K., Jesser, C., Li, X., & Fujimoto, J. (2001). Index matching to improve optical coherence tomography imaging through blood. *Circulation, 103*, 1999–2003.

Brezinski, M. E., Tearney, G. J., Bouma, B. E., Boppart, S. A., Hee, M. R., Swanson, E. A., et al. (1996). Imaging of coronary artery microstructure (in vitro) with optical coherence tomography. *American Journal of Cardiology, 77*, 92–93.

Brouwers, J. E. M., van Rietbergen, B., & Huiskes, R. (2007). No effects of in vivo micro-CT radation on proximal tibia in Wistar rats detected after eight weekly scans. *Journal for Ortho-paedic Research, 25*, 1325–1332.

Buchholz, B. M., Kaczorowski, D. J., Sugimoto, R., Yang, R., Wang, Y., Billiar, T. R., et al. (2008). Hydrogen inhalation ameliorates oxidative stress in transplantation induced intestinal graft injury. *American Journal of Transplantation, 8*, 2015–2024.

Buijs, G. J., Van, D. B., Stegenga, B., Bos, R. R., & Verkerke, G. J. (2007). Mechanical strength and stiffness of biodegradable and titanium osteofixation systems. *Journal of Oral and Maxillofacial Surgery, 11*, 2148–2158.

Butz, F., Ogawa, T., Chang, T. L., & Nishimura, I. (2006). Three-dimensional bone-implant integration profiling using micro-computed tomography. *The International Journal of Oral & Maxillofacial Implants, 21*, 687–695.

Carranza-Bencano, A., Perez-Tinao, M., Ballesteros-Vazquez, P., Armas-Padron, J. R., Hevia-Alonso, A., & Martos, C. F. (1999). Comparative study of the reconstruction of articular cartilage defects with free costal perichondrial grafts and free tibial periosteal grafts: an experimental study on rabbits. *Calcified Tissue International, 65*, 402–407.

Castellani, C., Lindtner, R. A., Hausbrandt, P., Tschegg, E., Stanzl-Tschegg, S. E., Zanoni, G., et al. (2011). Bone-implant interface strength and osseointegration: biodegradable magnesium alloy versus standard titanium control. *Acta Biomaterialia, 7*, 432–440.

Chang, T.-C., Wang, J.-Y. O. C.-M., & Lee, S. (2003). Grain refining of magnesium alloy AZ31 by rolling. *Proceedings of the 6th Asia Pacific Conference on Materials Processing, 140*, 588–591.

Chen, B., Lin, D.-L., Jin, L., Zeng, X.-Q., & Lu, C. (2008). Equal-channel angular pressing of magnesium alloy AZ91 and its effects on microstructure and mechanical properties. *Materials Science and Engineering A, 483–484*, 113–116.

Chen, Y., Song, Y., Zhang, S., Li, J., Zhao, C., & Zhang, X. (2011). Interaction between a high purity magnesium surface and PCL and PLA coatings during dynamic degradation. *Biomedical Materials, 6*, 25005.

Chiu, K., Wong, M., Cheng, F., & Man, H. (2007). Characterization and corrosion studies of fluoride conversion coating on degradable Mg implants. *Surface and Coatings Technology, 202*, 590–598.

Civitelli, R., & Ziambaras, K. (2011). Calcium and phosphate homeostasis: concerted interplay of new regulators. *Journal of Endocrinological Investigation, 34*, 3–7.

Claes, L., & Ignatius, A. (1998). *Hefte zur Zeitschrift"Der Unfallchirurg"*. Berlin: Springer.

Costa, M. A., & Simon, D. I. (2005). Molecular basis of restenosis and drug-eluting stents. *Circulation, 111*, 2257–2273.

Danckwardt-Lilliestroem, G. (1969). Reaming of the medullary cavity and its effect on diaphyseal bone. A fluorochromic, microangiographic and histologic study on the rabbit tibia and dog femur. *Acta Orthopaedica Scandinavica Supplementum, 128*, 1–153.

Del Valle, J. A., Carreño, F., & Ruano, O. A. (2006). Influence of texture and grain size on work hardening and ductility in magnesium based alloys processed by ECAP and rolling. *Acta Materialia, 54*, 4247–4259.

Disegi, J. A. (2000). Titanium alloys for fracture fixation implants. *Injury, S31*, 14–17.

Disegi, J. A., & Eschbach, L. (2000). Stainless steel in bone surgery. *Injury, S31*, 2–6.

Doernberg, M.-C. von, Rechenberg, B. von, Bohner, M., Grünenfelder, S., Van, L. H., Müller, R., et al. (2006). In vivo behaviour of calcium phosphate scaffolds with four different pore sizes. *Biomaterials, 27*, 5186–5198.

van Domburg, R. T., Foley, D. P., de Jaegere, P. P., de Feyter, P., van den Brand, M., van der Giessen, W., et al. (1999). Long term outcome after coronary stent implantation: a 10 year single centre experience of 1000 patients. *Heart, 82*(Suppl. 2), II27–II34.

Donath, K. (1988). Die Trenn-Dünnschliff-Technik zur Herstellung histologischer Präparate von nicht schneidbaren Geweben und Materialien—Apparate- und Methodenbeschreibung. *Der Präparator, 34*, 197–206.

Drynda, A., Hassel, T., Hoehn, R., Perz, A., Bach, F.-W., & Peuster, M. (2009). Development and biocompatibility of a novel corrodible fluoride-coated magnesium-calcium alloy with improved degradation kinetics and adequate mechanical properties for cardiovascular applications. *Journal of Biomedical Materials Research, 93a*, 763–775.

Duygulu, O., Kaya, R. A., Oktay, G., & Kaya, A. A. (2007). Investigation on the potential of magnesium alloy AZ31 as a bone implant. *Materials Science Forum, 546–549*, 421–424.

Dziuba, D., Meyer-Lindenberg, A., Seitz, J. M., Waizy, H., Angrisani, N., & Reifenrath, J. (2013). Long-term in vivo degradation behaviour and biocompatibility of the magnesium alloy ZEK100 for use as a biodegradable bone implant. *Acta Biomaterialia, 9*, 8548–8560.

Eggebrecht, H., Rodermann, J., Hunold, P., Schmermund, A., Böse, D., Haude, M., et al. (2005). Images in cardiovascular medicine. Novel magnetic resonance-compatible coronary stent: the absorbable magnesium alloy stent. *Circulation, 112*, 303–304.

El-Rahman, S. S. (2003). Neuropathology of aluminum toxicity in rats (glutamate and GABA impairment). *Pharmacological Research, 47*, 189–194.

Epple, M. (2003). *Biomaterialien und Biomineralisation* (1. Aufl). Stuttgart: Teubner.

Erbel, R., Bose, D., Haude, M., Kordish, I., Churzidze, S., Malyar, N., et al. (2007). Absorbable coronary stents. New promising technology. *Herz, 32*, 308–319.

Erbel, R., Di Mario, C., Bartunek, J., Bonnier, J., Bruyne de, B., Eberli, F. R., et al. (2007). Temporary scaffolding of coronary arteries with bioabsorbable magnesium stents: a prospective, non-randomised multicentre trial. *The Lancet, 369*, 1869–1875.

Erdmann, N., Angrisani, N., Reifenrath, J., Lucas, A., Thorey, F., Bormann, D., et al. (2011). Biomechanical testing and degradation analysis of MgCa0.8 alloy screws: a comparative in vivo study in rabbits. *Acta Biomaterials, 7*, 1421–1428.

Erne, P., Schier, M., & Resink, T. J. (2006). The road to bioabsorbable stents: reaching clinical reality? *Cardiovascular and Interventional Radiology, 29*, 11–16.

Exner, E., Müller, C., & Schmidt, H. (2004). Titan in der Gelenk- und Zahnprothetik: Verschleiß und Ermüdung als lebensdauerbegrenzende Faktoren. In T. Rossmann, & C. Tropea (Eds.), *Bionik: Aktuelle Forschungsergebnisse in Natur-, Ingenieur- und Geisteswissenschaft* (pp. 351–372). Berlin: Springer.

Feyerabend, F., Fischer, J., Holtz, J., Witte, F., Willumeit, R., Drücker, H., et al. (2010). Evaluation of short-term effects of rare earth and other elements used in magnesium alloys on primary cells and cell lines. *Acta Biomaterialia, 6*, 1834–1842.

Franke, H. (1934). Magnesium und Kohlehydratstoffwechsel, Naunyn-Schmiedeberg's. *Arch-Pharmacol, 174*, 727–741.

Freeman, S. B., & Blom, E. D. (2000). Frontal sinus stents. *Laryngoscope, 110*, 1179–1182.

Freeman, C. O., & Brook, I. M. (2006). Bone response to a titanium aluminium nitride coating on metallic implants. *Journal of Materials Science Materials in Medicine, 17*, 465–470.

Gabet, Y., Müller, R., Levy, J., Dimarchi, R., Chorev, M., Bab, I., et al. (2006). Parathyroid hormone 1-34 enhances titanium implant anchorage in low-density trabecular bone: a correlative micro-computed tomographic and biomechanical analysis. *Bone, 39*, 276–282.

Genesca, J., Betancourt, L., & Rodriguez, C. (1996). Electrochemical behaviour of a magnesium galvanic anode under ASTM test method G 97-89 conditions. *Corrosion Science, 52*, 502–507.

Geng, F., Tan, L. L., Jin, X. X., Yang, J. Y., & Yang, K. (2009). The preparation, cytocompatibility, and in vitro biodegradation study of pure beta-TCP on magnesium. *Journal of Materials Science Materials in Medicine, 20*, 1149–1157.

Ghali, E., Dietzel, W., & Kainer, K.-U. (2004). General and localized corrosion of magnesium alloys: acritical review. *Journal of Materials Engineering and Performance, 13*, 7–23.

Ghimire, G., Spiro, J., Kharbanda, R., Roughton, M., Barlis, P., Mason, M., et al. (2009). Initial evidence for the return of coronary vasoreactivity following the absorption of bio-absorbable magnesium alloy coronary stents. *Euro Intervention, 4*, 481–484.

Giles, J. J., & Bannigan, J. G. (2006). Teratogenic and developmental effects of lithium. *Current Pharmaceutical Design, 12*, 1531–1541.

Gogolewski, S. (2000). Bioresorbable polymers in trauma and bone surgery. *Injury, 31*, 28–32.

Gradinger, R., & Gollwitzer, H. (2006). *Ossäre integration*. Berlin, Heidelberg: Springer Medizin.

Gray, J., & Luan, B. (2002). Protective coatings on magnesium and its alloys — a critical review. *Journal of Alloys and Compounds, 336*, 88–113.

Grubb, B. R., Chadburn, J. L., & Boucher, R. C. (2002). In vivo microdialysis for determination of nasal liquid ion composition. *American Journal of Physiology: Cell Physiology, 282*, C1423–C1431.

Gu, X. N., Li, N., Zhou, W. R., Zheng, Y. F., Zhao, X., Cai, Q. Z., et al. (2011). Corrosion resistance and surface biocompatibility of a microarc oxidation coating on a Mg–Ca alloy. *Acta Biomaterialia, 7*, 1880–1889.

Gu, X.-N., & Zheng, Y.-F. (2010). A review on magnesium alloys as biodegradable materials. *Frontiers of Materials Science in China, 4*, 111–115.

Gu, X., Zheng, Y., Cheng, Y., Zhong, S., & Xi, T. (2009). In vitro corrosion and biocompatibility of binary magnesium alloys. *Biomaterials, 30*, 484–498.

Gu, X. N., Zhou, W. R., Zheng, Y. F., Cheng, Y., Wei, S. C., Zhong, S. P., et al. (2010). Corrosion fatigue behaviours of two biomedical Mg alloys—AZ91D and WE43—in simulated body fluid. *Acta Biomaterialia, 6*, 4605–4613.

Gunther, T., Vormann, J., & Forster, R. (1984). Regulation of intracellular magnesium by $Mg2^{+}$ efflux. *Biochemical and Biophysical Research Communications, 119*, 124–131.

Hallab, N., Jacobs, J. J., & Black, J. (2000). Hypersensitivity to metallic biomaterials: a review of leukocyte migration inhibition assays. *Biomaterials, 21*, 1301–1314.

Hampp, C., Ullmann, B., Reifenrath, J., Angrisani, N., Dziuba, D., Bormann, D., et al. (2012). Research on the biocompatibility of the new magnesium alloy LANd442-an in vivo study in the rabbit tibia over 26 weeks. *Advanced Engineering Materials, 14*, B28–B37.

Hänzi, A. C., Gunde, P., Schinhammer, M., & Uggowitzer, P. J. (2009). On the biodegradation performance of an Mg–Y–RE alloy with various surface conditions in simulated body fluid. *Acta Biomaterialia, 5*, 162–171.

Hayakawa, T., Yoshinari, M., Nemoto, K., Wolke, J. G., & Jansen, J. A. (2000). Effect of surface roughness and calcium phosphate coating on the implant/bone response. *Clinical Oral Implants Research, 11*, 296–304.

Hee, M. R., Izatt, J. A., Swanson, E. A., Huang, D., Schuman, J. S., Lin, C. P., et al. (1995). Optical coherence tomography of the human retina. *Archives of Ophthalmology, 113*, 325–332.

Hehrlein, C. (2007). Promises of biodegradable stents. *Catheterization and Cardiovascular Interventions, 69*, 739.

Heimann, R. B., Itiravivong, P., & Promasa, A. (2004). In vivo- Untersuchungen zur Osseointegration von Hydroxylapatitbeschichteten Ti6Al4V Implantaten mit und ohne bioinerter Titanoxid-Haftvermittlerschicht. *Biomaterialien, 5*, 38–43.

Henslee, A. M., Spicer, P. P., Yoon, D. M., Nair, M. B., Meretoja, V. V., Witherel, K. E., et al. (2011). Biodegradable composite scaffolds incorporating an intramedullary rod and delivering bone morphogenetic protein-2 for stabilization and bone regeneration in segmental long bone defects. *Acta Biomaterialia, 7*, 3627–3637.

Hermawan, H., Dube, D., & Mantovani, D. (2010). Developments in metallic biodegradable stents. *Acta Biomaterialia, 6*, 1693–1697.

Herrmann, B. W., Citardi, M. J., Vogler, G., Gardner, L., Smith, G., Javer, A. R., et al. (2004). A preliminary report on the effects of paclitaxel-impregnated stents on sheep nasal mucosa. *American Journal of Rhinology, 18*, 119–124.

Heublein, B., Rohde, R., Kaese, V., Niemeyer, M., Hartung, W., & Haverich, A. (2003). Biocorrosion of magnesium alloys: a new principle in cardiovascular implant technology? *Heart, 89*, 651–656.

Hirano, S., & Suzuki, K. T. (1996). Exposure, metabolism, and toxicity of rare earths and related compounds. *Environmental Health Perspectives, 104*, 85–95.

Hofmann, G. O. (1995). Biodegradable implants in traumatology: a review on the state-of-the-art. *Archives of Orthopaedic and Trauma Surgery, 114*, 123–132.

Hoffmann, R., & Mintz, G. S. (2000). Coronary in-stent restenosis - predictors, treatment and prevention. *European Heart Journal, 21*, 1739–1749.

Hoffmann, R., Weller, A., Helling, H. J., Krettek, C., & Rehm, K. E. (1997). Local foreign body reactions to biodegradable implants. A classification. *Unfallchirurg, 100*, 658–666.

von der Höh, N., Bormann, D., Lucas, A., Denkena, B., Hackenbroich, C., & Meyer-Lindenberg, A. (2009). Influence of different surface machining treatments of magnesium-based resorbable implants on the degradation behaviour in rabbits. *Advanced Engineering Materials, 11*, B47–B54.

von der Höh, N., Krause, A., Hackenbroich, C., Bormann, D., Lucas, A., & Meyer-Lindenberg, A. (2006). Influence of different surface machining treatments of resorbable implants made from different magnesium-calcium alloys on their degradation—a pilot study in rabbit models. *Dtsch Tierärztl Wochenschrift, 12*, 439–446.

von der Höh, N., von Rechenberg, B., Bormann, D., Lucas, A., & Meyer-Linderberg, A. (2009). Influence of different surface machining treatments of resorbable magnesium alloy implants on degradation—EDX-analysis and histology results. *Materialwissenchaft und Werkstofftechnik, 40*, 88–93.

Holdsworth, D. W., & Thornton, M. M. (2002). Micro-CT in small animal and specimen imaging. *Trends in Biotechnology, 20*, 34–39.

Horal, M., Ungerstedt, U., Persson, B., Westgren, M., & Marcus, C. (1995). Metabolic adaptation in IUGR neonates determined with microdialysis—a pilot study. *Early Human Development, 42*, 1–14.

Hosemann, W., Schindler, E., Wiegrebe, E., & Gopferich, A. (2003). Innovative frontal sinus stent acting as a local drug-releasing system. *European Archives of Otorhinolaryngology, 260*, 131–134.

Huang, C.-S., Kawamura, T., Toyoda, Y., & Nakao, A. (2010). Recent advances in hydrogen research as a therapeutic medical gas: free radical research. *Free Radical Research, 44*, 971–982.

Huang, J., Ren, Y., Jiang, Y., Zhang, B., & Yang, K. (2007). In vivo study of degradable magnesium and magnesium alloy as bone implant. *Frontiers of Materials Science in China, 1*, 405–409.

Huang, D., Swanson, E. A., Lin, C. P., Schuman, J. S., Stinson, W. G., Chang, W., et al. (1991). Optical coherence tomography. *Science, 254*, 1178–1181.

Huffer, W. E., Ruegg, P., Zhu, J. M., & Lepoff, R. B. (1994). Semiautomated methods for cancellous bone histomorphometry using a general-purpose video image analysis system. *Journal of Microscopy, 173*, 53–66.

Husby, O. S., Gjerdet, N. R., Erichsen, E. S., Rykkje, J. M., & Molster, A. O. (1989). Porosity of rat femora following intramedullary reaming and nailing. *Clinical Orthopaedics and Related Research, 246*, 305–312.

Hutmacher, D. W. (2000). Scaffolds in tissue engineering bone and cartilage. *Biomaterials, 21*, 2529–2543.

Iannello, S., & Belfiore, F. (2001). Hypomagnesemia. A review of pathophysiological, clinical and therapeutical aspects. *Panminerva Medica, 43*, 177–209.

Inagaki, K., & Haraguchi, H. (2000). Determination of rare earth elements in human blood serum by inductively coupled plasma mass spectrometry after chelating resin preconcentration. *Analyst, 125*, 191–196.

ISO10993–5:2009. (2009). *Biological evaluation of medical devices—Part 5: tests for in vitro cytotoxicity*. ANSI/AAMI. http://www.iso.org/iso/catalogue_detail.htm?csnumber=36406.

ISO10993–6:2009. (2009). *Biological evaluation of medical devices—Part 6: tests local effects after implantation*. ANSI/AAMI. http://www.iso.org/iso/catalogue_detail.htm?csnumber=36406.

Iwamoto, J., Takeda, T., Sato, Y., & Yeh, J. K. (2004). Response of cortical and cancellous bones to mild calcium deficiency in young growing female rats: a bone histomorphometry study. *Experimental Animals, 53*, 347–354.

Jacobson, I., Sandberg, M., & Hamberger, A. (1985). Mass transfer in brain dialysis devices - a new method for the estimation of extracellular amino acids concentration. *Journal of Neuroscience Methods, 15*, 263–268.

Jain, R., Podworny, N., Hearn, T., Anderson, G. I., & Schemitsch, E. H. (1997). Effect of stainless steel and titanium low-contact dynamic compression plate application on the vascularity and mechanical properties of cortical bone after fracture. *Journal of Orthopaedic Trauma, 11*, 490–495.

Jäkel, O., & Reiss, P. (2007). The influence of metal artefacts on the range of ion beams. *Physics in Medicine and Biology, 52*, 635–644.

Janning, C., Willbold, E., Vogt, C., Nellesen, J., Meyer-Lindenberg, A., Windhagen, H., et al. (2010). Magnesium hydroxide temporarily enhancing osteoblast activity and decreasing the osteoclast number in peri-implant bone remodelling. *Acta Biomaterialia, 6*, 1861–1868.

Jo, J.-H., Kang, B.-G., Shin, K.-S., Kim, H.-E., Hahn, B.-D., Park, D.-S., et al. (2011). Hydroxyapatite coating on magnesium with MgF_2 interlayer for enhanced corrosion resistance and biocompatibility. *Journal of Materials Science: Materials in Medicine, 22*, 2437–2447.

Kaartinen, E., Paavolainen, P., Holmstroem, T., & Slaetis, P. (1985). Loss of bone strength after intramedullary nailing: torsion tests of tibial osteotomies in rabbits. *Acta Orthopaedica Scandinavica, 56*, 487–490.

Kaese, V. (2002). *Beitragzum korrosionsschützenden Legieren von Magnesiumwerkstoffen*. Düsseldorf: VDI.

Kainer, K. U., & von Buch, F. (2006). The current state of technology and potential for further development of magnesium applications. In K. Kainer (Ed.), *Magnesium alloys and technologies* (pp. 1–22). Weinheim: Wiley-VCH.

Kannan, M. B., & Raman, R. K. (2008). In vitro degradation and mechanical integrity of calcium-containing magnesium alloys in modified-simulated body fluid. *Biomaterials, 29*, 2306–2314.

Kehr, J. (1993). A survey on quantitative microdialysis: theoretical models and practical implications. *Journal of Neuroscience Methods, 48*, 251–261.

Kiba, H., Hayakawa, T., Oba, S., Kuwabara, M., Haabata, I., & Yamamoto, H. (2003). Potential application of high-resolution microfocus X-ray techniques for observation of bone structure and bone-implant interface. *The International Journal of Oral & Maxillofacial Implants, 18*, 279–285.

Kietzmann, M., Löscher, W., Arens, D., Maass, P., & Lubach, D. (1993). The isolated perfused bovine udder as an in vitro model of percutaneous drug absorption. Skin viability and percutaneous absorption of dexamethasone, benzoyl peroxide, and etofenamate. *Journal of Pharmacological and Toxicology Methods, 30*, 75–84.

Kirkland, N. T., Birbilis, N., Walker, J., Woodfield, T., Dias, G. J., & Staiger, M. P. (2010). In-vitro dissolution of magnesium-calcium binary alloys: clarifying the unique role of calcium additions in bioresorbable magnesium implant alloys. *Journal of Biomedical Materials Research Part B: Applied Biomaterials, 95*, 91–100.

Koike, J., Ohyama, R., Kobayashi, T., Suzuki, M., & Maruyama, K. (2003). Grain-boundary sliding in AZ31 magnesium alloys at room temperature to 523 K. *Materials Transactions, 44*, 445–451.

Kopp, V., Lefebvre, W., & Pareige, C. (2011). Determination of the Mg-rich phase boundary of the binary mg-Nd phase diagram by means of atomprobetomography. *Journal of Phase Equilibria and Diffusion, 32*, 298–301.

Kostova, I., Momekov, G., & Stancheva, P. (2007). New Samarium(III), Gadolinium(III), and Dysprosium(III) complexes of coumarin-3-carboxylic acid as antiproliferative agents. *Metal Based Drugs*, 1–8.

Kovar, A., Nolting, A., & Derendorf, H. (1997). Microdialysis for the determination of free drugs in tissue. *Pharm Unserer Zeit, 26*, 17–23.

Kraus, T., Fischerauer, S. F., Hänzi, A. C., Uggowitzer, P. J., Löffler, J. F., & Weinberg, A. M. (2012). Magnesium alloys for temporary implants in osteosynthesis: in vivo studies of their degradation and interaction with bone. *Acta Biomaterialia, 8*, 1230–1238.

Krause, A., von der Höh, N., Bormann, D., Krause, C., Bach, F.-W., Windhagen, H., et al. (2010). Degradation behaviour and mechanical properties of magnesium implants in rabbit tibiae. *Journal of Materials Science, 45*, 624–632.

Kretz, F.-J., & Schäffer, J. (2008). *Anästhesie, Intensivmedizin, Notfallmedizin, Schmerztherapie* (5th ed.). Berlin: Springer.

Kubota, K., Mabuchi, M., & Higashi, K. (1999). Review processing and mechanical properties of fine-grained magnesium alloys. *Journal of Materials Science, 34*, 2255–2262.

Kuwahara, H., Al-Abdullat, Y., Mazaki, N., Tsutsumi, S., & Aizawa, T. (2001). Precipitation of magnesium apatite on pure magnesium surface during immersing in Hank's solution. *Materials Transactions, 42*, 1317–1321.

Lambotte, A. (1932). L'utilisation du magnesium comme matériel perdu dans l'osteosynthèse. *Bulletins et Mémoires de la Société Nationale de Chirurgie, 28*, 1325–1334.

Lang, G. (2006). *Histotechnik: Praxislehrbuch für die Biomedizinische Analytik*. Wien: Springer.

de Lange, E. C., de Boer, A. G., & Breimer, D. D. (2000). Methodological issues in microdialysis sampling for pharmacokinetic studies. *Advanced Drug Delivery Reviews, 45*, 125–148.

Lastra, M. D., Pastelin, R., Camacho, A., Monroy, B., & Aguilar, A. E. (2001). Zinc intervention on macrophages and lymphocytes response. *Journal of Trace Elements in Medicine and Biology, 15*, 5–10.

Lespinasse, V. (1910). A practical mechanical method of end-to-end anastomosis of blood-vessels: using absorbable magnesium rings. *JAMA, 55*, 1785–1790.

Levesque, J., Hermawan, H., Dube, D., & Mantovani, D. (2008). Design of a pseudo-physiological test bench specific to the development of biodegradable metallic biomaterials. *Acta Biomaterialia, 4*, 284–295.

Levin, D. C., & Fallon, J. T. (1982). Significance of the angiographic morphology of localized coronary stenoses: histopathologic correlations. *Circulation, 66*, 316–320.

Li, J., Forberg, S., & Hermansson, L. (1991). Evaluation of the mechanical properties of hot isostatically pressed titania and titania-calcium phosphate composites. *Biomaterials, 12*, 438–440.

Li, L., Gao, J., & Wang, Y. (2004). Evaluation of cyto-toxicity and corrosion behaviour of alkali-heat-treated magnesium in simulated body fluid. *Surface and Coatings Technology, 185*, 92–98.

Li, Z., Gu, X., Lou, S., & Zheng, Y. (2008). The development of binary Mg–Ca alloys for use as biodegradable materials within bone. *Biomaterials, 29*, 1329–1344.

Lind, A. Y., Eggebrecht, H., & Erbel, R. (2005). Images in cardiology: the invisible stent: imaging of an absorbable metal stent with multislice spiral computed tomography. *Heart, 91*, 1604.

Link, T. M., Berning, W., Scherf, S., Joosten, U., Joist, A., Engelke, K., et al. (2000). CT of metal implants: reduction of artifacts using an extended CT scale technique. *Journal of Computer Assisted Tomography, 24*, 165–172.

Liu, C., Huang, Y., Shen, W., & Cui, J. (2001). Kinetics of hydroxyapatite precipitation at pH 10 to 11. *Biomaterials, 22*, 301–306.

Liu, C., Xin, Y., Tian, X., & Chu, P. K. (2007). Corrosion behaviour of AZ91 magnesium alloy treated by plasma immersion ion implantation and deposition in artificial physiological fluids. *Thin Solid Films, 516*, 422–427.

Lonnroth, P., Jansson, P. A., & Smith, U. (1987). A microdialysis method allowing characterization of intercellular water space in humans. *American Journal of Physiology, 253*, E228–E231.

Lorenz, C., Brunner, J. G., Kollmannsberger, P., Jaafar, L., Fabry, B., & Virtanen, S. (2009). Effect of surface pre-treatments on biocompatibility of magnesium. *Acta Biomaterialia, 5*, 2783–2789.

Ma, A., Jiang, J., Saito, N., Shigematsu, I., Yuan, Y., Yang, D., et al. (2009). Improving both strength and ductility of a Mg alloy through a large number of ECAP passes. *Materials Science and Engineering A, 513–514*, 122–127.

Mahnken, A. H., Raupach, R., Wildberger, J. E., Jung, B., Heussen, N., Flohr, T. G., et al. (2003). A new algorithm for metal artifact reduction in computed tomography: in vitro and in vivo evaluation after total hip replacement. *Investigative Radiology, 38*, 769–775.

Makar, G. L., & Kruger, J. (1993). Corrosion of magnesium. *International Materials Reviews, 38*, 138–153.

di Mario, C., Griffiths, H., Goktekin, O., Peeters, N., Verbist, J., Bosiers, M., et al. (2004). Drug-eluting bioabsorbable magnesium stent. *Journal of Interventional Cardiology, 17*, 391–395.

Martin, I., Mastrogiacomo, M., de Leo, G., Muraglia, A., Beltrame, F., Cancedda, R., et al. (2002). Fluorescence microscopy imaging of bone for automated histomorphometry. *Tissue Engineering, 8*, 847–852.

McBride, E. D. (1938a). Absorbable metal in bone surgery. *The Journal of American Medical Association, 111*, 2464–2467.

McBride, E. D. (1938b). Magnesium screw and nail transfixion in fractures. *Southern Medical Journal, 31*, 508–515.

Mcdonald, D. J., Enneking, W. F., & Sundaram, M. (2002). Metal-associated angiosarcoma of bone: report of two cases and review of the literature. *Clinical Orthopaedics and Related Research, 396*, 206–214.

Meads, C., Cummins, C., Jolly, K., Stevens, A., Burls, A., & Hyde, C. (2000). Coronary artery stents in the treatment of ischaemic heart disease: a rapid and systematic review. *Health Technology Assessment, 4*, 1–153.

Meyer-Lindenderg, A., Krause, A., Krause, C., Bormann, D., & Windhagen, H. (2007). Entwicklung der mechanischen Eigenschaften von degradablen intramedullären Implantaten auf Magnesiumbasis nach unterschiedlicher Implantationsdauer. *Biomaterialien, 8*, 180.

Mjöberg, B., Hellquist, E., Mallmin, H., & Lindh, U. (1997). Aluminum, Alzheimer's disease and bone fragility. *Acta Orthopaedica, 68*, 511–514.

Mordike, B. L., & Ebert, T. (2001). Magnesium: properties—applications—potential. *Materials Science and Engineering A, 302*, 37–45.

Morice, M. C., Serruys, P. W., Sousa, J. E., Fajadet, J., Ban Hayashi, E., Perin, M., et al. (2002). A randomized comparison of a sirolimus eluting stent with a standard stent for coronary revascularization. *The New England Journal of Medicine, 346*, 1773–1780.

Moses, P. A., Lewis, D. D., Lanz, O. I., Stubbs, W. P., Cross, A. R., & Smith, K. R. (2002). Intramedullary interlocking nail stabilisation of 21 humeral fractures in 19 dogs and one cat. *Australian Veterinary Journal, 80*, 336–343.

Mueller, W. D., de Mele, M. F., Nascimento, M. L., & Zeddies, M. (2009). Degradation of magnesium and its alloys: dependence on the composition of the synthetic biological media. *Journal of Biomedical Materials Research Part A, 90*, 487–495.

Müller, M. J., Westenhöfer, J., Bosy-Westphal, A., Loser, C., & Selberg, O. (2007). Ernährungsmedizinische Untersuchungen. In M. J. Müller (Ed.), *Ernährungsmedizinische Praxis: Methoden—Prävention—Behandlung* (2nd ed.). (pp. 1–195). Heidelberg: Springer.

Muni, N. I., Califf, R. M., Foy, J. R., Boam, A. B., Zuckerman, B. D., & Kuntz, R. E. (2005). Coronary drug-eluting stent development: issues in trial design. *American Heart Journal, 149*, 415–433.

Nakamura, Y., Tsumura, Y., Tonogai, Y., Shibata, T., & Ito, Y. (1997). Differences in behaviour among the chlorides of seven rare earth elements administered intravenously to rats. *Toxicological Sciences, 37*, 106–116.

Nakano, Y., Toyosawa, S., & Takano, Y. (2004). Eccentric localization of osteocytes expressing enzymatic activities, protein, and mRNA signals for type 5 tartrate-resistant acid phosphatase (TRAP). *Journal of Histochemistry & Cytochemistry, 52*, 1475–1482.

Nemirovsky, D. (2003). Imaging of high-risk plaque. *Cardiology, 100*, 160–175.

Nicole, R. (1947). Metallschädigung bei Osteosynthesen. *Helvetica Chirurgica Acta, 14*(Suppl. 3), 1–74.

Niestroj, I. (2000). *Praxis der orthomolekularen Medizin: Physiologische Grundlagen; Therapie mit Mikro-Nährstoffen* (2nd ed.). Stuttgart: Hippokrates.

Ong, A. T., Mcfadden, E. P., Regar, E., De Jaegere, P. P., Van Domburg, R. T., & Serruys, P. W. (2005). Late angiographic stent thrombosis (LAST) events with drug-eluting stents. *Journal of the American College of Cardiology, 45*, 2088–2092.

van Oosterwyck, H., Duyck, J., Van der Sloten, J., Van der Perre, G., Jansen, J., Wevers, M., et al. (2000). The use of microfocus computerized tomography as a new technique for characterizing bone tissue around oral implants. *Journal of Oral Implantology, 26*, 5–12.

Pardo, A., Merino, M. C., Coy, A. E., Arrabal, R., Viejo, F., & Matykina, E. (2008). Corrosion behaviour of magnesium/aluminium alloys in 3.5 wt.% NaCl. *Corrosion Science, 50*, 823–834.

Parfitt, A. M. (1988). Bone histomorphometry: proposed system for standardization of nomenclature, symbols, and units. *Calcified Tissue International, 42*, 284.

Pasnik, J., Moll, J., Cywinska-Bernas, A., Sysa, A., & Zeman, K. (2007). Proteolytic and cytokine balance abnormalities in children with congenital heart disease undergoing cardiac surgery with cardiopulmonary bypass. *Kardiologica Polska, 65*, 1208–1214.

Pasterkamp, G., Falk, E., Woutman, H., & Borst, C. (2000). Techniques characterizing the coronary atherosclerotic plaque: influence on clinical decision making? *Journal of the American College of Cardiology, 36*, 13–21.

Paulus, M. J., Gleason, S. S., Easterly, M. E., & Foltz, C. J. (2001). A review of high resolution X-ray computed tomography and other imaging modalities for small animal research. *Laboratory Animals, 30*, 36–45.

Perloff, J. R., & Palmer, J. N. (2004). Evidence of bacterial biofilms on frontal recess stents in patients with chronic rhinosinusitis. *American Journal of Rhinology, 18*, 377–380.

Peuster, M., Beerbaum, P., Bach, F. W., & Hauser, H. (2006). Are resorbable implants about to become a reality? *Cardiology in the Young, 16*, 107–116.

Pietak, A., Mahoney, P., Dias, G. J., & Staiger, M. P. (2008). Bone-like matrix formation on magnesium and magnesium alloys. *Journal of Materials Science: Materials in Medicine, 19*, 407–415.

Pinto Slottow, T. L., Pakala, R., & Waksman, R. (2008). Serial imaging and histology illustrating the degradation of a bioabsorbable magnesium stent in a porcine coronary artery. *European Heart Journal, 29*, 314.

Pizzoferrato, A., Vespucci, A., Ciapetti, G., & Stea, S. (1985). Biocompatibility testing of prosthetic implant materials by cell cultures. *Biomaterials, 6*, 346–351.

Pohler, O. E. (2000). Unalloyed titanium for implants in bone surgery. *Injury, 31*, 7–13.

Post, J. I., Eibl, J. K., & Ross, G. M. (2008). Zinc induces motor neuron death via a selective inhibition of brain-derived neurotrophic factor activity. *Amyotrophic Lateral Sclerosis, 9*, 149–155.

Purnama, A., Hermawan, H., Couet, J., & Mantovani, D. (2010). Assessing the biocompatibility of degradable metallic materials: state-of-the-art and focus on the potential of genetic regulation. *Acta Biomaterialia, 6*, 1800–1807.

Qi, G., Zhang, S., Khor, K. A., Lye, S. W., Zeng, X., Weng, W., et al. (2008). Osteoblastic cell response on magnesium-incorporated apatite coatings. *Applied Surface Science, 255*, 304–307.

Radhi, J. M., Ibrahiem, K., & Al-Tweigeri, T. (1998). Soft tissue malignant lymphoma at sites of previous surgery. *Journal of Clinical Pathology, 51*, 629–632.

Rahn, B. V., Bacellar, F. C., Trapp, L., & Perren, S. M. (1980). Methode zur Fluoreszenz - Morphometrie des Knochenanbaus. *Akt Traumatol, 10*, 109–115.

Regar, E., Werner, F., Klauss, V., Siebert, U., Henneke, K. H., Rieber, J., et al. (1999). IVUS analysis of the acute and long-term stent result using motorized pullback: intraobserver and interobserver variability. *Catheterization and Cardiovascular Interventions, 48*, 245–250.

Regar, E., Werner, F., Siebert, U., Rieber, J., Theisen, K., Mudra, H., et al. (2000). Reproducibility of neointima quantification with motorized intravascular ultrasound pullback in stented coronary arteries. *American Heart Journal, 139*, 632–637.

Rehm, K. E., Helling, H. J., & Claes, L. (1994). Report of the biodegradable implants study group. *Akt Traumatol, 24*, 70–73.

Rehm, K. E., Helling, H. J., & Gatzka, C. (1997). New developments in the application of biodegradable implant. *Orthopäde, 26*, 489–497.

Reifenrath, J., Willbold, E., Ulrich, H., Palm, C., Nellesen, J., Crostack, H. A., et al. (2007). Der Einfluss von subchondralen Magnesiumimplantaten auf das peri-implantäre Gewebe. *Biomaterialien, 3*, 188.

Ren, Y., Wang, H., Huang, J., Zhang, B., & Yang, K. (2007). Study of biodegradation of pure magnesium. *Key Engineering Materials, 342–343*, 601–604.

Rettig, R., & Virtanen, S. (2009). Composition of corrosion layers on a magnesium rare-earth alloy in simulated body fluids. *Journal of Biomedical Materials Research Part A, 88*, 359–369.

Revell, P., Damien, E., Zhang, X., Evans, P., & Howlett, C. (2004). The effect of magnesium ions on bone bonding to hydroxyapatite. *Key Engineering Materials, 254–256*, 447–450.

Rogers, T. A., Haven, F. L., & Mahan, P. E. (1960). Exchange of radio active magnesium in Walker carcinosarcoma 256: anote. *Journal of the National Cancer Institute, 25*, 887–888.

Rogers, T. A., & Parker, E. M. (1959). Exchange of radioactive magnesium in the rat. *Proceedings of the Society for Experimental Biology and Medicine, 100*, 235–239.

Rokhlin, L. L. (2003). *Magnesium alloys containing rare earth metals: structure and properties*. London, New York: Taylor & Francis.

Romeis, B. (1989). *Mikroskopische Technik* (17. Aufl). München, Wien: Urban und Schwarzenberg.

Rosengren, A., Bjursten, L. M., Danielsen, N., Persson, H., & Kober, M. (1999). Tissue reactions to polyethylene implants with different surface topography. *Journal of Materials Science: Materials in Medicine, 10*, 75–82.

Rubin, A. H., Terasaki, M., & Sanui, H. (1979). Major intracellular cations and growth control: correspondence among magnesium content, protein synthesis, and the onset of DNA synthesis in BALB/c3T3 cells. *Proceedings of the National Academy of Sciences, 76*, 3917–3921.

Rude, R., Gruber, H., Norton, H., Wei, L., Frausto, A., & Kilburn, J. (2006). Reduction of dietary magnesium by only 50% in the rat disrupts bone and mineral metabolism. *Osteoporos International, 17*, 1022–1032.

Rudert, M. (2002). Histological evaluation of osteochondraldefects: consideration of animal models with emphasis on the rabbit, experimental setup, follow-up and applied methods. *Cells Tissues Organs, 171*, 229–240.

Ruegsegger, P., Koller, B., & Mueller, R. (1996). A microtomographic system for the nondestructive evaluation of bone architecture. *Calcified Tissue International, 58*, 24–29.

Saris, N.-E. L., Mervaala, E., Kappanan, H., Khawaja, J., & Lewenstam, A. (2000). Magnesium. An update on physiological, clinical and analytical aspects. *Clinica Chimica Acta, 294*, 1–26.

Schauer, A., & Scheibe, G. (1959). Über die Verwendung der reversibel polymeren Metachromasie von Pseudoisocyanin zur Gewebsfärbung. *Histochemie, 1*, 190–195.

Schilling, T., Cebotari, S., Tudorache, I., Hilfiker, A., Meyer, T., Biskup, C., et al. (2010). Stabilizing autologous vascularized intestine patch for myocardial substitution by magnesium scaffolds. *The Thoracic and cardiovascular Surgeon, 58*, 01.

Schoenhagen, P., & Nissen, S. (2002). Understanding coronary artery disease: tomographic imaging with intravascular ultrasound. *Heart, 88*, 91–96.

Schomig, A., Dibra, A., Windecker, S., Mehilli, J., Suarez de Lezo, J., Kaiser, C., et al. (2007). A meta-analysis of 16 randomized trials of sirolimus-eluting stents versus paclitaxel-eluting stents in patients with coronary artery disease. *Journal of the American College of Cardiology, 50*, 1373–1380.

Schranz, D., Zartner, P., Michel-Behnke, I., & Akintürk, H. (2006). Bioabsorbable metal stents for percutaneous treatment of critical recoarctation of the aorta in a newborn. *Catheterization and Cardiovascular Interventions, 67*, 671–673.

Schumacher, S., Stahl, J., Bäumer, W., Seitz, J. M., Bach, F. W., Petersen, L. J., et al. (2011). Ex vivo examination of biocompatibility of biodegradable magnesium via microdialysis in the isolated perfused bovine udder model. *The International Journal of Arificial Organs, 34*, 33–43.

Schwarz, F., Herten, M., Sager, M., Wieland, M., Dard, M., & Becker, J. (2007). Histological and immunohistochemical analysis of initial and early subepithelial connective tissue attachment at chemically modified and conventional SLA®titanium implants. A pilot study in dogs. *Clinical Oral Investigation Research, 18*, 481–488.

Seelig, M. (1924). A study of magnesium wire as an absorbable suture and ligature material. *Archives of Surgery, 8*, 669–680.

Seitz, J.-M., Collier, K., Wulf, E., Bormann, D., Angrisani, N., Meyer-Lindenberg, A., et al. (2011). The effect of different sterilization methods on the mechanical strength of magnesium based implant materials. *Advanced Engineering Materials, 13*, 1146–1151.

Serre, C. M., Papillard, M., Chavassieux, P., Voegel, J. C., & Boivin, G. (1998). Influence of magnesium substitution on a collagen-apatite biomaterial on the production of a calcifying matrix by human osteoblasts. *Journal of Biomedical Materials Research, 42*, 626–633.

Shadanbaz, S., & Dias, G. J. (2012). Calcium phosphate coatings on magnesium alloys for biomedical applications: a review. *Acta Biomaterialia, 8*, 20–30.

Shalabi, M. M., Wolke, J. G., Cuijpers, V. M., & Jansen, J. A. (2007). Evaluation of bone response to titanium-coated polymethyl methacrylate resin (PMMA) implants by X-ray tomography. *Journal of Materials Science: Materials in Medicine, 18*, 2033–2039.

Shaw, B. A. (2003). Corrosion resistance of magnesium alloys. In D. Stephen, & S. Bernard (Eds.), *ASM handbook volume 13A: corrosion: fundamentals, testing, and protection*. UK: ASM International.

Singh, R., & Dahotre, N. B. (2007). Corrosion degradation and prevention by surface modification of biometallic materials. *Journal of Materials Science: Materials in Medicine, 18*, 725–751.

Skalsky, H., & Carchman, R. (1983). Aluminum homeostasis in man. *International Journal of Toxicology, 1983*(2), 405–423.

Smet, E. D., Siegfried, S., Jaecques, V. N., Wevers, M., Jansen, J. A., Jacobs, R., et al. (2006). Effect of controlled early implant loading on bone healing and bone mass in guinea pigs, as assessed by micro-CT and histology. *European Journal of Oral Sciences, 114*, 232–242.

Somekawa, H., & Mukai, T. (2005). Effect of grain refinement on fracture toughness in extruded pure magnesium. *Scripta Materialia, 53*, 1059–1064.

Song, G. L. (2007). Control of biodegradation of biocompatable magnesium alloys. *Corrosion Science, 49*, 1696–1701.

Song, G. L., & Atrens, A. (1999). Corrosion mechanisms of magnesium alloys. *Advanced Engineering Materials, 1*, 11–33.

Song, G. L., & Atrens, A. (2003). Understanding magnesium corrosion—a framework for improved alloy performance. *Advanced Engineering Materials, 5*, 837–858.

Song, G., Bowles, L. A., & Stjohn, D. H. (2004). Corrosion resistance of aged die cast magnesium alloy AZ91D. *Materials Science and Engineering A, 366*, 74–86.

Song, Y. W., Shan, D. Y., & Han, E. H. (2008). Electrodeposition of hydroxyapatite coating on AZ91D magnesium alloy for biomaterial application. *Materials Letters, 62*, 3276–3279.

StJohn, D. H., Qian, M., Easton, M. A., Cao, P., & Hildebrand, Z. (2005). Grain refinement of magnesium alloys. *Metall and Mat Trans A, 36*, 1669–1679.

Staesche, M. (1948). Über die chemische Erzeugungeiner dickeren Magnesium fluorid-Schutzschicht auf Magnesium-Legierungen. *Arch Metallk, 3*, 99–102.

Staiger, M. P., Pietak, A. M., Huadmai, J., & Dias, G. (2006). Magnesium and its alloys as orthopedic biomaterials: a review. *Biomaterials, 27*, 1728–1734.

Stoppie, N., Wevers, M., & Naert, I. (2007). Feasibility of detecting trabecular bone around percutaneous titanium implants in rabbits by in vivo microfocus computed tomography. *Journal of Microscopy, 228*, 55–61.

Suganuma, J. A., & Alexander, H. (1993). Biological response of intramedullary bone to poly-L-lactic acid. *J. Appl. Biomater, 4*, 13–27.

Thomann, M., Krause, C., Angrisani, N., Bormann, D., Hassel, T., Windhagen, H., et al. (2010). Influence of a magnesium-fluoride coating of magnesium-based implants (MgCa0.8) on degradation in a rabbit model. *Journal of Biomedical Materials Research A, 93*, 1609–1619.

Thull, R. (2003). Oberflächenmodifikationen zurVerbesserung von Biokompatibilität und mechanischen Eigenschaften von orthopädischen Implantaten. *Der Orthopäde, 32*, 51–59.

Timmer, R. T., & Sands, J. M. (1999). Lithium intoxication. *JASN, 10*, 666–674.

Topf, J., & Murray, P. (2003). Hypomagnesemia and hypermagnesemia. *Reviews in Endocrine & Metabolic Disorders, 4*, 195–206.

Törmala, P., Pohjonen, T., & Rokkanen, P. (1998). Bioabsorbable polymers: materials technology and surgical applications. Proceedings of the institution of mechanical engineers, part H. *Journal of Engineering Medicine, 212*, 101–111.

Tsai, C., Ruey-Mo, L., Chien-Ping, J., & Jiin-Huey Chern, L. (2008). Bioresorption behaviour of tetracalcium phosphate-derived calcium phosphate cement implanted in femur of rabbits. *Biomaterials, 29*, 984–993.

Ullmann, B., Reifenrath, J., Dziuba, D., Seitz, J.-M., Bormann, D., & Meyer-Lindenberg, A. (2011). In vivo degradation behaviour of the magnesium alloy LANd442 in rabbit tibiae. *Materials, 4*, 2197–2218.

Ungethuem, M., & Winkler-Gniewek, W. (1984). Toxikologie der Metalleund Biokompatibilität metallischer Implantatwerkstoffe. *Zeitschrift fur Orthopadie und Ihre Grenzgeb, 122*, 99–105.

Verbrugge, J. (1934). Le Material Metaliqueresorbable en Chirurgieosseuse. *La Presse Médicale, 42*, 460–465.

Vormann, J. (2003). Magnesium: nutrition and metabolism. *Molecular Aspects of Medicine, 24*, 27–37.

Wachter, N. J., Augat, P., Krischak, G. D., Mentzel, M., Kinzl, L., & Claes, L. (2001). Prediction of cortical bone porosity in vitro by microcomputed tomography. *Calcified Tissue International, 68*, 38–42.

Wacker, W. E. C. (1980). *Magnesium and man*. Cambridge MA: Harvard University Press.

Waksman, R. (2006). Update on bioabsorbable stents: from bench to clinical. *Journal of Interventional Cardiology, 19*, 414–421.

Wang, H., Guan, S., Wang, Y., Liu, H., Wang, H., Wang, L., et al. (2011). In vivo degradation behaviour of Ca-deficient hydroxyapatite coated Mg–Zn–Ca alloy for bone implant application. *Colloids and Surface B: Biointerfaces, 88*, 254–259.

Wang, Y., Wei, M., Gao, J., Hu, J., & Zhang, Y. (2008). Corrosion process of pure magnesium in simulated body fluid. *Materials Letters, 62*, 2181–2184.

Ward, J. J., Thornbury, D. D., Lemons, J. E., & Dunham, W. K. (1990). Metal-induced sarcoma. A case report and literature review. *Clinical Orthopaedics and Related Research, 252*, 299–306.

Westengen, H., & Aune, T. (2006). Magnesium casting alloys. In H. Friedrich, & B. Mordike (Eds.), *Magnesium technology: metallurgy, design data, applications* (pp. 145–166). Berlin: Springer-Verlag.

Wiltfang, J., Merten, H. A., Schlegel, K. A., Schultze-Mosgau, S., Kloss, F. R., Rupprecht, S., et al. (2002). Degradation characteristics of alpha and beta tri-calcium-phosphate (TCP) in minipigs. *Journal of Biomedical Materials Research, 63*, 115–121.
Wintermantel, E. (2002). *Medizintechnik mitbiokompatiblen Werkstoffen und Verfahren* (3rd ed.). Berlin: Springer.
Wintermantel, E., & Ha, S.-W. (1996). *Biokompatible Werkstoffe und Bauweisen: Implantate für Medizinund Umwelt*. Berlin: Springer.
Wintermantel, E., Shah-Derler, B., Bruinink, A., Petitmermet, M., Blum, J., & Ha, S.-W. (2002). Biokompatibilität. In E. Wintermantelx (Ed.), *Medizintechnik mit biokompatiblen Werkstoffen und Verfahren* (3rd ed.) (pp. 5–43). Berlin: Springer.
Wintermantel, E., Shah-Derler, B., Bruinink, A., Petitmermet, M., Blum, J., & Ha, S.-W. (2009). Biokompatibilität. In E. Wintermantel (Ed.), *Medizintechnik: life science engineering* (5th ed.) (pp. 67–105). Berlin: Springer.
Witte, F., Abeln, I., Switzer, E., Kaese, V., Meyer-Lindenberg, A., & Windhagen, H. (2008b). Evaluation of the skin sensitizing potential of biodegradable magnesium alloys. *Journal of Biomedical Materials Research Part A, 86*, 1041–1047.
Witte, F., Fischer, J., Nellesen, J., Crostackl, H. A., Kaese, V., Pisch, A., et al. (2006). In vitro and in vivo corrosion measurements of magnesium alloys. *Biomaterials, 27*, 1013–1018.
Witte, F., Fischer, J., Nellesen, J., Vogt, C., Vogt, J., Donath, T., et al. (2010). In vivo corrosion and corrosion protection of magnesium alloy LAE442. *Acta Biomaterialia, 6*, 1792–1799.
Witte, F., Hort, N., Vogt, C., Cohen, S., Kainer, K. U., Willumeit, T.,R., et al. (2008a). Degradable biomaterials based on magnesium corrosion. *Current Opinion in Solid State & Materials Science, 12*, 63–72.
Witte, F., Kaese, V., Haferkamp, H., Switzer, E., Meyer-Lindenberg, A., Wirth, C. J., et al. (2005). In vivo corrosion of four magnesium alloys and the associated bone response. *Biomaterials, 26*, 3557–3563.
Witte, F., Nellesen, J., Crostack, H. A., & Beckmann, F. (2002). Subtraction-XTM as a sensitive method for detecting rare earth elements in magnesium implants. *Hasylab Annual Report, 2*, 825–826.
Witte, F., Ulrich, H., Palm, C., & Willbold, E. (2007a). Biodegradable magnesium scaffolds: Part II: peri-implant bone remodelling. *Journal of Biomedical Materials Research Part A, 81*, 757–765.
Witte, F., Ulrich, H., Rudert, M., & Willbold, E. (2007b). Biodegradable magnesium scaffolds: Part 1: appropriate inflammatory response. *Journal of Biomedical Materials Research Part A, 81*, 748–756.
Wolf, E., Röser, K., Hahn, M., Welkerling, H., & Delling, G. (1992). Enzyme and immunhistochemistry on undecalcified bone and bone marrow biopsies after embedding in plastic: a new embedding method for routine application. *Virchows Archiv A Pathological Anatomy, 420*, 17–24.
Wong, H. M., Yeung, K. W., Lam, K. O., Tam, V., Chu, P. K., Luk, K. D., et al. (2010). A biodegradable polymer-based coating to control the performance of magnesium alloy orthopaedic Implants. *Biomaterials, 31*, 2084–2096.
Wu, G., Fan, H., Zhai, C., & Zhou, Y. P. (2005). The effect of Ca and rare earth elements on the microstructure, mechanical properties and corrosion behaviour of AZ91D. *Materials Science and Engineering: A, 408*, 255–263.
Xin, Y., Hu, T., & Chu, P. K. (2011). In vitro studies of biomedical magnesium alloys in a simulated physiological environment: a review. *Acta Biomaterialia, 7*, 1452–1459.

Xu, L., Pan, F., Yu, G., Yang, L., Zhang, E., & Yang, K. (2009). In vitro and in vivo evaluation of the surface bioactivity of a calcium phosphate coated magnesium alloy. *Biomaterials, 30*, 1512–1523.

Xu, L., Yu, G., Zhang, E., Pan, F., & Yang, K. (2007). In vivo corrosion behaviour of Mg–Mn–Zn alloy for bone implant application. *Journal of Biomedical Materials Research Part A, 83*, 703–711.

Xu, L., Zhang, E., Yin, D., Zeng, S., & Yang, K. (2008). In vitro corrosion behaviour of Mg alloys in a phosphate buffered solution for bone implant application. *Journal of Materials Science: Materials in Medicine, 19*, 1017–1025.

Yabushita, H., Bouma, B. E., Houser, S. L., Aretz, H. T., Jang, I. K., Schlendorf, K. H., et al. (2002). Characterization of human atherosclerosis by optical coherence tomography. *Circulation, 106*, 1640–1645.

Yamamoto, A., & Hiromoto, S. (2009). Effect of inorganic salts, amino acids and proteins on the degradation of pure magnesium in vitro. *Materials Science and Engineering: C Biomimetic Supramolecular Systems, 29*, 1559–1568.

Yamamoto, A., Watanabe, A., Sugahara, K., Tsubakino, H., & Fukumoto, S. (2001). Improvement of corrosion resistance of magnesium alloys by vapor deposition. *Scripta Materialia, 44*, 1039–1042.

Yamasaki, Y., Yoshida, Y., Okazaki, M., Shimazu, A., Kubo, T., Akagawa, Y., et al. (2003). Action of $FGMgCO_3Ap$-collagen composite in promoting bone formation. *Biomaterials, 24*, 4913–4920.

Yamasaki, Y., Yoshida, Y., Okazaki, M., Shimazu, A., Uchida, T., Kubo, T., et al. (2002). Synthesis of functionally graded $MgCO_3$ apatite accelerating osteoblast adhesion. *Journal of Biomedical Materials Research, 62*, 99–105.

Yuen, C. K., & Ip, W. Y. (2010). Theoretical risk assessment of magnesium alloys as degradable biomedical implants. *Acta Biomaterialia, 6*, 1808–1812.

Zartner, P., Cesnjevar, R., Singer, H., & Weyand, M. (2005). First successful implantation of a biodegradable metal stent into the left pulmonary artery of a preterm baby. *Catheterization and Cardiovascular Interventions, 66*, 590–594.

Zelin, M. G., Yang, H. S., Valiev, R. Z., & Mukherjee, A. K. (1992). Interaction of high-temperature deformation mechanisms. *Metallurgical Transaction A, 23*, 3135.

Zhang, E., Xu, L., & Yang, K. (2005). Formation by ion plating of Ti-coating on pure Mg for biomedical applications. *Scripta Materialia, 53*, 523–527.

Zhang, E., Xu, L., Yu, G., Pan, F., & Yang, K. (2009). In vivo evaluation of biodegradable magnesium alloy bone implant in the first 6 months implantation. *Journal of Biomedical Materials Research Part A, 90*, 882–893.

Zhang, Y., Zhang, G., & Wei, M. (2009b). Controlling the biodegradation rate of magnesium using biomimetic apatite coating. *Journal of Biomedical Materials Research, 89*, 408–414.

Zhang, S., Zhang, X., Zhao, C., Li, J., Song, Y., Xie, C., et al. (2010). Research on an Mg–Zn alloy as a degradable biomaterial. *Acta Biomaterialia, 6*, 626–640.

Zöllner, C., Büsing, C. M., & Strutz, J. (1984). TCP-Implantate in der Mittelohrchirurgie. *Laryngo-Rhino-Otologie, 63*, 220–225.

Zreiqat, H., Howlett, C. R., Zannettino, A., Evans, P., Schulzetanzil, G., Knabe, C., et al. (2002). Mechanisms of magnesium-stimulated adhesion of osteoblastic cells to commonly used orthopaedic implants. *Journal of Biomedical Materials Research, 62*, 175–184.

Mechanical integrity of magnesium alloys for biomedical applications

5

Lokesh Choudhary, R.K. Singh Raman
Monash University, VIC, Australia

5.1 Magnesium and its alloys as implant materials

Human longevity considerably improved in the last century with significant advances in medical science. The most notable improvements include development of implant devices, since clinical cases requiring medical implants are increasing with desire for an active lifestyle and ageing population. Implants are the artificial devices used in the human body for repairing fractured bones and restoration of any impaired physical functions (Niinomi, 2002; Ratner, Hoffman, Schoen, & Lemons, 1996). Common use of implant devices can be varied based on the requirements, such as: (1) to aid healing process by using plates, rods, and screws for stabilization of fractured bones, (2) to improve the function of an organ or other parts of the body, and (3) to replace a damaged, diseased, or worn part of the anatomy. Since implant devices are used in the physiological environment, they must satisfy certain requirements, which include excellent biocompatibility, sufficient mechanical properties to hold or support the injured body part/tissues, and desired resistance to degradation due to corrosion and fatigue in aggressive human body fluid (Grimm, 2007, pp. 1165–1194; Nielsen, 1987).

In the last two decades, several research groups have investigated the idea that biodegradable metallic materials might be used for construction of implant devices. These temporary biomaterials will support tissue regeneration and healing process and then completely degrade away harmlessly. Traditional implants of stainless steels, cobalt–chromium alloys, and titanium alloys have been widely used because of their strength, ductility, and resistance to corrosion (Staiger, Pietak, Huadmai, & Dias, 2006). However, when these traditional alloys are used as temporary implant devices, such as plates, screws, and wires, a second surgery is required to remove the implant after tissues have healed. Besides the cost of these surgical procedures and inconvenience to patients, the traditional alloys also cause local inflammation due to potential release of cytotoxic ions (Granchi et al., 1999; Puleo & Huh, 1995; Wang et al., 2002). Therefore, it would be very advantageous if an implant material could be identified that degrades away in the physiological environment after completion of the healing process (Staiger et al., 2006; Witte et al., 2009). This approach will eliminate the need for the second surgical procedure. In this context, magnesium (Mg)-based

Surface Modification of Magnesium and its Alloys for Biomedical Applications. http://dx.doi.org/10.1016/B978-1-78242-077-4.00005-X

metallic alloys are potential candidates for use as biodegradable temporary implant devices because they can naturally degrade in the human body environment due to their high electrochemical activity (standard reduction potential of Mg is −2.4 V vs standard hydrogen electrode (SHE) (Makar & Kruger, 1993)). Mg, which is nontoxic to the human body, is also vital for human metabolic processes, including for the stabilization of DNA and RNA (Hartwig, 2001; Saris, Mervaala, Karppanen, Khawaja, & Lewenstam, 2000; Wolf & Cittadini, 2003). Any excess Mg is harmlessly excreted with urine. Unlike traditional implant materials, the mechanical properties of Mg and its alloys are also very suitable for implant applications: they possess low density $(\rho) = 1.74-2.0$ g cm^{-3} and elastic modulus $(E) = 41-45$ GPa, both of which are similar to the properties of human bones (Staiger et al., 2006). Hence the problem of stress shielding, which is caused by a mismatch in the elastic modulus between natural bone and implant, can be mitigated if Mg alloys are deployed as implant materials (Staiger et al., 2006; Witte, 2011; Xin, Hu, & Chu, 2011). Despite these highly advantageous properties of Mg, its alloys have rarely been used as human body implants. The major drawback in the use of Mg alloys as implants is that they tend to corrode very quickly in the corrosive human body fluid (that has a pH = 7.4−7.6) (Bobby Kannan & Singh Raman, 2008; Heublein et al., 2003; Kirkland, Lespagnol, Birbilis, & Staiger, 2010; Kirkland, Staiger, Nisbet, Davies, & Birbilis, 2011; Song, 2007; Staiger et al., 2006; Xin et al., 2011), thereby losing their mechanical integrity before the expected service life. Although these limitations have ruled out any effective use of Mg alloys as permanent implants, it is interesting that the Mg alloys could still be used as biodegradable temporary implant devices such as plates, wires, stents, pins, and screws.

5.2 Corrosion and assisted cracking of magnesium alloys

Mg and its alloys have been reported to suffer considerably rapid corrosion in the body environment that contains aggressive ions such as chlorides, carbonates, phosphates, and sulphates, in addition to the presence of proteins and glucose (Xin, Huo, Tao, Tang, & Chu, 2008). When used as biomaterials, Mg alloys will be susceptible to a few different forms of corrosion.

5.2.1 Galvanic corrosion

When two metals with different electrochemical potentials are in physical contact (electrically) and are exposed to an ion-conducting fluid, such as serum or physiological solution, galvanic corrosion will occur (Jones, 1996). Biodegradability is the major advantage in using Mg alloys as temporary implants. However, the problem of severe galvanic corrosion is one of the major obstacles in the successful use of Mg alloys as implants. Mg makes an active anode when in contact with most metals because of its high electrochemical activity, and it corrodes preferentially (Makar & Kruger, 1993; Song & Atrens, 1999).

Mg alloys are also not uniform in terms of their composition, microstructure, and even crystalline orientation. These differences can result in various electrochemical activities within a Mg alloy, thereby generating microgalvanic couples. The following two factors generally influence the formation of micro-galvanic cells in Mg alloys.

5.2.1.1 Impurities

Mg alloys contain iron (Fe), nickel (Ni), cobalt (Co), and copper (Cu) as impurities (Baker & Avedesian, 1999; Makar & Kruger, 1993; Song & Atrens, 1999) that are usually introduced during casting processes. These metallic impurities are highly cathodic to Mg and cause severe galvanic corrosion (Makar & Kruger, 1993; Song & Atrens, 1999), even when present in small amounts (as shown in Figure 5.1 (Makar & Kruger, 1993)).

Decreasing contents of these impurities below threshold levels can dramatically improve the alloy's corrosion resistance (Makar & Kruger, 1993). When these impurity contents are below critical concentrations (e.g., their respective solubility in the matrix phase), they are present in the form of solutes in Mg solid solutions, and therefore, no microgalvanic cells form between the impurities and Mg matrix. The tolerance limits for Fe, Cu, and Ni are 170, 1000, and 5 ppm, respectively (Makar & Kruger, 1993).

5.2.1.2 Secondary phases

Most of secondary phases that form in Mg alloys are highly cathodic to the primary alloy matrix phase, thereby capable of causing severe localized corrosion. However, when present in large amounts, these secondary phases can form a continuous network that can provide a barrier to the corrosion front and retard corrosion, as shown in Figure 5.2 (Song, 2011).

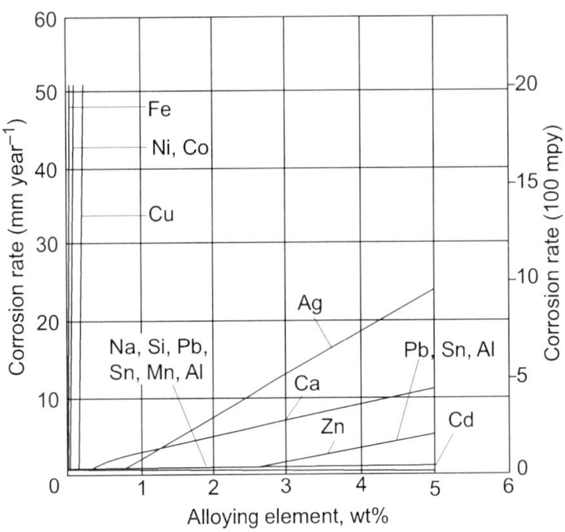

Figure 5.1 Effect of impurities and alloying elements on corrosion rate of Mg alloys in 3 wt% NaCl (Makar & Kruger, 1993).

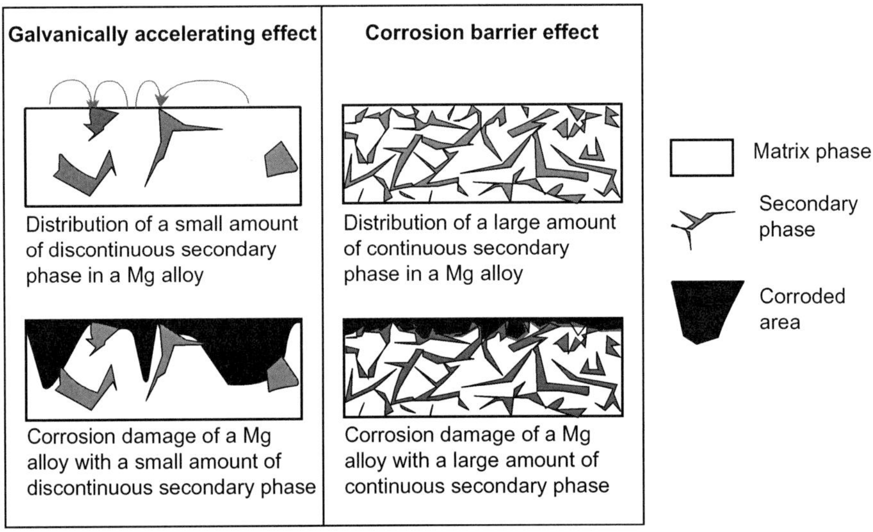

Figure 5.2 Schematic representation of dual role of secondary phase of Mg alloys in corrosion (Song, 2011).

For example, β-phase in common AZ alloys can either play a role in forming a corrosion barrier to retard corrosion or a galvanic cathode to accelerate corrosion, depending on the amount, distribution, and continuity of the β-phase (Figure 5.2). A fine distribution of β-phase can form a continuous layer and be efficient in providing the corrosion barrier, whereas the presence of a small amount of discontinuous β-phase accelerates the microgalvanic corrosion (Song, 2011). In the case of biodegradable Mg alloys, the microgalvanic corrosion is a serious problem and has been reported in several previous studies (Bobby Kannan & Singh Raman, 2008; Kirkland et al., 2010, 2011; Li, Gu, Lou, & Zheng, 2008; Xu, Yu, Zhang, Pan, & Yang, 2007; Xu, Zhang, Yin, Zeng, & Yang, 2008; Zainal Abidin, Atrens, Martin, & Atrens, 2011).

5.2.2 Localized/pitting corrosion

Mg alloys suffer localized or nonuniform attack as a result of microgalvanic effect (as described earlier) or due to the electrochemical inhomogeneity at the alloy surface (Singh Raman, 2004; Singh Raman, Birbilis, & Efthimiadis, 2004; Song & Atrens, 1999). Mg can undergo pitting corrosion at its free corrosion potential in chloride-containing medium (Tunold, Holtan, Berge, Lasson, & Steen-Hansen, 1977). Besides the roles of impurities and secondary phases, the uneven distribution of alloying elements in the matrix can cause development of microgalvanic cells and localized corrosion.

The localized corrosion and pitting have also been observed during *in vitro* and *in vivo* corrosion studies of Mg alloys for biomedical applications (Witte et al., 2005). Witte et al. (2005, 2006) reported the pitting corrosion of both LAE442 (Mg—4 wt% Li—4 wt% Al—2 wt% rare earth) and AZ91D (Mg—9 wt% Al—1 wt% Zn)

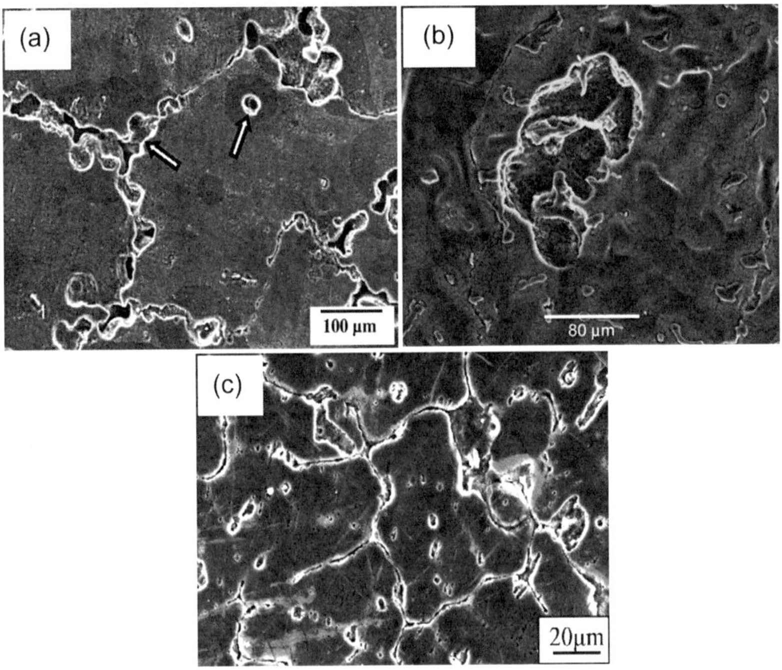

Figure 5.3 Localized/pitting corrosion observed in different Mg alloys: (a) AZ61 (Zhou et al., 2010), (b) AZ91 (Bobby Kannan & Singh Raman, 2008), and (c) Mg-Zn-Mn-Ca (Zhang & Yang, 2008).

during *in vitro* and *in vivo* tests (AZ91D showing a much greater susceptibility to the localized corrosion/pitting). Several other studies have also reported the susceptibility of AZ series Mg alloys to localized corrosion/pitting in simulated body fluid (SBF). Zhou, Shen, and Aung (2010) and Kannan and Raman (2008) observed the localized corrosion along the residual β-$Mg_{17}Al_{12}$ phase at the grain boundaries and found some pits within the grains (Figure 5.3(a) and (b)). Localized corrosion and pitting were observed in the case of some alloys more suitable for implant application, e.g., Mg-Zn-Mn-Ca alloys (Zhang & Yang, 2008), which was attributed to the tendency of Mg_2Ca network at the grain boundary to dissolve preferentially during exposure to Hanks' solution (Figure 5.3(c)). Recently, severe localized corrosion/pitting was reported for a high-strength Mg alloy, ZX50 (specifically designed for temporary implant applications in osteosynthesis), during both *in vitro* and *in vivo* testing (Fischerauer et al., 2012; Hänzi, Gerber, Schinhammer, Löffler, & Uggowitzer, 2010; Kraus et al., 2012).

5.2.3 Stress corrosion cracking

A considerable fraction of research on Mg alloys has focused on improving their general mechanical properties, creep resistance, and corrosion properties. The aspect that has received relatively less attention is their resistance to stress corrosion cracking

(SCC). SCC is a dangerous, complicated, and insidious form of corrosion-assisted failure, and it can cause catastrophic and unexpected fracture of structural components (Jones, 1992). It is a subcritical crack propagation in a brittle manner due to combined and interdependent effects of corrosion and tensile stress. SCC involves an optimum interaction of three parameters: (1) mechanical loading providing tensile stress, (2) a susceptible alloy, and (3) an environment causing corrosion at the required rate (Parkins, 1980; Sieradzki & Newman, 1985; Winzer et al., 2005). Consequently, SCC is relatively less frequent, although failures can be very costly and destructive when they occur. Unlike other types of corrosion, in the case of SCC, the surface may appear virtually unattacked, and only a few fine cracks (invisible to naked eye) initiate and propagate, leading to the brittle failures (Winzer et al., 2005). The most fundamental and detrimental feature of SCC is that a ductile material that would have undergone considerable elongation before fracture may suffer embrittlement in the presence of the corrosive environment.

Mg and its alloys are also reported to be susceptible to SCC, fracturing at stresses as low as 50% of the yield stress in common service environments (Winzer et al., 2008; Winzer et al., 2008a, 2008b). It is estimated that approximately 10–60 Mg alloy components in aerospace applications alone suffered SCC failures each year during 1960 and 1970 (Winzer et al., 2005). It is also suggested that the occurrence of SCC may be on increase because Mg alloys are now increasingly used in structural and automotive applications under load-bearing and harsh environmental conditions. Increased SCC incidences may also be expected because of increased stresses, as a result of smaller section sizes of the components in a bid to decrease weight.

5.3 SCC of magnesium alloys: a biomedical perspective

The biomaterials based on Mg alloys suffer from the problems of excessive corrosion rate and also concomitant hydrogen evolution (Ma & Xu, 2009; Staiger et al., 2006; Zberg, Uggowitzer, & Loffler, 2009). Mechanical integrity of implanted structures in the physiological environment is of fundamental importance in effective fracture fixations and cardiovascular surgeries. Metallic implant devices exposed to corrosive physiological environment are often also subjected to complex loadings during service. For example, a permanent hip implant can experience a load up to approximately four times of the body weight during a normal walk, whereas a temporary cardiovascular stent is continuously subjected to a cyclic loading due to heart beats (Gu et al., 2010; McCormack, Walsh, Wilson, & Prendergast, 1998). For an adult, loads in the spine during certain activities may exceed 3500 N (Kirkland, 2012). The synergistic presence of mechanical loading along with the corrosive environment may pose complication of sudden fracture of implants due to the phenomenon of SCC, which can occur even at stresses considerably below yield and design stresses (Jones, 1992; Singh Raman, 2005), and lead to a premature brittle failure. These brittle cracks generally initiate at locations of sharp contours such as root of a corrosion pit. Pre-existing macroscopic/microscopic flaws (e.g., micro-cracks, sharp corners, and protrusions in implant devices) can also markedly increase the probability of premature SCC failures and can be locations of SCC crack initiation. Moreover, the load-bearing implant devices are becoming relatively complex in design, thus also increasing the probability of high stress intensity points. Furthermore, galvanic, localized, or intergranular corrosion, and even thinning by uniform corrosion, could initiate SCC in biomaterials (Winzer et al., 2005). In this respect, it is advantageous to use simple shapes, with an aim of avoiding localized areas of high stress intensities. Bolted and riveted joints can produce a point of high stress concentrations. It may not always be not possible to avoid having such locations; however, it may still be possible to reduce stress by some surface treatment such as electropolishing.

Stress corrosion cracks may propagate undetected to a sudden catastrophic failure of implant devices during service, which may have many serious consequences such as painful irritation or inflammation of surrounding tissues, and troublesome removal of failed devices. In the past, several incidents involving SCC and corrosion fatigue of traditional implant devices of stainless steels and titanium alloys have been reported (Bombara & Cavallini, 1977; Bundy, Marek, & Hochman, 1983; Bundy, Vogelbaum, & Desai, 1986; Hughes & Jordan, 1972; Sivakumar, Kamachi Mudali, & Rajeswari, 1994; Sivakumar & Rajeswari, 1992; Yokoyama, Ichikawa, Murakami, Miyamoto, & Asaoka, 2002). In this context, it is also important to note that the Mg alloys are reported to be susceptible to SCC in aqueous environments, including distilled water and chloride solutions (Winzer et al., 2005). Hence, the characterization of SCC of biomedical Mg alloys in physiological conditions is a necessity before their actual service use. However, a very limited attention has been paid on the evaluation of mechanical integrity/SCC of Mg alloys in body fluid (Bobby Kannan & Singh Raman, 2008; Bobby Kannan, Singh Raman, Witte, Blawert, & Dietzel, 2011; Choudhary & Singh Raman, 2012).

5.4 Phenomenology of SCC of magnesium alloys

SCC mechanisms are generally dominated by mechanical or electrochemical processes and often specific to certain combinations of alloy and environment. Transgranular stress corrosion cracking (TGSCC) is the common mode of SCC in Mg alloys (Ben-Hamu, Eliezer, Dietzel, & Shin, 2008; Chen, Ai, Wang, Han, & Ke, 2009; Winzer et al., 2005, 2008a, 2008b), but instances of intergranular stress corrosion cracking (IGSCC) has also been reported (Bobby Kannan, Dietzel, Blawert, Atrens, & Lyon, 2008). The fundamental understanding of TGSCC and IGSCC can be found elsewhere (Winzer et al., 2008, 2008a). In the case of Mg alloys, it is generally accepted that the SCC processes usually involve hydrogen embrittlement (HE), with the source of hydrogen being cathodic reaction of the bare metal (Ebtehaj, Hardie, & Parkins, 1988; Stampella, Procter, & Ashworth, 1984), but there is a little consensus on a unified SCC mechanism. This section briefly discusses possible mechanisms of SCC of Mg alloys, influence of alloying elements on SCC, techniques for characterization of SCC in common service environments, and in SBF.

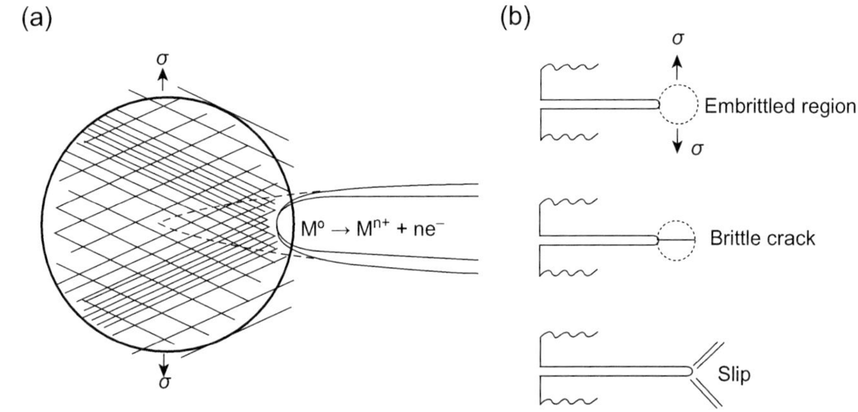

Figure 5.4 (a) Continuous crack propagation by dissolution following film rupture, and (b) cleavage-type fracture ahead of embrittled zone (Winzer et al., 2005). σ is tensile stress.

5.4.1 Mechanism of SCC in magnesium alloys

SCC of Mg and its alloys has been generally attributed to (1) continuous crack propagation by dissolution at the crack tip (Figure 5.4(a)) or (2) cleavage-type mechanism by discontinuous crack propagation (Figure 5.4(b)) (Winzer et al., 2005).

5.4.1.1 Dissolution mechanisms

The IGSCC observed in Mg-Al alloys is often attributed to the preferential dissolution of the metal matrix adjacent to the secondary phase precipitates ($Mg_{17}Al_{12}$ in Mg-Al alloy). Therefore, a network of grain boundary precipitates facilitates continuous crack propagation (Winzer et al., 2005). Microgalvanic corrosion causing IGSCC was also reported by Kannan et al. (Bobby Kannan et al., 2008) for an Al-free rare earth containing alloy, ZE41 (Figure 5.5).

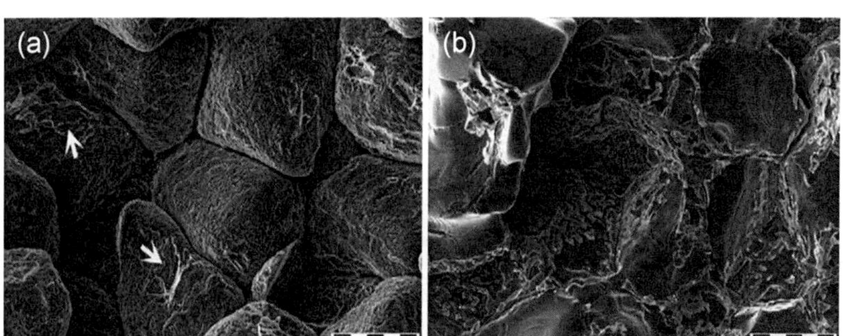

Figure 5.5 Fracture surfaces of ZE41 in: (a) 0.5 wt% NaCl—predominant intergranular and isolated transgranular (arrows) cracking, and (b) distilled water—corrosion along the grain boundaries (Bobby Kannan et al., 2008).

A film rupture model for SCC of Mg alloys has also been proposed (Winzer et al., 2005). The localized plastic deformation at the crack tip causes the rupture of passive film (Jones, 1992), exposing the bare metal to corrosive environment that causes a rapid dissolution, which leads to crack extension or propagation. The subsequent crack propagation is governed by the competing processes of film rupture and repassivation, as suggested also for Mg alloys by Ebtehaj et al. (1988) and Wearmouth, Dean, and Parkins (1973).

5.4.1.2 Mechanical fracture mechanisms

The limited slip systems in Mg alloys (hexagonal close-packed (HCP) structure) make them susceptible to cleavage fracture. A discontinuous cleavage results in the TGSCC of Mg alloys (Winzer et al., 2005). Various researchers (Chakrapani & Pugh, 1975, 1976; Meletis & Hochman, 1984) have reported that TGSCC in Mg alloys is a result of alternating processes of electrochemical dissolution and discontinuous crack advancement due to crystallographic constraints of HCP system. TGSCC resulted in fracture surfaces consisting of flat and parallel facets separated by perpendicular steps, which is consistent with cleavage mechanism. The matching and interlocking opposite fracture surfaces confirmed the occurrence of cleavage, which was difficult to explain by a dissolution model (Chakrapani & Pugh, 1975, 1976).

Mg alloys are reported to evolve considerable amount of hydrogen even at open circuit potential (OCP) as well as undergo localized corrosion. Hence, combined effects of hydrogen-assisted stress corrosion cracking (HASCC) and localized dissolution have been suggested to play an important role in deterioration of mechanical properties of AZ91D in modified simulated body fluid (*m*-SBF) (Choudhary & Singh Raman, 2012). In the case of magnesium alloys, an anodic polarization (which would normally not facilitate hydrogen generation in other alloy systems) is found to accelerate the SCC because hydrogen is generated even at such potentials due to the negative difference effect which is exclusive to Mg alloys (Stampella et al., 1984; Uematsu, Kakiuchi, & Nakajima, 2012; Winzer et al., 2005).

The TGSCC observed in Mg alloys has been widely attributed to mechanism involving hydrogen (H). Uematsu et al. (2012) showed that SCC of wrought AZ31 magnesium alloy was dominated by hydrogen embrittlement (HE). They reported higher crack propagation rate with increasing magnitude of cathodic potential that facilitated generation of hydrogen. Meletis and Hochman (1984) suggested that the presence of cleavage features at fracture surfaces could also be attributed to HE.

5.4.2 Influence of alloying elements

Most pure metals are immune to SCC; however, pure Mg has been reported to be susceptible to SCC (Fairman & Bray, 1971; Lynch & Trevena, 1988; Meletis & Hochman, 1984; Stampella et al., 1984). Stampella et al. (1984) performed a range of slow strain rate testing experiments on pure and commercial Mg in Na_2SO_4 under different environmental conditions to establish the mechanism of SCC. Both commercially pure and high-purity Mg were reported to be susceptible to SCC at room temperature in deaerated, pH 10, 10^{-3} M Na_2SO_4 solution. They also reported corrosion pits to be the necessary precursor for hydrogen embrittlement, and cracking. Pits provided bare, active, film-free sites that permit cathodically generated hydrogen to enter into the metal matrix. In both commercial and high-purity Mg, fracture surfaces appeared to be predominantly brittle, with the transgranular regions exhibiting typical quasi cleavage morphology. In contrast, when tested in air, both materials exhibited dimpled fracture surfaces as characteristic of ductile failure by microvoid nucleation, growth, and coalescence. Meletis and Hochman (1984) studied the SCC of 99.9% pure Mg in 3.3 wt% NaCl + 2 wt% K_2CrO_4, and reported failure in solution with 2.3% of strain in comparison with 5.1% strain in air, i.e., 55% reduction in strain in solution as compared to air (Figure 5.6). Fracture surfaces revealed TGSCC with specific features of cleavage-like fracture, which initiated at corrosion pits and was accompanied by H evolution.

Magnesium alloys are also reported to be susceptible to SCC in aqueous environments (Ben-Hamu et al., 2008; Bobby Kannan, Dietzel, Raman, & Lyon, 2007; Bobby Kannan et al., 2008; Makar, Kruger, & Sieradzki, 1993; Uematsu et al., 2012; Winzer et al., 2008, 2008a, 2008b). The alloying elements play a crucial role in either beneficial or detrimental manner on SCC of an alloy. The roles of most common alloying elements (aluminium (Al), zinc (Zn), and manganese (Mn)) and impurity (iron (Fe)) on the SCC of Mg alloys are discussed below.

5.4.2.1 Influence of aluminium

All Al-containing magnesium alloys (such as AZxx, containing Al and Zn, and AMxx, containing Al and Mn) have been reported to be susceptible to SCC to some extent in distilled water and chloride-containing solutions (Winzer et al., 2005). The susceptibility to SCC (represented as the minimum stress to cause SCC) was generally found to

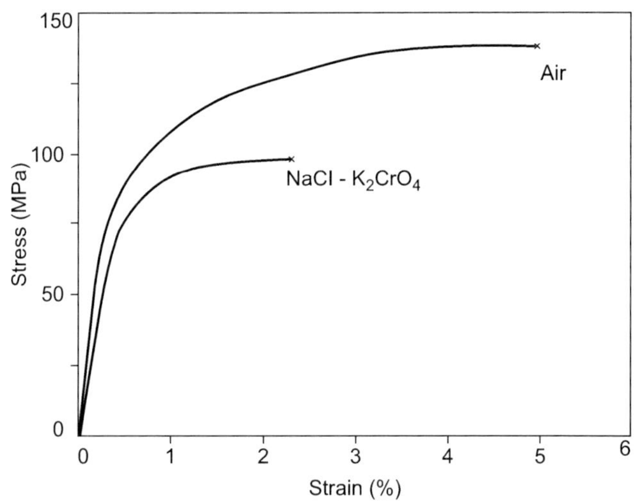

Figure 5.6 Stress versus strain curves for pure Mg specimens tested in air and corrosive solution (Meletis & Hochman, 1984).

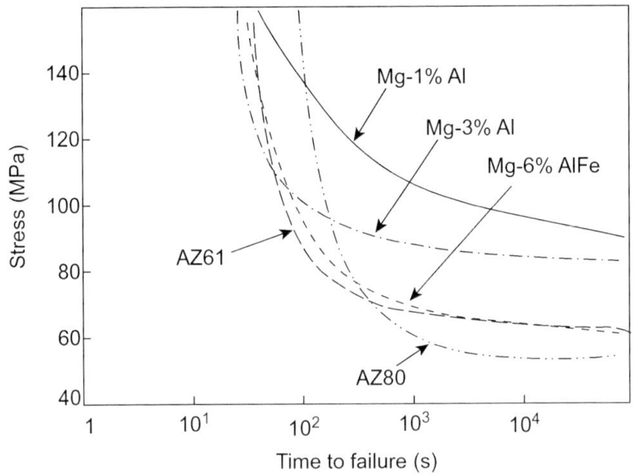

Figure 5.7 Stress versus time-to-failure curves for various Mg-Al alloys in 40 g L^{-1} NaCl + 40 g L^{-1} Na_2CrO_4 (Fairman & Bray, 1971).

increase with increasing Al concentration, as shown in Figure 5.7 (Fairman & Bray, 1971).

It should be noted that the Al is reported to be a neurotoxicant and to cause neurological disorders such as Alzheimer's disease and dementia (Crapper, Krishnan, & Quittkat, 1976; Forbes, Gentleman, & Maxwell, 1995). However, a few *in vivo* studies and *in vitro* cytocompatibility results have suggested the use of Al-containing Mg alloys (e.g., AZ91) to be appropriate for implant applications (Witte et al., 2005; Wong et al., 2010; Yuen & Ip, 2010). Nevertheless, the avoidance of Al in the Mg-alloy biomaterials may be quite useful in not only reducing the SCC susceptibility of the alloys in chloride-containing environments (such as human body fluid), but also in addressing the commonly perceived toxicity issues due to Al addition.

5.4.2.2 Influence of zinc

Zn also induces SCC susceptibility in Mg alloys (Baker & Avedesian, 1999). It is also reported that Mg-Zn alloys when alloyed with zirconium or rare earths (but no aluminium) have intermediate SCC resistance (Winzer et al., 2005).

5.4.2.3 Influence of manganese

Mg-Mn alloys are among the alloys with high resistance to SCC, and they are generally considered to be immune when loaded up to the yield strength in common test environments (Baker & Avedesian, 1999).

5.4.2.4 Influence of iron

Fe that is found in commercial Mg alloys as an impurity is known to reduce the corrosion resistance by promoting microgalvanic corrosion. The effect of Fe on SCC of Mg alloys remains unclear. Perryman (1951) reported that higher Fe content in Mg alloys decreased the SCC resistance in distilled water, whereas Timonova (1962) reported that Fe had no effect on SCC of Mg-Al-Zn-Mn alloys.

5.4.2.5 Influence of rare-earth elements

Rare-earth (RE) elements are generally added to Mg alloys for improving the creep resistance and corrosion resistance. It has been reported that Mg alloys containing RE may have intermediate SCC resistance (Baker & Avedesian, 1999). Rokhlin has also reported the addition of neodymium or cadmium to Mg—Zn—zirconium (Zr) alloy to increase the SCC resistance (Rokhlin, 2003). In a very recent study, Padekar, Singh Raman, Raja, and Paul (2013) showed that an RE-containing EV31A alloy possesses improved SCC resistance as compared to an RE-free alloy, AZ91E, particularly during constant load testing (CLT), which was attributed to the formation of a robust oxide layer on EV31A.

Influence of the surface corrosion film in SCC susceptibility is highly dependent on the alloy composition and/or the nature of the environment, such as the beneficial role of the robust films formed due to RE elements (as discussed above Padekar et al. (2013)), or due to chromate-containing aqueous solution (Ebtehaj et al., 1988).

5.4.3 Characterization of SCC using different techniques

There is no generalized approach that could provide a complete insight into prediction and mechanistic understanding of SCC. Combinations of a few techniques can qualitatively and quantitatively establish the susceptibility of different metals/alloys to SCC. The main techniques for characterization of SCC are slow strain rate tensile (SSRT) testing, U-bend testing, C-bend testing, compact tension (CT) testing, double cantilever beam (DCB) testing (Jones, 1992), circumferential notch tensile (CNT) testing (Singh Raman, Rihan, & Ibrahim, 2006), etc. A brief description of some of these testing methods is provided in this section.

5.4.3.1 Slow strain rate tensile testing

SSRT testing is a uniaxial tensile test conducted at extremely low cross-head speeds in the presence of corrosive environment (Dietzel, Srinivasan, & Atrens, 2011; Jones, 1992). These tests will invariably produce fracture either by SCC or mechanical failure or both. Slow straining ensures the opportunity for the environment to interact with crack tip and also the presence of continuing plastic strain, which encourages the initiation and growth of stress corrosion cracks. The SCC susceptibility is evaluated in terms of the time taken for failure to occur, the extension at failure, and/or morphology features of the fracture surface. SSRT is an accelerated laboratory testing and generates data in a relatively short time. Also, SSRT usually gives conservative results because of the severity of the tests.

Strain rate is a critical parameter in SSRT testing of Mg alloys. For a specific combination of alloy and environment, SCC occurs in a narrow window of strain rates

(Singh Raman, 2005). Wearmouth et al. (1973) investigated the role of strain rate for Mg-7Al alloy in aqueous chloride–chromate solution and attributed the role of strain to rupture of any surface film to allow localized dissolution or H ingress. Ebtehaj et al. (1988) also investigated the influence of strain rate on the SCC susceptibility of as-cast Mg-9Al alloy. They proposed that the film integrity was maintained at low strain rates, which prevented H ingress into the matrix, and failure occurred in ductile fashion. As the strain rate was increased, the film repassivation decreased, allowing H ingress more freely, causing embrittlement of the matrix. At relatively higher strain rates, ductile tearing occurred before embrittlement because of insufficient time available for H ingress. This explained the maximum SCC susceptibility at intermediate strain rates, as shown in Figure 5.8 (Ebtehaj et al., 1988).

In contrast to the maximum SCC susceptibility of Mg-Al alloy at intermediate strain rates in NaCl + K_2CrO_4 solution (Figure 5.8) reported by Ebtehaj et al. (1988), Winzer et al. (2008a) found the maximum susceptibility of a Mg-Al alloy tested in distilled water at lowest strain rate (Figure 5.9). This was characterized by the greater decrease in ultimate tensile strength (UTS) and elongation-to-failure in distilled water as compared to laboratory air in low strain rate regime (Figure 5.9). The occurrence of maximum SCC susceptibility at intermediate strain rates, as reported by Ebtehaj et al. (1988), is attributed to the passivating characteristic of the alloy in the chromate-containing solution used in their study. The corrosion film developed in this passivating environment required higher strain rate to sustain SCC, whereas a

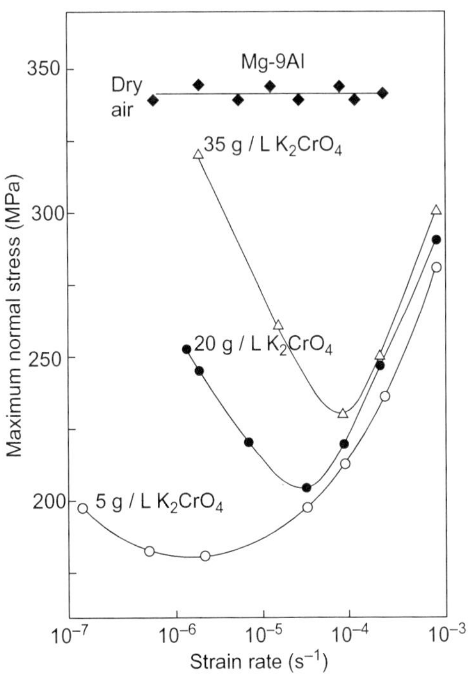

Figure 5.8 The effect of strain rate on the maximum nominal stress to fracture the specimens exposed to dry air or solutions containing 5 g L^{-1} NaCl and various amounts of K_2CrO_4 (Ebtehaj et al., 1988).

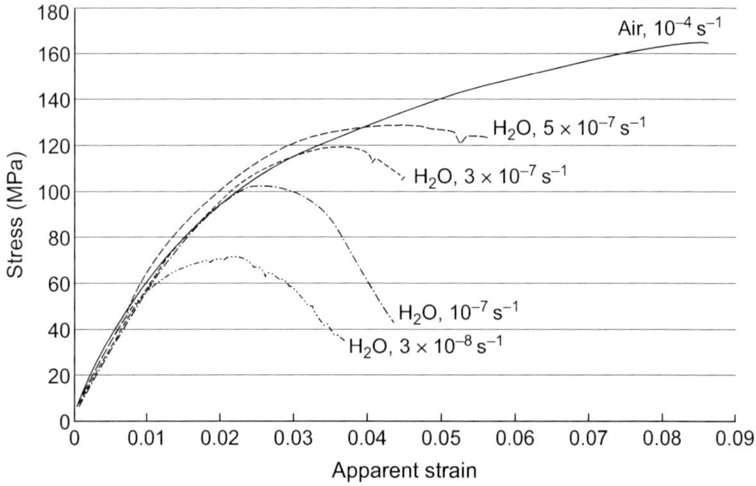

Figure 5.9 Stress versus apparent strain curves for AZ91 in distilled water and air (Winzer et al., 2008a).

less passivating solution in Winzer and co-workers' study (Winzer et al., 2008a) produced corrosion film that could be disrupted at lower strain rate and caused SCC.

In a recent study (Choudhary & Singh Raman, 2012) on SCC of AZ91D in *m*-SBF, SSRT tests were conducted at a strain rate of $2.2 \times 10^{-7}\ s^{-1}$ under different conditions: (a) strained in air, (b) strained in *m*-SBF solution, (c) immediately strained in air after preimmersion in *m*-SBF solution for 28 h, and (d) continuously cathodically charged and simultaneously pulled in *m*-SBF. The purpose of these tests were to investigate: (1) the occurrence of SCC and its mechanism and (2) whether the loss of the mechanical property is indeed a result of the synergistic effect of stress and corrosive environment (i.e., SCC) or it can simply be attributed to the continuously reducing cross-sectional area of the specimen due to high corrosion rate of Mg alloys. The stress versus time curves for AZ91D alloy under these conditions are shown in Figure 5.10. AZ91D tensile specimen, which was strained continuously in *m*-SBF at strain rate of $2.2 \times 10^{-7}\ s^{-1}$ (Figure 5.10(b)), failed in 28 h. Accordingly, a preimmersion time of 28 h was selected for straining the preimmersed AZ91D tensile specimen in air, assuming the stress-independent corrosion damage to be similar to that of specimen strained in *m*-SBF solution. The specimen pulled in *m*-SBF solution (Figure 5.10(b)) showed a considerable reduction in mechanical properties as compared to the specimen pulled in air (Figure 5.10(a)). The stress–time curve for the specimen preimmersed in *m*-SBF for 28 h and then strained in air (Figure 5.10(c)) was very similar to the specimen pulled in air (Figure 5.10(a)). Therefore, it is fair to infer that the simultaneous effect of stress and corrosive environment (i.e., SCC) was the primary cause of the loss of mechanical property in the case of specimen pulled in *m*-SBF (Figure 5.10(b)), and the stress-independent corrosion contributed only marginally (as the plots suggest). Also, a considerable loss of the mechanical strength and a decrease in the time-to-failure were observed in continuous

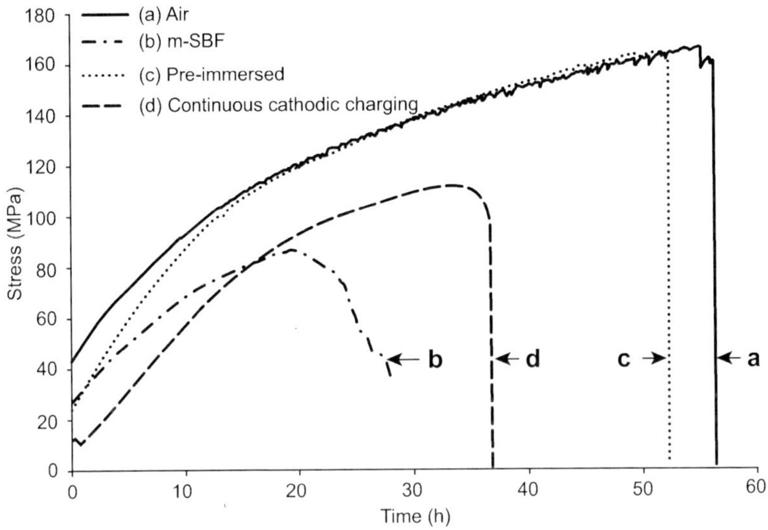

Figure 5.10 Stress versus time plots of AZ91D tested at a strain rate of 2.2×10^{-7} s^{-1}: (a) in air, (b) in *m*-SBF solution, (c) preimmersed in *m*-SBF for 28 h and then immediately strained in air, and (d) continuously charged at 200 mV cathodic to open circuit potential in *m*-SBF during the test (Choudhary & Singh Raman, 2012).

cathodic charging conditions (i.e., when anodic dissolution is minimal, Figure 5.10(d)). Thus, the cathodic charging experiment inferred hydrogen-assisted stress corrosion cracking (HASCC) as one of the SCC mechanisms. However, the maximum loss in mechanical property was observed in the case of the specimen strained in *m*-SBF solution at open circuit conditions (Figure 5.10(b)), which indicated possibly of a combined effect of HASCC and anodic dissolution. This combined mechanism may be plausible for Mg and its alloys that are known to invariably produce considerable amount of hydrogen during anodic dissolution (Song & Atrens, 1999; Song, Atrens, John, et al., 1997; Song, Atrens, Stjohn, et al., 1997).

The SSRT results of AZ91D tested under different conditions were further examined by fractography. The overall fracture surface of the specimen tested in air is shown in Figure 5.11(a). The entire surface of the specimen tested in air revealed dimples at higher magnification (Figure 5.11(b)), confirming the mechanical overload failure. There were also occasional features of brittle fracture, which may be associated with secondary phase particles of the alloy (predominantly β-phase). Mix-mode fracture observed in air was consistent with the fractographic features for the Mg alloy tested in air reported in the literature (Ben-Hamu et al., 2008; Bobby Kannan et al., 2008). In contrast, the overall fracture surface of the specimen tested in *m*-SBF showed the localized attack at the specimen circumference (Figure 5.11(c)). At higher magnification, the fractography of this specimen revealed distinctive feature of SCC such as transgranular cracking and localized cracks (Figure 5.11(d)).

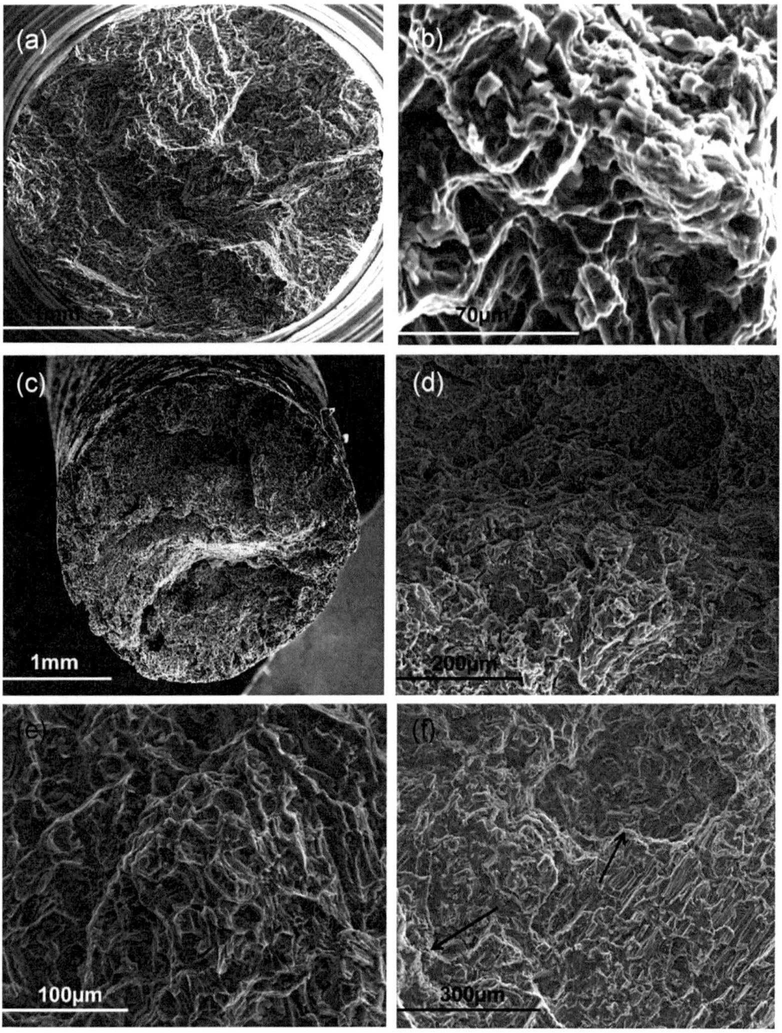

Figure 5.11 Fractograph of the specimen tested in air (a, b), fractograph of the specimen failed in *m*-SBF at a strain rate of $2.2 \times 10^{-7}\ s^{-1}$ (c, d), fractographs for the specimens: (e) immersed in *m*-SBF solution for 28 h and then strained in air, and (f) continuously cathodically charged and simultaneously pulled in *m*-SBF (Choudhary & Singh Raman, 2012).

The fractograph for the specimen immersed in *m*-SBF for 28 h followed by straining in air showed dimples formation (Figure 5.11(e)) similar to that observed for the specimen tested in air, suggesting the ductile failure. The specimen that was cathodically charged and simultaneously pulled in *m*-SBF solution exclusively showed the evidences of the transgranular cracking (arrows, Figure 5.11(f)). Transgranular cracking observed in the case of cathodically charged conditions was attributed to the mechanism involving hydrogen. In fact, the hydrogen generated during the cathodic charging diffuses ahead of the crack tip and embrittles the matrix (Winzer et al., 2005).

5.4.3.2 Fracture mechanics-based approach to SCC

The fracture mechanics-based approach is used for quantitative determination of important design parameters for the components susceptible to SCC. This type of study is of particular importance for the medical implant components and materials for the determination of safe service stress level in physiological conditions. In general, damage-tolerant design methods are applied for performing these studies, which assume that structure/component contains initial cracks. The concept of fracture mechanics is then used to characterize the conditions for propagation of cracks from these initial flaws. Usually, SCC can occur at stresses below general yield strength and propagate in elastic body; hence, linear elastic fracture mechanics (LEFM) principles are applied for determining the crack tip stress intensity factor (K_I) (Dietzel et al., 2011).

According to LEFM, the K_I at the crack tip can be given by Eqn (5.1):

$$K_I = Y\sigma(\pi a)^{1/2} \tag{5.1}$$

where Y is a geometrical factor, σ is the applied stress, and a is the crack length.

For a given material/environmental condition, the typical schematic of relationship between stress intensity and crack growth rate is shown in Figure 5.12 (Dietzel et al., 2011). The value of K_I at which the first measurable crack extension occurs is characterized by threshold stress intensity for SCC (K_{ISCC}). Consequently, there will not be any crack propagation at stress intensities lower than the K_{ISCC}. The practical meaning

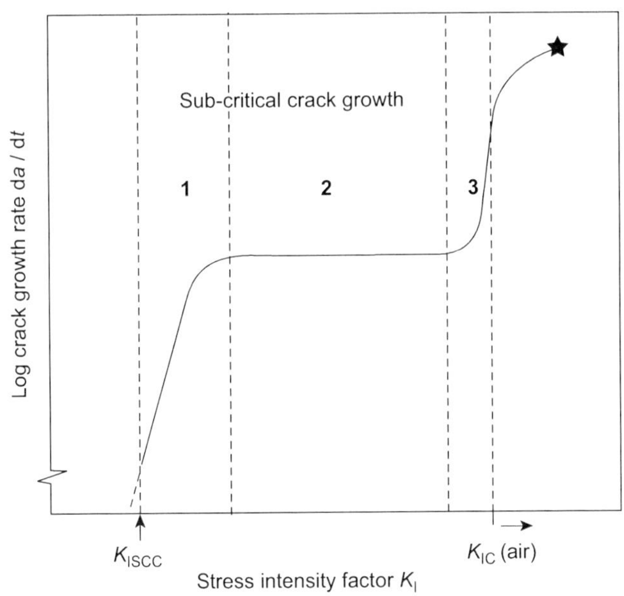

Figure 5.12 Typical plot of stress intensity factor versus crack growth rate (Dietzel et al., 2011).

of K_{ISCC} lies in the fact that below this stress intensity factor, crack growth rates become insignificantly low (i.e., $<10^{-12}$ m s^{-1}) (Singh Raman & Pal, 2011; Winzer et al., 2005).

The plot in Figure 5.12 is specifically characterized by three zones: region I (crack initiation and crack propagation), region II (steady state crack propagation), and region III (failure by overloading). During region I, the crack growth increases rapidly as the stress intensity is increased from the K_{ISCC}. The crack velocity is reasonably constant in region II and largely independent of the stress intensity factor. As the stress intensity factor approaches K_{IC} (i.e., fracture toughness, the critical stress intensity of the specimen tested in air at which it fails), pure mechanical rupture (ductile) dominates over subcritical crack extension caused by SCC, and rapid crack growth occurs. Determinations of K_{ISCC} and crack growth rate data for life prediction of components/structure are of prime importance for various applications, such as implants.

K_{ISCC} of AZ91D alloy in *m*-SBF was determined by the authors' research group (Choudhary & Singh Raman, 2012), using circumferential notch tensile (CNT) testing. The fatigue precracked CNT specimens of AZ91D alloy were loaded at different stress intensities in *m*-SBF at 37 °C. Specimens failed after different time durations depending on the applied stress intensities. The K_I at each applied load was calculated (following the procedure as described in references Choudhary and Singh Raman (2012), Pal, Ibrahim, and Singh Raman (2011), Rihan, Raman, and Ibrahim (2005)) and plotted against the time-to-failure (Figure 5.13). As evident from Figure 5.13, time-to-failure increased exponentially with the decreasing stress intensity. For example, a CNT specimen loaded at a stress intensity of 11.7 MPa $m^{1/2}$ failed in only 16 h, whereas it took 119 h for a specimen loaded at a stress intensity of 6.8 MPa $m^{1/2}$ to fail. The threshold stress intensity was determined by drawing an

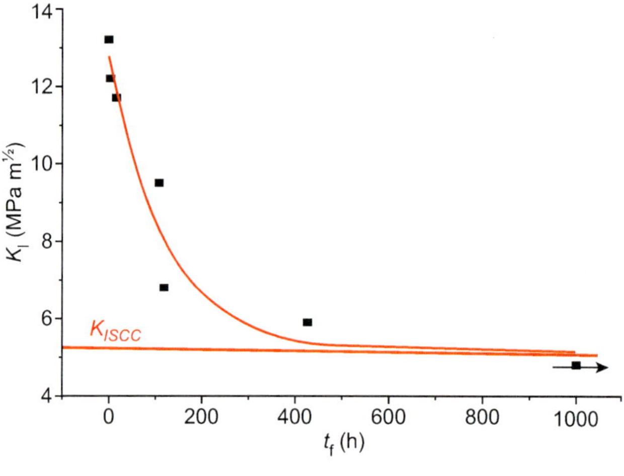

Figure 5.13 K_I versus t_f plot of AZ91D alloy tested in *m*-SBF at 37 °C (Choudhary and Singh Raman (2012)).

asymptote to time-to-failure axis in the plot of stress intensity versus time-to-failure (Figure 5.13).

The equation of the fitted curve can be given by Eqn (5.2):

$$K_{\mathrm{I}} = 7.58 \exp\left(-\frac{t_{\mathrm{f}}}{123.58}\right) + 5.18 \tag{5.2}$$

According to Eqn (5.2), the value of K_{I} when producing a horizontal asymptote to time-to-failure axis in Figure 5.13 (i.e., K_{ISCC}), was 5.18 MPa $\mathrm{m}^{1/2}$ for AZ91D alloy in *m*-SBF. Fracture surfaces of CNT specimens were observed under the scanning electron microscope (SEM) after cleaning of corrosion products. Figure 5.14(a) showed the overall fractograph of specimen that failed at a stress intensity of 6.8 MPa $\mathrm{m}^{1/2}$. Fracture surface consists of four regions: machined notch, fatigue precracked region, SCC zone, and mechanical overload failure zone. At higher magnifications, distinctive features of each of these regions become clearer. Figure 5.14(b) shows the transition from the fatigue precracked region to the SCC zone. The fatigue precracked region

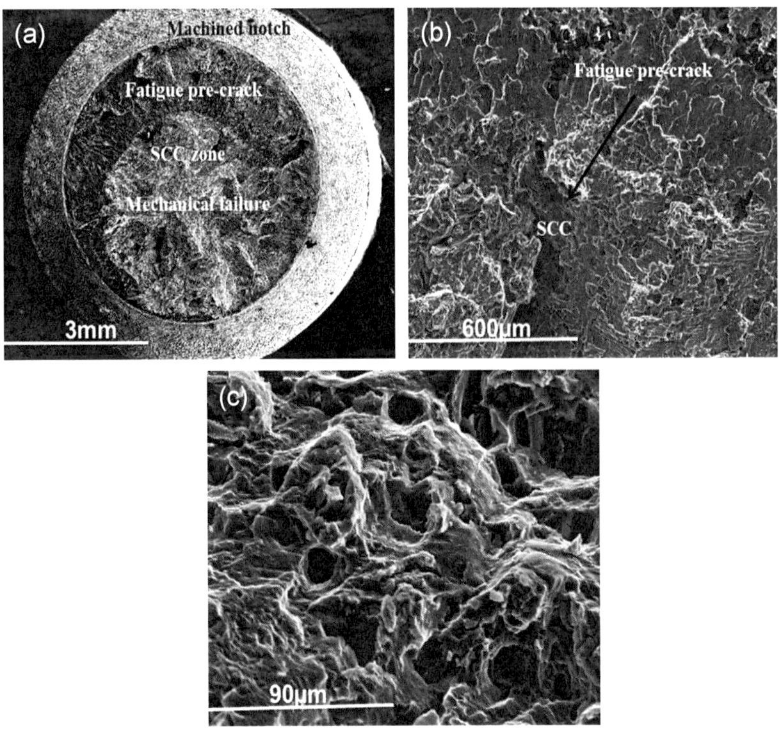

Figure 5.14 SEM fractographs of the specimen tested at $K_{\mathrm{I}} = 6.8$ MPa $\mathrm{m}^{1/2}$: (a) the overall fracture surface showing machined notch, fatigue precrack, SCC and mechanical failure zones, (b) fractograph showing fatigue precracked and SCC zones (arrow indicating the crack propagation direction), and (c) mechanical failure zone showing the dimples (Choudhary & Singh Raman, 2012).

was characterized as beach marks running from the circumference. SCC region mainly consisted of transgranular cracking and a few cleavage marks, which may be indicative of hydrogen-assisted cracking. Apart from transgranular cracking, a few localized pits were also visible. Stress corrosion crack propagated transgranularly until the critical stress intensity (i.e., fracture toughness) was attained. At this stage, the mechanical overloading takes over, and the central part of the fracture surface revealed ductile failure characterized by dimple formation at a higher magnification (Figure 5.14(c)).

Biodegradable temporary implants are required to serve at least for a definite period depending on their applications. For the implant devices with sharp corners and protrusions, it is essential to determine the K_{ISCC} and stress corrosion crack growth rate for predicting the service life of the component in the physiological environment. The implant devices generally have locations of stress concentrations such as the fastener holes, etc. that often show crack-like behaviour. The in-service inspection of the crack growth of such implants will be a nontrivial task. However, the flaw-tolerant design assessment of the implant devices (that may have a pre-existing fine crack), by the CNT testing as demonstrated in the study by Choudhary and Singh Raman (2012), will provide crucial information on whether the existing crack will be tolerable for the given service time of the device (James & Sire, 2010).

5.5 Prevention of SCC

There are a few recommendations to avoid SCC in practices, which are discussed below.

1. Selection of SCC resistant alloy

The SCC resistant alloys can be produced by judicious selection of alloying additions. For example, Mg alloys with higher concentration of Al are more susceptible to SCC, while the addition of Zr and rare earths as alloying elements have been reported to increase the SCC resistance (Baker & Avedesian, 1999).

2. Control of stress

SCC propagates when the stress intensities are above a critical level, which has been reported to be attained at 30–50% of the tensile yield strength of the material (Winzer et al., 2005). Hence, it is important to design Mg alloys having the threshold stress intensities reasonably lower than those presented by the service conditions. In this vein, a great importance should be given to the use of simple shapes that will provide fewer local high stress generating points (Winzer et al., 2005).

3. Surface modifications and coatings

Coatings may be an appropriate measure for preventing the SCC to some extent. As discussed, localized corrosion such as pitting is a common cause for initiation of SCC in Mg alloys. Accordingly, the coating, which decreases the pitting susceptibility of Mg alloys in human body fluid, could help in avoiding/retarding the SCC (Baker & Avedesian, 1999). Kannan and Lynnley (2011) recently studied the mechanical integrity of the hydroxyapatite (HA)-coated Mg alloy while exposed to body fluid and observed a significant improvement in mechanical integrity due to the coating. The enhanced mechanical integrity of the HA-coated specimen was attributed to the decrease in susceptibility of the coated alloy to localized corrosion/pitting.

An effective modification of stresses or environment is not practical in biomedical applications because implants are expected to work under given stress conditions in the corrosive human body fluid. So, the development of SCC-resistant Mg alloys appears to be one of the viable solutions. But the alloying elements must be nontoxic, which may limit the choices of alloying elements. Therefore, there is a value in exploring SCC mitigation with the use of biocompatible coatings on Mg alloys (Chen, Birbilis, & Abbott, 2011; Ng, Wong, & Cheng, 2010; Yu, Pan, & Uan, 2010; Zhou, Shan, Han, & Ke, 2008).

5.6 Conclusions

This chapter emphasizes the need for characterization of SCC of biodegradable Mg alloys for temporary implant applications.

Mg alloys have been consistently shown to be susceptible to SCC in SBF (that contains considerable amount of chloride) and as a result may suffer premature cracking during service. The localized corrosion is the common cause of pit initiation. The role of pits is primarily to provide a film-free and active surface through which atomic hydrogen can enter into the matrix and can subsequently cause embrittlement and cracking. The predominant mode for SCC crack propagation in Mg alloys was transgranular cracking. The slow strain rate testing experiments under different electrochemical conditions have suggested the SCC mechanism for a common Mg alloy in *m*-SBF to be the combined effect of HASCC and localized dissolution.

To prevent SCC initiation, the most promising recommendation is the use of appropriate biocompatible coatings that retard the localized corrosion/pitting.

The chapter also describes the importance of determination of threshold stress intensity for stress corrosion cracking (i.e., K_{ISCC}) as the critical design parameter for quantitative characterization of SCC before their use as implants and provides some typical data that were generated using a simple and inexpensive testing approach.

References

Baker, H., & Avedesian, M. M. (1999). In M. Avedesian, & H. Baker (Eds.), *Magnesium and magnesium alloys*. Materials Park, OH: ASM International.

Ben-Hamu, G., Eliezer, D., Dietzel, W., & Shin, K. S. (2008). Stress corrosion cracking of new Mg-Zn-Mn wrought alloys containing Si. *Corrosion Science, 50*, 1505–1517.

Bobby Kannan, M., Dietzel, W., Blawert, C., Atrens, A., & Lyon, P. (2008). Stress corrosion cracking of rare-earth containing magnesium alloys ZE41, QE22 and Elektron 21 (EV31A) compared with AZ80. *Materials Science and Engineering: A, 480*, 529–539.

Bobby Kannan, M., Dietzel, W., Raman, R. K. S., & Lyon, P. (2007). Hydrogen-induced-cracking in magnesium alloy under cathodic polarization. *Scripta Materialia, 57*, 579–581.

Bobby Kannan, M., & Singh Raman, R. K. (2008). Evaluating the stress corrosion cracking susceptibility of Mg-Al-Zn alloy in modified-simulated body fluid for orthopaedic implant application. *Scripta Materialia, 59*, 175–178.

Bobby Kannan, M., Singh Raman, R. K., Witte, F., Blawert, C., & Dietzel, W. (2011). Influence of circumferential notch and fatigue crack on the mechanical integrity of biodegradable magnesium-based alloy in simulated body fluid. *Journal of Biomedical Materials Research Part B: Applied Biomaterials, 96B*, 303–309.

Bombara, G., & Cavallini, M. (1977). Stress corrosion cracking of bone implants. *Corrosion Science, 17*, 77–85.

Bundy, K. J., Marek, M., & Hochman, R. F. (1983). In vivo and in vitro studies of the stress-corrosion cracking behavior of surgical implant alloys. *Journal of Biomedical Materials Research, 17*, 467–487.

Bundy, K. J., Vogelbaum, M. A., & Desai, V. H. (1986). The influence of static stress on the corrosion behavior of 316L stainless steel in Ringer's solution. *Journal of Biomedical Materials Research, 20*, 493–505.

Chakrapani, D., & Pugh, E. (1975). The transgranular SCC of a Mg-Al alloy: crystallographic, fractographic and acoustic-emission studies. *Metallurgical and Materials Transactions A, 6*, 1155–1163.

Chakrapani, D. G., & Pugh, E. N. (1976). Hydrogen embrittlement in a Mg-Al alloy. *Metallurgical Transactions A, 7*, 173–178.

Chen, J., Ai, M., Wang, J., Han, E.-H., & Ke, W. (2009). Stress corrosion cracking behaviors of AZ91 magnesium alloy in deicer solutions using constant load. *Materials Science and Engineering: A, 515*, 79–84.

Chen, X. B., Birbilis, N., & Abbott, T. B. (2011). A simple route towards a hydroxyapatite-$Mg(OH)_2$ conversion coating for magnesium. *Corrosion Science, 53*, 2263–2268.

Choudhary, L., & Singh Raman, R. K. (2012). Magnesium alloys as body implants: fracture mechanism under dynamic and static loadings in a physiological environment. *Acta Biomaterialia, 8*, 916–923.

Crapper, D. R., Krishnan, S. S., & Quittkat, S. (1976). Aluminium, neurofibrillary degeneration and Alzheimer's disease. *Brain: A Journal of Neurology, 99*, 67–80.

Dietzel, W., Srinivasan, P. B., & Atrens, A. (2011). Testing and evaluation methods for stress corrosion cracking (SSC) in metals. In V. S. Raja, & T. Shoji (Eds.), *Stress corrosion cracking: Theory and practice*. Woodhead Publishing Limited.

Ebtehaj, K., Hardie, D., & Parkins, R. N. (1988). The influence of chloride-chromate solution composition on the stress corrosion cracking of a Mg-Al alloy. *Corrosion Science, 28*, 811–821.

Fairman, L., & Bray, H. J. (1971). Transgranular see in Mg-Al alloys. *Corrosion Science, 11*, 533–541.

Fischerauer, S. F., Kraus, T., Wu, X., Tangl, S., Sorantin, E., Hanzi, A. C., et al. (2012). In vivo degradation performance of micro-arc-oxidized magnesium implants: a micro-CT study in rats. *Acta Biomaterialia, 9*, 5411–5420.

Forbes, W. F., Gentleman, J. F., & Maxwell, C. J. (1995). Concerning the role of aluminum in causing dementia. *Experimental Gerontology, 30*, 23–32.

Granchi, D., Ciapetti, G., Stea, S., Savarino, L., Filippini, F., Sudanese, A., et al. (1999). Cytokine release in mononuclear cells of patients with Co-Cr hip prosthesis. *Biomaterials, 20*, 1079–1086.

Grimm, M. J. (2007). *Selection of materials for biomedical applications. Handbook of materials selection*. John Wiley & Sons, Inc.

Gu, X. N., Zhou, W. R., Zheng, Y. F., Cheng, Y., Wei, S. C., Zhong, S. P., et al. (2010). Corrosion fatigue behaviors of two biomedical Mg alloys – AZ91D and WE43 – in simulated body fluid. *Acta Biomaterialia, 6*, 4605–4613.

Hänzi, A. C., Gerber, I., Schinhammer, M., Löffler, J. F., & Uggowitzer, P. J. (2010). On the in vitro and in vivo degradation performance and biological response of new biodegradable Mg-Y-Zn alloys. *Acta Biomaterialia, 6*, 1824–1833.

Hartwig, A. (2001). Role of magnesium in genomic stability. *Mutation Research/Fundamental and Molecular Mechanisms of Mutagenesis, 475*, 113–121.

Heublein, B., Rohde, R., Kaese, V., Niemeyer, M., Hartung, W., & Haverich, A. (2003). Biocorrosion of magnesium alloys: a new principle in cardiovascular implant technology? *Heart, 89*, 651–656.

Hughes, A. N., & Jordan, B. A. (1972). Metallurgical observations on some metallic surgical implants which failed in vivo. *Journal of Biomedical Materials Research, 6*, 33–48.

James, B. A., & Sire, R. A. (2010). Fatigue-life assessment and validation techniques for metallic vascular implants. *Biomaterials, 31*, 181–186.

Jones, R. H. (1992). *Stress-corrosion cracking*. ASM International.

Jones, D. A. (1996). *Principles and prevention of corrosion*. Prentice Hall.

Kannan, M. B., & Lynnley, O. (2011). In vitro mechanical integrity of hydroxyapatite coated magnesium alloy. *Biomedical Materials, 6*, 045003.

Kannan, M. B., & Raman, R. K. S. (2008). In vitro degradation and mechanical integrity of calcium-containing magnesium alloys in modified-simulated body fluid. *Biomaterials, 29*, 2306–2314.

Kirkland, N. T. (2012). Magnesium biomaterials: past, present and future. *Corrosion Engineering, Science and Technology, 47*, 322–328.

Kirkland, N. T., Lespagnol, J., Birbilis, N., & Staiger, M. P. (2010). A survey of bio-corrosion rates of magnesium alloys. *Corrosion Science, 52*, 287–291.

Kirkland, N., Staiger, M., Nisbet, D., Davies, C., & Birbilis, N. (2011). Performance-driven design of biocompatible Mg alloys. *JOM Journal of the Minerals, Metals and Materials Society, 63*, 28–34.

Kraus, T., Fischerauer, S. F., Hänzi, A. C., Uggowitzer, P. J., Löffler, J. F., & Weinberg, A. M. (2012). Magnesium alloys for temporary implants in osteosynthesis: in vivo studies of their degradation and interaction with bone. *Acta Biomaterialia, 8*, 1230–1238.

Li, Z., Gu, X., Lou, S., & Zheng, Y. (2008). The development of binary Mg-Ca alloys for use as biodegradable materials within bone. *Biomaterials, 29*, 1329–1344.

Lynch, S. P., & Trevena, P. (1988). Stress corrosion cracking and liquid metal embrittlement in pure magnesium. *Corrosion, 44*, 113–124.

Makar, G. L., & Kruger, J. (1993). Corrosion of magnesium. *International Materials Reviews, 38*, 138–153.

Makar, G. L., Kruger, J., & Sieradzki, K. (1993). Stress corrosion cracking of rapidly solidified magnesium-aluminum alloys. *Corrosion Science, 34*, 1311–1342.

Ma, E., & Xu, J. (2009). Biodegradable alloys: the glass window of opportunities. *Nature Materials, 8*, 855–857.

McCormack, B. A. O., Walsh, C. D., Wilson, S. P., & Prendergast, P. J. (1998). A statistical analysis of microcrack accumulation in PMMA under fatigue loading: applications to orthopaedic implant fixation. *International Journal of Fatigue, 20*, 581–593.

Meletis, E. I., & Hochman, R. F. (1984). Crystallography of stress corrosion cracking in pure magnesium. *Corrosion, 40*, 39–45.

Ng, W. F., Wong, M. H., & Cheng, F. T. (2010). Stearic acid coating on magnesium for enhancing corrosion resistance in Hanks' solution. *Surface and Coatings Technology, 204*, 1823–1830.

Nielsen, K. (1987). Corrosion of metallic implants. *British Corrosion Journal, 22*, 272–278.

Niinomi, M. (2002). Recent metallic materials for biomedical applications. *Metallurgical and Materials Transactions A, 33*, 477–486.

Padekar, B. S., Singh Raman, R. K., Raja, V. S., & Paul, L. (2013). Stress corrosion cracking of a recent rare-earth containing magnesium alloy, EV31A, and a common Al-containing alloy, AZ91E. *Corrosion Science, 71*, 1–9.

Pal, S., Ibrahim, R. N., & Singh Raman, R. K. (2011). Threshold stress intensity factor and crack growth rate for stress corrosion cracking of simulated heat affected zone in caustic solution. *Engineering Fracture Mechanics, 78*, 13–26.

Parkins, R. N. (1980). Predictive approaches to stress corrosion cracking failure. *Corrosion Science, 20*, 147–166.

Perryman, E. C. W. (1951). Stress-corrosion of magnesium alloys. *Journal of Institute of Metals, 78*, 621–642.

Puleo, D. A., & Huh, W. W. (1995). Acute toxicity of metal ions in cultures of osteogenic cells derived from bone marrow stromal cells. *Journal of Applied Biomaterials, 6*, 109–116.

Ratner, B. D., Hoffman, A. S., Schoen, F. J., & Lemons, J. E. (1996). *Biomaterials science: An introduction to materials in medicine*. Elsevier Science.

Rihan, R., Raman, R. K. S., & Ibrahim, R. N. (2005). Circumferential notched tensile (CNT) testing of cast iron for determination of threshold (K_{ISCC}) for caustic crack propagation. *Materials Science and Engineering: A, 407*, 207–212.

Rokhlin, L. L. (2003). *Magnesium alloys containing rare earth metals: Structure and properties*. Taylor & Francis.

Saris, N.-E. L., Mervaala, E., Karppanen, H., Khawaja, J. A., & Lewenstam, A. (2000). Magnesium: an update on physiological, clinical and analytical aspects. *Clinica Chimica Acta, 294*, 1–26.

Sieradzki, K., & Newman, R. C. (1985). Brittle behavior of ductile metals during stress-corrosion cracking. *Philosophical Magazine A, 51*, 95–132.

Singh Raman, R. (2004). The role of microstructure in localized corrosion of magnesium alloys. *Metallurgical and Materials Transactions A, 35*, 2525–2531.

Singh Raman, R. (2005). Evaluation of caustic embrittlement susceptibility of steels by slow strain rate testing. *Metallurgical and Materials Transactions A, 36*, 1817–1823.

Singh Raman, R. K., Birbilis, N., & Efthimiadis, J. (2004). Corrosion of Mg alloy AZ91 the role of microstructure. Corrosion Engineering. *Science and Technology, 39*, 346–350.

Singh Raman, R. K., & Pal, S. (2011). A simple approach to the determination of threshold stress intensity for stress corrosion cracking (K_{ISCC}) and crack growth of sensitized austenitic stainless steel. *Metallurgical and Materials Transactions A, 42*, 2643–2651.

Singh Raman, R. K., Rihan, R., & Ibrahim, R. N. (2006). A novel approach to the determination of the threshold for stress corrosion cracking (K_{ISCC}) using round tensile specimens. *Metallurgical and Materials Transactions A, 37*, 2963–2973.

Sivakumar, M., Kamachi Mudali, U., & Rajeswari, S. (1994). Investigation of failures in stainless steel orthopaedic implant devices: fatigue failure due to improper fixation of a compression bone plate. *Journal of Materials Science Letters, 13*, 142–145.

Sivakumar, M., & Rajeswari, S. (1992). Investigation of failures in stainless steel orthopaedic implant devices: pit-induced stress corrosion cracking. *Journal of Materials Science Letters, 11*, 1039–1042.

Song, G. (2007). Control of biodegradation of biocompatable magnesium alloys. *Corrosion Science, 49*, 1696–1701.

Song, G. L. (2011). Corrosion electrochemistry of magnesium and its alloys. In G. L. Song (Ed.), *Corrosion of magnesium alloys*. Cambridge, UK: Woodhead Publishing Limited.

Song, G. L., & Atrens, A. (1999). Corrosion mechanisms of magnesium alloys. *Advanced Engineering Materials, 1*, 11–33.

Song, G., Atrens, A., John, D. S., Wu, X., & Nairn, J. (1997). The anodic dissolution of magnesium in chloride and sulphate solutions. *Corrosion Science, 39*, 1981–2004.

Song, G., Atrens, A., Stjohn, D., Nairn, J., & Li, Y. (1997). The electrochemical corrosion of pure magnesium in 1 N NaCl. *Corrosion Science, 39*, 855–875.

Staiger, M. P., Pietak, A. M., Huadmai, J., & Dias, G. (2006). Magnesium and its alloys as orthopedic biomaterials: a review. *Biomaterials, 27*, 1728–1734.

Stampella, R. S., Procter, R. P. M., & Ashworth, V. (1984). Environmentally-induced cracking of magnesium. *Corrosion Science, 24*, 325–341.

Timonova, M. A. (1962). *Intercrystalline corrosion and corrosion of metals under stress*. Great Britain: Consultants Bureau.

Tunold, R., Holtan, H., Berge, M.-B. H., Lasson, A., & Steen-Hansen, R. (1977). The corrosion of magnesium in aqueous solution containing chloride ions. *Corrosion Science, 17*, 353–365.

Uematsu, Y., Kakiuchi, T., & Nakajima, M. (2012). Stress corrosion cracking behavior of the wrought magnesium alloy AZ31 under controlled cathodic potentials. *Materials Science and Engineering: A, 531*, 171–177.

Wang, M. L., Nesti, L. J., Tuli, R., Lazatin, J., Danielson, K. G., Sharkey, P. F., et al. (2002). Titanium particles suppress expression of osteoblastic phenotype in human mesenchymal stem cells. *Journal of Orthopaedic Research, 20*, 1175–1184.

Wearmouth, W. R., Dean, G. P., & Parkins, R. N. (1973). Role of stress in the stress corrosion cracking of a Mg-Al Alloy. *Corrosion, 29*, 251–258.

Winzer, N., Atrens, A., Dietzel, W., Raja, V. S., Song, G., & Kainer, K. U. (2008). Characterisation of stress corrosion cracking (SCC) of Mg-Al alloys. *Materials Science and Engineering: A, 488*, 339–351.

Winzer, N., Atrens, A., Dietzel, W., Song, G., & Kainer, K. U. (2008a). Comparison of the linearly increasing stress test and the constant extension rate test in the evaluation of transgranular stress corrosion cracking of magnesium. *Materials Science and Engineering: A, 472*, 97–106.

Winzer, N., Atrens, A., Dietzel, W., Song, G., & Kainer, K. U. (2008b). Fractography of stress corrosion cracking of Mg-Al alloys. *Metallurgical and Materials Transactions A, 39*, 1157–1173.

Winzer, N., Atrens, A., Song, G., Ghali, E., Dietzel, W., Kainer, K., et al. (2005). A critical review of the stress corrosion cracking (SCC) of magnesium alloys. *Advanced Engineering Materials, 7*, 659–693.

Witte, F. (2011). The history of biodegradable magnesium implants: a review. *Acta Biomaterialia, 6*, 1680–1692.

Witte, F., Fischer, J., Nellesen, J., Crostack, H.-A., Kaese, V., Pisch, A., et al. (2006). In vitro and in vivo corrosion measurements of magnesium alloys. *Biomaterials, 27*, 1013–1018.

Witte, F., Hort, N., Vogt, C., Cohen, S., Kainer, K. U., Willumeit, R., et al. (2009). Degradable biomaterials based on magnesium corrosion. *Current Opinion in Solid State and Materials Science, 12*, 63–72.

Witte, F., Kaese, V., Haferkamp, H., Switzer, E., Meyer-Lindenberg, A., Wirth, C. J., et al. (2005). In vivo corrosion of four magnesium alloys and the associated bone response. *Biomaterials, 26*, 3557–3563.

Wolf, F. I., & Cittadini, A. (2003). Chemistry and biochemistry of magnesium. *Molecular Aspects of Medicine, 24*, 3–9.

Wong, H. M., Yeung, K. W. K., Lam, K. O., Tam, V., Chu, P. K., Luk, K. D. K., et al. (2010). A biodegradable polymer-based coating to control the performance of magnesium alloy orthopaedic implants. *Biomaterials, 31*, 2084–2096.

Xin, Y., Hu, T., & Chu, P. K. (2011). In vitro studies of biomedical magnesium alloys in a simulated physiological environment: a review. *Acta Biomaterialia, 7*, 1452–1459.

Xin, Y., Huo, K., Tao, H., Tang, G., & Chu, P. K. (2008). Influence of aggressive ions on the degradation behavior of biomedical magnesium alloy in physiological environment. *Acta Biomaterialia, 4*, 2008–2015.

Xu, L., Yu, G., Zhang, E., Pan, F., & Yang, K. (2007). In vivo corrosion behavior of Mg-Mn-Zn alloy for bone implant application. *Journal of Biomedical Materials Research Part A, 83A*, 703–711.

Xu, L., Zhang, E., Yin, D., Zeng, S., & Yang, K. (2008). In vitro corrosion behaviour of Mg alloys in a phosphate buffered solution for bone implant application. *Journal of Materials Science: Materials in Medicine, 19*, 1017–1025.

Yokoyama, K., Ichikawa, T., Murakami, H., Miyamoto, Y., & Asaoka, K. (2002). Fracture mechanisms of retrieved titanium screw thread in dental implant. *Biomaterials, 23*, 2459–2465.

Yuen, C. K., & Ip, W. Y. (2010). Theoretical risk assessment of magnesium alloys as degradable biomedical implants. *Acta Biomaterialia, 6*, 1808–1812.

Yu, B. L., Pan, X. L., & Uan, J. Y. (2010). Enhancement of corrosion resistance of Mg-9 wt.% Al-1 wt.% Zn alloy by a calcite ($CaCO_3$) conversion hard coating. *Corrosion Science, 52*, 1874–1878.

Zainal Abidin, N. I., Atrens, A. D., Martin, D., & Atrens, A. (2011). Corrosion of high purity Mg, Mg2Zn0.2Mn, ZE41 and AZ91 in Hank's solution at 37 °C. *Corrosion Science, 53*, 3542–3556.

Zberg, B., Uggowitzer, P. J., & Loffler, J. F. (2009). MgZnCa glasses without clinically observable hydrogen evolution for biodegradable implants. *Nature Materials, 8*, 887–891.

Zhang, E., & Yang, L. (2008). Microstructure, mechanical properties and bio-corrosion properties of Mg-Zn-Mn-Ca alloy for biomedical application. *Materials Science and Engineering: A, 497*, 111–118.

Zhou, W., Shan, D., Han, E.-H., & Ke, W. (2008). Structure and formation mechanism of phosphate conversion coating on die-cast AZ91D magnesium alloy. *Corrosion Science, 50*, 329–337.

Zhou, W., Shen, T., & Aung, N. N. (2010). Effect of heat treatment on corrosion behaviour of magnesium alloy AZ91D in simulated body fluid. *Corrosion Science, 52*, 1035–1041.

Effect of amino acids and proteins on the *in vitro* performance of coated magnesium for biomedical applications

6

Nicholas Travis Kirkland[1], *Jay Waterman*[2]
[1]Nagasaki University, Nagasaki, Japan; [2]University of Canterbury, Christchurch, New Zealand

6.1 Introduction

Magnesium corrosion rates and mechanisms depend heavily on the corrosion environment. The complexity of the *in vivo* environment precludes that the corrosion may be influenced in many different ways. In vitro corrosion tests, ideally, simplify the complicated environment of the body to make measuring corrosion rates and performance easier and cheaper to perform. To use an *in vitro* approximation that will accurately predict *in vivo* performance, it is necessary to understand the effect different components have on the corrosion mechanisms. Unfortunately, these effects are often interdependent on other aspects of the system, making clear relationships difficult to obtain. Alloys, coatings, and choice of buffer are just a few of the many system facets that can interact with other parts of the solution, affecting results in difficult-to-predict ways. Consequently, the final corrosion mechanisms in a body environment are not necessarily the same as the sum of the individual components of an *in vitro* test. To that end, it is critical to understand the role and effects of *in vivo* fluid components on a wide spectrum of test cases/conditions to effectively choose an *in vitro* test.

The organic components of the *in vivo* environment are characteristic of the complex interactions that can affect corrosion rates of Mg in unexpected ways. Although large molecules found in the body, such as amino acids and proteins, may not react directly with the electrochemical corrosion reactions of Mg, they have been shown to affect the ultimate corrosion rates *in vitro* (Virtanen, 2011). These molecules, too often neglected in more simple *in vitro* experiments, can affect the rates of adsorption to the surface of corroding Mg, affecting properties such as buffer capacity and interacting with the corrosion products. These secondary processes in turn affect the corrosion rates and performance of the material *in vitro*. Before any *in vitro* test may be accurately used to determine the correct degradation rate of Mg-based implants *in vivo*, it is necessary to understand and fully characterize these effects and understand how these molecules affect corrosion behavior.

Surface Modification of Magnesium and its Alloys for Biomedical Applications. http://dx.doi.org/10.1016/B978-1-78242-077-4.00006-1

The use of Mg and Mg alloys as biomedical implant materials is primarily limited by the corrosion rate requirements of the implant. The corrosive nature of the body means that, for many implant applications, corrosion rates of pure Mg and Mg compatible alloys are sufficiently high that additional corrosion protection must be applied. Biocompatible coatings are one of the most promising solutions to this problem (Zeng, Dietzel, Witte, Hort, & Blawert, 2008). Calcium phosphate (CaP) coatings have been widely used to improve the biocompatibility of traditional permanent metal implants (León & Jansen, 2008). CaP coatings are insoluble enough in physiological systems to provide corrosion protection as well as biocompatibility. However, due to the complex nature of the *in vivo* environment, it is necessary to understand how the addition of these coatings will interact with more complex molecules in solution. Therefore, an investigation into the effects of such a coating on the corrosion layers, behavior, and ultimate corrosion rate in both amino acids and protein solutions are presented here.

6.2 The role of amino acids and proteins in biocorrosion

6.2.1 Amino acids—bridging the physiological gap

Amino acids (AA), molecules containing C, H, O, and N, are essentially the building blocks of proteins and perform countless functions in the body's metabolism (Puleo & Bizios, 2009). In a biocorrosion sense, this includes acting as a pH regulator, affecting the buffering capacity of *in vitro* solutions, although not as effectively as carbonic buffering (Malda et al., 2008). Further, they may influence corrosion behavior of Mg alloys by adsorbing onto the sample surface, following the Langmuir isotherm (Ashassi-Sorkhabi, Ghasemi, & Seifzadeh, 2005; Ashassi-Sorkhabi, Majidi, & Seyyedi, 2004). They have also been found to form a complex with metal cations, potentially encouraging the dissolution of metal (Bruneel & Helsen, 1988; William & William, 2004). They may chelate (metal atom attached to large molecule) with Mg and inhibit the formation of the insoluble Mg compounds that form the passive layer, decreasing the corrosion resistance (Yamamoto & Hiromoto, 2009). It has been suggested that other organic compounds may have a similar effect to reduce the corrosion resistance of any insoluble salt layer on Mg (Yamamoto & Hiromoto, 2009).

Conversely, it has also been found that AA absorbed onto the surface of Mg alloys increased the resistance to polarization and reduced the corrosion current density (Gu, Zheng, & Chen, 2009). A similar increase in corrosion resistance has also been reported for steel and aluminum (Ashassi-Sorkhabi et al., 2004, 2005; El-Shafei, Moussa, & El-Far, 1997). To date it appears the specific role of amino acids on Mg biomaterials is unclear, and it certainly warrants further investigation.

The addition of the amino acids to measure corrosion is essential to bridge the gap between simple ionic solutions and more complicated protein solutions used in biocorrosion experiments. Perhaps the most commonly used solution used in the literature is Eagle's Minimum Essential Medium (MEM), a formula containing ionic compounds and amino acids in amounts similar to those found in the body, providing a suitable environment for cell cultures. However, the number of Mg-based experiments that have used an AA solution is still relatively small (Gu et al., 2009; Mueller, Lucia Nascimento, & Lorenzo De Mele, 2010; Witte et al., 2007; Xu et al., 2009; Yamamoto & Hiromoto, 2009; Yang & Zhang, 2009; Yun et al., 2009; Zheng, Gu, Xi, & Chai, 2010). Others have used a variation of MEM, namely, Dulbecco's Modified Eagle Medium (DMEM), a version of MEM with additional amino acids present (Carboneras, Garcã-A-Alonso, & Escudero, 2011). Others have also tried McCoy's medium (Yun et al., 2009), PRMI1640 (Xu et al., 2009), and a simulated blood plasma (Yang & Zhang, 2009).

The reason for the lack of widespread use of MEM in the bio-Mg literature may stem from its traditional function as a base media for protein of cell cultures rather than corrosion tests. However, MEM and similar solutions provide a link between a completely inorganic salt solution and one containing proteins and is a necessary solution in the understanding of both coated and uncoated Mg corrosion *in vivo*.

6.2.2 Proteins—an important biological addition

The importance of proteins in the body is well understood. For the sake of conciseness, their salient properties will not be discussed in great detail herein, and instead the reader is directed toward some of the many excellent tomes on the subject (Horbett, 2004; Latour, 2008). In short, they are known to be a vital part to the success of any implant in the body, and their connection to the surface is crucial because they provide attachment sites on which cells bind (Miller, Fainerman, Leser, & Michel, 2004).

In a biocorrosion sense, proteins may directly influence the corrosion properties of metals (Clark & Williams, 1982), as well as being important to their perceived biocompatibility (Latour, 2008). Built of chains of amino acids, the structures of proteins typically have functional groups on the outside that can be polar, nonpolar, hydrophilic, or hydrophobic, thus creating a complex surface. For a given protein with a set isoelectric point, the overall charge depends on the pH of the environment (Latour, 2008). This charge and surface composition will affect the adsorption rates of the proteins onto the surface. It is the adsorption of these proteins that can affect the corrosion surface, and any interactions are crucial to the overall corrosion behavior.

Fetal bovine serum (FBS) is most commonly used to provide proteins for *in vitro* SBFs for corrosion tests (Eliezer & Witte, 2010; Keim, Brunner, Fabry, & Virtanen, 2011; Kirkland & Birbilis, 2013; Salunke, Shanov, & Witte, 2011). Taken from the blood plasma of a calf fetus, it is often used for cell culture work due to the amount of growth factors it contains as well as the low number of antibodies (Willmer, 1965). FBS typically contains proteins amounting to 30–45 g/L (Equitech-Bio Inc., 2010), with the majority of the protein composition made of bovine serum albumin (BSA) (Lenter, 1981). BSA is effectively the bovine equivalent of human serum albumin (HAS), which accounts for about half of the blood serum protein in humans.

The biocorrosion properties of Mg and Mg alloys have been shown to be strongly affected by the addition and concentration of proteins in solution. Several studies have shown that proteins decreased the corrosion of various Mg alloys (Eliezer & Witte,

2010; Liu, Xin, Tian, & Chu, 2007; Liu et al., 2010; Yamamoto & Hiromoto, 2009; Yang, Hort, Willumeit, & Feyerabend, 2012); however, the reason for this effect is not always clear due to the complex nature of the interactions. For example, Mueller et al. reported that corrosion rates of pure Mg in PBS increase when BSA is added to 0.1%, but then decrease as further BSA is added (up to 10%) (Mueller, De Mele, Nascimento, & Zeddies, 2009). Further, this trend was dependent on alloy, with AZ31 displaying very little change in corrosion rate across BSA concentrations, while LAE442 steadily increased. Kirkland et al. found that the addition of 10% FBS slowed the corrosion rate of a number of alloys as well as pure Mg in MEM (Kirkland & Birbilis, 2013). However, increasing additions of BSA to MEM solutions from 20 to 60 g/L led to increasing corrosion rates for pure Mg, although all were ultimately lower than in MEM alone (Kirkland & Birbilis, 2013). These varying trends suggest the corrosion influence of proteins is more complicated than the simplistic adsorption model suggests, because proteins in solution have a complicated relationship on the passivation layers that form (Willumeit et al., 2011).

A number of tests that have been performed on Mg with the addition of proteins to a base simulated body fluid are summarized in Table 6.1.

In the literature, the amount of protein added to solutions has varied widely, from approximately 0.1 (Mueller et al., 2009) to 40 g/L (Rettig & Virtanen, 2009). It is important to consider that, when added in FBS form, the proteins are not solely BSA, and a 10% FBS solution would be equivalent to adding a total of ~4 g/L of protein. Consequently, all FBS experiments had between 4 and 6 g/L of protein. Although none of the experiments justified why this amount was chosen, it is a commonly used amount in cell cultures, and FBS is expensive (Helgason & Miller, 2005).

Directly adding BSA proteins results in the same amount of protein in solution, as it is in a pure form. Most experiments that used BSA additions also did not specify why a certain amount was chosen. However, Mueller et al. chose three different amounts to make comparisons between them (Mueller et al., 2009), and Rettig and Virtanen chose to use 40 g/L in two separate experiments, the most realistic amount investigated to date (Rettig & Virtanen, 2008, 2009). This was justified as it would equivalent to the physiological amount of HSA in the body.

Although the effect of proteins has been widely investigated for a number of biomaterials, studies of their influence on Mg alloys has still been rather limited. Apart from the tests mentioned above, very few studies have looked at how they might affect the corrosion properties of coated alloys *in vitro*. As one of the key parameters determining the success of an implant in the body, protein interaction with the surface of Mg alloys is still not well understood.

It is important to consider that the initial surface of the implant will be the surface the proteins immediately come into contact with, and consequently it is this surface that warrants the greatest study. Although CaP coatings do not chemically react strongly with proteins, the changes in effective surface area, corrosion rate, local pH, and diffusion will heavily affect the interaction with proteins, further altering the corrosion behavior.

The biomimetic process uses solutions similar in ionic composition to physiological fluids, with the aim of creating a coating that is similar in properties to the layer that

Table 6.1 Magnesium in vitro experiments involving proteins

Primary author (reference)	Base media	Protein addition
Gu (Gu et al., 2009)	MEM	10% FBS
Witte (Witte et al., 2007)	MEM	10% FBS
Yamamoto (Yamamoto & Hiromoto, 2009)	E-MEM	10% FBS
Pietak (Pietak, Mahoney, Dias, & Staiger, 2007)	MEM	15% FBS
Liu (Liu et al., 2007)	Basic SBF (not specified)	1 g/L BSA
Mueller (Mueller et al., 2007)	PBS	1/10 g/L BSA
Mueller (Mueller et al., 2010)	PBS	1/10 g/L BSA
Mueller (Mueller et al., 2009)	PBS	0.1/1/10 g/L BSA
Rettig (Rettig & Virtanen, 2008)	Oyane m-SBF[a]	40 g/L BSA
Eliezer (Eliezer & Witte, 2010)	0.9% NaCl, HBSS	Not stated
Willumeit (Willumeit et al., 2011)	DMEM, HBSS, Water	10/20% FBS
Salunke (Salunke et al., 2011)	DMEM	10% FBS
Rettig (Rettig & Virtanen, 2009)	Oyane m-SBF[a]	40 g/L BSA
Liu (Liu et al., 2010)	Water, 0.9% NaCl	1/10 g/L BSA
Keim (Keim et al., 2011)	"SBF," DMEM, 100% FBS	10% FBS

[a]Solution based on modified SBF by Oyane (Oyane et al., 2003).

would form *in vivo* (Barrere, 2002). The similarity of a biomimetic coating to natural bone mineral can increase bioactivity of the surface (Baker et al., 2006). These coatings do not contain any elements not found in bone or body fluids and have been used in various implant applications (Barrère, 2003; Kokubo, 1996; Lin & Li, 2006; Zhu & Song, 2005). Biomimetic coatings can be created easily using the process described in (Waterman et al., 2011). These coatings will be used to measure how an applied coating can affect the corrosion response of amino acids and proteins in solution on pure Mg. Due to its ubiquity, unique advantages, and ease of application, biomimetic coatings were chosen in this work as the most ideal for further investigation.

6.3 Effect of amino acids on corrosion performance of magnesium

6.3.1 Uncoated magnesium performance

Before proceeding to coated samples, it is first important to consider the impact of amino acids on the corrosion of uncoated Mg. This provides a base for comparison and further insight to the difference and protection coatings can offer.

The electrochemical behavior of pure Mg in a simple salt solution (Hanks Balanced Salt Solution, HBSS) is shown in comparison to MEM, which contains amino acids. Although the compositions of both media are similar, containing the same Mg^{2+} ion content, MEM has a lower Cl^- as well as higher Ca^{2+} and PO_4^{3-} concentrations (Table 6.2). Theoretically, such a difference would suggest slightly decreased corrosion rates and increased CaP formation on the surface.

The polarization potential (E_{Corr}) of uncoated Mg was found to be higher for samples in MEM than in HBSS (Figure 6.1). The corrosion current density (i_{Corr}) was also found to be higher in MEM, which at a basic level is counterintuitive, given the aforementioned higher E_{Corr} and lower Cl^- content in MEM. It is possible that this increased

Table 6.2 Composition of corrosion media

Component	Human plasma (HP)	Hank's balanced salt solution (HBSS)[a]	Minimum essential medium (MEM)[b]	MEM + bovine serum albumin (MEM + BSA)[c]
Na^+	142	145	117.4	117.4
Cl^-	103	144.6	123.5	123.5
K^+	5.0	5.8	5.4	5.4
Ca^{2+}	2.5	1.3	1.8	1.8
Mg^{2+}	1.5	0.4	0.4	0.4
HPO_4^{2-}	1.0	0.8	1	1
SO_4^{2-}	0.5	0.4	0.4	0.4
D-Glucose	5	5.5	5.5	5.5
Bicarbonate (HCO_3^-)	22–30	±26.2	±26.2	±26.2
HEPES	–	±25	±25	±25
Phenol red	–	0.03	0.03	0.03
Albumin (g/L)	34–54	–	–	40

All concentrations in mmol/L unless otherwise stated.
Concentrations of inorganic blood contents given as in Warrel (2003).
[a]H1641, Sigma-Aldrich.
[b]56414C, Sigma-Aldrich.
[c]MP Biomedical NZ Ltd.

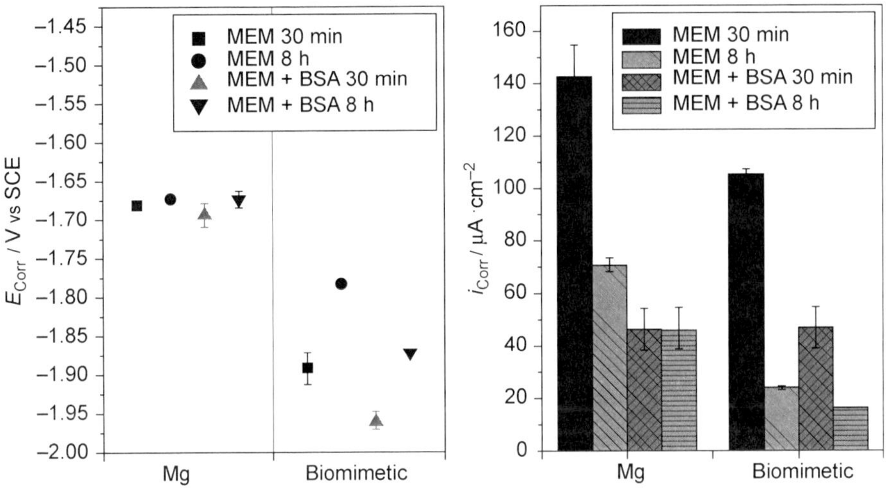

Figure 6.1 Corrosion potential and current density for uncoated samples with and without amino acids and proteins.

corrosion rate was observed due to the increased buffer capacity provided by amino acids (Malda et al., 2008), as well as the proposed mechanism that amino acids chelate with Mg^{2+} and inhibit passive layer formation (Yamamoto & Hiromoto, 2009).

Materials and methods

The pure Mg used for this work was a high-purity form (99.99%) obtained specifically for this study (Timmenco Ltd., Toronto, Canada). The chemical compositions were determined independently via inductively coupled plasma atomic emission spectroscopy (results not shown herein). Unless otherwise stated, all experiments presented in this chapter were performed using a three-electrode flat-cell (K0235, Princeton Applied Research, TN, USA) with a volume of 300 mL of media and an exposed working area of 1 cm^2. Results were recorded on a Biologic SP-150 using EC-Lab 10.2 software (BioLogic Inc., TN, USA).

Although MEM contains a similar composition of inorganic ions as HBSS, the amino acids it contains may act as charge carriers, resulting in the possibility that the higher current measurements were an artefact of solution conductivity. To investigate this, the conductivities of the solutions were measured, with MEM found to possess a slightly lower conductivity than HBSS with HEPES, indicating that this was not the case (Figure 6.2).

6.3.2 Effect on coated magnesium

Analyzing the polarization behavior of coated Mg samples in HBSS and MEM revealed some interesting results (Figure 6.3). A similar trend was observed across the samples, with an increase in both E_{Corr} and i_{Corr} in MEM over HBSS, primarily due to a large increase in the cathodic kinetics (i.e., increase in cathodic reactions

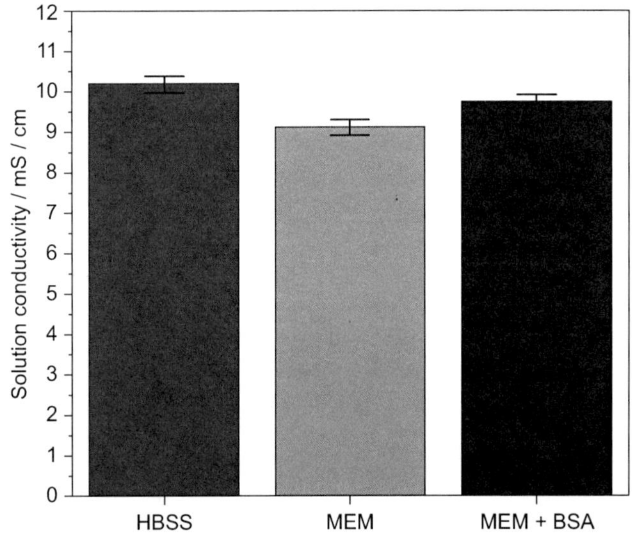

Figure 6.2 Solution conductivity at 37 °C with the HEPES buffer.

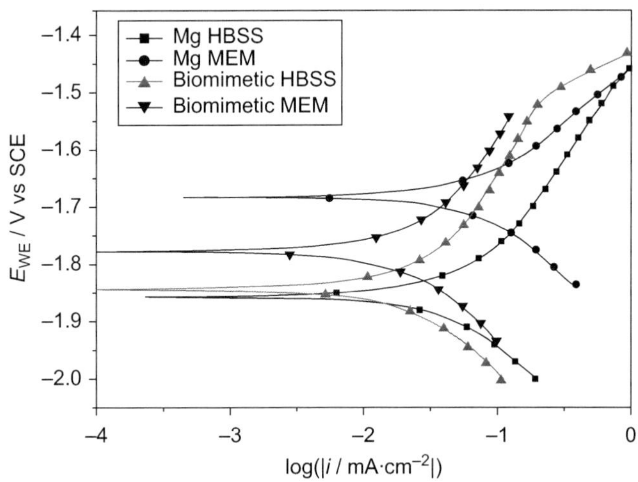

Figure 6.3 PDP in HBSS versus MEM buffered with HEPES at 8 h immersion.

taking place). The nature of the corrosion layers can be determined through impedance testing. Although it has been reported elsewhere that, for steel and aluminum, amino acids increase polarization resistance (Ashassi-Sorkhabi et al., 2004, 2005), the opposite appears true for uncoated Mg in MEM (at least for the first few hours) (Figure 6.4). From the Nyquist plots, it can be seen that, for each time point, both the film and EDL resistance was decreased in the presence of the amino acids. Consequently, instead of slowing short-term corrosion, amino acids are increasing the rate at which it proceeds. This is supported by the findings of Yamamoto et al. (Yamamoto & Hiromoto, 2009).

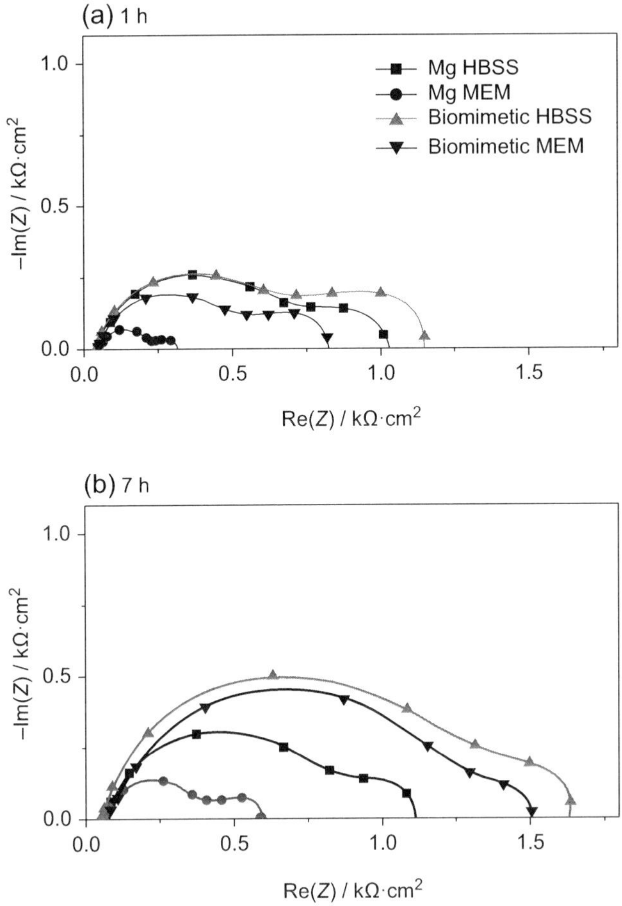

Figure 6.4 EIS of all samples in HBSS and MEM buffered with HEPES at 1 h (a) and 7 h (b) immersion.

Coated samples exhibited similar behavior in the presence of amino acids, with the polarization resistance remaining primarily cathodically controlled and increased in MEM compared with HBSS. However, unlike uncoated Mg, this shift was accompanied by a decrease in the anodic kinetics, making the biomimetic-coated sample display less corrosion in MEM than HBSS after 6 h. A similar behavior can be seen when analyzing the polarization resistance, where the coated samples initially have lower resistance in MEM, but become almost equal after 7 h immersion. Coupled with the more noble potential, this leads to an overall reduction in the corrosion rate.

From these data, we can see that the effect of MEM is altered when the sample is coated versus uncoated. For uncoated samples, the additional buffer capacity of the MEM and/or the inhibition of the passive layers increased corrosion rates. But when the coatings are applied and MEM is added, the effect of these passivation layers is less important, because the bulk of the protection comes from the CaP coating.

Also, because the effective corrosion rate is reduced, pH rise due to corrosion drops, and the buffering capacity of the amino acids becomes less important. The net result is that the accelerating effects on corrosion one might expect amino acids to have is reduced by the application of a coating.

6.4 Effect of proteins on magnesium biocorrosion

6.4.1 Uncoated magnesium

Uncoated Mg displayed highly consistent corrosion potential in both MEM and MEM-containing 40 g/L BSA (MEM + BSA) throughout the duration of the tests Figure 6.1(a). Proteins do not appear to directly interact with the corrosion of pure Mg in solution. For uncoated Mg, the proteins decreased the cathodic and anodic kinetics for scans at both 30 min (Figure 6.5) and 8 h immersion (Figure 6.6) without changing the corrosion potential significantly. The effect of proteins is reported to be focused around the adsorption to form a protective layer (Liu et al., 2007, 2010). The total i_{Corr} for Mg in MEM was initially high, more than 140 μA/cm^2, but rapidly decreased over the 8 h of investigation Figure 6.1(b). This behavior may be attributed to the formation of the hydroxide layer, charge separation, and a rise in local pH, with longer immersion times decreasing the corrosion rate. Looking at the polarization plots for uncoated samples, it can be seen that there was a slight drop in both anodic and cathodic branches, consistent with the formation of a corrosion layer on Mg (Figure 6.7(a)). This is confirmed via EIS (Figure 6.8(a) and (b)), as Mg in MEM displays the two-time constant system that commonly characterizes semiprotective hydroxide layers. After 8 h immersion, the total impedance of both the oxide layer and the EDL has increased, indicating the corrosion film became slightly more protective as corrosion progressed and the layer formed, which can be seen later in Figure 6.9.

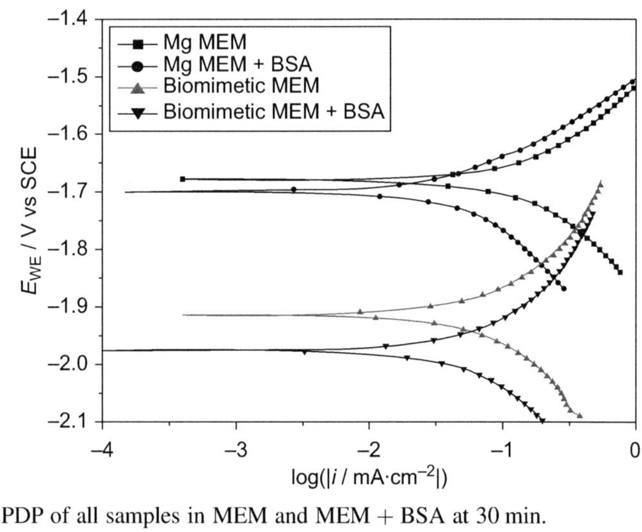

Figure 6.5 PDP of all samples in MEM and MEM + BSA at 30 min.

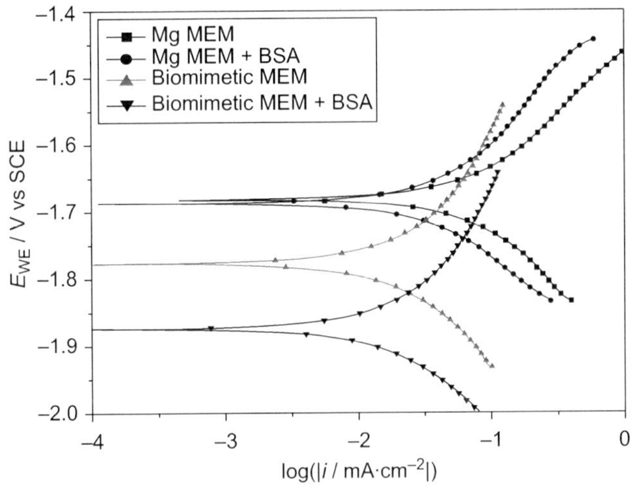

Figure 6.6 PDP of all samples in MEM and MEM + BSA at 8 h.

The addition of the protein albumin to the test solution strongly affects the corrosion behavior of uncoated Mg. The corrosion potential, E_{Corr}, does not significantly change with the addition of albumin; however, the i_{Corr} was found to drop considerably (Figure 6.1). BSA addition was found to decrease both the cathodic and anodic reactions (Figure 6.7(a)). When corroding in BSA, the total corrosion current density did not drop much from 30 min to 8 h, suggesting that the initial layers that formed within 30 min did not change much over the duration of the test. The Nyquist plots confirm this, showing similar impedance behavior for Mg in BSA at both time points (Figure 6.8). The relative shape of the Nyquist plots remain similar to MEM solutions alone. This is indicative of a greater film and double-layer resistance due to proteins impeding corrosion by adsorbing to the surface and forming a protective layer (Liu et al., 2007, 2010; Rettig & Virtanen, 2008, 2009).

Previously, albumin has been shown to increase E_{Corr} of Mg by decreasing the anodic reaction rate on AZ91 (Liu et al., 2007), as well as increase the anodic reaction rate on WE43 and LAE442 (Mueller et al., 2010). Over the 8 h tests reported here, both anodic and cathodic reactions decreased with the addition of BSA. This matches work previously reported by (Kirkland et al., 2010). Adsorption of proteins to the Mg surface creates a layer that partially protects the underlying metal by reducing the effective exposed surface area. While it has been reported that chelating metal ions with the proteins would increase the corrosion rate of other metals (Clark & Williams, 1982), the data here support previous works that do not show this effect on pure Mg (Willumeit et al., 2011). However, it is clear from the literature that different alloys, surfaces, and corrosion conditions can affect the measured results. To understand the effects of surface condition and corrosion rate on the amino acid and protein modified solutions, the data presented above will be used as a baseline from which to compare coated samples.

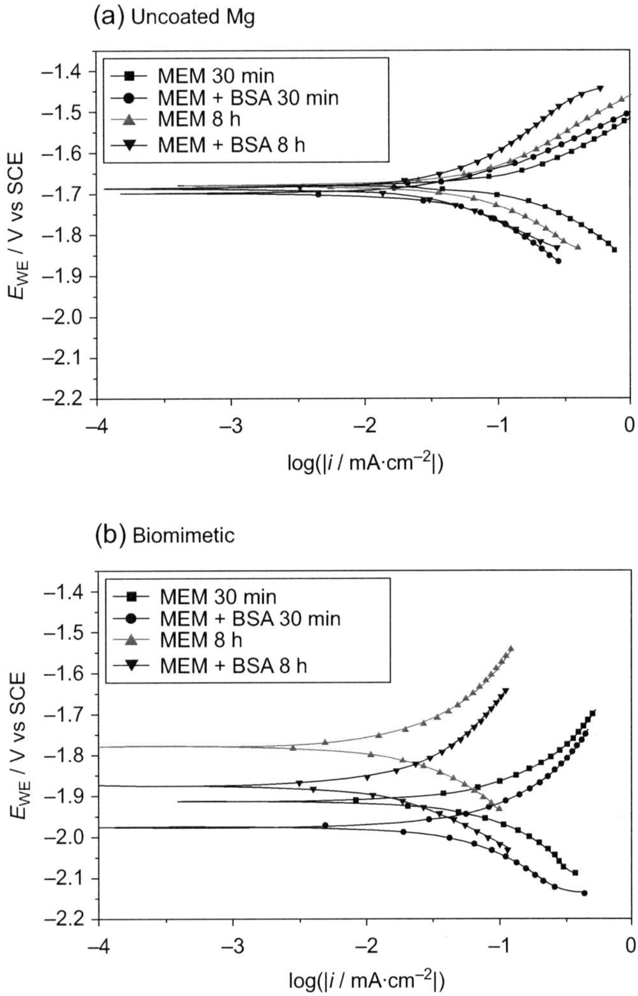

Figure 6.7 PDP over time in MEM and MEM + BSA for (a) uncoated Mg, (b) biomimetic coated.

6.4.2 Proteins on biomimetically coated magnesium

The electrochemical behavior of biomimetically coated samples was tested under the same conditions as the bare Mg substrates. The coatings have the effect of changing the surface, reducing corrosion and diffusion rates from the corrosion sites to the bulk solution. The coatings caused a decrease in the measured E_{Corr} for all samples. Although reduced in the short term, the corrosion potential of the biomimetic-coated Mg increased toward, but not reaching, the potential of the uncoated samples in both solutions over the 8 h of testing (Figure 6.1). Although the addition of proteins did not significantly affect the corrosion potential of uncoated Mg, when the coating is

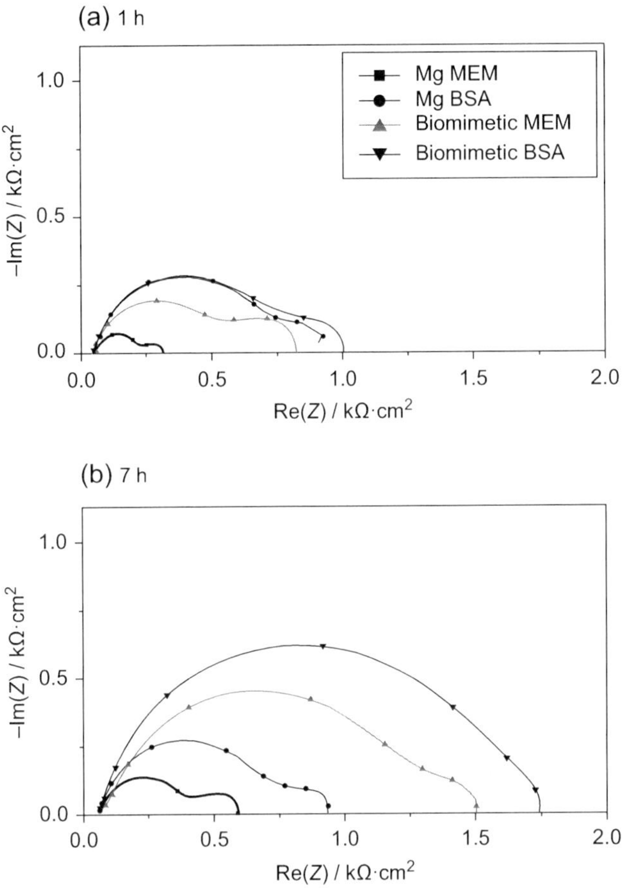

Figure 6.8 EIS of solutions with and without proteins at 1 h (a) and 7 h (b).

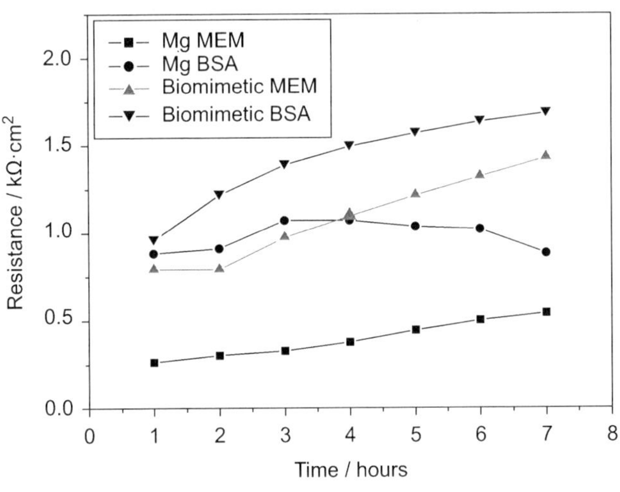

Figure 6.9 EIS over time total polarization resistance.

present proteins in solution lowered E_{Corr} at both 30 min and 8 h. The effect of coatings on corrosion current density was about what would be expected for these samples. High initial corrosion rates after immersion quickly drop once the corrosion layers begin to form. The protective coating prevents diffusion and leads to local pH rises that help form protective layers underneath (Waterman et al., 2011). The proteins enhance the coatings protective effects, leading to a decrease in the corrosion rate early on, and further improving over the 8 h test. The proteins caused a decrease in the cathodic kinetics of the samples (Figure 6.5). This is again the case with the biomimetic coatings after 8 h (Figure 6.6), when the lower cathodic kinetics provides a lower corrosion current density despite an increase in anodic kinetics due to the more negative potential. The proteins here had a synergistic effect with the coatings. While protein solutions led to lower corrosion rates on uncoated Mg, there was little change in i_{Corr} from 30 min to 8 h. However, the coated samples continued to become more protective, leading to a corrosion rate lower than the coating applied alone.

There were no large changes in cathodic kinetics over time for coated samples in each solution (Figure 6.7(b)). The drop in corrosion current density is therefore due to the decrease in anodic kinetics as immersion time increases. This is due to the formation of corrosion layers impeding the oxidation of Mg. For the coated samples, BSA decreased the amount of anodic shift seen in MEM. These corrosion layers also exhibited the two-time constant system typical of a semiporous coating (Figure 6.8). The proteins adsorbed to the surface and reduced the effective area vulnerable to corrosion. This effect is seen by the increased film resistance. Note that the magnitude of the impedance of the biomimetic-coated sample in BSA is comparable to the uncoated Mg at this time in BSA, which is larger than the impedance of the biomimetic-coated sample in MEM alone. The presence of proteins in this solution affects the early corrosion rates as much as the coating. Indeed, the initial corrosion current density in MEM and MEM + BSA is higher and nearly equal to (within error) the corrosion rates of uncoated Mg in MEM + BSA. Thus, the reduction of corrosion rate (at least initially) due to proteins is significant. However, the barrier to corrosion that the proteins provide does not appear to offer complete protection, as the impedance of the uncoated Mg in protein-containing solutions stays constant or rises over the test period.

The biomimetic coatings slow the corrosion rates initially, but the porous nature of the coating allows the corrosion to continue through the pores until the corrosion layers reach steady state. Thus, over the 8 h of testing, the coated samples show increased protection as time goes on. When corrosion occurs in the pores of the coating, the local pH rise promotes formation of $Mg(OH)_2$ layers as well as deposition of other calcium and phosphate compounds on the Mg (Waterman et al., 2011). This accounts for the decrease in the anodic reaction rate over time and the improvement in the film resistance. When proteins are added to the solutions, the film resistance improves even further. The corrosion potential is lower, leading to a small increase in anodic kinetics, but despite this, the overall corrosion current density is lowered by proteins for biomimetic-coated samples. The uncoated samples in protein solutions show large reductions in corrosion initially, but past 30 min the gains are less than for the coated samples.

6.5 Effect of buffer and atmosphere on amino acids/protein-induced corrosion

Because Mg corrosion depends heavily on pH, the type and amount of buffering agent has been shown to have a great effect on Mg corrosion behavior (Kirkland et al., 2011; Yamamoto & Hiromoto, 2009). Differences in buffer capacity and chemical composition affect the local pH and therefore the corrosion layer. Proteins and protein–surface interactions are highly sensitive to the pH (Latour, 2008). The pH will affect the binding of albumin to divalent ions such as Ca^{2+} (Kragh-Hansen & Vorum, 1993). The binding reaction of Mg^{2+} is similar and competes at the same sites (Pedersen, 1972). Albumin-Mg^{2+} binding has been shown to exhibit changes in binding behavior between pH 7 and 7.5 (Guillaume, Guinchard, & Berthelot, 2000). A change in local pH due to the corrosion rate can therefore affect the protein behavior. As the binding increases with pH (Kragh-Hansen & Vorum, 1993), the greater variation in pH with a CO_2/HCO_3 buffer due to its lower capacity could affect the corrosion potential and the formation of the passive layer. Additionally, the changes in surface charges with pH may affect the adsorption rate and therefore the protection provided by the protein layer (Latour, 2008). When testing samples *in vitro*, the effect of these complex molecules on corrosion rates can vary when different coatings and buffer systems change the corrosion environment.

E_{Corr} and i_{Corr} were measured for bare and coated samples in both buffer systems (Figure 6.10). BSA addition does not greatly affect E_{Corr} when the HEPES buffer is used on uncoated samples. However, these proteins lower E_{Corr} when used with a HCO_3 buffered system. Uncoated Mg in HEPES buffered MEM solutions have

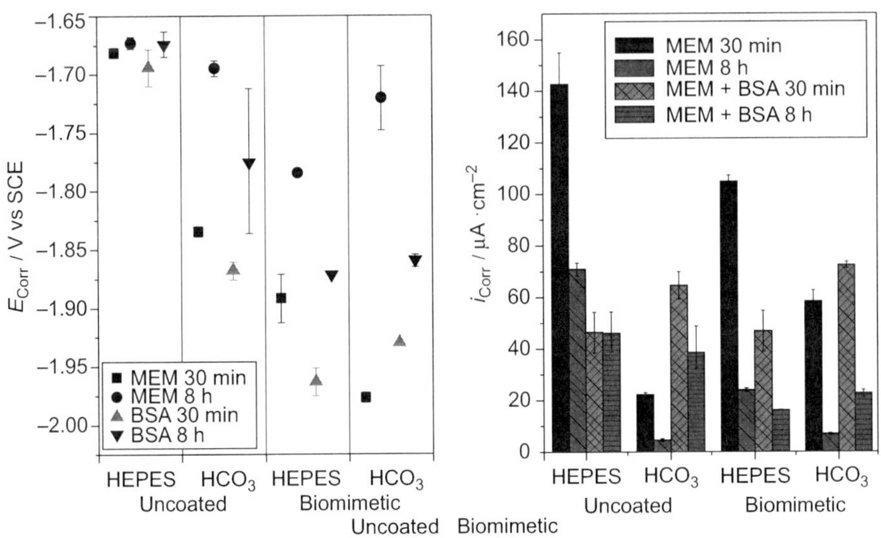

Figure 6.10 E_{Corr} and i_{Corr} of coated samples with HEPES and HCO_3 buffers.

high initial i_{Corr} that falls as the corrosion progresses. For MEM + HCO_3, i_{Corr} is lower after 30 min and continues to drop over 8 h as the passive layer forms. This is similar to the corrosion behavior in HBSS and HCO_3, where the passive hydroxide and carbonate layer forms to impede corrosion (Xin, Huo, Tao, Tang, & Chu, 2008).

Samples in HCO_3/CO_2 buffered solutions displayed a much greater corrosion current densities once proteins are added, between 40 and 60 μA/cm^2. This suggests the corrosion-resistant layer that formed in the HCO_3 buffer was impeded by the proteins. Uncoated Mg at 30 min shows that the carbonate buffer exhibits large decreases in cathodic kinetics and a smaller decrease in anodic kinetics (Figure 6.11). With the addition of BSA, the decrease in anodic kinetics is reduced, possibly due to the proteins interfering with the protective passivation layer that forms in the HCO_3 buffer. After 8 h, the HCO_3 buffer in MEM results in a significant drop in both anodic and cathodic kinetics of roughly equal magnitude, leading to the very small change in E_{Corr} (Figure 6.12). When proteins are present, the cathodic kinetics are further decreased, but the anodic kinetics remain similar to the HEPES buffered BSA solution. The explanation for this is found in the limitation of the formation of the passive layer, perhaps by proteins taking up Ca^{2+} ions and preventing the calcium carbonates from forming on the surface. The reduced E_{Corr} might also be explained by binding of Mg^{2+} ions, which would occur in the presence of the higher pH (Kragh-Hansen & Vorum, 1993). The fact that the HEPES buffered reaction kinetics drop only slightly with BSA addition is in line with other findings that the influence of the buffer, specifically CO_2/HCO_3 buffer, is crucial to a reduction in the corrosion rate. (Willumeit et al., 2011).

Nyquist plots of uncoated Mg in both MEM and MEM + BSA over the first 7 h displayed a large initial EDL resistance in MEM buffered with HCO_3 (Figure 6.13). This effect increases as time passes, corrosion occurs, the pH rises, and the layer

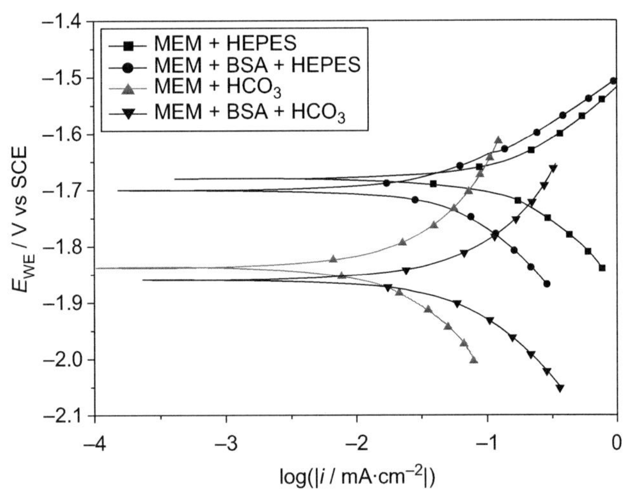

Figure 6.11 PDP of uncoated Mg after 30 min in HEPES and HCO_3 buffered protein solutions.

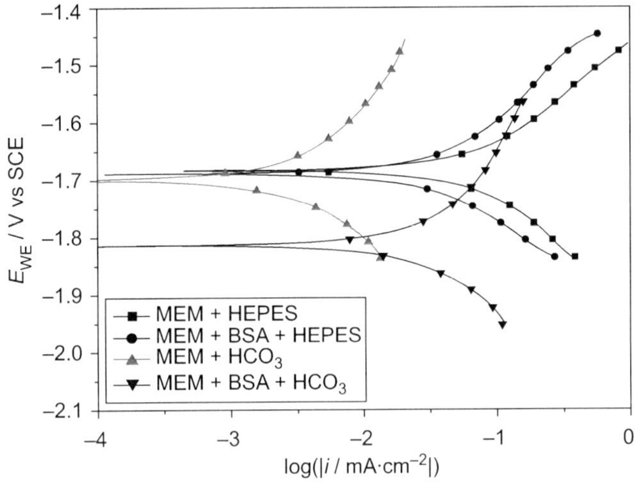

Figure 6.12 PDP of protein solutions and uncoated Mg after 8 h.

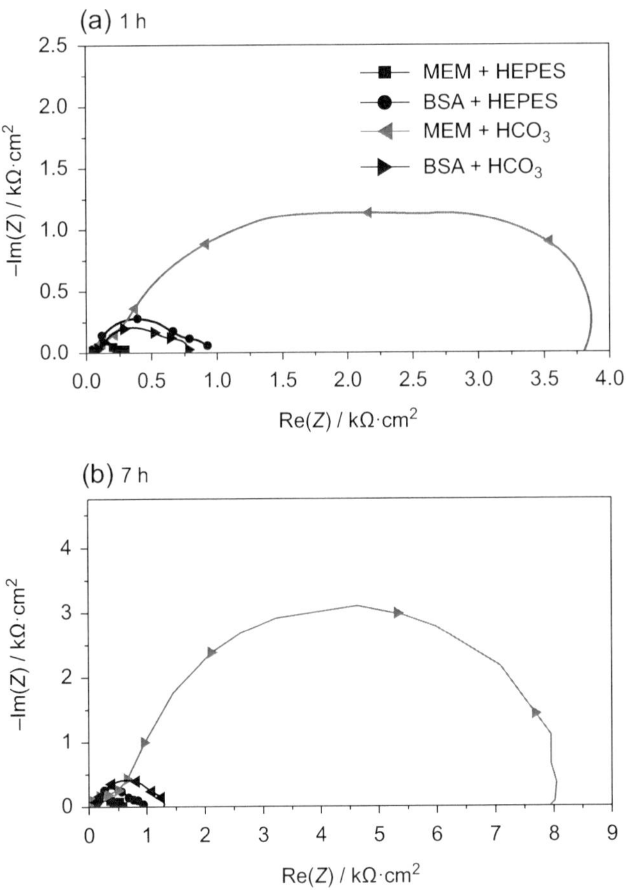

Figure 6.13 Nyquist plots of uncoated Mg in protein solutions buffered with HCO_3 after 1 hour (a) and 7 hours (b).

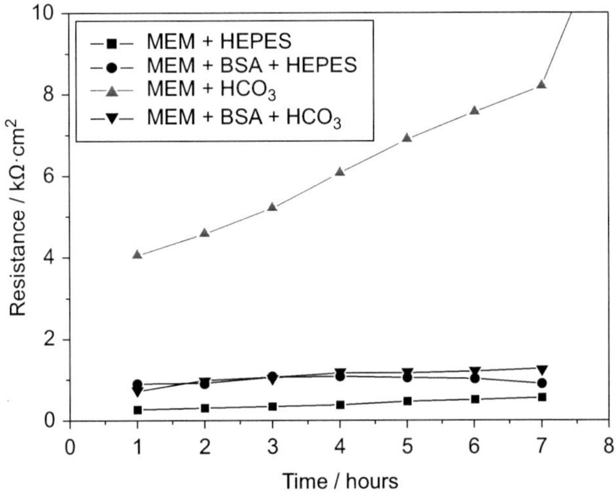

Figure 6.14 Polarization resistance in MEM and MEM + BSA of uncoated Mg samples with buffers.

becomes increasingly passive (Figure 6.14). Both solutions containing albumin display similar initial impedance layers in both buffers. The larger initial time constant appears as the result of the semipermeable protein layer resisting charge transfer, which causes the drop in the anodic and cathodic reactions. Closer examination of the Nyquist plot (Figure 6.15) shows that the uncoated Mg in MEM + BSA buffered with HCO_3 actually displays three distinct time constants. The first is the small passivation layer presented by the carbonate and hydroxide layer, followed by a layer of proteins, and finally the smaller charge transfer resistance from the electrolytic double layer itself. The smaller size of the first and third time constants is evidence of the proteins inhibiting the passivation that provides such protection in MEM and HCO_3 alone.

For biomimetic coated samples, the HCO_3/CO_2 buffer resulted in reduced anodic and cathodic kinetics for MEM solutions at 30 min (Figure 6.16). The addition of proteins altered the anodic kinetics to decrease, but the cathodic kinetics are concomitantly increased by the HCO_3/CO_2 buffer. This follows the trend seen before for uncoated Mg. After 8 h, the difference between these two buffer systems is more pronounced (Figure 6.17). At this point, both layers have had a chance to settle, and HCO_3 buffer reduces both reactions in MEM. However, when albumin is present, the cathodic kinetics remain increased, and the anodic kinetics are also slightly greater. The inhibitory effects of the proteins on the passivation layers cause the same effect on these coated samples, blocking the passivation of the pores in the coatings. Beyond 8 h of immersion, the protective layer that formed in MEM on bare Mg does not appear to persist, as the corrosion resistance dropped significantly (Figure 6.18). Consequently, in the presence of amino acids, the layer does not seem to last.

The biomimetic coatings in MEM + HCO_3 also displayed a significant increase in the passive properties of the coating, lasting over the 24 h experiment. However, with the addition of BSA, the protective effect of the low-capacity HCO_3 buffer

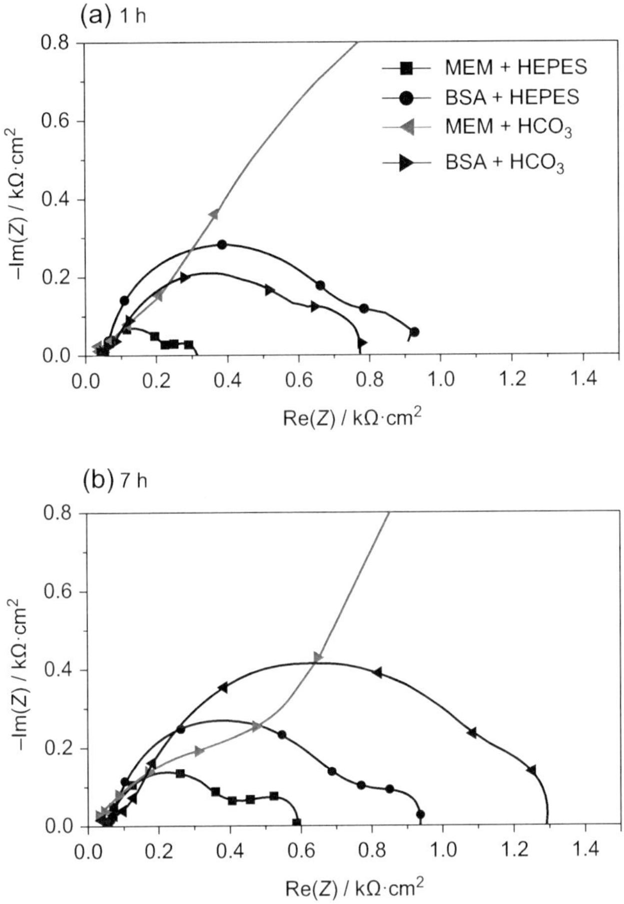

Figure 6.15 Nyquist plots of uncoated Mg in protein solutions buffered with HCO_3 after 1 h (a) and 7 h (b).

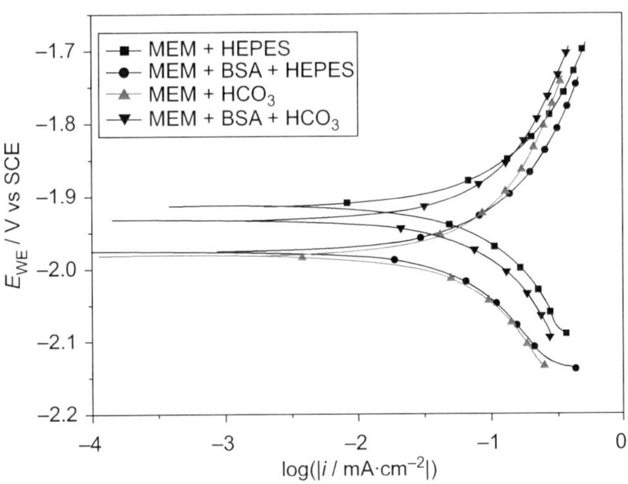

Figure 6.16 PDP of biomimetic coated samples in protein solutions at 30 min.

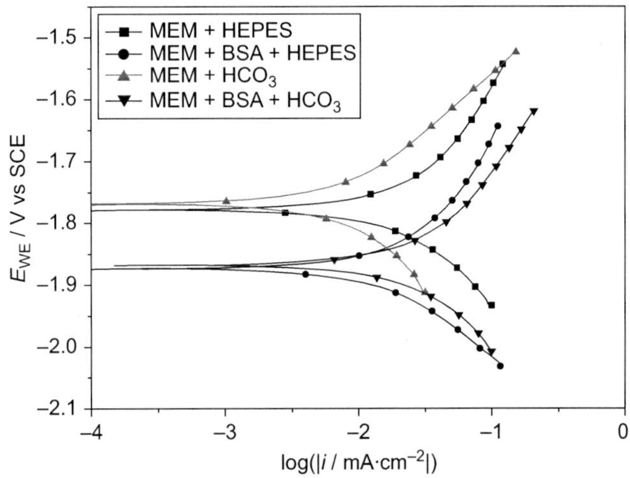

Figure 6.17 PDP of protein solutions on biomimetic-coated samples after 8 h.

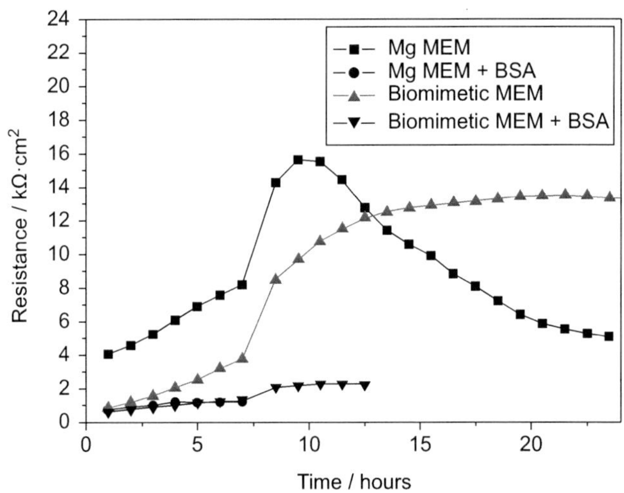

Figure 6.18 Polarization resistance of biomimetic-coated samples in HCO_3 buffered protein solutions.

was much less pronounced, and the corrosion rates were more typical of what one would expect at pH 7.4.

The Nyquist plots show the coating resistances increasing with time (Figure 6.19), and the bicarbonate buffered sample in MEM displays the greatest increases due to the layer that forms. The lower resistance of the biomimetic-coated sample will be an artefact of the lower corrosion rate than uncoated Mg alone, in which the pH rise quickly outpaces the buffer capacity, forming the insoluble calcium and Mg carbonates that supplement the magnesium hydroxide layer. When the coating is added, the pH

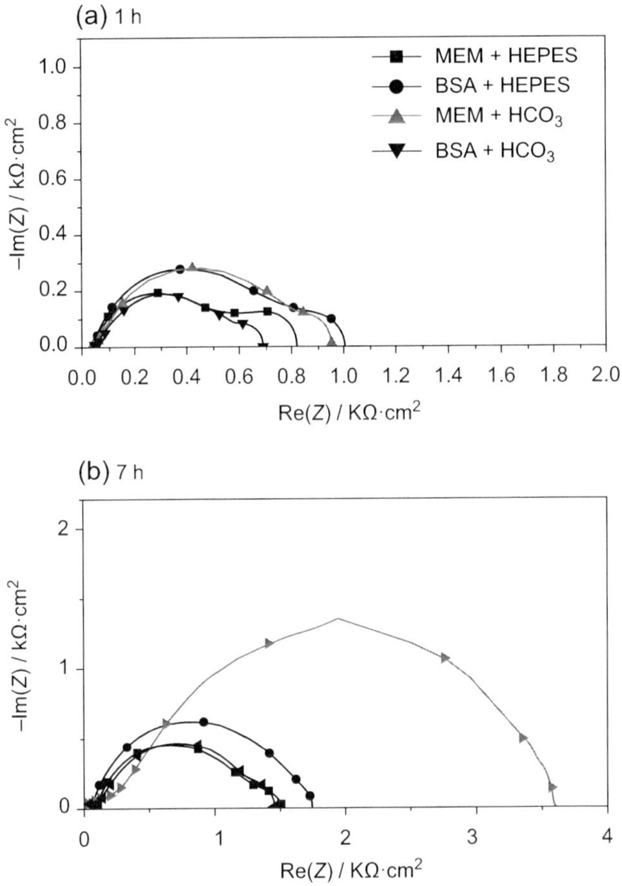

Figure 6.19 Nyquist plots of biomimetic-coated samples in protein solutions buffered with HCO_3 after 1 h (a) and 7 h (b).

changes far mores gradually, giving the bicarbonate ion and CO_2 atmosphere time to equilibrate, resulting in a less protective layer. With proteins binding to free Ca^{2+} and Mg^{2+} ions in solution (Clark & Williams, 1982; Kragh-Hansen & Vorum, 1993), these ions are effectively not as available to form the insoluble salts that prevent the corrosion. Therefore, the kinetics of the reactions differs depending on the buffer choice for each system. With a complicated environment such as the body, it will be important to understand the effects each component will have on the ultimate corrosion reactions and passivation properties expected at the implant site.

The relationship between E_{Corr} and i_{Corr} for uncoated and coated samples in the protein solutions versus buffer type is compared in Figure 6.20. The uncoated Mg in HEPES, displaying no real great passive behavior, did not have a pronounced correlation with regard to E_{Corr} and i_{Corr}. All other samples tested displayed the "expected" trend of decreasing i_{Corr} with increasing E_{Corr}. As the equilibrium

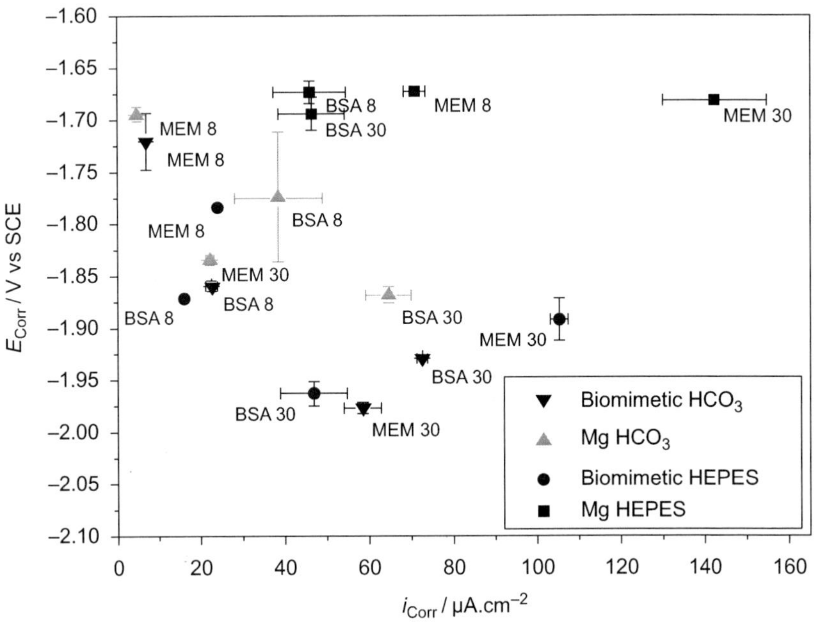

Figure 6.20 E_{Corr} versus i_{Corr} plots for uncoated Mg and biomimetic-coated samples for different buffer types on MEM and MEM + BSA.

of the layers settled in the solution, the corrosion rates dropped over time for all samples, the only difference being the impeding corrosion layer that forms, whether due to proteins, the buffer, or a combination of both.

6.6 Conclusions

From the results reported herein, it is clear that both amino acids and proteins have a strong effect on the corrosion mechanisms and rates for coated (and uncoated) Mg. This chapter goes some way to looking at the deeper causes the organic components have, which is important when attempting to understand the individual effects *in vitro* if effective models and predictions are to be developed by *in vivo* testing.

Key findings include:

- Early corrosion rates of uncoated Mg are accelerated by the addition of amino acids, despite lower levels of Cl^-, decreased conductivity, and a more positive open circuit potential. The more positive E_{Corr} measured for Mg suggests the chelation of Mg^{2+} ions is not the dominant factor in the early corrosion mechanisms as has been proposed previously (Yamamoto & Hiromoto, 2009). The increased buffer capacity and inhibition of passive layer formation are therefore more likely to be responsible (Malda et al., 2008).
- Adsorption of amino acids did not slow the early rates of corrosion for uncoated Mg like they do on other metals (Ashassi-Sorkhabi et al., 2004, 2005). It can be concluded that the overall

effect of the amino acids is to increase the corrosion rate *in vitro*, and it would be expected to contribute to the corrosion rate *in vivo* as well.

- The corrosion environment and thus the overall corrosion rate were affected in a complex fashion by the proteins. It was found that the proteins reduced the corrosion of uncoated Mg in solution by the adsorption of proteins to the surface (Liu et al., 2007, 2010). The effect of increasing corrosion on uncoated Mg by chelating of metal ions with proteins was not a significant contribution to the early corrosion kinetics in this study (Willumeit et al., 2011). The corrosion rates were dominated by the pH and Cl^- ions and the relative area available for corrosion.
- The proteins affected the corrosion through biomimetic coatings as well. With the addition of proteins, the extra impedance provided by the adsorption to the surface led to better passive properties of the exposed metal. The reduction in corrosion rate shows that the proteins facilitate the barrier the biomimetic coating provides. The corrosion through the defects was lessened when the proteins were present. Therefore, in a high-protein solution such as the body, it could be expected that this effect will be strengthened, further reducing the corrosion rate of the biomimetic-coated samples.
- The behavior of the protein solutions with different buffer systems was also found to be relevant to the corrosion reactions. While the presence of the carbonate buffer led to rapid carbonate film formation in amino acids, the proteins appear to interfere with the layer formation by binding to salts (Clark & Williams, 1982; Kragh-Hansen & Vorum, 1993). This is evidence that the HCO_3 buffered systems may not be the most physiologically relevant just because of the chemical equivalency. Of course, it must be considered that this is only a single type of protein addition, but corrosion behavior of protein solutions depends on the buffer choice.

The actual physiological system has many different types of proteins, and their behavior is complex. Therefore, further *in vitro* and *in vivo* tests will be needed to understand the effects of each different type of protein. The different solutions and buffers all agreed on the relative corrosion protection provided by each coating, but for the exact rates and mechanisms, the effect of the proteins is not trivial and may result in greater or lesser protection than expected from simple SBFs, depending on the solution and the buffering agent. These factors will be important to consider when designing a proper *in vitro* test to test the protection of coatings on biodegradable Mg.

There is a critical need to establish a good correlation between corrosion behavior *in vitro* and *in vivo* for both coated and uncoated Mg alloys, an area that the authors have previously addressed (Kirkland & Birbilis, 2013). This area requires significant further research and development, but presents problems that are not insurmountable. By establishing the effects of different biological components and testing their relevance under different types of tests (e.g., coated vs uncoated), standard practices for *in vitro* screening can be developed that accurately predict *in vivo* behavior. Understanding the predicted effects of the *in vivo* system will also allow tailoring of properties of the implant materials to be better suited to the environment. The establishment of standards derived from and leading to a better understanding of the degradation behavior of Mg alloys is a crucial next step in the realization of this next generation of implant materials.

References

Ashassi-Sorkhabi, H., Ghasemi, Z., & Seifzadeh, D. (2005). The inhibition effect of some amino acids towards the corrosion of aluminum in 1 M HCl + 1 M H_2SO_4 solution. *Applied Surface Science, 249*, 408–418.

Ashassi-Sorkhabi, H., Majidi, M. R., & Seyyedi, K. (2004). Investigation of inhibition effect of some amino acids against steel corrosion in HCl solution. *Applied Surface Science, 225*, 176–185.

Baker, K. C., Anderson, M. A., Oehlke, S. A., Astashkina, A. I., Haikio, D. C., Drelich, J., et al. (2006). Growth, characterization and biocompatibility of bone-like calcium phosphate layers biomimetically deposited on metallic substrata. *Materials Science and Engineering: C, 26*, 1351–1360.

Barrère, F. (2003). Osteointegration of biomimetic apatite coating applied onto dense and porous metal implants in femurs of goats. *Journal of Biomedical Materials Research Part B: Applied Biomaterials, 67B*, 655–665.

Barrere, F. (2002). *Biomimetic calcium phosphate coatings: physicochemistry and biological activity* (Doctor's Degree Ph.D.). University of Twente.

Bruneel, N., & Helsen, J. A. (1988). In vitro simulation of biocompatibility of Ti–Al–V. *Journal of Biomedical Materials Research, 22*, 203–214.

Carboneras, M., Garcã-A-Alonso, M. C., & Escudero, M. L. (2011). Biodegradation kinetics of modified magnesium-based materials in cell culture medium. *Corrosion Science, 53*, 1433–1439.

Clark, G. C. F., & Williams, D. F. (1982). The effects of proteins on metallic corrosion. *Journal of Biomedical Materials Research, 16*, 125–134.

El-Shafei, A. A., Moussa, M. N. H., & El-Far, A. A. (1997). Inhibitory effect of amino acids on Al pitting corrosion in 0.1 M NaCl. *Journal of Applied Electrochemistry, 27*, 1075–1078.

Eliezer, A., & Witte, F. (2010). Corrosion behaviour of magnesium alloys in biomedical environments. *Advanced Materials Research, 95*, 17–20.

Equitech-Bio Inc. (2010). *Sterile filtered fetal bovine serum, US origin [online]*. Equitech-Bio Inc. Available https://www.equitech-bio.com/Sterile-filtered-fetal-bovine-serum-US-origin.html Accessed 13.05.10.

Gu, X. N., Zheng, Y. F., & Chen, L. J. (2009). Influence of artificial biological fluid composition on the biocorrosion of potential orthopedic Mg–Ca, AZ31, AZ91 alloys. *Biomedical Materials, 4*, 8.

Guillaume, Y. C., Guinchard, C., & Berthelot, A. (2000). Affinity chromatography study of magnesium and calcium binding to human serum albumin: pH and temperature variations. *Talanta, 53*, 561–569.

Helgason, C. D., & Miller, C. L. (2005). *Basic cell culture protocols*. Totowa, NJ: Humana Press.

Horbett, T. A. (2004). The role of adsorbed proteins in tissue response to biomaterials. In B. D. Ratner, A. S. Hoffman, F. J. Schoen, & J. E. Lemons (Eds.), *Biomaterials science: An introduction to materials in medicine*. San Diego: Elsevier Academic Press.

Keim, S., Brunner, J. G., Fabry, B., & Virtanen, S. (2011). Control of magnesium corrosion and biocompatibility with biomimetic coatings. *Journal of Biomedical Materials Research – Part B Applied Biomaterials, 96B*, 84–90.

Kirkland, N., Waterman, J., Birbilis, N., Dias, G., Woodfield, T., Hartshorn, R., et al. (2011). Buffer-regulated biocorrosion of pure magnesium. *Journal of Materials Science: Materials in Medicine, 23*, 283–291.

Kirkland, N. T., & Birbilis, N. (2013). *Magnesium biomaterials – design, testing and best practice*. New York: Springer.
Kirkland, N. T., Birbilis, N., Walker, J., Woodfield, T., Dias, G. J., & Staiger, M. P. (2010). In-vitro dissolution of magnesium–calcium binary alloys: clarifying the unique role of calcium additions in bioresorbable magnesium implant alloys. *Journal of Biomedical Materials Research Part B: Applied Biomaterials, 95B*, 91–100.
Kokubo, T. (1996). Formation of biologically active bone-like apatite on metals and polymers by a biomimetic process. *Thermochimica Acta, 280–281*, 479–490.
Kragh-Hansen, U., & Vorum, H. (1993). Quantitative analyses of the interaction between calcium ions and human serum albumin. *Clinical Chemistry, 39*, 202–208.
Latour, R. A. (2008). Biomaterials: protein-surface interactions. In G. L. Bowlin (Ed.), *Encyclopedia of biomaterials and biomedical engineering*. New York: Taylor & Francis.
Lenter, C. (1981). *Geigy scientific tables*. Basle, Switzerland: Ciba-Geigy.
León, B., & Jansen, J. A. (Eds.). (2008). *Thin calcium phosphate coatings for medical implants*. New York: Springer.
Lin, C., & Li, X. (2006). Role of CO_2 in the initial stage of atmospheric corrosion of AZ91 magnesium alloy in the presence of NaCl. *Rare Metals, 25*, 190–196.
Liu, C., Xin, Y., Tian, X., & Chu, P. K. (2007). Degradation susceptibility of surgical magnesium alloy in artificial biological fluid containing albumin. *Journal of Materials Research, 22*, 1806–1814.
Liu, C. L., Zhang, X. M., Wang, Y. J., Huang, W. J., Zeng, R. Z., & Chu, P. K. (2010). In vitro corrosion degradation behaviour of Mg–Ca alloy in the presence of albumin. *Corrosion Science, 52*, 3341–3347.
Malda, J., Woodfield, T. B. F., Radisic, M., Levenberg, S., Oomens, C., Baaijens, F. P., et al. (2008). Cell nutrition: in vitro and in vivo. *Tissue Engineering: A Textbook*, 327–362.
Miller, R., Fainerman, V. B., Leser, M. E., & Michel, M. (2004). Kinetics of adsorption of proteins and surfactants. *Current Opinion in Colloid & Interface Science, 9*, 350–356.
Mueller, W. D., De Mele, M. F. L., Nascimento, M. L., & Zeddies, M. (2009). Degradation of magnesium and its alloys: dependence on the composition of the synthetic biological media. *Journal of Biomedical Materials Research Part A, 90A*, 487–495.
Mueller, W. D., Lucia Nascimento, M., & Lorenzo De Mele, M. F. (2010). Critical discussion of the results from different corrosion studies of Mg and Mg alloys for biomaterial applications. *Acta Biomaterialia, 6*, 1749–1755.
Mueller, W. D., Nascimento, M. L., Zeddies, M., Córsico, M., Gassa, L. M., & De Mele, M. A. F. L. (2007). Magnesium and its alloys as degradable biomaterials: corrosion studies using potentiodynamic and EIS electrochemical techniques. *Materials Research, 10*, 5–10.
Oyane, A., Kim, H.-M., Furuya, T., Kokubo, T., Miyazaki, T., & Nakamura, T. (2003). Preparation and assessment of revised simulated body fluids. *Journal of Biomedical Materials Research Part A, 65A*, 188–195.
Pedersen, K. O. (1972). Binding of calcium to serum albumin III. Influence of ionic strength and ionic medium. *Scandinavian Journal of Clinical & Laboratory Investigation, 29*, 427–432.
Pietak, A. M., Mahoney, T., Dias, G., & Staiger, M. P. (2007). Bone-like matrix formation on magnesium and magnesium alloys. *Journal of Biomedical Materials Research, 19*, 407–415.
Puleo, D. A., & Bizios, R. (2009). *Biological interactions on materials surfaces: understanding and controlling protein, cell and tissue responses*. London: Springer.
Rettig, R., & Virtanen, S. (2008). Time-dependent electrochemical characterization of the corrosion of a magnesium rare-earth alloy in simulated body fluids. *Journal of Biomedical Materials Research Part A, 85A*, 167–175.

Rettig, R., & Virtanen, S. (2009). Composition of corrosion layers on a magnesium rare-earth alloy in simulated body fluids. *Journal of Biomedical Materials Research Part A, 88*, 359–369.

Salunke, P., Shanov, V., & Witte, F. (2011). High purity biodegradable magnesium coating for implant application. *Materials Science and Engineering: B, 176*, 1711–1717.

Virtanen, S. (2011). Biodegradable Mg and Mg alloys: corrosion and biocompatibility. *Materials Science and Engineering: B, 176*, 1600–1608.

Warrel, D. A. (2003). *Oxford textbook of medicine*. New York: Oxford University Press.

Waterman, J., Pietak, A., Birbilis, N., Woodfield, T., Dias, G., & Staiger, M. P. (2011). Corrosion resistance of biomimetic calcium phosphate coatings on magnesium due to varying pretreatment time. *Materials Science and Engineering: B, 176*, 1756–1760.

William, D. F., & William, R. L. (2004). Degradative effects of the biological environment on metals and ceramics. In B. D. Ratner, A. S. Hoffman, F. J. Schoen, & J. E. Lemons (Eds.), *Biomaterials science: An introduction to materials in medicine*. San Diego: Elsevier Academic Press.

Willmer, E. N. (1965). *Cells and tissues in culture: Methods, biology, and physiology*. London: Academic Press.

Willumeit, R., Fischer, J., Feyerabend, F., Hort, N., Bismayer, U., Heidrich, S., et al. (2011). Chemical surface alteration of biodegradable magnesium exposed to corrosion media. *Acta Biomaterialia, 7*, 2704–2715.

Witte, F., Feyerabend, F., Maier, P., Fischer, J., Stormer, M., Blawert, C., et al. (2007). Biodegradable magnesium-hydroxyapatite metal matrix composites. *Biomaterials, 28*, 2163–2174.

Xin, Y., Huo, K., Tao, H., Tang, G., & Chu, P. K. (2008). Influence of aggressive ions on the degradation behavior of biomedical magnesium alloy in physiological environment. *Acta Biomaterialia, 4*, 2008–2015.

Xu, L., Pan, F., Yu, G., Yang, L., Zhang, E., & Yang, K. (2009). In vitro and in vivo evaluation of the surface bioactivity of a calcium phosphate coated magnesium alloy. *Biomaterials, 30*, 1512–1523.

Yamamoto, A., & Hiromoto, S. (2009). Effect of inorganic salts, amino acids and proteins on the degradation of pure magnesium in vitro. *Materials Science and Engineering: C, 29*, 1559–1568.

Yang, L., Hort, N., Willumeit, R., & Feyerabend, F. (2012). Effects of corrosion environment and proteins on magnesium corrosion. *Corrosion Engineering, Science and Technology, 47*, 335–339.

Yang, L., & Zhang, E. (2009). Biocorrosion behavior of magnesium alloy in different simulated fluids for biomedical application. *Materials Science and Engineering: C, 29*, 1691–1696.

Yun, Y., Dong, Z., Yang, D., Schulz, M. J., Shanov, V. N., Yarmolenko, S., et al. (2009). Biodegradable Mg corrosion and osteoblast cell culture studies. *Materials Science and Engineering: C, 29*, 1814–1821.

Zeng, R., Dietzel, W., Witte, F., Hort, N., & Blawert, C. (2008). Progress and challenge for magnesium alloys as biomaterials. *Advanced Engineering Materials, 10*, B3–B14.

Zheng, Y. F., Gu, X. N., Xi, Y. L., & Chai, D. L. (2010). In vitro degradation and cytotoxicity of Mg/Ca composites produced by powder metallurgy. *Acta Biomaterialia, 6*, 1783–1791.

Zhu, L., & Song, G. (2005). Improved corrosion resistance of AZ91D magnesium alloy by an aluminium-alloyed coating. *Surface and Coatings Technology, 200*, 2834–2840.

Biocompatibility of surface-modified magnesium and magnesium alloys

7

Ke Yang, Xiao Lin
Chinese Academy of Sciences, Shenyang, China

7.1 Introduction

Biocompatibility is a critical factor for successful applications of biomedical materials. During the years between 1940 and 1980, biomaterials were thought to be bioinert. Therefore, it was considered that biocompatibility would be achieved by designing material to be extremely chemically stable. However, with the development of biomaterials, it is gradually being recognized that it will be better that a biomaterial can properly react with the surrounding tissues rather than being totally inert. More recently, it has become widely accepted that, for some purposes, biomaterials that can finally degrade in the physiological environment are better than those that exist permanently. In 2008, David F. Williams proposed a concept of biocompatibility as follows: Biocompatibility refers to the ability of a biomaterial to perform its desired function with respect to a medical therapy, without eliciting any undesirable local or systemic effects in the recipient or beneficiary of that therapy, but generating the most appropriate beneficial cellular or tissue response in that specific situation, and optimizing the clinically relevant performance of that therapy (Williams, 2008).

Magnesium (Mg)-based metals include pure Mg and Mg alloys. They are degradable in the body environment, which is an advantage compared with permanently implanted metallic biomaterials. For surface-modified Mg-based metals, not only the surface should generate "the most appropriate beneficial cellular or tissue response," but also the degradation products of the alloys and the surface modification layers should not elicit "any undesirable local or systemic effects." In this chapter, we will analyze the medical suitability of the materials introduced by the surface modifications on Mg-based metals. Then, we will introduce both the *in vitro* and *in vivo* biocompatibility studies of surface-modified Mg-based metals and discuss the influencing factors on biocompatibility. Lastly, we will give a likely trend of the future development of the biocompatible surface-modified Mg-based metals.

7.2 Biocompatibility of the materials introduced by surface modification

Since the surface-modified Mg-based metals are biodegradable in the body environment, the alloying elements in the alloys and the materials introduced onto alloy

Surface Modification of Magnesium and its Alloys for Biomedical Applications. http://dx.doi.org/10.1016/B978-1-78242-077-4.00007-3

surfaces by different surface modifications should be controlled to ensure that the degradation products from the Mg implants will not elicit any undesirable local or systemic effects. Information on the biocompatibility of alloying elements can be found elsewhere (Habibovic & Barralet, 2011; Li & Zheng, 2013; Witte et al., 2008). Here we give several recommended alloy elements from the biocompatibility consideration in Table 7.1. In this section, we mainly focus on the biocompatibility of the materials introduced by surface modification on Mg-based metals. The general properties of commonly adopted coating materials, which are introduced in detail in the following sections, are listed in Table 7.2.

7.2.1 Calcium phosphates

Many works have studied calcium phosphate-containing coatings on the biodegradable Mg-based metals (Chang, Tian, Liu, & Duan, 2013; Niu et al., 2013; Shadanbaz & Dias, 2012; Wang, Guan, Wang, Ren, & Wang, .; Wen et al., 2009). The main inorganic component in bones is exactly a biological apatite (CaP). Thus, CaP has always been considered to have an excellent biocompatibility, especially for the bone tissues. There are many kinds of CaPs, including calcium phosphate dehydrate (DCP), octacalcium phosphate (OCP), tricalcium phosphate (TCP), and hydroxyapatite (HA). The biocompatibilities of CaPs have been clearly revealed, and they are all biodegradable, although they have different degradation rates (Shadanbaz & Dias, 2012). CaPs are suggested as the coating materials on biodegradable Mg-based metals.

7.2.2 Magnesium phosphate

Magnesium phosphate-containing coatings are mainly fabricated by the substrate-involving chemical reactions through either chemical conversion (Ishizaki, Shigematsu, & Saito, 2009; Zhou, Shan, Han, & Ke, 2008) or microarc oxidation (MAO) (Bai et al., 2012; Bala Srinivasan, Liang, Blawert, St Rmer, & Dietzel, 2009; Bala Srinivasan, Liang, Blawert, St Rmer, & Dietzel, 2010; Gao, Yerokhin, & Matthews, 2013; Lin et al., 2014) methods. Recently, magnesium phosphate cement (MPC) has attracted much attention as a potential biodegradable bone-implant material (Mestres & Ginebra, 2011; Waselau, Samii, Weisbrode, Litsky, & Bertone, 2007; Wu et al., 2008; Yu et al., 2010). The *in vitro* and *in vivo* studies showed that MPC has excellent biocompatibility. It is believed that magnesium phosphate can be transformed into CaP *in vivo*, which has higher chemical stability.

7.2.3 Magnesium silicate

Magnesium silicate-containing coatings on the biodegradable Mg-based metals have been fabricated through anodization (Wu et al., 2007; Xue, Yun, Schulz, & Shanov, 2011) or MAO (Gu et al., 2011; Lin, Tan, Zhang et al., 2013; Wan et al., 2013; Yao, Li, & Jiang, 2009) methods. Magnesium silicate was mainly in the form of Mg_2SiO_4, which has been widely studied as a biodegradable material for the scaffold in bone-tissue engineering and/or load-bearing devices due to its biodegradability and good

Table 7.1 **Recommended alloy elements for biodegradable magnesium alloy**

Elements	Biofunctionality	References
Calcium	Major component in human bone; essential in chemical signaling with cells; elevated calcium has a number of stimulating effects on osteoblasts	(Adams, Mansfield, Perlot, & Shapiro, 2001; Dvorak et al., 2004; Ilich & Kerstetter, 2000)
Zinc	Plays an important role in various physiological processes; one of the constituents of the antioxidant system; involved in the synthesis of a large number of proteins and is required for their stability; important to the bone tissues health	(Prasad, Bao, Beck, Kucuk, & Sarkar, 2004; Tapiero & Tew, 2003; Yamaguchi, Oishi, & Suketa, 1988)
Silicon	An essential trace element in the skeletal development; associated with calcium in an early stage of calcification; induce proliferation of osteoblast-like cells and angiogenesis during bone regeneration	(Carlisle, 1970; Carlisle, 1982; Shie, Ding, Chang et al., 2011; Zhai et al., 2012)
Strontium	About 98% in the human body is in the skeleton; strontium renalate is used clinically in the treatment of osteoporotic patients in Europe; positive effects on ostegenic differentiation and mineral formation; inhibitory effects on osteoclastic differentiation and resorption; dose-dependent effect	(Bonnelye, Chabadel, Saltel, & Jurdic, 2008; Grynpas & Marie, 1990; Marie, Ammann, Boivin, & Rey, 2001; Skoryna, 1984)
Copper	Osteoporosis associates with copper deficiency, while a supplement of copper may have a potential therapeutic application in the treatment and prevention of involutional osteoporosis; essential for blood vessel formation	(Rico et al., 2000; Yee, Kubena, Walker, Champney, & Sampson, 1995; De Lima et al., 2008)

Table 7.2 Biocompatibility and biodegradability of coating materials on biodegradable magnesium alloy

Materials	Biocompatibility	Biodegradability
Calcium phosphates	Excellent biocompatibility	Degradable; degradation rate varies significantly between different calcium phosphates
Magnesium phosphate	Biocompatible	Degradable
Magnesium silicate	Biocompatible	Degradable; degradation rate is very low
Magnesium fluoride	Dose-dependent biocompatibility	Degradable
Titania	Questionable; depend on its size and existing state	Not degradable
Alumina	Biocompatible	Not degradable
Degradable polymers	Biocompatible	Degradable; degradation products may elicit slight inflammatory reaction

biocompatibility (Kharaziha & Fathi, 2009; Kharaziha & Fathi, 2010; Ni & Chang, 2009; Ni, Chou, & Chang, 2007). It has been reported that silicon (Si) is an essential trace element in skeletal development (Carlisle, 1982; Jugdaohsingh et al., 2004; Yamada et al., 2003). It was also found that an appropriate Si concentration was effective in supporting the proliferation of osteoblast-like cells (Shie et al., 2011). Thus, magnesium silicate is a biocompatible coating material. However, it should be noted that its degradation rate is very low (Ni & Chang, 2009; Tavangarian & Emadi, 2011).

7.2.4 Magnesium fluoride

Magnesium fluoride (MgF_2)-containing coatings were prepared on biodegradable Mg-based metals through hydrofluoric acid-involving chemical conversion (Chiu, Wong, Cheng, & Man, 2007; Lin, Tan, Wan et al., 2013; Pereda et al., 2010; Thomann et al., 2010; Yan et al., 2010) or MAO (Pan, Chen, Wang, & Lin, 2013; Pan, Chen, Wang, & Zhao, 2013; Seyfoori, Mirdamadi, Khavandi, & Raufi, 2012) methods. Fluorine (F) is a natural component in human bones and teeth (Zheng, Wu, Ng, Wang, & Lian, 2002). A proper release of F will not cause harm to the organisms (Thomann et al., 2010). The bone response to fluoride is dose-dependent: a low-dose release of F will facilitate the bone formation; however, a high dose of it will cause the formation of poorly mineralized osteoid (Ellingsen, 1995). Thus, MgF_2 is an acceptable coating material, although the degradation rate should be controlled.

7.2.5 Titania

Some researchers fabricated titania-containing coatings on biodegradable Mg-based metals through one of the methods of sol–gel (Hu et al., 2011; Li, Chen, Zhang, Hu, & Su, 2010), direct current reactive magnetron sputtering (Chen et al., 2011), and ion implantation (Liu, Xin, Tian, & Chu, 2007; Zhao et al., 2013). The titania coatings have been prepared on the permanently implanted metallic materials, showing excellent biocompatibility and bioactivity (Chen et al., 2004; Huang et al., 2003; Jokinen et al., 1998). A porous titania ceramic has been studied as a bone graft substitute, which demonstrated excellent biocompatibility (Sabetrasekh, Tiainen, Lyngstadaas, Reseland, & Haugen, 2011). However, titania is an inert and poorly soluble material (Wang et al., 2007). The use of titania as coating material on biodegradable Mg-based metals is controversial. In addition, the titania particles from the coating might elicit some undesirable local or systemic effects (Wang et al., 2007).

7.2.6 Alumina

Alumina-containing coatings have been prepared on biodegradable Mg-based metals through plasma immersion ion implantation (Liu et al., 2007; Wong et al., 2013; Zhao, Wu, Pan, Yeung, & Chu, 2012) or filtered cathodic arc deposition (Xin et al., 2008) methods. Alumina is frequently used for medical implants and prostheses because of its excellent biocompatibility (Fischer et al., 2005; Thomas, Barnstorf, Summer, Willmann, & Przybilla, 2003). However, alumina is a bioinert ceramic with high chemical stability, and its degradability could be a concern when applied on biodegradable Mg-based metals.

7.2.7 Organic materials

Many researchers have fabricated organic materials containing coatings on biodegradable Mg-based metals (Abdal-hay, Dewidar, & Lim, 2012; Abdal-hay, Amna, & Lim, 2013a; Abdal-hay, Barakat, & Lim, 2013b; Abdal-hay, Barakat, & Lim, 2013c; Alabbasi, Liyanaarachchi, & Kannan, 2012; Bai et al., 2012; Cui, Beniash, Gawalt, Xu, & Sfeir, 2013; Degner, Singer, Cordero, Boccaccini, & Virtanen, 2013; Fekry, Ghoneim, & Ameer, 2014; Gray-Munro, Seguin, & Strong, 2009; Killian, Wagener, Schmuki, & Virtanen, 2010; Ko, Choe, Jung, & Kim, 2013; Kunjukunju et al., 2013; Li, Cao, Zhang, Zhang, & He, 2010b; Liu et al., 2013; Oosterbeek, Seal, Seitz, & Hyland, 2013; Ostrowski et al., 2013; Shanshan, Lili, Yingxue, Bingchun, & Ke, 2013; Srinivasan, Ranjani, & Rajendran, 2013; Wagener, Killian, Turhan, & Virtanen, 2013; Wang, Zhao, Chen, Li, & Zhang, 2012; Wang et al., 2013; Wong et al., 2010; Xu & Yamamoto, 2012; Xue et al., 2012; Zomorodian et al., 2013). The coating materials include the well-known biodegradable polymers, such as poly(L-lactic acid) (PLA) (Abdal-hay et al., 2013c; Alabbasi et al., 2012; Gray-Munro et al., 2009; Xu & Yamamoto, 2012), poly (lactide-co-glycolide) (PLGA) (Li, Cao et al., 2010; Shanshan et al., 2013), poly (e-caprolactone) (PCL) (Degner et al.,

2013; Wang et al., 2012; Wong et al., 2010), polyether imide (PEI) (Zomorodian et al., 2013), poly(1,3-trimethylene carbonate) (PTMC) (Wang et al., 2013), and silane (Ko et al., 2013; Liu et al., 2013; Xue et al., 2012). Protein (Killian et al., 2010; Wagener et al., 2013) and peptide (Cui et al., 2013) were also used as the coating materials. The biocompatibility of the above materials has been well addressed (Francis Suh & Matthew, 2000; Gunatillake & Adhikari, 2003; Mohanty, Misra, & Hinrichsen, 2000; Rezwan, Chen, Blaker, & Boccaccini, 2006). However, the degradation of some polymers usually produces a high quantity of acidic degradation products in a short time, which may induce a slight inflammatory response (Bostman & Pihlajamaki, 2000; Taylor, Daniels, Andriano, & Heller, 1994).

7.3 Biocompatibility of surface-modified magnesium-based metals – *in vitro* results

Various surface modifications have been applied to Mg-based metals to control their degradation rates. The coatings can be divided into three classes according to their compositions: inorganic coatings, organic coatings, and inorganic/organic composite coatings. In this section, *in vitro* biocompatibility studies of these three classes of coatings are introduced.

7.3.1 Inorganic coating

A number of methods have been used to modify the surfaces of biodegradable Mg-based metals by forming inorganic coatings, as illustrated in Figure 7.1. The coating mechanisms and formed coating materials are summarized in Table 7.3. Although the compositions of coatings prepared by the same method could vary from each other, their physicochemical properties were similar. To a great extent, the surface biocompatibility of the surface-modified Mg-based metals should depend on the physicochemical properties of the surface layers. Thus in the following part, *in vitro* biocompatibility of the inorganic coatings is introduced according to the modification methods.

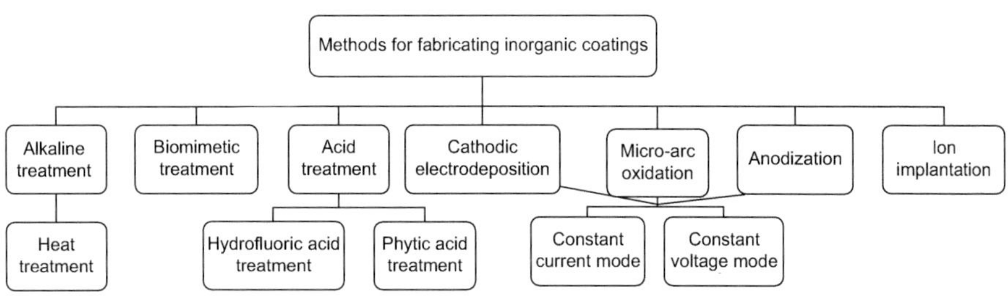

Figure 7.1 Main inorganic coating methods for biodegradable magnesium alloy. Sometimes more than one method is used to fabricate inorganic–inorganic composite coatings.

Table 7.3 Coating mechanisms and formed coating materials of several inorganic coating methods for biodegradable magnesium alloy

Method	Coating mechanism	Coating materials
Alkaline treatment	Substrate-involved chemical reaction	MgO; $Mg(OH)_2$
Biomimetic treatment	Substrate-involved chemical reaction; pH variation-induced chemical deposition	MgO; calcium phosphates; magnesium phosphates
Acid treatment	Substrate-involved chemical reaction	Phytic-Mg phytate complex; MgF_2; MgO; $Mg(OH)_2$
Cathodic electrodeposition	Chemical reaction; electrochemical reaction; pH variation-induced chemical deposition	Calcium phosphates, especially HA; bioactive element doped calcium phosphates
Micro-arc oxidation	Substrate-involved chemical reaction; electrochemical reaction	MgO, calcium phosphates; magnesium phosphates; CaO, MgF_2, magnesium silicates

7.3.1.1 Alkaline treatment

In this treatment, Mg-based metals were immersed in an alkaline solution. A coating will be formed through a chemical reaction. A heat treatment is always followed to densify the coating. Gu et al. (Gu, Zheng, Cheng, & Zheng, 2009) soaked a Mg1.4 wt% Ca alloy in three different alkaline solutions (Na_2HPO_4, Na_2CO_3, and $NaHCO_3$) for 24 h and subsequently annealed the alloy at 773 K for 12 h. Magnesium oxide layers were formed on the surface of the Mg-Ca alloy after different alkaline treatments. All the heat-treated Mg alloy samples showed no cell toxicity to L929 cells during 7 d of co-culture. Lorenz et al. (Lorenz et al., 2009) passivated a pure Mg (99.9% purity) by soaking it in 1 M NaOH for 24 h. The formed passive film, which was mainly composed of $MgO/Mg(OH)_2$, was in the order of nanometers in size. The initial dissolution rate of the pure Mg in the cell culture medium was decreased by approximately a factor of 20 due to the passivation. The human HeLa cell density on the treated sample was significantly higher than the polished one, where only single isolated cells could be found and the cell spreading was poor. For the passivated sample, however, strong formation of the actin cytoskeleton could be observed, and the cells were well spread, covering most of the surface.

7.3.1.2 Biomimetic treatment

This method refers to immersing the Mg-based metals in a Ca and P coexisting solution, especially the simulated body fluid (SBF) under the near-physiological condition, in which CaP-containing coating will be formed. Lorenz et al. (Lorenz et al., 2009) soaked pure Mg (99.9% purity) in the M−SBF, leading to the formation of mixed Ca/Mg-phosphate layers on the surface, with thickness of tens of micrometers. The roughness was strongly increased, which led to a strongly hydrophilic behavior of the surface. However, the increase of pH for such treated Mg was even larger than the polished one. The density of human HeLa cells cultured on the treated sample was significantly higher than that on the polished one. However, the cell spreading was poor on the treated sample, and the actin cytoskeleton was not well formed. Furthermore, no filopodia formation could be detected.

In that study, it was also shown that the treated sample facilitated the cell adhesion initially. However, the poor protection of the layer limited the long-term survival of the cells. Keim et al. (Keim, Brunner, Fabry, & Virtanen, 2011) soaked the above surface-modified pure Mg in 1 M NaOH at room temperature for 24 h or heat-treated it in a steam for 1 h to improve the protective ability of the surface. The treatments led to reductions of the cell densities. They also treated the pure Mg by soaking it in a cell culture medium, and a carbon-rich layer was formed. The surface morphology was similar to the layer formed in SBF, and the corrosion protection ability of the surface layer was 10 times higher than that formed in the SBF. The density of cells growing for 24 h on the Mg samples pretreated in the cell culture medium was greatly increased compared with the SBF-treated samples. Xu et al. (Xu et al., 2009) prepared flake-like $CaHPO_4 \cdot 2H_2O$ crystals containing coating on the Mg-manganese (Mn)-zinc (Zn) Mg alloy by immersing the alloy in a solution based on calcium dihydro phosphate. The number of L929 cells on the coated Mg sample was significantly larger than that on the uncoated Mg sample, indicating that the surface of coated Mg alloy is better for cell attachment, growth, and proliferation. Geng et al. (Geng, Tan, Jin, Yang, & Yang, 2009) introduced β-TCP coating on the pure Mg by biomimetic method. MG63 cells grew well on the coating, while the bare Mg showed some cell toxicity.

7.3.1.3 Acid treatment

The acid treatment is to immerse the Mg-based metals in an acidic solution, forming a protective layer through chemical reactions. Ye et al. (Ye, Zheng, Wang, Xi, & Li, 2012) modified WE43 Mg alloy in phytic acid with different pH values. The corrosion resistance of the alloy was increased. All the modified groups showed better cell viability than the unmodified control. The coating with highest corrosion resistance behaved the best cell compatibility and the lowest hemolytic potential. Another commonly used acid for the treatment is hydrofluoric acid (Lozano et al., 2013; Mao, Yuan, Niu, Zong, & Ding, 2013; Seitz, Eifler, Stahl, Kietzmann, & Bach, 2012). Seitz et al. (Seitz et al., 2012) studied the cell compatibility of MgNd2 alloy with and without hydrofluoric acid treatment. The cell viability of murine fibroblasts or keratinocytes was not significantly affected by the bare and the treated MgNd2 extracts. However, the cell proliferation of murine fibroblasts and keratinocytes seemed to be inhibited by the bare alloy, but not in the case of the treated one. Lozano (Lozano et al., 2013) examined the cell compatibility of a hydrofluoric acid-treated AZ31 Mg alloy (AZ31HF) to the MC3T3-E1 osteoblasts and L929 fibroblasts. The AZ31HF hindered the proliferation of the two cell lines. The cell viability of the two cell lines was not affected by the AZ31HF. The respiratory activity rates of cell lines cultured with the AZ31 and the AZ31HF were similar. However, the AZ31HF induced a harmful effect on the plasma membrane of MC3T3-E1 osteoblasts. The alkaline phosphatase (ALP) expression of MC3T3-E1 osteoblasts was enhanced by the interaction with AZ31HF. Meanwhile, the study of macrophages indicated that the AZ31HF might lead to an undesirable inflammatory response in the body. Mao et al. (Mao et al., 2013) studied the cell- and hemo-compatibility of a hydrofluoric acid-treated Mg-Nd-Zn-Zr alloy. The results showed that the treated alloy showed no cytotoxicity to the EA.hy 926 cells. Although the hemolysis ratio of alloy could be reduced by the HF treatment, it still did not meet the requirement for implant materials. The treated alloy showed a good anti-platelets adhesion performance.

7.3.1.4 Cathodic electrodeposition

This method could fabricate a CaP coating on Mg-based metals. The solution is similar to that used in the biomimetic method. Differently, this method worked under the assistance of electric field. Guan et al. (Guan et al., 2012) fabricated an HA coating on a Mg-4.0Zn-1.0Ca-0.6Zr alloy. The hemolysis ratio of both HA-coated and uncoated alloy samples was lower than 5%, which meets the requirement for implant materials. The L929 fibroblasts cultured with extracts of the HA-coated samples had slightly greater relative growth rate (RGR%) than those cultured with the uncoated sample extract. Li et al. (Li, Song et al., 2010) studied the *in vitro* response of human bone marrow stem cells (BMSCs) to a fluoridated hydroxyapatite coated biodegradable Mg-Zn alloy. The human BMSCs grew well on the sample, and the cell viability (MTT results) in the coated sample group was significantly higher than that in the uncoated group after 2 and 3 days of cultivation. The ALP activity of the coated group was lower than that of the uncoated group at 7 days. After longer time of incubation, higher ALP activity for the coated group was shown. The real time polymerase chain reaction test showed that expression levels of main differentiation markers were significantly increased for the coated group.

7.3.1.5 Microarc oxidation

In the MAO treatment, the samples are immersed in an electrolyte bath, and a high bipolar pulse voltage is applied on the sample–electrolyte interface. MAO coatings can be formed on Mg-based metals through the chemical reaction under high pressure and in a high-temperature environment. MAO became the most likely commercially applicable protection method for Mg-based metals in the postchromate coating era (Hornberger, Virtanen, & Boccaccini, 2012). Yang et al. (Yang et al., 2013) evaluated the biocompatibility/bioactivity of a forsterite-containing MAO coating on ZK60 Mg alloy. It was found that the cell adhesion, spreading, and viability on the MAO coating were obviously better than those on the bare ZK60 alloy. The intracellular ALP activity of murine BMSCs cultured with the MAO-coated Mg alloy extract was significantly higher than that with the bare alloy extract. Gu et al. (Gu et al., 2011) prepared MAO coatings on a Mg–Ca alloy under three voltages (300, 360, and 400 V). The corrosion resistance of the alloy was increased by increasing the voltage. Both the indirect and direct cytotoxicity assays showed that the MAO coating greatly improved the cell compatibility of the Mg–Ca alloy. The number of MG63 cells on the 300-V treated sample was larger than those on the other two samples at day one and day three. However, the cell number on the 300-V treated sample decreased sharply at day five, lower than those on the other two samples. The MG63 cells attached and extended well on the 360- and 400-V treated samples; however, cells on the 300-V treated sample showed unhealthy morphology, with round shapes. The ALP activity of MG63 cells was significantly improved by the MAO coating, and an increased ALP activity was found by increasing the applied voltage. Lin et al. (Lin et al., 2014) prepared a strontium (Sr)-containing MAO coating on ZK60 Mg alloy. The viability of MC3T3-E1 cells and ALP activity of murine BMSCs were significantly improved for the Sr-incorporated coating.

7.3.1.6 Other treatments

Xin et al. (Xin et al., 2009) fabricated a hydrogenated amorphous silicon (a-Si:H) coating on a die-cast AZ91 Mg alloy by direct current magnetron sputtering deposition. The coating improved the corrosion resistance of the substrate by about two orders. The direct culture of hFOB 1.19 cells showed that the cells attached well on the coating and proliferated normally after both 1 and 3 days. No cells survived after seeding on the untreated alloy. Wong et al. (Wong et al., 2013) treated an AZ91 Mg alloy by means of dual aluminum and oxygen plasma immersion ion implantation and deposition (PIII&D). An alumina coating was formed on the surface of the alloy. The GFP osteoblasts grew well on the treated alloy. No cell growth was observed on the untreated sample. Li et al. (Li et al., 2013) prepared an amorphous Si coating on a WE43 Mg alloy by plasma enhanced chemical vapor deposition (PECVD). Higher cell viability exhibited for the coated sample compared with the uncoated sample. The hemolysis ratio reduced from $6.1\% \pm 0.4\%$ for the uncoated sample to $2.9\% \pm 0.7\%$ for the Si coated sample.

7.3.2 Organic coating

Recently, organic coated biodegradable Mg implants were extensively studied. The preparation methods include dip-coating, spray-coating, electrospinning, layer-by-layer, etc., which are shown in Figure 7.2. Li et al. (Li, Cao et al., 2010) fabricated PLGA coating on a Mg-6Zn alloy by the dip-coating method. The coated Mg alloy samples showed significantly enhanced ability of MC3T3-E1 cell attachment compared with the bare Mg alloy. Wong et al. (Wong et al., 2010) fabricated two kinds of PCL coatings on an AZ91 Mg alloy: low porosity membrane (LPM) and high porosity membrane (HPM). The PCL coatings improved the corrosion resistance of the alloy

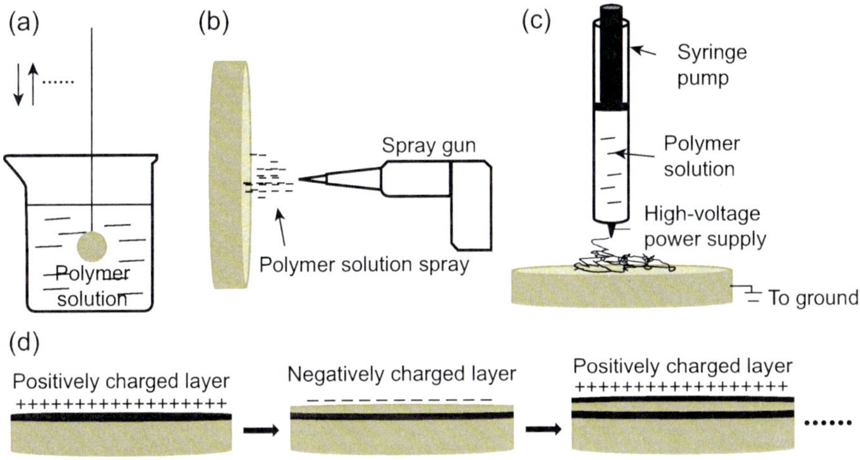

Figure 7.2 Common methods to fabricate organic coatings on the biodegradable magnesium alloy. (a) Dip-coating method. Sample is immersed in the polymer solution and then slowly withdrawn from the solution. (b) Spray coating. Polymer solution was sprayed on the sample. (c) Electrospinning method. Polymer solution with a proper viscosity emit from a spinneret to from nanofibers, which are coated on the sample under assistance of strong electric field. For all three above methods, organic coating forms on the sample after evaporation of solvent. (d) Layer-by-layer method. Films are formed by depositing alternating layers of oppositely charged materials with wash steps in between. Method used to fabricate each single layer of layer-by-layer coating could be any one of methods (a), (b) and (c).

and decreased the release rate of Mg ions and the increase of pH caused by the alloy degradation. The cells cultured with the coated alloy extract after long-term immersion grew better than those cultured with the bare alloy extract. The PCL coating facilitated the cell adhesion and spreading. Abdal-hay et al. (Abdal-hay et al., 2012) prepared both porous and compact poly vinyl acetate (PVAc) coatings on an AM50 Mg alloy by the dip-coating method. The osteoblasts grew better on the coated sample than on the uncoated one. It was also shown that the porous coating would better facilitate the osteoblasts' attachment, spreading, and proliferation compared with the compact one. Later, Abdal-hay et al. (Abdal-hay et al., 2013c) prepared PLA coating on an AM50 alloy by the dip-coating and electrospinning techniques, respectively. The MC3T3-E1 osteoblasts grew well on both coatings. Xu et al. (Xu & Yamamoto, 2012) prepared four kinds of organic coatings with PCL and PLLA on pure Mg by the spin-coating method. The four coatings all significantly enhanced the SaOS-2 cell compatibility of the alloy. Ostrowski et al. (Ostrowski et al., 2013) fabricated three kinds of organic coatings on an AZ31 alloy through the layer-by-layer method. The coatings were with the sequence of "ABCBCBCBCB", where A was PEI, C was poly allylamine hydrochloride (PAH), B was one of the 50:50 PLGA, 75:25 PLGA, or PCL. The coatings improved the adhesion, spreading, and proliferation of the MC3T3 preosteoblasts and human mesenchymal stem cells (hMSCs). Compared with the PCL coating, the two PLGA coatings showed better biocompatibility, although with inferior corrosion resistance.

7.3.3 Composite coating

Different types of inorganic–organic composite coatings were developed most recently (Figure 7.3). The base material of the coating could be inorganic or organic depending on the fabrication method as shown in Figure 7.3. Sometimes, more than one method is needed to fabricate the composite coating. Hahn et al. (Hahn et al., 2011) prepared an HA-chitosan composite coating on an AZ31 Mg alloy through the aerosol deposition method. Compared with the HA coating, the spreading of the cells having multiple filopodia on the surface of the HA-chitosan composite coating was enhanced, implying that the biocompatibility was improved by the incorporation of chitosan. Abdal-hay et al. (Abdal-hay, Amna et al., 2013a) designed and fabricated a porous HA/PCL hybrid nano-composite coating on an AM50 Mg alloy with either high or low porosity by the dipping method. Compared with the single PCL-coated alloy, the osteoblasts cultured in the extract of the alloy with the composite coating showed better cell proliferation, indicating that the incorporation of nano-hydroxyapatite particles (nHAp) into the polymer coating improved the biocompatibility. Abdal-hay et al. (Abdal-hay et al., 2013b) also prepared HA-doped PLA porous coating on an AZ31B Mg alloy by the spraying method. The composite coating had better compatibility to the MC3T3 cells than the pristine polymer membrane. Zomorodian et al. (Zomorodian et al., 2013) fabricated PEI coatings with and without nHAp on an AZ31 alloy via the dip-coating method. The results showed that the incorporation of HA improved the cytocompatibility of the PEI coating. With increase of the HA content, the corrosion resistance and cytocompatibility of the coating decreased. Gao et al. (Gao et al., 2011) prepared a composite coating on the Mg-2Zn-0.24Ca alloy by MAO followed by a sol–gel process. The coating composed of an inner ceramic MAO layer and an outer propolis layer. The corrosion resistance of composite coating was higher than that of the single MAO coating. Mesenchymal stem cells (MSCs) grew better on the composite coating than on the single MAO coating. The cell viability of MSCs was also higher in the composite coating group.

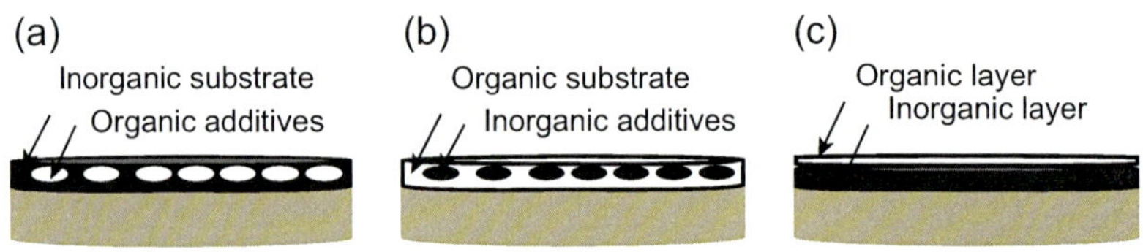

Figure 7.3 Different types of inorganic–organic composite coatings. (a) Inorganic substrate plus organic additives. The coating substrate is inorganic. Organic materials are added to enhance the corrosion resistance and/or biocompatibility. (b) Organic substrate plus inorganic additives. The coating substrate is organic. Inorganic substances are incorporated mainly to enhance the biocompatibility/bioactivity. The above two coatings are usually prepared by the one-step method. (c) Two-layer composite coating. The inner layer is an inorganic coating with high wear resistance and bonding strength. The outer layer is an organic one to enhance the corrosion resistance and biocompatibility. Two or more methods are needed to fabricate such composite coatings.

7.4 Biocompatibility of surface-modified magnesium-based metals – *in vivo* findings

The *in vitro* cell tests could give a first insight of the biocompatibility of surface-modified Mg-based metals. However, the *in vitro* results would not be totally consistent with the *in vivo* ones. In addition, the *in vivo* study could supply information about the tissue response, important organ alternation, and the long-term *in vivo* degradation profile of the implant. Thus the *in vivo* study is highly important for research on biodegradable Mg-based metals. Many *in vivo* studies have been conducted, and some important works are introduced here.

7.4.1 Inorganic coatings

Witte et al. (Witte et al., 2010) implanted an LAE442 Mg alloy with magnesium fluoride (MgF_2) coating into the femur condyles of female New Zealand white rabbits. The average corrosion rate of the implants during 12 weeks was 0.1 ± 0.03 mm/year. No appearance of subcutaneous gas cavities was found. Neither infections nor other adverse host reactions were clinically observed during the postoperative follow-up. The blood examinations were all in the physiological range for rabbits. The histopathology of the synovial tissue in the 2-week group showed granular cell infiltration in all cases, but in one case, a fibrotic tissue formation with local area of cartilage metaplasia was observed. In the 4-week group, only one case of inflammatory cell infiltration was observed. All other sections of synovial tissue showed a regular histology without sign of pathological lesion. In two cases (one at 2 weeks and one at 4 weeks), minor focal mineralization was found in the kidneys, while the surrounding parenchyma showed no sign of degenerative or inflammatory change.

Thomann et al. (Thomann et al., 2010) implanted MgCa 0.8 pins with MgF_2 coatings into the tibia marrow cavities of adult female New Zealand white rabbits. The mean losses in volume after 3 and 6 months were 8.54% and 25.33%, respectively. A total of 20% and 31% of the slices showed new bone formation on the surfaces after 3 and 6 months, respectively, which was mostly in the form of trabecula (44%); however, clasps of bone (7%) also occurred, which affected only a part of the surface. Endosteal new bone formation was shown after 3 and 6 months, which reached a maximum after 3 months of implantation. Periosteal new bone formation could be found in 20% and 9% of the evaluated slices after 3 and 6 months, respectively.

Sun et al. (Sun et al., 2013) studied the bioactivity of Mg-3Zn-0.8Zr alloys separately with MgF_2 and Ca-P coatings by implanting the alloys in rabbits' femoral shafts. The mean losses in volumes of the bare, CaP-coated, and MgF_2-coated alloy rods after 3 months were 37.02%, 22.67%, and 23.85%, respectively. No gas shadow was observed around any of the rods at 1, 2, or 3 months. No visible evidence of inflammation was seen in the local bone tissue of any of the samples after 3 months. New bone trabeculas gradually formed around the implants. The arrangement of the newly formed bone trabecula around the MgF_2 sample was more neat and compact than that around the bare and CaP samples. The amount of new bone, new bone volume fraction

(BVF), and tissue mineral density (TMD) around the MgF_2 implant were the highest, followed by CaP and bare implants, at 3 months. The degradation of all the implants had no influence on raising the serum Mg^{2+} or Ca^{2+} levels and the normal function of the kidney and liver.

Xu et al. (Xu et al., 2009) evaluated the surface bioactivity of Mg-Mn-Zn alloy with CaP coating by implanting rods into the femoral shaft of Japanese big-ear rabbits. Compared with the bare Mg alloy implant, the newly formed osteoid tissue around the CaP-coated Mg alloy implant was compact and uniform after 4 weeks of implantation. Unlike the bare Mg alloy, lymphocytic infiltration and plasmablastic infiltration were not seen around the coated alloy within the first 2 weeks after implantation. At all time intervals (1, 2, 3 and 4 weeks), the CaP-coated alloy group showed higher BMP-2, TGF-β1, and PDGF expression levels than the bare alloy group. Xiao et al. (Xiao et al., 2013) implanted AZ60 Mg alloys with and without CaP coating into the femoral shafts of adult New Zealand rabbits. All rabbits healed well, and no difference was observed between the two groups. The mean weight of rabbits increased for the CaP-coated sample group, whereas it decreased for the uncoated sample group. Wang et al. (Wang et al., 2011) studied the *in vivo* performance of the Ca-deficient HA-coated Mg-Zn-Ca alloy by implanting rods into rabbit femur bones. The *in vivo* degradation rate of the alloy was dramatically decreased by the coating. The fibroblastic band formation was suppressed, while new bone formation was promoted by the coating.

Fischerauer et al. (Fischerauer et al., 2013) implanted a spinel containing MAO-coated Mg-Zn-Ca-Mn alloy in the femoral mid-diaphyseal region of rats. No wound infection was observed after implantation. The MAO coating led to a delayed degradation process within the first 4 weeks. The decreased gas formation of the coated implant within the first 2 weeks minimized the callus formation in the cortical bone and allowed apposition of new bone along the MAO surface. No foreign-body reaction was visible at week four. At week 12, when the implants just fully corroded, a progressive bone restoration at the original drill-hole site was shown. Adjacent to the former implant interface, bone formation was exhibited, producing comparatively unstructured woven bone with large osteocyte lacunae. After 24 weeks, histological ground sections showed notable tissue regeneration. The bone marrow was also completely restored. Adipocytes predominated in the cell line of the bone marrow. A few scattered bone tissue formations remained in the medullary cavity at that point.

Lin et al. (Lin, Tan, Wang et al., 2013) implanted a forsterite-containing MAO-coated ZK60 alloy into the rabbit femoral bones. Within the first 2 weeks, the MAO coating could protect the alloy well. However, the protective ability was gradually lost. The bone tissue grew better around the coated alloy than around the bare one within 2 weeks. The morphology of the surrounding bone was altered by the hydrogen generation. The biological functionality of the surrounding tissues was not significantly affected during the entire serving period of the implant. Qi et al. (Qi et al., 2013) compared the bone response to the MAO-coated ZK60 alloy and PLLA with a distal femur model of 3-month-old male Sprague–Dawley mice. At 26 weeks postoperatively, direct bone contact between the Mg alloy implant and the new bone could be found. A new bone sheath in place of cancellous bone was observed in the bone marrow cavity, which did not differ from the cortical bone in microstructure. Fibrous tissue was also observed partly around the alloy implant. A cavity was detected around the Mg alloy. The blood biochemical examination results showed that some values of GPT and GOT and all tested values of serum K were out of the recommended values. However, no significant difference was observed.

Chen et al. (Chen et al., 2012) implanted a Mg-Zn-Ca alloy with MAO/ED composite coating into the femur shaft of an adult New Zealand white rabbit. The degradation rate of the coated samples was much smaller than that of the substrates. More newly formed bone was found around the coated samples at 8 and 18 weeks postoperatively. No lymphocytic and plasmablastic inflammations were observed around the coated sample at 8 weeks. The coated sample was surrounded by enormous bone trabeculars and osteoclasts at 12 weeks. The bone trabeculars aligned compactly and osteoid tissues were connected around the coated samples at 18 weeks. At 50 weeks, when the coated implant was completely degraded, the newly formed bone was the same as the original bone tissue.

Badar et al. (Badar et al., 2013) inserted a pure Mg wire with a partially protective magnesium hydroxide layer into the lower tail artery of female BALB/c mice. At 4 weeks, the entire surface of the Mg wire was coated by an organic layer. In the tissue containing the Mg implant, no evidence of tissue necrosis, inflammatory giant cells, or gas bubbles could be detected at any time. The histological analysis revealed that the pure Mg implant was essentially as tissue-compatible as the two conventional implant materials.

Wong et al. (Wong et al., 2013) implanted Al_2O_3-coated AZ91 Mg alloy implants, which were fabricated by aluminum and oxygen plasma implantation method, into the lateral femur epicondyle of a 2-month-old female Sprague–Dawley rat. The untreated implant showed significantly less new bone formation than the treated one at 8 weeks postoperative. The serum Mg^{2+} concentrations in the rats implanted with either the untreated or plasma-treated implants were in the normal range of physiological Mg levels. New bone tissues formed around the implants and showed direct bone/implant contact. More new bone was formed around the plasma-treated implant, where osteoblast-like cells could be observed.

7.4.2 Organic coatings

Wong et al. (Wong et al., 2010) studied the *in vivo* biocompatibility of the PCL-coated AZ91 Mg alloy. The low porosity membrane (LPM) owned a better *in vivo* corrosion resistance than that of the high porosity membrane (HPM). No inflammation or necrosis around the implants was observed. All the samples showed direct contact with the newly formed bone. The uncoated sample had the least amount of new bone formation, and the polymer-coated samples showed the new bone volume in order of LPM > HPM. The detected serum Mg^{2+} levels were within the normal range of physiological Mg levels. Wang et al. (Wang et al., 2013) implanted uncoated, PCL-coated, and PTMC-coated Mg-Zn-Mn wires subcutaneously in rats. During 16 weeks of implantation, the uncoated, PCL-coated, and PTMC-coated samples neither induced any local toxic effect nor caused any obvious local response of the tissue. From macroscopic examination, no significant signs of inflammation, encapsulation, hemorrhage, necrosis, or discoloration were observed. A few differences between the

PCL-coated group and the other two groups were detected from microscopic examination. For the uncoated and PTMC-coated samples, only a fibrocyte monolayer existed at the implant–tissue interface. For the PCL-coated sample, there was a thicker fibrous capsule ($\sim$40 μm) and a small number of macrophages and plasmocytes in the surrounding region, with a slight inflammation response. No gas bubbles were observed in the surrounding tissue by eye survey or histological analysis in any samples. Overall, the PTMC-coated sample displayed an excellent tissue response.

7.5 Influencing factors on the biocompatibility of the surface-modified magnesium-based metals

7.5.1 Morphology, composition, and physicochemical properties of the surface

As presented in the studies on other biomaterials (Costa et al., 2013; Huebsch et al., 2010; Mei et al., 2010; Minagar, Wang, Berndt, Ivanova, & Wen, 2013; Phillips, Petrie, Creighton, & Garcia, 2010; Tan & Teoh, 2007), the morphology, composition, and physicochemical properties of the surface coating on biodegradable Mg-based metals will greatly affect their initial surface biocompatibilities. In addition, the degradation products, which depend on the compositions of the modified layers, will also have influences on the biocompatibility, as discussed in Section 7.2.

Lorenz et al. (Lorenz et al., 2009) showed that the surface chemistry (presences of Ca and P on the surface) and/or surface roughness of coatings on pure Mg greatly influenced the initial cell adhesion. Seyfoori et al. (Seyfoori, Mirdamadi, Mehrjoo, & Khavandi, 2013) demonstrated that a phosphate coating with a rougher surface than the silicate coating on the AZ31 Mg alloy could provide better cell adhesion for osteoblastic cells. In the study by Abdal-hay et al. (Abdal-hay et al., 2012), it was found that the porous PVAc organic coating on AM50 Mg alloy had a better biocompatibility than the compact one. They also prepared PLA coatings with two different morphologies on AM50 alloy by electrospinning and dip-coating methods, respectively. The results showed that with a prolonging of the incubation time, the large surface area of the coated samples with wide nano-fiber membranes presented more cell adhesion and proliferation (Abdal-hay et al., 2013c).

Ostrowski et al. (Ostrowski et al., 2013) showed that even with lower corrosion resistance, the PLLA coating on AZ31 Mg alloy better facilitated the growth of osteoblastic cells and hMSCs than the PCL coated alloy. Liu et al. (Liu et al., 2013) prepared the anticorrosive silane coatings with and without heparin on an AZ31 alloy. The platelet adhesion test demonstrated a significantly lower platelet adhesion for the silane-heparin coating compared with the single silane coating. Hahn et al. (Hahn et al., 2011) demonstrated that the incorporation of chitosan into an HA coating on an AZ31 alloy could improve the biocompatibility of the coating. Abdal-hay et al. (Abdal-hay, Amna et al., 2013a) showed that the incorporation of nHAp within the polymer matrix on AM50 alloy could improve the cell viability and survival in comparison with the plain PCL coating. Similar results were obtained in a study on HA/PLA composite coating on AZ31 alloy (Abdal-hay et al., 2013b). Furthermore, Zomorodian et al. (Zomorodian et al., 2013) found that the HA addition into the PEI coating on AZ31 alloy enhanced the cellular response to the coating. Lin et al. (Lin et al., 2014) altered phosphate MAO coating by incorporating Sr. The Sr-containing coating demonstrated much better biocompatibility/bioactivity.

7.5.2 Degradation resistance of the surface modification layer

Many studies showed that the corrosion resistance of the coatings on Mg-based metals also greatly influenced the initial surface biocompatibility (Gu et al., 2011; Keim et al., 2011; Lorenz et al., 2009; Lu, Chen, Huang, & Yan, 2012; Wang et al., 2013; Wong et al., 2010; Xu & Yamamoto, 2012). The influence of corrosion resistance could be attributed to the different control of the Mg^{2+} release and pH elevation caused by the degradation of Mg substrate. Keim et al. (Keim et al., 2011) cultured HeLa cells on a glass and incubated them in the culture medium with increased Mg^{2+} concentrations ranging from 20 to 750 mg/L. The results showed that the Mg^{2+} concentration in the given range did not affect the cell adhesion and growth. Meanwhile, this study showed that the protective ability of a coating was critical for enhancing the cell adhesion and spreading by way of reducing the hydrogen evolution and pH elevation. Seuss et al. (Seuss, Seuss, Turhan, Fabry, & Virtanen, 2011) cultured HeLa cells on glass in culture mediums with different pH values. The results showed that the cell density decreased with increasing pH of the culture medium. However, a dramatic decline in cell density and spreading area was only observed at $pH > 10$. Wong et al. (Wong et al., 2010) showed the correlation between cell viability and the amount of released Mg ions. When the Mg ion concentration was higher than 750 ppm, the extract showed an obvious cytotoxicity.

It needs to be mentioned that the corrosion protect ability of the modified layer is not necessary to be designed as high as possible to achieve the surface biocompatibility. After all, the surface-modified Mg-based metals are developed as a kind of biodegradable implant material. The coating itself should be biodegradable with a proper rate. Some materials, such as titania and alumina, are not recommended as the coating materials although they have good biocompatibility as bioinert materials. In addition, if the Mg implant is not used for load-bearing application, a mild Mg^{2+} release from the substrate, which is beneficial for the surrounding bone tissue, is acceptable. Furthermore, the surface biocompatibility could be better for an Mg implant with surface layer that has proper composition and topography than that with surface layer that has inferior composition and topography, even if the corrosion resistance of the former one is lower than that of the latter. All in all, the composition, topography, and corrosion resistance should be optimized simultaneously to achieve good surface biocompatibility of the surface-modified Mg implant.

7.5.3 Average degradation rate of the surface-modified magnesium-based metals

The above two factors are related to the surface biocompatibility of the surface-modified Mg alloy. Once the surface layer fails and the substrate starts to be degraded, the average degradation rate rather than the surface could be critical to the biocompatibility of the Mg implant. The degradation rate of an Mg implant is directly related to the Mg^{2+} and hydrogen release rate. To obtain an activated new bone formation, a proper Mg^{2+} release rate is needed (Serre, Papillard, Chavassieux, Voegel, & Boivin, 1998; Staiger, Pietak, Huadmai, & Dias, 2006; Zreiqat et al., 2002). The hydrogen release rate could also affect the new bone growth (Fischerauer et al., 2013; Kraus et al., 2012; Lin, Tan, Wang et al., 2013; Qi et al., 2013). As expected, the tissue response was quite different to the Mg implants with different average degradation rates (Huehnerschulte et al., 2011; Kraus et al., 2012; Thomann et al., 2009; Witte et al., 2005; Wong et al., 2010). Thomann et al. (Thomann et al., 2009) implanted MgCa0.8 and LAE442 Mg alloy cylinders into tibiae of female adult New Zealand white rabbits. The MgCa0.8 implants degraded obviously faster than LAE442 implants. After 12 months, the μ-CT results showed that more bone-implant contact was found in the MgCa0.8 group. However, more new endosteal bone formation was found in the LAE442 group. Similar results were obtained in an *in vivo* comparative study on ZEK100 and AX30 Mg alloys (Huehnerschulte et al., 2011). The ZEK100 implants had a higher corrosion rate than those made of AX30. The periosteal, endosteal, and trabecular new bone formation were all different for both materials at 3 and 6 months. It was supposed that a relation existed between the bone response and the degradation rate of the alloy.

Kraus et al. (Kraus et al., 2012) carefully compared the cortical bone responses to ZX50 and WE21 Mg alloys in a growing rat skeleton model. The ZX50 alloy exhibited a significantly higher degradation rate than the WZ21 alloy. It was found that the ZX50 released large amount of gas bubbles. These gas bubbles impeded the connectivity of osteocytes with the ZX50 implant surface. The gas pressure induced some mechanical disturbance of bone regeneration, which resulted in a distinct callus formation. In addition, the fast Mg ion release of ZX50 led to an enhanced neo-formation of bone tissue around the implant. Gas formation around the WE21 implant occurred in an amount that did not affect the bone regeneration and was almost entirely absorbed by surrounding tissue. New bone formation around the WE21 pin occurred in an enhanced way. Also, new bone formation closely contacted to the implant at the cortical and medullary cavity site within the first 12 weeks.

The average degradation rate of surface-modified Mg-based metals can be decided by the corrosion resistances of both the modified layer and the base alloy. The corrosion resistance of Mg substrate and modified layer could be adjusted in a wide range. Thus, it is possible to design Mg implants with a proper *in vivo* degradation rate. It must be noted that the unexpected local fail of the surface-modified layer will lead to a local and accelerated degradation of the implant. The unexpected fails include local detachment of the coating under shear stress, local abrasion during implantation, and service process when in contact with the surgical instruments, hard tissues, or other implants. The defects in modified layers, such as cracks or pores in conversion coatings and MAO coating, will also lead to the failure of surface coating in a peeling-off way, as reported by Lin et al. (Lin, Tan, Zhang et al., 2013). The local and accelerated degradation of the implant will, on one hand, lead to undesired local burst release of Mg^{2+}, OH^-, and H_2; on the other hand, they result in the quick loss of mechanical strength of the implant. Both of them will finally cause the failure of the

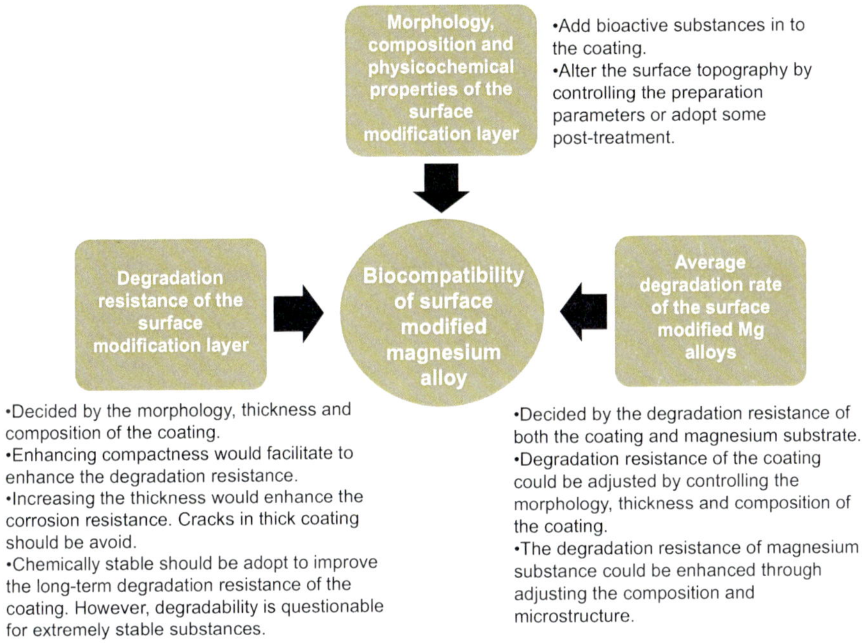

Figure 7.4 Influencing factors of biocompatibility of surface-modified magnesium alloy and how to control these factors.

implant. Thus the defect-free surface layer with high bonding strength and wear resistance should be prepared to avoid the unexpected failure.

The above three factors are schematically summarized in Figure 7.4, in which how to control the three factors is also addressed. The above three factors need to be considered simultaneously when designing biocompatible surface-modified Mg alloys with required properties for different applications.

7.6 Future trends

The biocompatibility of a surface-modified Mg implant is decided by both the surface coating and the alloy substrate. Several requirements for a biocompatible surface-modified Mg alloy are depicted in Figure 7.5.

An ideal Mg alloy substrate should contain no biotoxic alloying elements and no nondegradable second phases and have a proper corrosion resistance. Additionally, the base alloy should have enough mechanical strength for load-bearing applications. These properties can be obtained by alloy composition design, heat treatment, and deformation processing.

A biocompatible surface coating should facilitate the cell adhesion and proliferation and ideally should promote restoration of the surrounding tissue by releasing some bioactive substances. Meanwhile, the degradation products of the surface coating should not elicit any undesirable local or systemic effects. In addition, the surface coating should well-protect the Mg alloy substrate from fast degradation. For load-bearing applications,

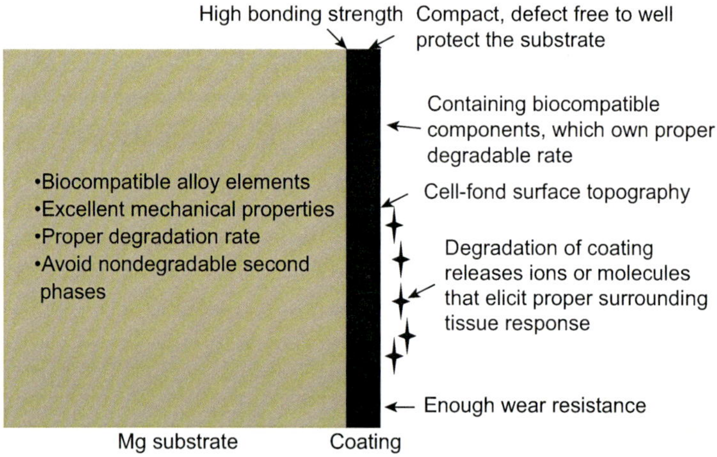

Figure 7.5 Requirements for a biocompatible surface-modified magnesium alloy.

the surface coating should totally prevent the degradation of Mg substrate within a proper period. Furthermore, the mechanical properties of the surface coating, such as bonding strength, and wear resistance, are also critical for the clinical, especially for orthopedic, applications. One single modification method is hardly able to fabricate a surface coating that fulfills all the above requirements. Two or more different modification methods need be used simultaneously to fabricate a surface coating to meet all the above requirements. An innermost layer is better formed by Mg substrate-involved chemical reactions, which will have an excellent bonding strength. The wear resistance is another consideration for preparation of the innermost layer. On the innermost layer, a biofunctional layer could be prepared to elicit the desired biological response. The multilayered coating should be compact and contain no weak spots to ensure good corrosion resistance. The inorganic/organic composite coating could be a future research hotspot. Besides the above requirements, the coating process should not change the properties of Mg substrate. For real clinical applications, the market price for the biodegradable Mg implants should be acceptable to most of the customers. Thus the Mg substrate and the surface treatment process should be cost-effective.

The biocompatibility test methods should be standardized to obtain comparable results for different surface-modified Mg-based metals. The test procedure should be normalized or the control material should be standardized. Widely accepted reference materials should be recommended for use in all the biocompatibility tests.

7.7 Sources of further information and advice

7.7.1 Book and journal

To our knowledge, no books specifically written about the biodegradable Mg-based metals are currently available. The relative research works could be found in the following journals: *Acta Biomaterialia, Biomaterials, Journal of Biomedical Materials*

Research, Materials Science and Engineering: C, Journal of Materials Science: Materials in Medicine, Colloids and Surface B: Biointerfaces, Surface and Coatings Technology, Applied Surface Science, Corrosion Science, Electrochimica Acta, etc.

7.7.2 Research body

Engineering Research Center for Revolutionizing Metallic Biomaterials, National Science Foundation, USA.

http://erc.ncat.edu/

7.7.3 Research centers

Institute of Metal Research, Chinese Academy of Sciences, China.
Peking University, China
Shanghai Jiao Tong University, China
Southwest Jiaotong University, China
Zhengzhou University, China
The Chinese University of Hong Kong, China
The University of Hong Kong, China
National Institute for Materials Science (NIMS), Japan
Chonbuk National University, South Korea
Hannover Medical School, Germany
Leibniz University of Hannover, Germany
Charité-Universitätsmedizin Berlin, Germany
Helmholtz-Zentrum Geesthacht, Germany
Medical University Graz, Austria
ETH Zurich, Switzerland
North Carolina A&T State University, USA
University of Pittsburgh, Pittsburgh, USA
University of Cincinnati, USA

7.7.4 Companies

Biotronik, Germany
http://www.biotronik.com/wps/wcm/connect/int_web/biotronik/home
aap Implantate, Germany
http://www.aap.de/
Syntellix, Germany
http://syntellix.com/en/
Trauson, China
http://www.trauson.com/index_e.asp
Eontec, China
http://www.e-ande.com/english/

References

Abdal-hay, A., Amna, T., & Lim, J. K. (2013a). Biocorrosion and osteoconductivity of PCL/nHAp composite porous film-based coating of magnesium alloy. *Solid State Sciences, 18*, 131–140.

Abdal-hay, A., Barakat, N. A. M., & Lim, J. K. (2013b). Hydroxyapatite-doped poly(lactic acid) porous film coating for enhanced bioactivity and corrosion behavior of AZ31 Mg alloy for orthopedic applications. *Ceramics International, 39*, 183−195.

Abdal-hay, A., Barakat, N. A. M., & Lim, J. K. (2013c). Influence of electrospinning and dip-coating techniques on the degradation and cytocompatibility of Mg-based alloy. *Colloids and Surfaces A: Physicochemical and Engineering Aspects, 420*, 37−45.

Abdal-hay, A., Dewidar, M., & Lim, J. K. (2012). Biocorrosion behavior and cell viability of adhesive polymer coated magnesium based alloys for medical implants. *Applied Surface Science, 261*, 536−546.

Adams, C. S., Mansfield, K., Perlot, R. L., & Shapiro, I. M. (2001). Matrix regulation of skeletal cell apoptosis role of calcium and phosphate ions. *Journal of Biological Chemistry, 276*, 20316−20322.

Alabbasi, A., Liyanaarachchi, S., & Kannan, M. B. (2012). Polylactic acid coating on a biodegradable magnesium alloy: an in vitro degradation study by electrochemical impedance spectroscopy. *Thin Solid Films, 520*, 6841−6844.

Badar, M., Lunsdorf, H., Evertz, F., Rahim, M. I., Glasmacher, B., Hauser, H., et al. (2013). The formation of an organic coat and the release of corrosion microparticles from metallic magnesium implants. *Acta Biomaterialia, 9*, 7580−7589.

Bai, K., Zhang, Y., Fu, Z., Zhang, C., Cui, X., Meng, E., et al. (2012). Fabrication of chitosan/magnesium phosphate composite coating and the in vitro degradation properties of coated magnesium alloy. *Materials Letters, 73*, 59−61.

Bala Srinivasan, P., Liang, J., Blawert, C., St Rmer, M., & Dietzel, W. (2009). A preliminary study of calcium containing plasma electrolytic oxidation coatings on AM50 magnesium alloy. *Journal of Materials Science, 45*, 1406−1410.

Bala Srinivasan, P., Liang, J., Blawert, C., St Rmer, M., & Dietzel, W. (2010). Characterization of calcium containing plasma electrolytic oxidation coatings on AM50 magnesium alloy. *Applied Surface Science, 256*, 4017−4022.

Bonnelye, E., Chabadel, A., Saltel, F., & Jurdic, P. (2008). Dual effect of strontium ranelate: stimulation of osteoblast differentiation and inhibition of osteoclast formation and resorption in vitro. *Bone, 42*, 129−138.

Bostman, O., & Pihlajamaki, H. (2000). Clinical biocompatibility of biodegradable orthopaedic implants for internal fixation: a review. *Biomaterials, 21*, 2615−2621.

Carlisle, E. M. (1970). Silicon: a possible factor in bone calcification. *Science, 167*, 279−280.

Carlisle, E. M. (1982). The nutritional essentiality of silicon. *Nutrition Reviews, 40*, 193−198.

Chang, L., Tian, L., Liu, W., & Duan, X. (2013). Formation of dicalcium phosphate dihydrate on magnesium alloy by micro-arc oxidation coupled with hydrothermal treatment. *Corrosion Science, 72*, 118−124.

Chen, S., Guan, S., Chen, B., Li, W., Wang, J., Wang, L., et al. (2011). Corrosion behavior of TiO_2 films on Mg−Zn alloy in simulated body fluid. *Applied Surface Science, 257*, 4464−4467.

Chen, S., Guan, S., Li, W., Wang, H., Chen, J., Wang, Y., et al. (2012). In vivo degradation and bone response of a composite coating on Mg−Zn−Ca alloy prepared by microarc oxidation and electrochemical deposition. *Journal of Biomedical Materials Research Part B: Applied Biomaterials, 100B*, 533−543.

Chen, J. Y., Wan, G. J., Leng, Y. X., Yang, P., Sun, H., Wang, J., et al. (2004). Behavior of cultured human umbilical vein endothelial cells on titanium oxide films fabricated by plasma immersion ion implantation and deposition. *Surface and Coatings Technology, 186*, 270−276.

Chiu, K., Wong, M., Cheng, F., & Man, H. (2007). Characterization and corrosion studies of fluoride conversion coating on degradable Mg implants. *Surface and Coatings Technology, 202*, 590−598.

Costa, D. O., Prowse, P. D., Chrones, T., Sims, S. M., Hamilton, D. W., Rizkalla, A. S., et al. (2013). The differential regulation of osteoblast and osteoclast activity by surface topography of hydroxyapatite coatings. *Biomaterials, 34*, 7215–7226.

Cui, W., Beniash, E., Gawalt, E., Xu, Z., & Sfeir, C. (2013). Biomimetic coating of magnesium alloy for enhanced corrosion resistance and calcium phosphate deposition. *Acta Biomaterialia, 9*, 8650–8659.

De Lima, M., Mcmannis, J., Gee, A., Komanduri, K., Couriel, D., Andersson, B. S., et al. (2008). Transplantation of ex vivo expanded cord blood cells using the copper chelator tetraethylenepentamine: a phase I/II clinical trial. *Bone Marrow Transplant, 41*, 771–778.

Degner, J., Singer, F., Cordero, L., Boccaccini, A. R., & Virtanen, S. (2013). Electrochemical investigations of magnesium in DMEM with biodegradable polycaprolactone coating as corrosion barrier. *Applied Surface Science, 282*, 264–270.

Dvorak, M. M., Siddiqua, A., Ward, D. T., Carter, D. H., Dallas, S. L., Nemeth, E. F., et al. (2004). Physiological changes in extracellular calcium concentration directly control osteoblast function in the absence of calciotropic hormones. *Proceedings of the National Academy of Sciences of the United States of America, 101*, 5140–5145.

Ellingsen, J. (1995). Pre-treatment of titanium implants with fluoride improves their retention in bone. *Journal of Materials Science: Materials in Medicine, 6*, 749–753.

Fekry, A. M., Ghoneim, A. A., & Ameer, M. A. (2014). Electrochemical impedance spectroscopy of chitosan coated magnesium alloys in a synthetic sweat medium. *Surface and Coatings Technology, 238*, 126–132.

Fischerauer, S. F., Kraus, T., Wu, X., Tangl, S., Sorantin, E., Hanzi, A. C., et al. (2013). In vivo degradation performance of micro-arc-oxidized magnesium implants: a micro-CT study in rats. *Acta Biomaterialia, 9*, 5411–5420.

Fischer, H., Niedhart, C., Kaltenborn, N., Prange, A., Marx, R., Niethard, F. U., et al. (2005). Bioactivation of inert alumina ceramics by hydroxylation. *Biomaterials, 26*, 6151–6157.

Francis Suh, J.-K., & Matthew, H. W. (2000). Application of chitosan-based polysaccharide biomaterials in cartilage tissue engineering: a review. *Biomaterials, 21*, 2589–2598.

Gao, J. H., Shi, X. Y., Yang, B., Hou, S. S., Meng, E. C., Guan, F. X., et al. (2011). Fabrication and characterization of bioactive composite coatings on Mg-Zn-Ca alloy by MAO/sol-gel. *Journal of Materials Science: Materials in Medicine, 22*, 1681–1687.

Gao, Y., Yerokhin, A., & Matthews, A. (2013). DC plasma electrolytic oxidation of biodegradable cp-Mg: in-vitro corrosion studies. *Surface and Coatings Technology, 234*, 132–142.

Geng, F., Tan, L. L., Jin, X. X., Yang, J. Y., & Yang, K. (2009). The preparation, cytocompatibility, and in vitro biodegradation study of pure beta-TCP on magnesium. *Journal of Materials Science: Materials in Medicine, 20*, 1149–1157.

Gray-Munro, J. E., Seguin, C., & Strong, M. (2009). Influence of surface modification on the in vitro corrosion rate of magnesium alloy AZ31. *Journal of Biomedical Materials Research. Part A, 91*, 221–230.

Grynpas, M. D., & Marie, P. J. (1990). Effects of low doses of strontium on bone quality and quantity in rats. *Bone, 11*, 313–319.

Guan, R. G., Johnson, I., Cui, T., Zhao, T., Zhao, Z. Y., Li, X., et al. (2012). Electrodeposition of hydroxyapatite coating on Mg-4.0Zn-1.0Ca-0.6Zr alloy and in vitro evaluation of degradation, hemolysis, and cytotoxicity. *Journal of Biomedical Materials Research. Part A, 100*, 999–1015.

Gu, X. N., Li, N., Zhou, W. R., Zheng, Y. F., Zhao, X., Cai, Q. Z., et al. (2011). Corrosion resistance and surface biocompatibility of a microarc oxidation coating on a Mg–Ca alloy. *Acta Biomaterialia, 7*, 1880–1889.

Gunatillake, P. A., & Adhikari, R. (2003). Biodegradable synthetic polymers for tissue engineering. *European Cells & Materials, 5*, 1–16. discussion 16.

Gu, X., Zheng, W., Cheng, Y., & Zheng, Y. (2009). A study on alkaline heat treated Mg-Ca alloy for the control of the biocorrosion rate. *Acta Biomaterialia, 5*, 2790–2799.

Habibovic, P., & Barralet, J. E. (2011). Bioinorganics and biomaterials: bone repair. *Acta Biomaterialia, 7*, 3013–3026.

Hahn, B.-D., Park, D.-S., Choi, J.-J., Ryu, J., Yoon, W.-H., Choi, J.-H., et al. (2011). Aerosol deposition of hydroxyapatite–chitosan composite coatings on biodegradable magnesium alloy. *Surface and Coatings Technology, 205*, 3112–3118.

Hornberger, H., Virtanen, S., & Boccaccini, A. R. (2012). Biomedical coatings on magnesium alloys—a review. *Acta Biomaterialia, 8*, 2442–2455.

Huang, N., Yang, P., Leng, Y. X., Chen, J. Y., Sun, H., Wang, J., et al. (2003). Hemocompatibility of titanium oxide films. *Biomaterials, 24*, 2177–2187.

Huebsch, N., Arany, P. R., Mao, A. S., Shvartsman, D., Ali, O. A., Bencherif, S. A., et al. (2010). Harnessing traction-mediated manipulation of the cell/matrix interface to control stem-cell fate. *Nature Materials, 9*, 518–526.

Huehnerschulte, T. A., Angrisani, N., Rittershaus, D., Bormann, D., Windhagen, H., & Meyer-Lindenberg, A. (2011). In vivo corrosion of two novel magnesium alloys ZEK100 and AX30 and their mechanical suitability as biodegradable implants. *Materials, 4*, 1144–1167.

Hu, J., Zhang, C., Cui, B., Bai, K., Guan, S., Wang, L., et al. (2011). In vitro degradation of AZ31 magnesium alloy coated with nano TiO_2 film by sol–gel method. *Applied Surface Science, 257*, 8772–8777.

Ilich, J. Z., & Kerstetter, J. E. (2000). Nutrition in bone health revisited: a story beyond calcium. *Journal of the American College of Nutrition, 19*, 715–737.

Ishizaki, T., Shigematsu, I., & Saito, N. (2009). Anticorrosive magnesium phosphate coating on AZ31 magnesium alloy. *Surface and Coatings Technology, 203*, 2288–2291.

Jokinen, M., P Tsi, M., Rahiala, H., Peltola, T., Ritala, M., & Rosenholm, J. B. (1998). Influence of sol and surface properties on in vitro bioactivity of sol-gel-derived TiO_2 and TiO_2-SiO_2 films deposited by dip-coating method. *Journal of Biomedical Materials Research, 42*, 295–302.

Jugdaohsingh, R., Tucker, K., Qiao, N., Cupples, L., Kiel, D., & Powell, J. (2004). Dietary silicon intake is positively associated with bone mineral density in men and premenopausal women of the Framingham offspring cohort. *Journal of Bone and Mineral Research, 19*, 297–307.

Keim, S., Brunner, J. G., Fabry, B., & Virtanen, S. (2011). Control of magnesium corrosion and biocompatibility with biomimetic coatings. *Journal of Biomedical Materials Research Part B: Applied Biomaterials, 96B*, 84–90.

Kharaziha, M., & Fathi, M. H. (2009). Synthesis and characterization of bioactive forsterite nanopowder. *Ceramics International, 35*, 2449–2454.

Kharaziha, M., & Fathi, M. H. (2010). Improvement of mechanical properties and biocompatibility of forsterite bioceramic addressed to bone tissue engineering materials. *Journal of the Mechanical Behavior of Biomedical Materials, 3*, 530–537.

Killian, M. S., Wagener, V., Schmuki, P., & Virtanen, S. (2010). Functionalization of metallic magnesium with protein layers via linker molecules. *Langmuir, 26*, 12044–12048.

Ko, Y.-M., Choe, H.-C., Jung, S.-C., & Kim, B.-H. (2013). Plasma deposition of a silicone-like layer for the corrosion protection of magnesium. *Progress in Organic Coatings, 76*, 1827–1832.

Kraus, T., Fischerauer, S. F., H Nzi, A. C., Uggowitzer, P. J., L Ffler, J. F., & Weinberg, A. M. (2012). Magnesium alloys for temporary implants in osteosynthesis: In vivo studies of their degradation and interaction with bone. *Acta Biomaterialia, 8*, 1230–1238.

Kunjukunju, S., Roy, A., Ramanathan, M., Lee, B., Candiello, J. E., & Kumta, P. N. (2013). A layer-by-layer approach to natural polymer-derived bioactive coatings on magnesium alloys. *Acta Biomaterialia, 9*, 8690–8703.

Li, J. N., Cao, P., Zhang, X. N., Zhang, S. X., & He, Y. H. (2010). In vitro degradation and cell attachment of a PLGA coated biodegradable Mg–6Zn based alloy. *Journal of Materials Science, 45*, 6038–6045.

Li, M., Cheng, Y., Zheng, Y. F., Zhang, X., Xi, T. F., & Wei, S. C. (2013). Plasma enhanced chemical vapor deposited silicon coatings on Mg alloy for biomedical application. *Surface and Coatings Technology, 228*, S262–S265.

Li, M., Chen, Q., Zhang, W., Hu, W., & Su, Y. (2010). Corrosion behavior in SBF for titania coatings on Mg–Ca alloy. *Journal of Materials Science, 46*, 2365–2369.

Lin, X., Tan, L., Wan, P., Yu, X., Yang, K., Hu, Z., et al. (2013). Characterization of micro-arc oxidation coating post-treated by hydrofluoric acid on biodegradable ZK60 magnesium alloy. *Surface and Coatings Technology, 232*, 899–905.

Lin, X., Tan, L., Wang, Q., Zhang, G., Zhang, B., & Yang, K. (2013). In vivo degradation and tissue compatibility of ZK60 magnesium alloy with micro-arc oxidation coating in a transcortical model. *Materials Science and Engineering: C, 33*, 3881–3888.

Lin, X., Tan, L., Zhang, Q., Yang, K., Hu, Z., Qiu, J., et al. (2013). The in vitro degradation process and biocompatibility of a ZK60 magnesium alloy with a forsterite-containing micro-arc oxidation coating. *Acta Biomaterialia, 9*, 8631–8642.

Lin, X., Yang, X., Tan, L., Li, M., Wang, X., Zhang, Y., et al. (2014). In vitro degradation and biocompatibility of a strontium-containing micro-arc oxidation coating on the biodegradable ZK60 magnesium alloy. *Applied Surface Science, 288*, 718–726.

Li, J., Song, Y., Zhang, S., Zhao, C., Zhang, F., Zhang, X., et al. (2010). In vitro responses of human bone marrow stromal cells to a fluoridated hydroxyapatite coated biodegradable Mg–Zn alloy. *Biomaterials, 31*, 5782–5788.

Liu, C., Xin, Y., Tian, X., & Chu, P. K. (2007). Corrosion behavior of AZ91 magnesium alloy treated by plasma immersion ion implantation and deposition in artificial physiological fluids. *Thin Solid Films, 516*, 422–427.

Liu, X., Yue, Z., Romeo, T., Weber, J., Scheuermann, T., Moulton, S., et al. (2013). Biofunctionalized anti-corrosive silane coatings for magnesium alloys. *Acta Biomaterialia, 9*, 8671–8677.

Li, N., & Zheng, Y. (2013). Novel magnesium alloys developed for biomedical application: a review. *Journal of Materials Science & Technology, 29*, 489–502.

Lorenz, C., Brunner, J. G., Kollmannsberger, P., Jaafar, L., Fabry, B., & Virtanen, S. (2009). Effect of surface pre-treatments on biocompatibility of magnesium. *Acta Biomaterialia, 5*, 2783–2789.

Lozano, R. M., Perez-Maceda, B. T., Carboneras, M., Onofre-Bustamante, E., Garcia-Alonso, M. C., & Escudero, M. L. (2013). Response of MC3T3-E1 osteoblasts, L929 fibroblasts, and J774 macrophages to fluoride surface-modified AZ31 magnesium alloy. *Journal of Biomedical Materials Research: Part A, 101*, 2753–2762.

Lu, W., Chen, Z., Huang, P., & Yan, B. (2012). Microstructure, corrosion resistance and biocompatibility of biomimetic HA-based Ca-P coatings on ZK60 magnesium alloy. *International Journal of Electrochemical Science, 7*, 12668–12679.

Mao, L., Yuan, G., Niu, J., Zong, Y., & Ding, W. (2013). In vitro degradation behavior and biocompatibility of Mg–Nd–Zn–Zr alloy by hydrofluoric acid treatment. *Materials Science and Engineering: C, 33*, 242–250.

Marie, P. J., Ammann, P., Boivin, G., & Rey, C. (2001). Mechanisms of action and therapeutic potential of strontium in bone. *Calcified Tissue International, 69*, 121–129.

Mei, Y., Saha, K., Bogatyrev, S. R., Yang, J., Hook, A. L., Kalcioglu, Z. I., et al. (2010). Combinatorial development of biomaterials for clonal growth of human pluripotent stem cells. *Nature Materials, 9*, 768–778.

Mestres, G., & Ginebra, M. P. (2011). Novel magnesium phosphate cements with high early strength and antibacterial properties. *Acta Biomaterialia, 7*, 1853–1861.

Minagar, S., Wang, J., Berndt, C. C., Ivanova, E. P., & Wen, C. (2013). Cell response of anodized nanotubes on titanium and titanium alloys. *Journal of Biomedical Materials Research. Part A, 101*, 2726–2739.

Mohanty, A., Misra, M., & Hinrichsen, G. (2000). Biofibres, biodegradable polymers and biocomposites: an overview. *Macromolecular Materials and Engineering, 276*, 1–24.

Ni, S., & Chang, J. (2009). In vitro degradation, bioactivity, and cytocompatibility of calcium silicate, dimagnesium silicate, and tricalcium phosphate bioceramics. *Journal of Biomaterials Applications, 24*, 139–158.

Ni, S., Chou, L., & Chang, J. (2007). Preparation and characterization of forsterite (Mg_2SiO_4) bioceramics. *Ceramics International, 33*, 83–88.

Niu, J., Yuan, G., Liao, Y., Mao, L., Zhang, J., Wang, Y., et al. (2013). Enhanced biocorrosion resistance and biocompatibility of degradable Mg-Nd-Zn-Zr alloy by brushite coating. *Materials Science & Engineering. C: Materials for Biological Applications, 33*, 4833–4841.

Oosterbeek, R. N., Seal, C. K., Seitz, J.-M., & Hyland, M. M. (2013). Polymer–bioceramic composite coatings on magnesium for biomaterial applications. *Surface and Coatings Technology, 236*, 420–428.

Ostrowski, N., Lee, B., Enick, N., Carlson, B., Kunjukunju, S., Roy, A., et al. (2013). Corrosion protection and improved cytocompatibility of biodegradable polymeric layer-by-layer coatings on AZ31 magnesium alloys. *Acta Biomaterialia, 9*, 8704–8713.

Pan, Y. K., Chen, C. Z., Wang, D. G., & Lin, Z. Q. (2013). Preparation and bioactivity of micro-arc oxidized calcium phosphate coatings. *Materials Chemistry and Physics, 141*, 842–849.

Pan, Y. K., Chen, C. Z., Wang, D. G., & Zhao, T. G. (2013). Effects of phosphates on microstructure and bioactivity of micro-arc oxidized calcium phosphate coatings on Mg-Zn-Zr magnesium alloy. *Colloids and Surface B: Biointerfaces, 109*, 1–9.

Pereda, M. D., Alonso, C., Burgos-Asperilla, L., Del Valle, J. A., Ruano, O. A., Perez, P., et al. (2010). Corrosion inhibition of powder metallurgy Mg by fluoride treatments. *Acta Biomaterialia, 6*, 1772–1782.

Phillips, J. E., Petrie, T. A., Creighton, F. P., & Garcia, A. J. (2010). Human mesenchymal stem cell differentiation on self-assembled monolayers presenting different surface chemistries. *Acta Biomaterialia, 6*, 12–20.

Prasad, A. S., Bao, B., Beck, F. W. J., Kucuk, O., & Sarkar, F. H. (2004). Antioxidant effect of zinc in humans. *Free Radical Biology and Medicine, 37*, 1182–1190.

Qi, Z. R., Zhang, Q., Tan, L. L., Lin, X., Yin, Y., Wang, X. L., et al. (2014). Comparison of degradation behavior and the associated bone response of ZK60 and PLLA in vivo. *Journal of Biomedical Materials Research. Part A, 102*, 1255–1263. http://dx.doi.org/10.1002/jbm.a.34795.

Rezwan, K., Chen, Q., Blaker, J., & Boccaccini, A. R. (2006). Biodegradable and bioactive porous polymer/inorganic composite scaffolds for bone tissue engineering. *Biomaterials, 27*, 3413–3431.

Rico, H., Roca-Botran, C., Hern Ndez, E. R., Seco, C., Paez, E., Valencia, M. J., et al. (2000). The effect of supplemental copper on osteopenia induced by ovariectomy in rats. *Menopause, 7*, 413–416.

Sabetrasekh, R., Tiainen, H., Lyngstadaas, S. P., Reseland, J., & Haugen, H. (2011). A novel ultra-porous titanium dioxide ceramic with excellent biocompatibility. *Journal of Biomaterials Applications, 25*, 559–580.

Seitz, J. M., Eifler, R., Stahl, J., Kietzmann, M., & Bach, F. W. (2012). Characterization of MgNd2 alloy for potential applications in bioresorbable implantable devices. *Acta Biomaterialia, 8*, 3852–3864.

Serre, C. M., Papillard, M., Chavassieux, P., Voegel, J. C., & Boivin, G. (1998). Influence of magnesium substitution on a collagen–apatite biomaterial on the production of a calcifying matrix by human osteoblasts. *Journal of Biomedical Materials Research, 42*, 626–633.

Seuss, F., Seuss, S., Turhan, M. C., Fabry, B., & Virtanen, S. (2011). Corrosion of Mg alloy AZ91D in the presence of living cells. *Journal of Biomedical Materials Research. Part B: Applied Biomaterials, 99*, 276–281.

Seyfoori, A., Mirdamadi, S., Khavandi, A., & Raufi, Z. S. (2012). Biodegradation behavior of micro-arc oxidized AZ31 magnesium alloys formed in two different electrolytes. *Applied Surface Science, 261*, 92–100.

Seyfoori, A., Mirdamadi, S., Mehrjoo, M., & Khavandi, A. (2013). In-vitro assessments of micro arc oxidized ceramic films on AZ31 magnesium implant: degradation and cell-surface response. *Progress in Natural Science: Materials International, 23*, 425–433.

Shadanbaz, S., & Dias, G. J. (2012). Calcium phosphate coatings on magnesium alloys for biomedical applications: a review. *Acta Biomaterialia, 8*, 20–30.

Shanshan, C., Lili, T., Yingxue, T., Bingchun, Z., & Ke, Y. (2013). Materials science & engineering. C: materials for biological applications. *Materials Science and Engineering C Materials for Biological Applications, 33*, 1476–1480.

Shie, M.-Y., Ding, S.-J., & Chang, H.-C. (2011). The role of silicon in osteoblast-like cell proliferation and apoptosis. *Acta Biomaterialia, 7*, 2604–2614.

Skoryna, S. C. (1984). Metabolic aspects of the pharmacologic use of trace elements in human subjects with specific reference to stable strontium. *Trace Substances in Environmental Health, 18*, 23.

Srinivasan, A., Ranjani, P., & Rajendran, N. (2013). Electrochemical polymerization of pyrrole over AZ31 Mg alloy for biomedical applications. *Electrochimica Acta, 88*, 310–321.

Staiger, M., Pietak, A., Huadmai, J., & Dias, G. (2006). Magnesium and its alloys as orthopedic biomaterials: a review. *Biomaterials, 27*, 1728–1734.

Sun, J., Wang, J., Jiang, H., Chen, M., Bi, Y., & Liu, D. (2013). In vivo comparative property study of the bioactivity of coated Mg-3Zn-0.8Zr alloy. *Materials Science & Engineering. C: Materials for Biological Applications, 33*, 3263–3272.

Tan, P. S., & Teoh, S. H. (2007). Effect of stiffness of polycaprolactone (PCL) membrane on cell proliferation. *Materials Science and Engineering: C, 27*, 304–308.

Tapiero, H., & Tew, K. D. (2003). Trace elements in human physiology and pathology: zinc and metallothioneins. *Biomedicine & Pharmacotherapy, 57*, 399–411.

Tavangarian, F., & Emadi, R. (2011). Improving degradation rate and apatite formation ability of nanostructure forsterite. *Ceramics International, 37*, 2275–2280.

Taylor, M. S., Daniels, A. U., Andriano, K. P., & Heller, J. (1994). Six bioabsorbable polymers: In vitro acute toxicity of accumulated degradation products. *Journal of Applied Biomaterials, 5*, 151–157.

Thomann, M., Krause, Ch., Angrisani, N., Bormann, D., Hassel, T., Windhagen, H., et al. (2010). Influence of a magnesium-fluoride coating of magnesium-based implants (MgCa0.8) on degradation in a rabbit model. *Journal of Biomedical Materials Research. Part A, 93*, 1609–1619.

Thomann, M., Krause, C., Bormann, D., Von Der Höh, N., Windhagen, H., & Meyer-Lindenberg, A. (2009). Comparison of the resorbable magnesium alloys LAE442 und MgCa0.8 concerning their mechanical properties, their progress of degradation and the bone-implant-contact after 12 months implantation duration in a rabbit model. *Materialwissenschaft und Werkstofftechnik, 40*, 82–87.

Thomas, P., Barnstorf, S., Summer, B., Willmann, G., & Przybilla, B. (2003). Immuno-allergological properties of aluminium oxide (Al_2O_3) ceramics and nickel sulfate in humans. *Biomaterials, 24*, 959–966.

Wagener, V., Killian, M. S., Turhan, C. M., & Virtanen, S. (2013). Albumin coating on magnesium via linker molecules–comparing different coating mechanisms. *Colloids and Surfaces. B: Biointerfaces, 103*, 586–594.

Wang, H., Guan, S., Wang, Y., Liu, H., Wang, H., Wang, L., et al. (2011). In vivo degradation behavior of Ca-deficient hydroxyapatite coated Mg-Zn-Ca alloy for bone implant application. *Colloids and Surfaces. B: Biointerfaces, 88*, 254–259.

Wang, H. X., Guan, S. K., Wang, X., Ren, C. X., & Wang, L. G. (2010). In vivo degradation behavior of Ca-deficient hydroxyapatite coated Mg-Zn-Ca alloy for bone implant application. *Colloids and Surfaces. B: Biointerfaces, 6*, 1743–1748.

Wang, J., He, Y., Maitz, M. F., Collins, B., Xiong, K., Guo, L., et al. (2013). A surface-eroding poly(1,3-trimethylene carbonate) coating for fully biodegradable magnesium-based stent applications: toward better biofunction, biodegradation and biocompatibility. *Acta Biomaterialia, 9*, 8678–8689.

Wang, H., Zhao, C., Chen, Y., Li, J., & Zhang, X. (2012). Electrochemical property and in vitro degradation of DCPD–PCL composite coating on the biodegradable Mg–Zn alloy. *Materials Letters, 68*, 435–438.

Wang, J., Zhou, G., Chen, C., Yu, H., Wang, T., Ma, Y., et al. (2007). Acute toxicity and biodistribution of different sized titanium dioxide particles in mice after oral administration. *Toxicology Letters, 168*, 176–185.

Wan, P., Lin, X., Tan, L., Li, L., Li, W., & Yang, K. (2013). Influence of albumin and inorganic ions on electrochemical corrosion behavior of plasma electrolytic oxidation coated magnesium for surgical implants. *Applied Surface Science, 282*, 186–194.

Waselau, M., Samii, V. F., Weisbrode, S. E., Litsky, A. S., & Bertone, A. L. (2007). Effects of a magnesium adhesive cement on bone stability and healing following a metatarsal osteotomy in horses. *American Journal of Veterinary Research, 68*, 370–378.

Wen, C., Guan, S., Peng, L., Ren, C., Wang, X., & Hu, Z. (2009). Characterization and degradation behavior of AZ31 alloy surface modified by bone-like hydroxyapatite for implant applications. *Applied Surface Science, 255*, 6433–6438.

Williams, D. F. (2008). On the mechanisms of biocompatibility. *Biomaterials, 29*, 2941–2953.

Witte, F., Fischer, J., Nellesen, J., Vogt, C., Vogt, J., Donath, T., et al. (2010). In vivo corrosion and corrosion protection of magnesium alloy LAE442. *Acta Biomaterialia, 6*, 1792–1799.

Witte, F., Hort, N., Vogt, C., Cohen, S., Kainer, K., Willumeit, R., et al. (2008). Degradable biomaterials based on magnesium corrosion. *Current Opinion in Solid State and Materials Science, 12*, 63–72.

Witte, F., Kaese, V., Haferkamp, H., Switzer, E., Meyer-Lindenberg, A., Wirth, C., et al. (2005). In vivo corrosion of four magnesium alloys and the associated bone response. *Biomaterials, 26*, 3557–3563.

Wong, H. M., Yeung, K. W., Lam, K. O., Tam, V., Chu, P. K., Luk, K. D., et al. (2010). A biodegradable polymer-based coating to control the performance of magnesium alloy orthopaedic implants. *Biomaterials, 31*, 2084–2096.

Wong, H. M., Zhao, Y., Tam, V., Wu, S., Chu, P. K., Zheng, Y., et al. (2013). In vivo stimulation of bone formation by aluminum and oxygen plasma surface-modified magnesium implants. *Biomaterials, 34*, 9863–9876.

Wu, H., Cheng, Y., Li, L., Chen, Z., Wang, H., & Zhang, Z. (2007). The anodization of ZK60 magnesium alloy in alkaline solution containing silicate and the corrosion properties of the anodized films. *Applied Surface Science, 253*, 9387–9394.

Wu, F., Wei, J., Guo, H., Chen, F., Hong, H., & Liu, C. (2008). Self-setting bioactive calcium-magnesium phosphate cement with high strength and degradability for bone regeneration. *Acta Biomaterialia, 4*, 1873–1884.

Xiao, X., Yu, H., Zhu, Q., Li, G., Qu, Y., & Gu, R. (2013). In vivo corrosion resistance of Ca-P coating on AZ60 magnesium alloy. *Journal of Bionic Engineering, 10*, 156–161.

Xin, Y., Jiang, J., Huo, K., Tang, G., Tian, X., & Chu, P. K. (2009). Corrosion resistance and cytocompatibility of biodegradable surgical magnesium alloy coated with hydrogenated amorphous silicon. *Journal of Biomedical Materials Research. Part A, 89*, 717–726.

Xin, Y., Liu, C., Zhang, W., Jiang, J., Tang, G., Tian, X., et al. (2008). Electrochemical behavior Al_2O_3/Al coated surgical AZ91 magnesium alloy in simulated body fluids. *Journal of the Electrochemical Society, 155*, C178–C182.

Xue, D., Tan, Z., Schulz, M. J., Vanooij, W. J., Sankar, J., Yun, Y., et al. (2012). Corrosion studies of modified organosilane coated magnesium–yttrium alloy in different environments. *Materials Science and Engineering: C, 32*, 1230–1236.

Xue, D., Yun, Y., Schulz, M. J., & Shanov, V. (2011). Corrosion protection of biodegradable magnesium implants using anodization. *Materials Science and Engineering: C, 31*, 215–223.

Xu, L., Pan, F., Yu, G., Yang, L., Zhang, E., & Yang, K. (2009). In vitro and in vivo evaluation of the surface bioactivity of a calcium phosphate coated magnesium alloy. *Biomaterials, 30*, 1512–1523.

Xu, L., & Yamamoto, A. (2012). Characteristics and cytocompatibility of biodegradable polymer film on magnesium by spin coating. *Colloids and Surfaces. B: Biointerfaces, 93*, 67–74.

Yamada, M. O., Tohno, Y., Tohno, S., Utsumi, M., Moriwake, Y., & Yamada, G. (2003). Silicon compatible with the height of human vertebral column. *Biological Trace Element Research, 95*, 113–121.

Yamaguchi, M., Oishi, H., & Suketa, Y. (1988). Zinc stimulation of bone protein synthesis in tissue culture: activation of aminoacyl-tRNA synthetase. *Biochemical Pharmacology, 37*, 4075–4080.

Yang, X., Li, M., Lin, X., Tan, L., Lan, G., Li, L., et al. (2013). Enhanced in vitro biocompatibility/bioactivity of biodegradable Mg–Zn–Zr alloy by micro-arc oxidation coating contained Mg_2SiO_4. *Surface and Coatings Technology, 233*, 65–73.

Yan, T., Tan, L., Xiong, D., Liu, X., Zhang, B., & Yang, K. (2010). Fluoride treatment and in vitro corrosion behavior of an AZ31B magnesium alloy. *Materials Science and Engineering: C, 30*, 740–748.

Yao, Z., Li, L., & Jiang, Z. (2009). Adjustment of the ratio of Ca/P in the ceramic coating on Mg alloy by plasma electrolytic oxidation. *Applied Surface Science, 255*, 6724–6728.

Yee, C., Kubena, K., Walker, M., Champney, T., & Sampson, H. W. (1995). The relationship of nutritional copper to the development of postmenopausal osteoporosis in rats. *Biological Trace Element Research, 48*, 1–11.

Ye, C. H., Zheng, Y. F., Wang, S. Q., Xi, T. F., & Li, Y. D. (2012). In vitro corrosion and biocompatibility study of phytic acid modified WE43 magnesium alloy. *Applied Surface Science, 258*, 3420–3427.

Yu, Y., Wang, J., Liu, C., Zhang, B., Chen, H., Guo, H., et al. (2010). Evaluation of inherent toxicology and biocompatibility of magnesium phosphate bone cement. *Colloids and Surfaces. B: Biointerfaces, 76*, 496–504.

Zhai, W., Lu, H., Chen, L., Lin, X., Huang, Y., Dai, K., et al. (2012). Silicate bioceramics induce angiogenesis during bone regeneration. *Acta Biomaterialia, 8*, 341–349.

Zhao, Y., Wu, G., Lu, Q., Wu, J., Xu, R., Yeung, K. W. K., et al. (2013). Improved surface corrosion resistance of WE43 magnesium alloy by dual titanium and oxygen ion implantation. *Thin Solid Films, 529*, 407–411.

Zhao, Y., Wu, G., Pan, H., Yeung, K. W. K., & Chu, P. K. (2012). Formation and electrochemical behavior of Al and O plasma-implanted biodegradable Mg-Y-RE alloy. *Materials Chemistry and Physics, 132*, 187–191.

Zheng, Y., Wu, J., Ng, J. C., Wang, G., & Lian, W. (2002). The absorption and excretion of fluoride and arsenic in humans. *Toxicology Letters, 133*, 77–82.

Zhou, W., Shan, D., Han, E.-H., & Ke, W. (2008). Structure and formation mechanism of phosphate conversion coating on die-cast AZ91D magnesium alloy. *Corrosion Science, 50*, 329–337.

Zomorodian, A., Garcia, M. P., Moura E Silva, T., Fernandes, J. C., Fernandes, M. H., & Montemor, M. F. (2013). Corrosion resistance of a composite polymeric coating applied on biodegradable AZ31 magnesium alloy. *Acta Biomaterialia, 9*, 8660–8670.

Zreiqat, H., Howlett, C. R., Zannettino, A., Evans, P., Schulze-Tanzil, G., Knabe, C., et al. (2002). Mechanisms of magnesium-stimulated adhesion of osteoblastic cells to commonly used orthopaedic implants. *Journal of Biomedical Materials Research, 62*, 175–184.

Part Three

Testing and characterization of surface-modified magnesium and magnesium alloys for biomedical applications

Characterization of modified magnesium and magnesium alloys for biomedical applications

8

W.R. Zhou, Y.F. Zheng
Peking University, Beijing, China

8.1 Introduction

Metals its alloys, and cobalt–chromium alloys are widely used in biomedical devices and components, particularly as osteosynthesis implant devices as well as stents and dental devices because of their mechanical features and fracture toughness (Li et al., 2012; Park et al., 2012; Yang, Cui, & Lee, 2011). However, these bio-inert metals require a secondary surgery for implant removal after satisfactory treatment of the affected tissue; otherwise, long-term implantation may cause local inflammation (Brar, Platt, Sarntinoranont, Martin, & Manuel, 2009; Staiger, Pietak, Huadmai, & Dias, 2006). Repeated surgeries not only increase the morbidity rate of patients, but also result in an increased cost for the health care system. In addition, stress shielding effects, which are caused by the mismatched elastic moduli, will decrease the stability of an implant or may even cause the implant to fail (Yan et al., 2010). To reduce such complications, the use of degradable and biocompatible metallic implants has been investigated.

In the recent decade, magnesium (Mg) and its alloys have been proposed as promising alternatives to implant materials because they can eliminate the adverse effects of neutral metallic implant materials. As illustrated in Table 8.1, Mg has mechanical properties similar to that of human natural bone, which avoids stress shielding effects and resulting osteopenia. Mg is essential to human metabolism as a cofactor for many enzymes, and the corrosion products of Mg are thought to be beneficial physiologically. Last, but most important, Mg and its alloys are perfectly degradable in body fluids, which makes them available for use as biodegradable metallic implants. Earlier research works on biodegradable magnesium alloys such as Mg–aluminum (Al) (Cheng, Qin, Wang, & Zhang, 2009; Wen, Wu, Dai, & Yang, 2009), Mg–calcium (Ca) (Liu et al., 2010; Zhou & Gong, 2012), Mg–zinc (Zn) (Chen et al., 2010; He et al., 2009; Zhang, Li et al., 2010), and Mg–rare earth (RE) alloys (Gunde, Furrer, Hanzi, Schmutz, & Uggowitzer, 2010; Hort et al., 2010) have shown the feasibility of Mg and its alloys to be used for biomedical applications. Figure 8.1 shows some of the implants made of biomedical Mg alloys. However, Mg and its alloys have not yet been applied extensively in the clinic setting, mainly as a result of the excessively high degradation rates in body fluids. During severe corrosive attacks, hydrogen

Surface Modification of Magnesium and its Alloys for Biomedical Applications. http://dx.doi.org/10.1016/B978-1-78242-077-4.00008-5

Table 8.1 Mechanical properties of the metals that are used as biomedical and related materials (Gu et al., 2009; Yang et al., 2010; Hermawan et al., 2010, Witte et al., 2008)

	Mg	Pure iron	Co–Cr alloy	Ti alloy	Nitinol Ni–Ti	SS 316L	Tantalum	Natural bone
Tensile strength (MPa)	86.8	180–210	951–1220	760	895	480–620	2057	42–109
Yield strength (MPa)	20.9	120–150	448–648	485	195–690 (austenite phase)	170–310	138	77–114
Modulus elasticity (GPa)	41	211.4	210	110	83 (austenite phase)	193	185	3–20
Density (g/cm^3)	1.74	7.87	9.2	4.5	6.7	8.0	16.6	1.8–2.1

Co, cobalt; Cr, chromium; Mg, magnesium; Ni, nickel; Ti, titanium.

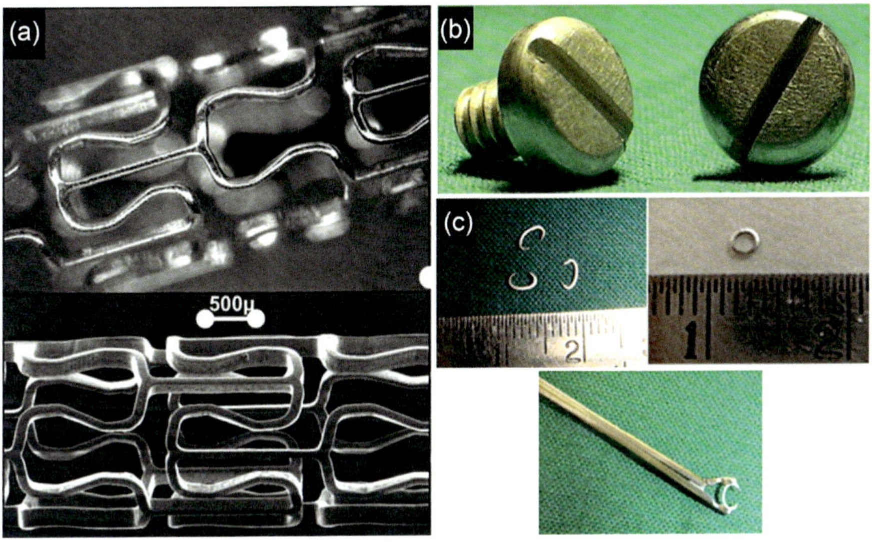

Figure 8.1 Photograph of the tubular slot balloon expandable stent (a) (Di Mario et al., 2004), bone screws (b) (Erdmann et al., 2010), and microclips (c) (Chng, Lau, Choo, & Chui, 2012) made of magnesium alloys.

is produced rapidly and the OH^- ion induces alkalization of the surrounding environments. The pH-dependent physiological processes around the implants are badly affected, leading to reduced biocompatibility. Simultaneously, mechanical integrity is lost before the formation of stable tissues. Therefore, reducing the degradation rate to a reasonable level is becoming a key issue to be addressed in the development of Mg and its alloys for biomedical applications.

In general, there are two possible ways to improve the corrosion resistance of Mg and its alloys. Element alloying is well known as one of the ways by which composition and microstructure can be tailored and, subsequently, improves corrosion resistance and mechanical properties. Throughout the years, a great number of novel Mg alloy systems have been developed by alloying and have undergone investigation for potential clinical applications, such as Mg–zirconium (Zr)–Ca (Zhou et al., 2012), Mg–strontium (Gu, Xie, Li, Zheng, & Qin, 2012), Mg-Zr–strontium (Li et al., 2012), Mg–silicon (Si)–Ca, and Mg-Si-Zn (Zhang, Yang, Xu, Chen, 2010) alloys. The alloying effect of Mg is sometimes not the first choice because of the low solubility of many elements in Mg (Hornberger, Virtanen, & Boccaccini, 2012), surface modification, as the other possible way, is of high significance and attraction to improve corrosion resistance. Surface modification can be considered in terms of surface treatment (modification of a surface by changing its composition or microstructure or both), surface coating (the addition of a distinct layer of material to the original surface), and a mixed treatment. In this way, a layer of protective ceramic, polymer, or composite coatings are generated on surfaces of substrate Mg materials. Accordingly, the characteristics of modified Mg and its alloys are altered with the purpose of reducing and controlling the corrosion rates and increasing their initial biocompatibilities. In this chapter, the characterization of modified Mg and its alloys for biomedical applications are reviewed, and are discussed in terms of surface characteristics, mechanical properties, anticorrosive properties, and biological response.

8.2 Characterization methods

An ideal device made of biodegradable Mg and its alloys requires durability during the process of tissue restoration and an active biological response. Although durability is governed mainly by the bulk properties of the material, a hard and protective coating preserves the substrate from a surrounding corrosive attack during the initial period. In addition, interactions between biological environments and biomedical materials take place on the material's surface, and the biological response from living tissues to these extrinsic biomaterials depends on the surface properties. Therefore, a layer of coating, which combines the near-surface region of a modifid coating and the substrate, also has an important influence on the biological performance of implant materials.

8.2.1 Surface characteristics

Surface characteristics consist of surface morphology, surface topography, surface roughness, wettability, and so on. Coatings with various surface characteristics have been produced via different surface-treating technologies. The differences in superficial features improve corrosive behavior and biocompatibility of the implant materials to different extents, which need to be designed in accordance with specific field applications.

8.2.1.1 Morphology

The morphology of modified Mg and its alloys comprises, in general, a surface pattern, coating thickness, and smoothness/roughness. Surfaces with various kinds of morphologies are usually visualized by scanning electron microscope. Zhu, Zhao, Zhang, and Wu (2012) have synthesized successfully a protective coating on the AZ31 Mg alloy using a hydrothermal method with deionized water as a mineralizer. The surface structure, morphology, and composition of the coatings were investigated by X-ray diffraction, field emission scanning electron microscopy, and energy-dispersing spectroscopy, respectively. The coating is obviously uniform and dense, and presents as large hexagonal flakes and small irregularly shaped platelets on the substrates. In contrast, plasma electrolytic oxidation, anodization, and microarc oxidation (MAO) always generate coatings with a porous structure, such as an irregular porous oxide film formed on Mg-35Zn-3Ca by using plasma anodization in an alkaline electrolyte (Park et al., 2012). In general, this kind of porous microstructure may be beneficial for the rapid adhesion and growth of cells, resulting in a signifiantly stronger bond to the parent tissue; however, it also results in a decrease in corrosion resistance (Gu et al., 2011). Therefore, a proper surface treatment under advisable conditions should be screened to combine appropriate corrosion resistance along with good biocompatibility of Mg and its alloys, especially during the initial implantation period. After fulfilling its task of tissue repair, a fast corrosion rate is required for total degradation of the surface-modified implants to avoid future inflammation.

Typical thickness characterization methods include scanning electron microscopy and a coating thickness gauge. The ranges for various surface modification technologies are shown in Figure 8.2 (Hornberger et al., 2012; Marin, Lanzutti, Guzman, & Fedrizzi, 2011; Yang, Cui, Lee, & Wang, 2010, Yang et al., 2011). Coating thickness is often highlighted during the surface treatment process. It is one of the important parameters in determining the performance of Mg and its alloys for biomedical usage. Gray and Luan (2002) have reported that the protection afforded by the coating is proportional to the coating thickness. Gu, Zheng, Cheng, & Zheng (2009) have studied the biocorrosive behavior of three alkaline, heat-treated Mg-Ca alloys in simulated body fluids. Their results indicate that the $NaHCO_3$-heated Mg-Ca alloy sample, which has a thicker and denser modified surface layer than the Na_2HPO_4-heated and Na_2CO_3-heated Mg-Ca alloys, exhibited the best corrosion resistance. However, Wu et al. (2007) pointed out that a thicker film does not always mean better corrosion resistance. When the thickness is too great, cracks appear on the film as a result of the increased internal stress of the film. As a consequence, the corrosion resistance of the film becomes worse than before the coating was applied. An appropriate thickness for each application has to be selected carefully to make an optimum compromise with coating quality.

Surface roughness with different values is needed for different targeted applications (e.g., stents, orthopedic implants, or tissue engineering scaffolds). For example, a modified implant material with a rough surface is not appropriate for stents, because a greater area is then exposed to blood, which results in blood coagulation. Therefore,

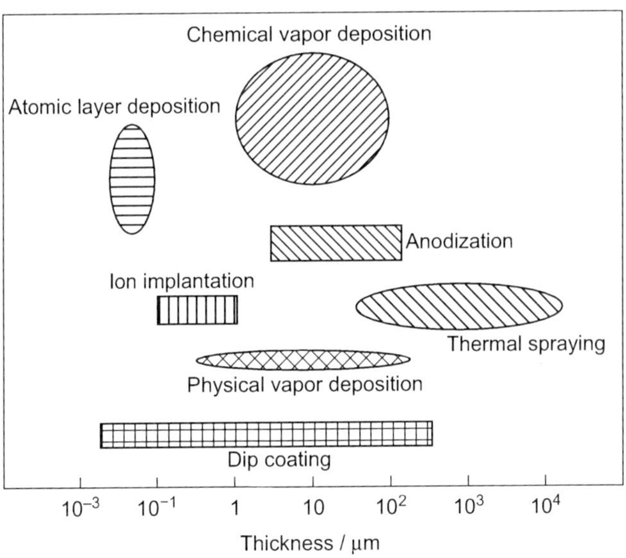

Figure 8.2 Typical thickness ranges for various surface modification technologies.

the majority of vascular stents are polished to a mirror finish after their manufacture. This process is believed to decrease platelet aggregation and reduce thrombus formation. However, a very different case is noted for a hard tissue implant. A rougher surface is considered to improve cell attachment, spreading, and proliferation. In addition, surface roughness also influences the corrosive behavior of substrate materials. It has been noted that a small surface roughness of the substrate material provides a smaller surface area for corrosive attack (Budke, Krempel-Hesse, Maidhof, & Sch Ssler, 1999). Therefore, suitable roughness can be achieved through surface modification that makes a compromise between biocompatibility and corrosion resistance. Atomic force microscopy (AFM), a roughness detector, or a surface profilometer are commonly used to measure surface roughness. In general, AFM obtains a detailed morphology of coatings and reveals the structure with very high resolution. Zhang, Yang, Cui et al. (2010) deposited a layer of diamondlike carbon (DLC) films on AZ31 and AM50 alloys by radio frequency plasma-assisted chemical vapor deposition. Based on AFM observation, deposition of DLC films is shown to improve the rough nature of the Mg alloy surface; the average roughness of AZ31 and AM50 alloys is reduced to 248 nm and 295 nm from original values of about 395 nm and 422 nm, respectively. By treating surfaces with various currents and frequencies, Park et al. (2012) developed coatings with a roughness ranging from 0.615 to 2.197 μm for Mg alloy substrates, which implies the roughness is controllable by changing the operating parameters to satisfy specific requirements of certain applications.

8.2.1.2 Surface energy

In general, surface energy is one of the key factors that affect biocompatibility of certain materials. For example, low surface energy is beneficial to achieve good hemocompatibility of stent material. In addition, it has been reported that low surface energy facilitates albumin adhesion and inhibits fibrinogen adhesion (Zheng et al., 2005). This behavior is a basic prerequisite for blood-contacting occasion for decreasing platelet adhesion. The contact angle, measured by a contact angle or interfacial tension measuring machine, is considered to be an indicator to evaluate wettability and to estimate surface energy. After undergoing a hydrothermal treatment, the contact angle of the AZ31 alloy was much lower after surface modification, 11° for 160 °C for 1 h, 8.0° for 160 °C for 2 h, 7.5° for 160 °C for 3 h, 6.5° for 160 °C for 4 h, and 13.5° for 120 °C 3 h, compared with 40.5° of the blank substrate (Zhu et al., 2012). These results suggest that the coating surfaces are more hydrophilic than the naked substrate. Although it is believed, in general, that a hydrophobic surface can reduce the electrolyte to penetrate through the surface and thus enhance corrosion resistance of the coating system, the synthesized coating can offer much greater corrosion resistance when the surface is highly hydrophilic, which is explained by the fact that hydrophobicity is not necessary for an anticorrosive surface, and the integrity, surface thickness, and density may have more influence on corrosion resistance. Moreover, a hydrophilic surface is usually beneficial to cell adhesion, and improvements in surface wettability may foster cell adhesion.

8.2.2 Mechanical properties

High microhardness, adhesion strength, and wear resistance are required mechanical features for a high-quality coating. It has been reported that poor adhesion between a coating and a substrate causes coating delamination, which affects the performance and reliability of a device. Normally, a tensile test using a mechanical testing system is performed to evaluate the adhesion strength between coatings and the substrate. Da Conceicao, Scharnagl, Dietzel, and Kainer (2011) prepared PEI solutions by dissolving the polymer in two different solvents—N′N′-dimethylacetamide (DMAc) and N-methylpyrollidone (NMP)—respectively. HF-treated and ground AZ31 alloy samples were modified by dipping them in the PEI solutions for 20 seconds to allow wetting of the surface, and then withdrawing them from the solution and drying. The dry and wet adhesion of AZ31 alloys with a PEI coating were measured, for which the wet condition is 12 h of immersion in distilled water. It was noted that the values decreased considerably after 12 h of exposure to distilled water, indicating that a liquid environment has an adverse impact on the interfacial stability of these coatings. Besides, the HF-treated NMP coatings have a greater stability compared with HF-treated DMAc coatings and ground NMP and DMAc coatings, which indicates that substrate roughness and the chemical composition of the substrate also contribute to coating stability. Pretreatment based mainly on chemical reactions is often performed on organic-based coatings (Hornberger et al., 2012) during which a bond between layers is created that improves surface stability. Moreover, surface stability can be adjusted by selecting a proper surface-treating technology, such as ion implantation, MAO, chemical vapor deposition (CVD), and ion beam-assisted deposition (Yang, Cui et al., 2008; Yang, Jiao et al., 2008; Yang et al., 2011; Zhao, Cui, Wang, & Bu, 2010).

Surface hardness is, in general, measured by a nanoindentation technique. The typical hardness curve of a surface-modified Mg alloy as a function of distance from the surface is illustrated schematically in Figure 8.3. From the outer layer to the interior of the coating, the hardness values increase gradually and reach a maximal value. Shortly after that, the hardness decreases with indentation depth and approaches a hardness value close to that of the bulk Mg alloy substrates. Zhang, Zhao, Wu, Wang, and Wu (2007) measured the hardness of MAO-treated AZ91D alloy samples by using a microhardness tester under an applied load of 200 g; a peak value of appropriately 1634 kgf/mm^2 was detected at about a 40-μm distance beneath the surface. With the increase of indentation depth, the hardness value decreased gradually to a value of about 100 kgf/mm^2, which is close to that of the AZ91D substrate. It is clear that the microhardness of an MAO coating is considerably greater than that of an AZ91D alloy. Therefore, the substrate AZ91D alloy is strengthened mechanically by MAO, which is also beneficial to improving wear resistance of the samples.

To date, only a few studies have reported the wear behavior of biomedical Mg and its alloys, and they are conducted only on a modified AZ91 Mg alloy. These evaluations of wear behavior usually use a friction and wear monitor unit under dry or lubricating conditions. Zhang et al. (2007) studied the wear resistance of the AZ91D Mg alloy with and without MAO treatment, with Hank's solution as the lubricant. Results show that MAO samples exhibit a lower mass loss and normalized wear rate compared

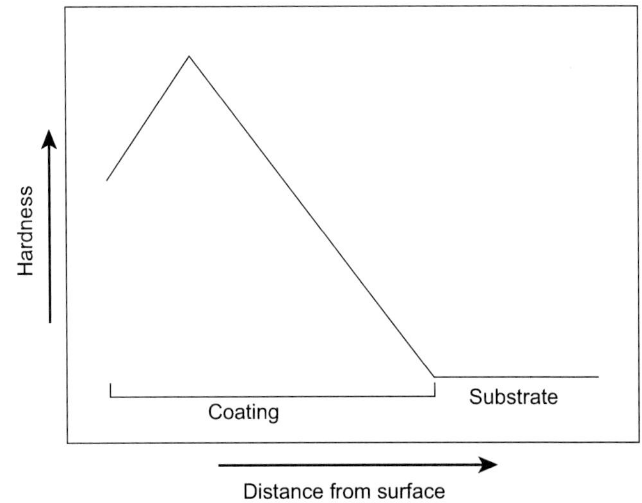

Figure 8.3 Schematic illustration for the typical cross-sectional hardness profile of a surface-modified magnesium alloy.

with the untreated samples, which can be attributed to the massive chimbs and holes on the surface of the MAO coating. In addition, the AZ91 alloy which has experienced a thermal oxidation at 200 °C for 25 h is also been investigated in the viewpoint of wear resistance by a ball on disc wear testing unit against a diamond surface (Majumdar, Bhattacharyya, Biswas, & Manna, 2008). It is relevant that in the thermally oxidized AZ91 alloy, both the magnitude and rate of wear are reduced significantly compared with the as-received AZ91 alloy. The improved wear resistance in the thermally oxidized AZ91 alloy is mainly the result of an increased surface hardness because of the presence of oxide scale. Zou et al. (2011) have the modified AZ91 alloy by depositing a DLC coating, and they studied the tribological behavior of uncoated and coated AZ91 alloys using a ball-on-disk tribotester. Compared with the uncoated AZ91 alloy, the Mg alloy coated with a DLC coating exhibited a lower friction coefficient and a narrow, shallow wear track. The wear resistance of the AZ91 Mg alloy can be improved significantly by adding a layer of DLC protective film, which is a result of its high hardness and low friction coefficient. To improve the poor wear resistance of Mg alloys, currently many surface treatment techniques have been developed with the intent of generating a protective hard coating, such as electroplating, chemical plating, anodic oxidation, chemical conversion coatings, physical vapor deposition, laser surface treatment, and so on (Zhang et al., 2007).

8.2.3 Anticorrosive behavior

When applying a surface-treating technology to Mg and its alloys, the primary concern about the coatings is whether they can maintain their corrosion resistance during degradation. Numerous corrosion studies have been carried out using immersion tests and electrochemical measurements. Changes in morphology and phase composition, weight loss, pH value, hydrogen evolution volume, and released ion concentrations during immersion are critical features for evaluating the anticorrosive behavior of surface-modified Mg alloys. For example, the coatings fail when $Mg(OH)_2$ X-ray diffraction peaks appear. However, for electrochemical measurements, corrosion protection in the presence of the coatings is usually determined on the basis of the corrosion current density I_{corr}. Different simulated body fluids, such as SBF (Gu et al., 2009; Shi, Qi, Chen, & Shi, 2011; Zhang, Zhang, & Wei, 2009), Hank's solution (Geng, Tan, Jin, Yang, & Yang, 2009; Hu et al., 2011), and Dulbecco's Modified Eagle's Medium without 10% fetal bovine serum (Roy et al., 2011; Singh, Roy, Lee, & Kumta, 2011), are often used for corrosion tests. Because of the differences in composition, concentration, and volume of these solutions, and the test time and the parameters used during tests, comparisons among results are very difficult to make. Overall, all coatings have been found to reduce the corrosion rate to a certain extent, as expected. For example, Gu et al. (2009) studied the biocorrosive behavior of an alkaline heat-treated Mg-Ca alloy. The immersion test and the electrochemical test were both carried out in SBF at 37 °C. During the first 200 hours of immersion, the pH values of the solutions corresponding to the heat-treated samples increased more slowly with immersion time than the untreated samples. Meanwhile, the hydrogen evolution volumes of the heat-treated samples were much less smaller than that of the untreated samples. Results of the electrochemical test showed that the corrosion current densities of heat-treated samples were approximately one order of magnitude less than those that of the untreated samples at the same potential range, meaning, an improved anti-corrosion property by alkaline heat treatment was noted. Table 8.2 presents polarization measurements for several coated Mg substrates tested under different conditions, as reported in the literature. When comparing the electrochemical parameters with those of the bare counterparts, the corrosion potentials of surface-modified Mg alloys were shifted toward the positive potential, whereas the corrosion densities were decreased to varying degrees, demonstrating better corrosion resistance.

8.2.4 Biocompatibility

A material with good biocompatibility should induce a minimal level of cytotoxicity, hemolysis, blood coagulation, and inflammation, and should have no risk of inducing mutagenic or carcinogenic reactions. Biological assessment items are usually divided into three steps: step 1 is performed at the cellular level, step 2 is performed in small animals such as rats and hamsters, along with step 3, a final test that is also called the *application test*. Safety and efficacy are estimated simultaneously in large animals at this stage, such as monkeys and dogs.

8.2.4.1 Cellular biocompatibility

A cytotoxicity test, which belongs to early testing, is an economical method, with the advantages of a relatively simple testing method, high replicability, accurate result, and large-scale assessment as a result of standardization. For the cytotoxicity assessment, various cell lines are used in accordance with the target application. Marrow cells and osteoblastlike cells are adopted commonly for the evaluation of bone implant

Table 8.2 **Polarization measurements on several coated Mg substrates tested under different conditions, as reported in the literature**

Conditions	Substrate	I_{corr} (A/cm^2)	E_{corr} (V)	Coating Method	I_{corr} (A/cm^2)	E_{corr} (V)	References
SBF 37 °C	AZ31	2.51×10^{-5}	−1.6 (SCE)	ED HA	2.51×10^{-8}	−1.42 (SCE)	Wen et al. (2009)
SBF 36.5 ± 0.5 °C	Mg-2Zn-0.2Ca	3.84×10^{-4}	−1.705 (SCE)	MAO	6.51×10^{-6}	−1.614 (SCE)	Li et al. (2011)
SBF 36.5 ± 0.5 °C	Mg-2Zn-0.2Ca	3.84×10^{-4}	−1.705 (SCE)	MAO and ED	9.13×10^{-7}	−1.495 (SCE)	Li et al. (2011)
Hank's SBF	AZ31	3.163×10^{-5}	−1.596 (SCE)	Sol-gel method TiO_2	1.26×10^{-7}	−1.419 (SCE)	Hu et al. (2011)
SBF 37 °C	WE43	5.701×10^{-4}	−1.681 (SCE)	MAO	5.741×10^{-5}	−1.441 (SCE)	Liu et al. (2012)
SBF 37 °C	WE43	5.701×10^{-4}	−1.681 (SCE)	MAO and LBL self-assembly process	2.796×10^{-6}	−1.295 (SCE)	Liu et al. (2012)
3.5% NaCl room temperature	AZ31	6.763×10^{-3}	−1.356 (SCE)	Hydrothermal method	5.421×10^{-6}	−1.324 (SCE)	Zhu et al. (2012)

SBF 37 °C	WE43	6.025×10^{-4}	−1.972 (SCE)	Al and O plasma implantation	4.468×10^{-5}	−1.586 (SCE)	Zhao et al. (2012)
SBF	Mg-Zn-Ca	1.43×10^{-4}	−1.771	MAO	1.57×10^{-5}	−1.687	Bai et al. (2012)
SBF	Mg-Zn-Ca	1.43×10^{-4}	−1.771	Chitosan/MAO	1.4×10^{-5}	−1.549	Bai et al. (2012)
SBF	AZ31	7.77×10^{-4}	−1.700	PEO and EDP MgO/HA composite	3.64×10^{-6}	−1.226	Sreekanth and Rameshbabu (2012)
SBF 36.5 ± 0.5 °C	AZ31	1.01×10^{-4}	−1.30	MAO	3.70×10^{-6}	−1.07	Gu et al. (2012)
0.9% NaCl 37 ± 0.5 °C	Mg-6Zn	2.65×10^{-5}	−1.46	Dip coating PLGA	8.5×10^{-8}	−1.44	Li et al. (2010)
SBF 37 °C	AZ31B	1.262×10^{-5}	−1.690	Fluoride treatment	1.785×10^{-6}	−1.670	Yan et al. (2010)
SBF 37 °C	Mg	1.68×10^{-4}	−1.98	Cr-O ion implantation	5.89×10^{-5}	−1.63	Xu et al. (2011)

ED, electrochemical deposition; MAO, microarc oxidation; PEO, plasma electrolytic oxidation.

materials, whereas the evaluation of materials used within blood vessels, endothelial cells and human smooth muscle cells are mostly used. Direct and indirect contact assays are the two major methods and are divided by culturing cells on the samples or in their extracts. After incubation for a prescribed period, a microscopy and/or a microplate reader are usually used to characterize cell morphology and viability/cytotoxicity. For instance, the cytotoxicity of the HA-coated Mg-4Zn-1Ca-0.6Zr alloy has been evaluated using an indirect assay (Guan et al., 2012). No significant difference in fibroblast morphology was observed between the HA-coated and the uncoated sample groups. However, the relative growth rate calculated based on MTT assay results demonstrated that the HA-coated samples had a slightly greater relative growth rate than the uncoated samples. Li, Gao, & Wang (2004) examined the cytotoxicity of alkaline heat-treated Mg using a direct method. Through microscopic observation, marrow cells in contact with alkali and heat-treated samples gave the evidence of cell morphology similar to that of bare Mg. No signs of cellular lysis were observed and no inhibitory effects on cell growth were detected as a result of the presence of Mg samples. In addition, the average cell numbers in the culture media exposed to heat-treated samples were greater than that of negative controls, suggesting an improved biocompatibility after surface modification. The cytocompatibility study to investigate the interactions between Mg and tissue cells is at a preliminary stage. Because Mg and its alloys are intended to be used as hard tissue substitutes and stent materials, complete *in vitro* biocompatibility studies, including hemocompatibility, cytotoxicity tests, and antibacterial effect, are required.

8.2.4.2 Hemocompatibility

Assessment of blood compatibility is essential when testing anticorrosion and surface biocompatibility properties of implant biomaterials. To determine the blood compatibility of implant materials, hemolysis and blood coagulation assay are the most commonly used methods. Hemolysis is the breakage or destruction of red corpuscles, which causes hemoglobin to be released into the surrounding medium. Interactions of red blood cells with biomaterials or with extracts of biomaterials in solution can cause hemolysis. Normally, a hemolytic ratio that is less than 5% is required for excellent blood compatibility (American Society for Testing and Materials, 2000). Samples or their extracts are cultured together with blood for 60 min and are then centrifuged. The hemolytic ratio is calculated on the basis of the optical density of the supernatant solution. Table 8.3 illustrates hemolytic ratios of modified Mg and several of its alloys, as reported in the literature. It shows that, most of the surface modification technologies provide enhanced antihemolysis. This enhanced blood compatibility of modified Mg alloy samples can be attributed to the reduced Mg^{2+} concentration as well as a lower pH value because of the improved corrosion resistance (Lu, Cao, Liu, Xu, & Wu, 2011). The *in vitro* blood coagulation times of an MAO-PLLA-modified WE42 Mg alloy were measured using an automated blood coagulation analyzer (Lu et al., 2011). The prothrombin times of the WE42 group, the WE42-MAO group, and the WE42-MAO/PLLA group were the same as those of the control group. These results suggest that WE42 and MAO/PLLA film do not interfere with the extrinsic

Table 8.3 **Hemolytic ratios of several Mg and Mg alloys after surface modification, as reported in the literature**

Substrate	Hemolytic ratio (%)	Coating method	Hemolytic ratio (%)	References
WE42	50.37	MAO	3.67	Lu et al. (2011)
WE42	50.37	MAO and dipping PLLA	1.79	Lu et al. (2011)
Mg-1Zn-1Ca	24.58	MAO	2.25	Wang et al. (2011)
Mg-35Zn-3Ca	1.062	Plasma anodization	0.16	Park et al. (2012)
Mg-4Zn-1Ca-0.6Zr	4.12	Alkaline heat, electrodeposition HA	4.35	Guan et al. (2012)
WE43	9.27	Immersion phytic acid	2.02	Ye et al. (2012)
AZ31	90	Chemical deposition Ca-P coating	4.3	Tan et al. (2010)
Mg	59.24	Heat–organic–film treatment	2.20	Gao et al. (2006)

MAO, microarc oxidation; Mg, magnesium.

pathway of coagulation. However, it is notable that the recalcification time of the WE42 group was shortened after surface modification. Mg^{2+} ions released during contact with blood act as a natural antagonist. Mg^{2+} ions combine with the extracellular binding site of Ca^{2+} competitively, interrupting Ca^{2+} channels and delaying platelet aggregation. Therefore, it can be concluded that after MAO-PLLA modification, WE42 alloys have a weakened anticoagulant property against the intrinsic pathway as a result of enhanced corrosion resistance.

8.2.4.3 *Antibacterial effect*

Bare magnesium metal is reported to present an antibacterial effect (Robinson, Griffith, Shechtman, Evans, & Conzemius, 2010), and the mechanism appears to be alkaline pH during its degradation in bacterial solution. This event is meaningful in clinical fields because the infections associated with surgical implants are currently a serious issue. Ren, Lin, Tan, and Yang (2011) have studied comparatively the antibacterial behaviors of Mg-based metals with different coatings. The tests were conducted by coculturing bacterial and other specimens for a prescribed period. The antibacterial rate was determined based on the number of bacterial colonies. Compared with bare metals, Mg with a porous Si-containing coating by MAO still maintains its antibacterial ability with a mild increase in pH value. However, pure Mg and the AZ31 alloy with fluoride-containing and Si-containing coatings by chemical conversion, respectively, lost their antibacterial ability, with nearly no change in pH, as a result of the much more dense coatings on the surfaces. These results indicate that antibacterial ability can be influenced by surface characteristics after surface modification.

8.2.4.4 *Biocompatibility assessment* in vivo

For biomedical applications, any protective coating on Mg should be nontoxic and should have an improved bioactivity. Animal studies, which belong to the late stage, are used to assess biocompatibility and degradation of implant materials. Samples are implanted into defective tissue in an animal, and the healing process is monitored by X-ray radiography until the animal is euthanized. Then, microcomputed tomography and histological observation are used to visualize the corrosion morphology and to quantify the *in vivo* corrosion rate and bone formation. Jo et al. (2011) studied *in vivo* degradation behavior of HA-MgF_2-coated Mg using the rabbit femoral defect model. In contrast to bare Mg, HA-MgF_2-coated samples maintained their implant shape because of reduced Mg corrosion, and a larger bone–implant contact area was observed. Furthermore, a greater bone–implant contact ratio was measured based on the histological images, thus confirming the positive effects of the HA-MgF_2 coating with regard to the bioactivity of the Mg implant. Wong et al. (2010) produced biodegradable polymer coatings on the AZ91 alloy. Their *in vivo* study indicated that polymer-coated samples had greater volumes of new bone compared with uncoated samples, as determined by microcomputed tomography, although histological analysis indicated no inflammation, necrosis, or hydrogen gas accumulation on either of the samples during degradation, which proves their biocompatibility. Besides, the measurement of serum Mg levels after implantation showed no significant differences between uncoated and coated samples, most likely as a result of homeostatic regulation by the kidney. Taking these data together, polymer-coated samples reduce the rate of Mg ion release and allow for homeostatic maintenance of physiological Mg levels. More important, the data indicate that, after the polymer-coating is degraded, leaving behind the uncoated implant, the release of Mg ions from the uncoated implant does not induce toxic levels of Mg. However, additional long-term, *in vivo* studies, lasting longer than the complete degradation of the implant, are needed for verification.

8.3 Future trends

Surface modification is regarded as one of the effective methods for controlling the corrosive behavior of Mg and its alloys. An appropriate surface-treating technique should be chosen on the basis of many factors, including the substrate material, coating component design and geometry, cost, and, most important, the end application. Afterward, a full-scale characterization of Mg and its alloys is demanded to assess the potential for clinical use.

In this chapter, types of surface modification methods conducted on different Mg-based materials were reviewed. It was noted that the critical factors of modified Mg materials to be used as implant materials are assessed in terms of coating morphology, surface chemistry, corrosion rate, adhesion between the coating and the substrate, and biocompatibility, along with the correlation between surface properties and resulting performance. For an ideal surface treatment technology, a combination of strong corrosion resistance during tissue healing and an active biological response should be provided to the Mg alloy substrate. In previous publications, most studies have set anticorrosion behavior as a concern, with only one or two surface properties evaluated, omitting the examination of the whole range of functionalities and coating properties.

Optimum characteristics of a protective coating change with the target application. For example, materials with a rough coating are not appropriate for stent applications because blood coagulation is easily caused by the more exposed area. For hard tissue repair, a rough surface is preferred because it is beneficial to the adhesion between osteoblasts and the implant. To date, accurate criteria are still lacking for determining the recommended range of a certain property for a specific field application. The various characterization methods and conditions used in previous work make exact comparisons between different surface-modified Mg materials difficult. For this reason, the development of appropriate and unified methods to study biocompatibility and degradation processes in detail and in the long term remains an important area of research.

8.4 Sources of additional information and advice

Surface modification of biomedical Mg and its alloys is being researched at an exponentially increasing rate. In the scientific literature, there are many publications of specific, relevant research work that covers different surface modification technologies and the resulting performance of modified Mg alloys. Several reviews concerning this work have been undertaken and published. Yang's review (2011), "Surface Modifications of Magnesium Alloys for Biomedical Applications," provides a general overview of the current research and development status of surface modification technologies of Mg alloys for biomedical materials research, and the advantages and disadvantages of the different methods and with regard to the most promising method for Mg alloys are also discussed. The application and use of modified Mg alloys, surface chemistry, corrosion rate, coating morphology, and cell adhesion—common issues—are discussed in another review: "Biomedical Coatings on Magnesium Alloys: A Review" (Hornberger et al., 2012).

Most of the publications are the results of various research groups and organizations, such as Peking University (http://lbmd.coe.pku.edu.cn/), the Institute of Metal Research (http://www.imr.cas.cn), the GKSS Materials Research Center (http://www.hzg.de/institute/materials_research/index.html.en), the McGowan Institute for Regenerative Medicine (http://www.mirm.pitt.edu), and the Institute of Materials Science (http://www.ims.uconn.edu). In addition, there are several conference series devoted to biomaterials, and useful information is available from the proceedings of these conferences. The World Biomaterials Congress and the Symposium on Biodegradable Metals for Biomedical applications have been in place for a number of years, and a significant amount of scientific information regarding surface modification of biomedical Mg and its alloys is available in the proceedings of the conferences.

References

American Society for Testing and Materials. (2000). *Standard practice for assessment of hemolytic properties of materials*. ASTM F 756–00. Philadelphia.

Bai, K., Zhang, Y., Fu, Z., Zhang, C., Cui, X., Meng, E., et al. (2012). Fabrication of chitosan/magnesium phosphate composite coating and the in vitro degradation properties of coated magnesium alloy. *Materials Letters, 73*, 59–61.

Brar, H. S., Platt, M. O., Sarntinoranont, M., Martin, P. I., & Manuel, M. V. (2009). Magnesium as a biodegradable and bioabsorbable material for medical implants. *JOM, 61*, 31–34.

Budke, E., Krempel-Hesse, J., Maidhof, H., & Sch Ssler, H. (1999). Decorative hard coatings with improved corrosion resistance. *Surface and Coatings Technology, 112*, 108–113.

Chen, Y., Zhang, S., Li, J., Song, Y., Zhao, C., & Zhang, X. (2010). Dynamic degradation behavior of MgZn alloy in circulating m-SBF. *Materials Letters, 64*, 1996–1999.

Cheng, Y.-L., Qin, T.-W., Wang, H.-M., & Zhang, Z. (2009). Comparison of corrosion behaviors of AZ31, AZ91, AM60 and ZK60 magnesium alloys. *Transactions of Nonferrous Metals Society of China, 19*, 517–524.

Chng, C. B., Lau, D. P., Choo, J. Q., & Chui, C. K. (2012). A bioabsorbable microclip for laryngeal microsurgery: design and evaluation. *Acta Biomaterialia, 8*, 2835–2844.

Da Conceicao, T. F., Scharnagl, N., Dietzel, W., & Kainer, K. U. (2011). Corrosion protection of magnesium AZ31 alloy using poly(ether imide) [PEI] coatings prepared by the dip coating method: Influence of solvent and substrate pre-treatment. *Corrosion Science, 53*, 338–346.

Di Mario, C., Griffiths, H., Goktekin, O., Peeters, N., Verbist, J., Bosiers, M., et al. (2004). Drug-eluting bioabsorbable magnesium stent. *Journal of Interventional Cardiology, 17*, 391–395.

Erdmann, N., Bondarenko, A., Hewicker-Trautwein, M., Angrisani, N., Reifenrath, J., Lucas, A., et al. (2010). Evaluation of the soft tissue biocompatibility of MgCa0.8 and surgical steel 316L in vivo: a comparative study in rabbits. *BioMedical Engineering OnLine, 9*, 63.

GAO, J-c., QIAO, L-y., LI, L-c., & Wang, Y. (2006). Hemolysis effect and calcium-phosphate precipitation of heat-organic-film treated magnesium. *Transactions of Nonferrous Metals Society of China, 16*, 539−544.

Geng, F., Tan, L. L., Jin, X. X., Yang, J. Y., & Yang, K. (2009). The preparation, cytocompatibility, and in vitro biodegradation study of pure β-TCP on magnesium. *Journal of Materials Science: Materials in Medicine, 20*, 1149−1157.

Gray, J., & Luan, B. (2002). Protective coatings on magnesium and its alloys—a critical review. *Journal of Alloys and Compounds, 336*, 88−113.

Gu, X. N., Li, N., Zhou, W. R., Zheng, Y. F., Zhao, X., Cai, Q. Z., et al. (2011). Corrosion resistance and surface biocompatibility of a microarc oxidation coating on a Mg-Ca alloy. *Acta Biomaterialia, 7*, 1880−1889.

Gu, X. N., Xie, X. H., Li, N., Zheng, Y. F., & Qin, L. (2012). In vitro and in vivo studies on a Mg-Sr binary alloy system developed as a new kind of biodegradable metal. *Acta Biomaterialia, 8*, 2360−2374.

Gu, X. N., Zheng, W., Cheng, Y., & Zheng, Y. F. (2009). A study on alkaline heat treated Mg-Ca alloy for the control of the biocorrosion rate. *Acta Biomaterialia, 5*, 2790−2799.

Guan, R.-G., Johnson, I., Cui, T., Zhao, T., Zhao, Z.-Y., Li, X., et al. (2012). Electrodeposition of hydroxyapatite coating on Mg-4.0Zn-1.0Ca-0.6Zr alloy and in vitro evaluation of degradation, hemolysis, and cytotoxicity. *Journal of Biomedical Materials Research Part A, 100A*, 999−1015.

Gunde, P., Furrer, A., Hanzi, A. C., Schmutz, P., & Uggowitzer, P. J. (2010). The influence of heat treatment and plastic deformation on the bio-degradation of a Mg-Y-RE alloy. *Journal of Biomedical Materials Research. Part A, 92*, 409−418.

He, Y., Tao, H., Zhang, Y., Jiang, Y., Zhang, S., Zhao, C., et al. (2009). Biocompatibility of bio-Mg-Zn alloy within bone with heart, liver, kidney and spleen. *Chinese Science Bulletin, 54*, 484−491.

Hermawan, H., Dube, D., & Mantovani, D. (2010). Developments in metallic biodegradable stents. *Acta Biomaterialia, 6*, 1693−1697.

Hornberger, H., Virtanen, S., & Boccaccini, A. R. (2012). Biomedical coatings on magnesium alloys - a review. *Acta Biomaterialia, 8*, 2442−2455.

Hort, N., Huang, Y., Fechner, D., Stormer, M., Blawert, C., Witte, F., et al. (2010). Magnesium alloys as implant materials−principles of property design for Mg-RE alloys. *Acta Biomaterialia, 6*, 1714−1725.

Hu, J., Zhang, C., Cui, B., Bai, K., Guan, S., Wang, L., et al. (2011). In vitro degradation of AZ31 magnesium alloy coated with nano TiO_2 film by sol−gel method. *Applied Surface Science, 257*, 8772−8777.

Jo, J. H., Kang, B. G., Shin, K. S., Kim, H. E., Hahn, B. D., Park, D. S., et al. (2011). Hydroxyapatite coating on magnesium with MgF(2) interlayer for enhanced corrosion resistance and biocompatibility. *Journal of Material Science: Materials in Medicine, 22*, 2437−2447.

Li, J. N., Cao, P., Zhang, X. N., Zhang, S. X., & He, Y. H. (2010). In vitro degradation and cell attachment of a PLGA coated biodegradable Mg -6Zn based alloy. *Journal of Materials Science, 45*, 6038−6045.

Li, L., Gao, J., & Wang, Y. (2004). Evaluation of cyto-toxicity and corrosion behavior of alkali-heat-treated magnesium in simulated body fluid. *Surface and Coatings Technology, 185*, 92−98.

Li, W., Guan, S., Chen, J., Hu, J., Chen, S., Wang, L., et al. (2011). Preparation and in vitro degradation of the composite coating with high adhesion strength on biodegradable Mg -Zn -Ca alloy. *Materials Characterization, 62*, 1158–1165.

Li, Y., Wen, C., Mushahary, D., Sravanthi, R., Harishankar, N., Pande, G., et al. (2012). Mg-Zr-Sr alloys as biodegradable implant materials. *Acta Biomaterialia, 8*, 3177–3188.

Liu, P., Pan, X., Yang, W., Cai, K., & Chen, Y. (2012). Improved anticorrosion of magnesium alloy via layer-by-layer self-assembly technique combined with micro-arc oxidation. *Materials Letters, 75*, 118–121.

Liu, C. L., Wang, Y. J., Zeng, R. C., Zhang, X. M., Huang, W. J., & Chu, P. K. (2010). In vitro corrosion degradation behaviour of Mg–Ca alloy in the presence of albumin. *Corrosion Science, 52*, 3341–3347.

Lu, P., Cao, L., Liu, Y., Xu, X., & Wu, X. (2011). Evaluation of magnesium ions release, biocorrosion, and hemocompatibility of MAO/PLLA-modified magnesium alloy WE42. *Journal of Biomedical Materials Research Part B: Applied Biomaterials, 96B*, 101–109.

Majumdar, J. D., Bhattacharyya, U., Biswas, A., & Manna, I. (2008). Studies on thermal oxidation of Mg-alloy (AZ91) for improving corrosion and wear resistance. *Surface and Coatings Technology, 202*, 3638–3642.

Marin, E., Lanzutti, A., Guzman, L., & Fedrizzi, L. (2011). Chemical and electrochemical characterization of TiO_2/Al_2O_3 atomic layer depositions on AZ-31 magnesium alloy. *Journal of Coatings Technology and Research, 9*, 347–355.

Park, R. S., Kim, Y. K., Lee, S. J., Jang, Y. S., Park, I. S., Yun, Y. H., et al. (2012). Corrosion behavior and cytotoxicity of Mg-35Zn-3Ca alloy for surface modified biodegradable implant material. *Journal of Biomedical Materials Research Part B: Applied Biomaterials, 100*, 911–923.

Ren, L., Lin, X., Tan, L., & Yang, K. (2011). Effect of surface coating on antibacterial behavior of magnesium based metals. *Materials Letters, 65*, 3509–3511.

Robinson, D. A., Griffith, R. W., Shechtman, D., Evans, R. B., & Conzemius, M. G. (2010). In vitro antibacterial properties of magnesium metal against *Escherichia coli, Pseudomonas aeruginosa* and *Staphylococcus aureus*. *Acta Biomaterialia, 6*, 1869–1877.

Roy, A., Singh, S. S., Datta, M. K., Lee, B., Ohodnicki, J., & Kumta, P. N. (2011). Novel sol–gel derived calcium phosphate coatings on Mg4Y alloy. *Materials Science and Engineering: B, 176*, 1679–1689.

Shi, Y., Qi, M., Chen, Y., & Shi, P. (2011). MAO-DCPD composite coating on Mg alloy for degradable implant applications. *Materials Letters, 65*, 2201–2204.

Singh, S. S., Roy, A., Lee, B., & Kumta, P. N. (2011). Aqueous deposition of calcium phosphates and silicate substituted calcium phosphates on magnesium alloys. *Materials Science and Engineering: B, 176*, 1695–1702.

Sreekanth, D., & Rameshbabu, N. (2012). Development and characterization of MgO/hydroxyapatite composite coating on AZ31 magnesium alloy by plasma electrolytic oxidation coupled with electrophoretic deposition. *Materials Letters, 68*, 439–442.

Staiger, M. P., Pietak, A. M., Huadmai, J., & Dias, G. (2006). Magnesium and its alloys as orthopedic biomaterials: a review. *Biomaterials, 27*, 1728–1734.

Tan, L.-l., Wang, Q., Geng, F., Xi, X.-s., Qiu, J.-h., & Yang, K. (2010). Preparation and characterization of Ca-P coating on AZ31 magnesium alloy. *T Nonferr Metal Soc, 20*, s648–s654.

Wang, D.-W., Cao, Y., Qiu, H., & Bi, Z.-G. (2011). Improved blood compatibility of Mg-1.0Zn-1.0Ca alloy by micro-arc oxidation. *Journal of Biomedical Materials Research Part A, 99A*, 166–172.

Wen, Z., Wu, C., Dai, C., & Yang, F. (2009). Corrosion behaviors of Mg and its alloys with different Al contents in a modified simulated body fluid. *Journal of Alloys and Compounds, 488*, 392–399.

Witte, F., Hort, N., Vogt, C., Cohen, S., Kainer, K. U., Willumeit, R., et al. (2008). Degradable biomaterials based on magnesium corrosion. *Current Opinion in Solid State and Materials Science, 12*, 63–72.

Wong, H. M., Yeung, K. W., Lam, K. O., Tam, V., Chu, P. K., Luk, K. D., et al. (2010). A biodegradable polymer-based coating to control the performance of magnesium alloy orthopaedic implants. *Biomaterials, 31*, 2084–2096.

Wu, H.-L., Cheng, Y.-L., Li, L.-L., Chen, Z.-H., Wang, H.-M., & Zhang, Z. (2007). The anodization of ZK60 magnesium alloy in alkaline solution containing silicate and the corrosion properties of the anodized films. *Applied Surface Science, 253*, 9387–9394.

Xu, R., Wu, G., Yang, X., Hu, T., Lu, Q., & Chu, P. K. (2011). Controllable degradation of biomedical magnesium by chromium and oxygen dual ion implantation. *Materials Letters, 65*, 2171–2173.

Yan, T., Tan, L., Xiong, D., Liu, X., Zhang, B., & Yang, K. (2010). Fluoride treatment and in vitro corrosion behavior of an AZ31B magnesium alloy. *Materials Science and Engineering: C, 30*, 740–748.

Yang, J., Cui, F.-Z., Lee, I. S., & Wang, X. (2010). Plasma surface modification of magnesium alloy for biomedical application. *Surface and Coatings Technology, 205*, S182–S187.

Yang, J., Cui, F., & Lee, I. S. (2011). Surface modifications of magnesium alloys for biomedical applications. *Annals of Biomedical Engineering, 39*, 1857–1871.

Yang, J. X., Cui, F. Z., Lee, I.-S., Jiao, Y. P., Yin, Q. S., & Zhang, Y. (2008). Ion-beam assisted deposited C–N coating on magnesium alloys. *Surface and Coatings Technology, 202*, 5737–5741.

Yang, J. X., Jiao, Y. P., Cui, F. Z., Lee, I.-S., Yin, Q. S., & Zhang, Y. (2008). Modification of degradation behavior of magnesium alloy by IBAD coating of calcium phosphate. *Surface and Coatings Technology, 202*, 5733–5736.

Ye, C. H., Zheng, Y. F., Wang, S. Q., Xi, T. F., & Li, Y. D. (2012). In vitro corrosion and biocompatibility study of phytic acid modified WE43 magnesium alloy. *Applied Surface Science, 258*, 3420–3427.

Zhang, S., Li, J., Song, Y., Zhao, C., Xie, C., & Zhang, X. (2010). Influence of heat treatments on in vitro degradation behavior of Mg-6Zn alloy studied by electrochemical measurements. *Advanced Engineering Materials, 12*, B170–B174.

Zhao, Y., Wu, G., Pan, H., Yeung, K. W. K., & Chu, P. K. (2012). Formation and electrochemical behavior of Al and O plasma-implanted biodegradable Mg-Y-RE alloy. *Materials Chemistry and Physics, 132*, 187–191.

Zhang, Y., Yang, J. X., Cui, F. Z., Lee, I. S., & Lee, G. H. (2010c). Characterization and degradation comparison of DLC film on different magnesium alloys. *Surface and Coatings Technology, 205*, S15–S20.

Zhang, E., Yang, L., Xu, J., & Chen, H. (2010). Microstructure, mechanical properties and bio-corrosion properties of Mg-Si(-Ca, Zn) alloy for biomedical application. *Acta Biomaterialia, 6*, 1756–1762.

Zhang, X. P., Zhao, Z. P., Wu, F. M., Wang, Y. L., & Wu, J. (2007). Corrosion and wear resistance of AZ91D magnesium alloy with and without microarc oxidation coating in Hank's solution. *Journal of Materials Science, 42*, 8523–8528.

Zhang, Y., Zhang, G., & Wei, M. (2009). Controlling the biodegradation rate of magnesium using biomimetic apatite coating. *Journal of Biomedical Materials Research Part B: Applied Biomaterials, 89B*, 408–414.

Zhao, L., Cui, C., Wang, Q., & Bu, S. (2010). Growth characteristics and corrosion resistance of micro-arc oxidation coating on pure magnesium for biomedical applications. *Corrosion Science, 52*, 2228–2234.

Zheng, C., Cui, F., Meng, B., Ge, J., Liu, D., & Lee, I.-S. (2005). Hemocompatibility of C–N films fabricated by ion beam assisted deposition. *Surface and Coatings Technology, 193*, 361–365.

Zhou, P., & Gong, H. R. (2012). Phase stability, mechanical property, and electronic structure of an Mg–Ca system. *Journal of the Mechanical Behavior of Biomedical Materials, 8*, 154–164.

Zhou, Y. L., An, J., Luo, D. M., Hu, W. Y., Li, Y. C., Hodgson, P., et al. (2012). Microstructures and mechanical properties of as cast Mg-Zr-Ca alloys for biomedical applications. *Materials Technology: Advanced Performance Materials, 27*, 52–54.

Zhu, Y., Zhao, Q., Zhang, Y.-H., & Wu, G. (2012). Hydrothermal synthesis of protective coating on magnesium alloy using de-ionized water. *Surface and Coatings Technology, 206*, 2961–2966.

Zou, Y. S., Wu, Y. F., Yang, H., Cang, K., Song, G. H., Li, Z. X., et al. (2011). The microstructure, mechanical and friction properties of protective diamond like carbon films on magnesium alloy. *Applied Surface Science, 258*, 1624–1629.

Effect of surface treatments on the fatigue life of magnesium and its alloys for biomedical applications

9

R.A. Antunes[1], *M.C.L. de Oliveira*[2]
[1]Universidade Federal do ABC, São Paulo, Brazil; [2]Electrocell Ind. Com. Equip. Elet. LTDA, Technology, Entrepreneurship and Innovation Center (CIETEC), São Paulo, Brazil

9.1 Introduction

Cyclic stresses account for the vast majority of mechanical failures in biomedical devices for load-bearing applications (Teoh, 2000). The relevance of fatigue properties of implant materials has been highlighted by many authors (Majumdar, Singh, & Chakraborty, 2010; Niinomi, 2007). Fatigue failure has a catastrophic nature and often occurs at stress levels below the threshold to produce failure under monotonic loading. For orthopedic alloys, the ultimate failure is associated frequently with the synergistic effect between corrosion and cyclic mechanical loading (Magnissalis, Zinelis, Karachalios, & Hartofilakidis, 2003), which is called corrosion fatigue. This phenomenon arises when a metallic material is subjected to cyclic stresses in contact with an aggressive environment in which it is prone to electrochemical corrosion. Body fluids are an active medium that have been reported to cause severe pitting corrosion of magnesium alloys (Witte et al., 2005). Hence, the corrosion fatigue of these materials and other load-bearing biomedical alloys has received much attention (Antunes & De Oliveira, 2012).

In light of the scenario just described, it is clear the successful performance of magnesium biomedical alloys depends on their surface properties related to both fatigue and corrosion resistance. To prevent premature failure resulting from fatigue or corrosion fatigue, the designer of magnesium implants has to use mitigation methods. Surface treatments play a pivotal role in this process. The fatigue life of metallic materials depends on surface conditions such as roughness, residual stresses, and the presence of stress raisers (Arola & Williams, 2002; Zhang, Lindemann, & Leyens, 2010a). Pazos, Corengia, and Svoboda (2010) found a strong dependence of fatigue crack nucleation on the surface conditions of titanium implants. Khan et al. (2011) observed that shot blasting improved the fatigue strength of the as-cast AM60 Mg alloy, probably as a result of alterations of the surface residual stress profile. Mechanical surface treatments are reported to have direct action on the fatigue crack nucleation and propagation behavior of metallic materials (Rodopoulos, Curtis, De los Rios, & Romero, 2004). Examples of such treatments are shot peening, laser

Surface Modification of Magnesium and its Alloys for Biomedical Applications. http://dx.doi.org/10.1016/B978-1-78242-077-4.00009-7

peening, burnishing, and severe plastic deformation (SPD) processes. It is also of prime importance to consider the effect of different coatings on the fatigue properties of biomedical alloys. The well-known low-corrosion resistance of magnesium alloys (Witte et al., 2008) constitutes a critical issue within this context. This characteristic can be exploited advantageously to produce temporary implant devices such as pins, screws, plates, and stents. The degradable nature of magnesium alloys eliminates the need for a revision surgery to remove these devices from the patient, which occurs when traditional stainless steel or titanium materials are used (Witte, 2010). The main obstacle to consolidate this attractive attribute into a practical engineering application is to guarantee the proper control of the corrosion rate so the magnesium implant can sustain mechanical loadings without failure during the healing period (Staiger, Pietak, Huadmai, & Dias, 2006). To overcome this technological barrier, different coating strategies have proliferated. The aim of this chapter is to provide a deep overview of the current literature and future trends regarding the surface treatments of magnesium alloys for biomedical applications. The chapter is divided into three parts; in the first, relevant concepts on the fatigue mechanisms of magnesium alloys are reviewed. In the second, different surface treatments of magnesium alloys are detailed, focusing the analysis of their effects on the fatigue properties of these materials. Third and last, future trends in the development of effective surface treatments for magnesium biomedical alloys are discussed.

9.2 Fatigue behavior of magnesium alloys

The fatigue behavior of engineering materials can be investigated according to two fundamental concepts. The first concept is cumulative fatigue damage (CFD), which relates to total fatigue life, safe life, or damage-intolerant life design. The fatigue data are represented as Wöhler's or $S-N$ curves. In this representation stress, S, is plotted on the ordinates, whereas the number of stress cycles to failure, N, is plotted on the abscissas (Lados & Apelian, 2004). When fatigue behavior is evaluated using this approach, the material strain level is concentrated in the elastic regime. The stress amplitudes are relatively low, and the fatigue regime is known as high-cycle fatigue (HCF) (for $N > 10^5$ cycles to failure). Although there is no gross plastic deformation, the material undergoes localized plastic deformation. A detailed description of the fatigue failure mechanisms under high-cycle conditions can be found in the excellent book by Suresh (2004). The fatigue strength of biomedical metallic alloys is commonly evaluated using this approach (Bernard, Balla, Bose, & Bandyopadhyay, 2011; Rubitschek, Niendorf, Karaman, & Maier, 2012). The CFD approach has also been used to evaluate the fatigue life of magnesium alloys for biomedical purposes or others (Gu et al., 2010; Nascimento et al., 2010). Yang and Liu (2008) studied the HCF behavior of AZ91D and AZ91D with 1 wt% cerium magnesium alloys. The $S-N$ curves obtained by the authors are reproduced in Figure 9.1. It is observed that the addition of cerium to the AZ91D alloy increased its fatigue strength. According to the authors, cerium addition decreased the grain size of the alloy. As a consequence, dislocation movement was pinned by the augmented grain boundary area,

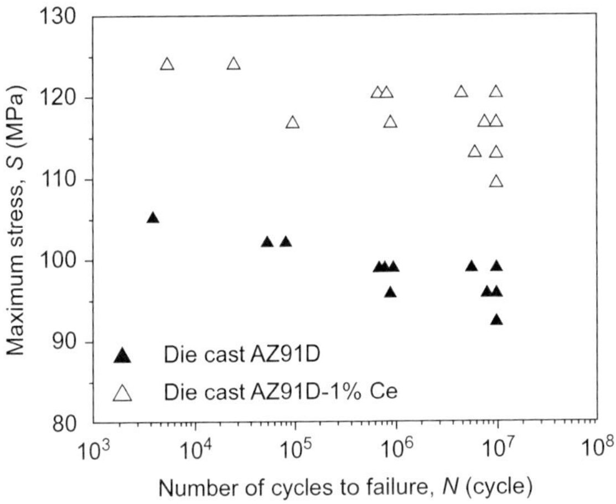

Figure 9.1 *S–N* curves for die-cast AZ91D and AZ91D with addition of 1 wt% Ce. From Yang and Liu (2008).

thus enhancing fatigue properties of the refined microstructure. The main result obtained from this type of curve is the endurance limit or the fatigue limit of the alloy. The endurance limit is defined as the stress range below which there is no crack growth and the material presents an infinite life under cyclic stresses. It is generally accepted that the endurance limit corresponds to a fatigue life of 10^7 cycles, and this is considered to be "infinite life." Figure 9.2 shows *S–N* curves for the AZ31 Mg alloy. The curves were obtained for specimens exposed to air and to NaCl 3 wt.% solution. The curve obtained for specimens exposed to air presents an endurance limit, whereas it disappeared when the fatigue tests were conducted in NaCl solution. The decrease of fatigue strength in saline solution is evident. This behavior has been reported by several authors for different metallic materials and is often associated with the formation of corrosion pits that act as preferential sites for nucleation of fatigue cracks (Bhuiyan, Mutoh, Murai, & Iwakami, 2008). Biomedical alloys are affected markedly

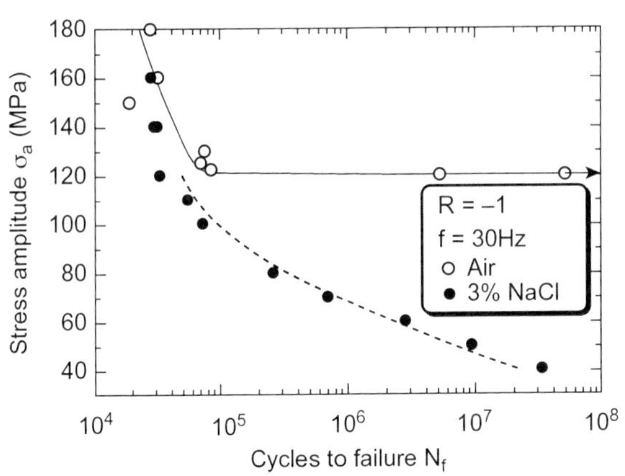

Figure 9.2 *S–N* curves for AZ31 Mg alloy in air and in NaCl 3.0 wt% solution.
From Nan et al. (2008).

by corrosion fatigue because physiological fluids consist of a complex mixture of saline species. In the case when the endurance limit is not defined and the stress decreases continuously for an increasing number of cycles to failure, the fatigue strength is based on the concept of "fatigue limit." This parameter is defined arbitrarily as the stress at a specific number of cycles, typically 10^8 or 5×10^8 cycles (Lados & Apelian, 2004).

Another approach is based on strain-controlled fatigue tests when there is interest in evaluating the low-cycle fatigue behavior ($N < 10^5$ cycles) of metallic alloys. In this regime, plastic deformation is more intense and the cyclic strain becomes more relevant than in the high-cycle regime. The total strain amplitude ($\Delta\varepsilon/2$) can be expressed by the sum of the elastic ($\Delta\varepsilon_e/2$) and plastic ($\Delta\varepsilon_p/2$) strain amplitudes, according to Eqn (9.1):

$$\frac{\Delta\varepsilon_e}{2} = \frac{\sigma_f'}{E}(2N)^b \tag{9.1}$$

Elastic strain amplitude can be modeled by Basquin's Eqn (9.2), whereas Coffin-Manson's Eqn (9.3) is usually used to model the plastic strain amplitude of metallic materials under low-cycle conditions. The parameters in these equations are defined as follows: σ_f' is the fatigue strength coefficient, E is the Young's modulus, b is the fatigue strength exponent, ε_f' is the fatigue ductility coefficient, and c is the fatigue ductility exponent.

$$\frac{\Delta\varepsilon_e}{2} = \frac{\sigma_f'}{E}(2N)^b \tag{9.2}$$

$$\frac{\Delta\varepsilon_p}{2} = \varepsilon_f'(2N)^c \tag{9.3}$$

Strain-controlled fatigue tests of magnesium alloys have been performed by several authors. The main concerns relate to engineering applications for aerospace and automotive applications for which more severe cyclic loading makes the study of low-cycle fatigue a critical issue. In this context, conventional magnesium alloys such as AZ31B, AZ91, and AZ61 are investigated frequently, either under symmetric or asymmetric stress–strain cycles (Lin, Chen, Liu, & Chen, 2013; Shiozawa, Kitajima, Kaminashi, Murai, & Takahashi, 2011; Wu et al., 2010). Examples of low-cycle fatigue data of the AZ31 alloy are shown in Figure 9.3 (Begum, Chen, Xu, & Luo, 2009). Figure 9.3(a) displays a representation of elastic strain amplitude as a function of the number of reversals to failure (2N) according to Basquin's model. Figure 9.3(b) displays a representation of plastic strain amplitude as a function of the number of reversals to failure (2N) according to Coffin-Manson model. Despite the validity of this approach to some biomedical applications, such as stents (Vojtech et al., 2011), it is often neglected for most biomedical metallic alloys. This is particularly true for the newest biodegradable magnesium alloys based on the systems Mg–Zn–Ca, Mg–Sr, and Mg–Nd (Brar, Wong, & Manuel, 2012; González et al., 2012; Seitz, Eifler, Stahl, Kietzmann,

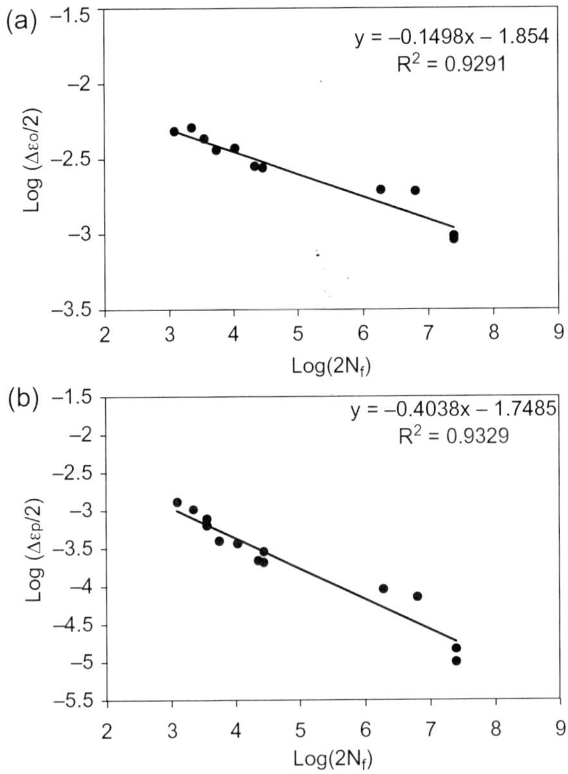

Figure 9.3 Low-cycle fatigue data of the AZ31 alloy. (a) Elastic strain amplitude versus number of reversals to failure. (b) Plastic strain amplitude versus number of reversals to failure.
From Begum et al. (2009).

& Bach, 2012). It is often reported that the loading path or texture affects the strain-controlled fatigue response of magnesium alloys strongly (Geng et al., 2013). Moreover, extruded alloys have better low-cycle fatigue properties than cast alloys (Wu et al., 2010) as a result of their extended plasticity. The validity of these general indications to biodegradable magnesium alloys has yet to be investigated.

Both stress life and strain life approaches are part of CFD concepts. Another aspect of fatigue failure is fatigue crack propagation (FCP). The CFD approach is often criticized because the presence of flaws in the material is not considered. Thus, it is of limited applicability when a mechanistic understanding of the fatigue behavior is to be addressed (Lados & Apelian, 2004). The FCP philosophy is based on a "damage-tolerant" design in which the presence of flaws in the material is recognized. This allows for the determination of safe loads and flaw sizes to prevent fatigue failure under specific conditions during service life.

The basic relationships between fatigue crack growth rate and stress level are derived from linear elastic fracture mechanics, based on the pioneering work by Paris, Gomez, and Anderson (1961). Paris's law (Eqn (9.4)) is a well-known power–law relationship used to interpret experimental fatigue crack growth rate (da/dN) data of engineering materials.

$$\frac{\mathrm{d}a}{\mathrm{d}N} = C \cdot \Delta K^{\mathrm{m}} \tag{9.4}$$

In Eqn (9.4), C and m are the Paris coefficient and exponent, respectively, that depend on the material, environment, frequency, and stress ratio ($R = \sigma_{\min}/\sigma_{\max}$). ΔK ($K_{\max} - K_{\min}$) is the stress intensity factor range expressed as

$$\Delta K = Y \cdot \Delta\sigma \cdot \sqrt{\pi \cdot a} \tag{9.5}$$

where Y is a dimensionless parameter that depends on the geometry and size of an existing crack of length a in the material. A typical da/dN versus ΔK plot has a sigmoidal shape, as shown in Figure 9.4. This plot can be characterized by three different regions that describe the fatigue crack growth behavior of the material. Region I corresponds to the regime in which the average crack growth increment per cycle is smaller than a lattice spacing (Suresh, 2004). A threshold value of the stress intensity factor range (ΔK_{th}) denotes the end of region I and the beginning of region II. This parameter is related closely to the concept of fatigue limit (Carpinteri & Paggi, 2009). The conventional definition of ΔK_{th} is taken as the value of ΔK below which the crack growth rate is 10^{-9} mm/cycle or less. Paris's law is valid in region II, where long cracks propagate (stage II cracks). Region III is the regime of fast crack propagation, leading to final failure at very high values of ΔK, approaching K_{IC}. K_{IC} is the material's fracture toughness, an intrinsic property related to its resistance to brittle fracture. FCP philosophy has been used by several authors to investigate the crack growth rate of magnesium alloys (Rozali, Mutoh, & Nagata, 2011; Tokaji, Nakajima, & Uematsu, 2009; Zeng, Xu, Ke, & Han, 2009).

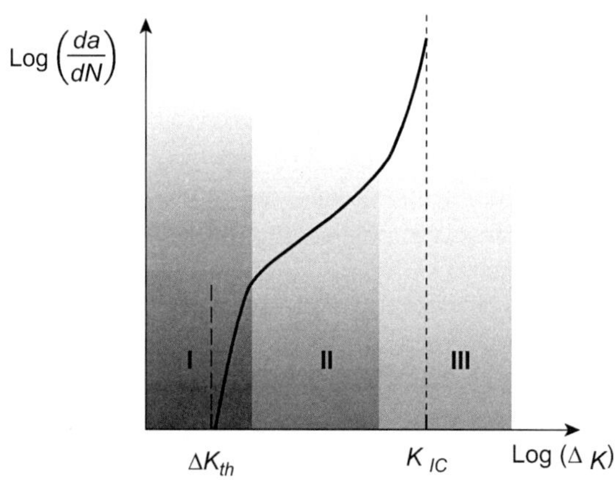

Figure 9.4 Schematic representation of a typical da/dN versus ΔK plot. From Jones et al. (2008).

However, the validity of the linear elastic fracture mechanics concepts to the interpretation of fatigue data has well-established limitations, which have been outlined by Pugno, Ciavarella, Cornetti, and Carpinteri (2006). Despite these limitations, the fatigue propagation rate behavior of metallic materials has been widely characterized through d*a*/d*N* versus ΔK plots. Several factors can affect the FCP behavior of magnesium alloys, such as temperature, humidity, microstructure, heat treatments, corrosive species, and loading frequency. The influence of loading frequency on the FCP behavior of the AZ80 and the AZ61 alloys can be observed in Figure 9.5 (Zeng, Han, & Ke, 2012). The fatigue crack growth rate was found to decrease with increasing loading frequencies. Despite the relevance of such data for the characterization of fatigue behavior of biomedical alloys, the FCP approach is often disregarded for magnesium alloys specifically developed for biomedical purposes. The gap in this area is yet to be filled.

The basic concepts addressed in this section are useful in supporting the reasoning developed in next section, which focuses on the influence of surface treatments on the fatigue behavior of biomedical magnesium alloys. This analysis can be based on the evaluation of the stress life or strain life behavior of metallic alloys (CFD approach), or on the relationship between fatigue crack growth rate and the stress intensity factor range (FCP approach). In this context, an overview of basic aspects of mechanical treatments and coating methods is given, along with their influence on surface properties of the treated alloys and the mechanisms involved in fatigue failure.

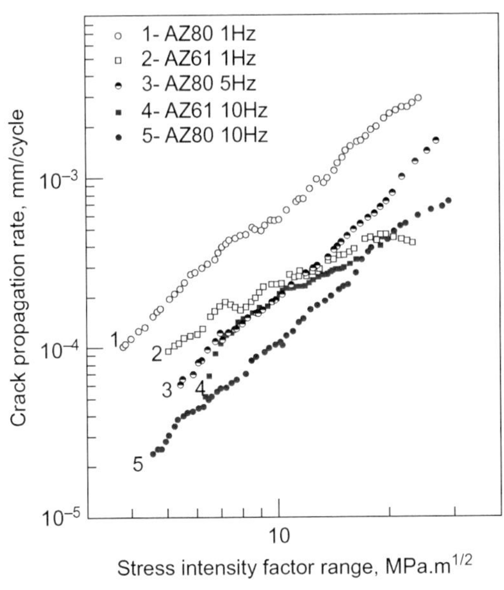

Figure 9.5 Fatigue crack growth data of AZ61 and AZ80 alloys at different loading frequencies. From Zeng et al. (2012).

9.3 Surface treatments

9.3.1 Mechanical treatments

The fatigue mechanism of metallic alloys is associated with the nucleation of surface cracks assisted by the presence of stress raisers such as holes, machining marks, weld beads, corrosion pits, and the like. Increasing fatigue life can be achieved by designing surface treatments properly that act directly in retarding both the nucleation and propagation of fatigue cracks. Mechanical methods are traditional options to achieve these goals. The fatigue strength of structural magnesium alloys can be improved by proper mechanical treatments.

9.3.1.1 Shot peening

Shot peening is a mechanical treatment that produces a near-surface compressive residual stress profile in metallic alloys. As a result, fatigue crack nucleation and/or propagation can be retarded, thus improving the fatigue strength of peened surfaces (Webster & Ezeilo, 2001). Wagner (1999) showed that the fatigue performance of the Mg–8Al–0.5Zn (AZ80) alloy can be improved by shot peening. However, the peening intensity (quantified by the Almen intensity) is critical. Overpeening can lead to detrimental effects resulting from surface degradation, thus imposing practical limitations on the widespread use of this technique for improving fatigue performance of magnesium alloys. Despite this drawback, several authors have studied the fatigue behavior of these materials after shot peening. Zhang, Lindemann, and Leyens (2010b) investigated the effect of different peening media on the residual stress profile, surface roughness, and fatigue strength of the wrought AZ80 alloy. Peening media included glass beads, Zirblast B30 (essentially a mixture of ZrO_2 and SiO_2) and Ce–ZrO_2 (80–90% ZrO_2, 10–20% CeO_2). The fatigue strength of the AZ80 alloy was improved by 60–75% at optimum peening conditions. The best performance was achieved when Ce–ZrO_2 was used as the peening medium. The overpeening effect could be suppressed completely for this type of shot, thus widening the process window. The beneficial response of the surfaces treated with Ce–ZrO_2 shots originated from its relatively high density and coarse grain size. According to the authors, shots of small size and low density need to travel at greater velocities to achieve the same peening intensity of shots with greater density and larger particle size. Thus, the strain rate of small and light shots should be high, leading to more surface defects on the treated material and, consequently, to a more restricted process window. Barry, Hainsworth, and Fitzpatrick (2009) investigated the effect of shot peening on the fatigue properties of a cast magnesium alloy (A8). They observed the endurance limit was increased up to 30% and the fatigue life was improved five times depending on the applied stress amplitude. Shot peening was effective at retarding the initiation and growth of cracks, which initiated at regions of residual porosity (stress raisers). Shot peening led to increased surface roughness in comparison with as-machined surfaces. However, this was not detrimental to the fatigue behavior of the cast alloy. Despite the insensitivity of the fatigue initiation sites to the surface roughness resulting from the shot peening process of the A8 alloy, it should not be disregarded that this effect can be deleterious to the fatigue performance of metallic alloys. As stated by Curtis, De los Rios, Rodopoulos, and Levers (2003), surface roughening is expected to impair the fatigue life of metallic components because it facilitates both initiation and propagation of short fatigue cracks. The reduced ductility of the hexagonal, closely packed crystal structure of magnesium alloys makes them more sensitive to surface damage induced by shot peening, thus aggravating this effect (Liu, Dong, Zhang, Zhai, & Ding, 2009). Liu et al. (2011) studied the effect of shot peening on the fatigue performance of cast and extruded Mg−10Gd−3Y alloys in the as-processed condition and submitted to precipitation-hardening heat treatments. The improvement of fatigue performance obtained by shot peening depends on the work-hardening and compressive residual stresses in the near-surface region. There is a tendency toward relaxation of such stresses during cyclic loading, making the material softer at the surface, losing the work-hardening characteristics gained after shot peening (Jujierm, Altenberger, & Scholtes, 2006). In this regard, precipitates formed after heat treatment can help to sustain the work-hardening level during cyclic loading, acting as barriers to dislocation movement. This effect was more pronounced for the extruded Mg−10Gd−3Y alloy than for the cast alloy. The smaller grain size of the extruded material in comparison with the cast one contributes to diminishing the relaxation phenomenon during cyclic loading. Hence, the fatigue properties of the extruded and heat-treated alloy were improved, as shown in Figure 9.6. Furthermore, the intensity of the shot peening process also affects the roughness and residual stress profile of the component, as shown in Figure 9.7. The condition that presented the lowest surface roughness and greatest compressive residual stress (extruded with the T5 precipitation-hardening heat treatment) yielded the best fatigue performance,

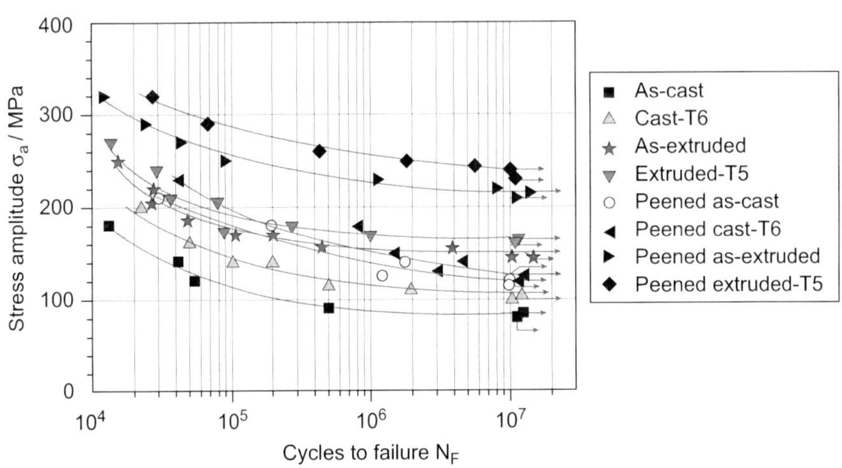

Figure 9.6 *S−N* curves showing the effect of shot peening on the fatigue behavior of the as-cast and extruded Mg−10Gd−3Y alloy without heat treatment and submitted to precipitation-hardening heat treatments (T5 and T6).
From Liu et al. (2011).

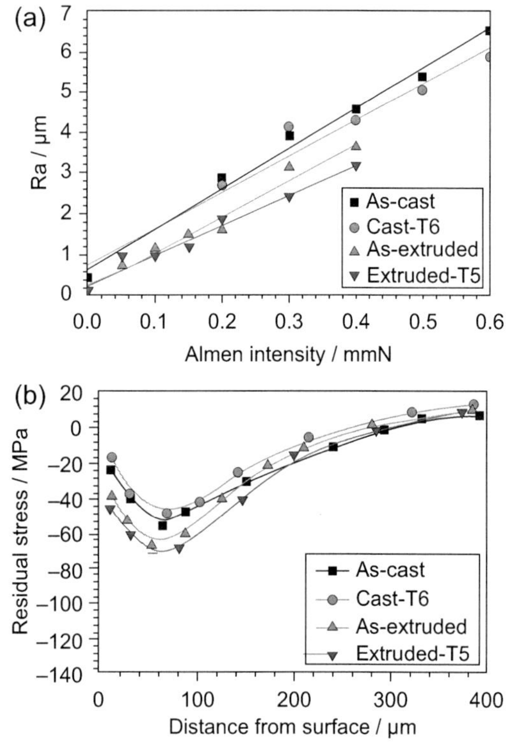

Figure 9.7 Surface roughness (Ra) (a) and residual stress profile (b) of the Mg−10Gd−3Y alloy under different surface finishing conditions after shot peening.
From Liu et al. (2011).

according to the results presented in Figure 9.6. These effects can be explored advantageously to drive the development of fatigue-resistant biomedical magnesium alloys. Based on the literature, extruded alloys should be preferred over cast materials. Moreover, heat-treatable alloys can present a further improvement of the fatigue properties as a result of the blockage of dislocation slip by precipitates. Shot peening should be designed judiciously to avoid the overpeening effect, in which excessive surface defects such as microcracks undermine fatigue strength. The best compromise between surface roughness and compressive residual stresses should be pursued to optimize the fatigue properties of the shot-peened alloy. Despite the benefits of shot peening to the fatigue behavior of magnesium alloys, one should not neglect its influence on the corrosion behavior of these materials. If, on the one hand, several reports point toward an enhancement of fatigue properties after shot peening, on the other, there are indications that the corrosion resistance of metallic materials can be damaged by this process (Azar, Hashemi, & Yazdi, 2010).

9.3.1.2 Laser peening

Laser shock peening has emerged as an alternative surface treatment capable of overcoming the limitations of conventional shot peening. During this process, a solid-state laser beam is pulsed on a metallic surface. Shock waves are then generated, traveling throughout the component and causing plastic deformation (Hill et al., 2003). According to Montross, Wei, Ye, Clark, and Mai (2002), the plastic deformation caused by shock waves propagating through a metallic structure can generate compressive residual stresses that have a deeper penetration than those obtained by conventional shot peening (Hatamleh, Lyons, & Forman, 2007). An improvement of fatigue life of laser-peened alloys is, therefore, expected. Indeed, this expectation has been confirmed by several authors (Gao, 2011; Ganesh et al., 2011). Figure 9.8 shows the increased fatigue strength of an aluminum alloy after laser shock peening. Table 9.1 summarizes the major aspects of a comparison between conventional shot peening and laser shock peening. The increased fatigue strength imparted by laser shock peening can be torn down as a result of the relatively low stability of the compressive residual stresses during cyclic loading (Liao, Suslov, Ye, & Cheng, 2012). This relaxation effect is more pronounced at high temperatures (Ye, Suslov, Kim, Stach, & Cheng, 2011). Despite this drawback, laser shock peening has been shown to be a promising treatment to improve both the mechanical and corrosion properties of magnesium implants (Sealy & Guo, 2010). Guo, Sealy, and Guo (2012) have shown that the surface integrity of Mg–Ca biodegradable alloys submitted to laser shock peening can be tailored to produce an adequate corrosion rate for biomedical purposes. Surface roughness, residual stresses, microhardness, and subsurface microstructure affect the electrochemical response of the treated alloy. Studies focusing the fatigue behavior of magnesium alloys treated by laser shock peening, either using the CFD or the FCP approach are scarce (Singh & Harimkar, 2012). Hence, there is a great demand for investigations regarding the effects of laser shock peening on the fatigue performance of biomedical metallic alloys.

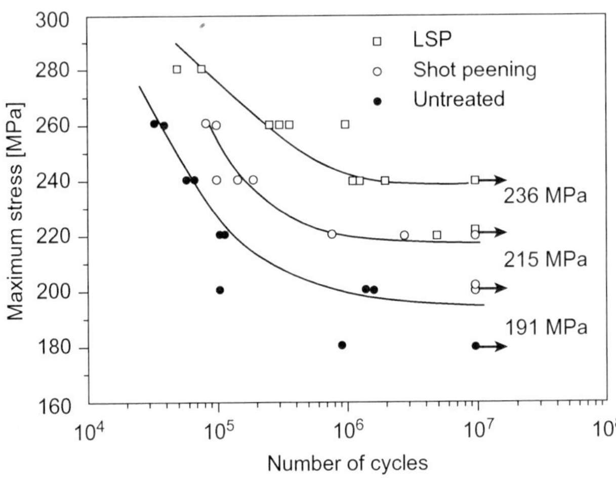

Figure 9.8 Fatigue strength of a 7075-T7351 alloy in three different conditions: untreated, shot peened, and laser shock peened (LSP).
From Montross et al. (2002).

Table 9.1 Advantages and drawbacks of laser shock peening over conventional shot peening

Processing characteristic	Advantage	Drawback
Microstructural alterations	Increased dislocation density and twinning (increased hardness)	–
Compressive residual stresses	Deeper compressive residual stresses	–
Process control	Laser power controlled easily to tailor the final residual stress profile	–
Surface roughness	Smoother surface of laser-peened parts	–
Stability of the residual stress layer	–	More intense stress relaxation effects during cyclic loading

9.3.1.3 Roller burnishing

Roller burnishing has been proposed as an alternative to shot peening to improve the fatigue performance of metallic materials. This process can be defined as a mechanical surface treatment based on the cold-working of metallic surfaces by the action of rotating cylindrical or spherical rollers (Balland, Tabourot, Degre, & Moreau, 2013). The surface finishing of burnished parts is improved by the removal of scratches, pits, and machining marks (Hassan & Al-Bsharat, 1996). Moreover, the burnishing tool introduces compressive residual stresses on the treated surface up to a depth of a few micrometers. This effect is expected to strengthen the fatigue properties of burnished surfaces (Zhang & Lindemann, 2005). Zhang and Lindemann (2005) used roller burnishing to improve the high-cycle fatigue performance of the wrought AZ80 magnesium alloy. They observed an improvement of 110% in the fatigue limit of the burnished alloy in comparison with the untreated one. The fatigue behavior was even better than that of the same alloy submitted to conventional shot peening as a result of the lower surface roughness and greater compressive residual stresses introduced by roller burnishing. Zhang, Lindemann, Ding, and Leyens (2010) showed that roller burnishing is effective at improving the fatigue strength of the Mg–10Gd–3Y alloy in hot-rolled and aged conditions. The improvement in the hot-rolled condition was superior to that obtained for the aged alloy as a result of the formation of a thicker deformation layer in the more ductile hot-rolled material in comparison with the harder aged alloy. Furthermore, more intense compressive residual stresses developed on the surface of the hot-rolled material, giving rise to a beneficial effect on the retardation of fatigue crack nucleation (stage I fatigue cracks). Another important finding was related to the influence of the rolling force on the fatigue life of the alloy. The authors identified an optimum value, above which the excess of surface defects led to a decrease in fatigue life. This reveals the relevance of designing processing parameters carefully—such as ball diameter, pressure, number of passes, and speed—to avoid excessive deformation of the metallic surface during burnishing.

In addition to effects on fatigue behavior of conventional magnesium alloys, roller burnishing has been also investigated as a method of controlling the degradation rate of biodegradable Mg–Ca alloys. A recent example can be found in the report by Salahshoor and Guo (2011). These authors investigated the effect of burnishing on the surface characteristics of biodegradable Mg–Ca alloys. The aim was to determine the influence of processing parameters (burnishing speed, number of passes, and burnishing pressure) on surface topography, roughness, microhardness, microstructure, and residual stresses. These surface characteristics were, then, associated qualitatively with the degradation rate of the alloys. However, no direct measurement of degradation rate has been performed. The burnishing parameters were found to affect surface roughness, microhardness, and residual stresses, whereas the microstructure of the alloy was little affected. In comparison with machined surfaces, burnished ones were smoother and softer. Nevertheless, the hardness in subsurface regions was deeper than that measured for machined surfaces. Low burnishing pressure and high speed in one pass favored the achievement of high compressive residual stresses. The authors hypothesize that burnishing could improve the corrosion performance of the Mg–Ca alloy as a result of the positive effect on surface roughness (a rougher surface is expected to increase the corrosion rate) and on the production of high compressive residual stresses, which are considered to slow down the kinetics of corrosion in comparison with tensile stresses. This hypothesis was confirmed by Denkena and Lucas (2007). They found that the corrosion resistance of the Mg–0.3Ca alloy was improved by a factor of 100 after burnishing.

The concomitant action of fatigue and corrosion on burnished magnesium alloys has not yet been reported in the literature. Despite this lack of investigations, it is possible to infer that it could be used advantageously to improve corrosion fatigue performance of biomedical magnesium alloys based on the beneficial effect of burnishing on the isolated phenomena. However, one major drawback of this process is that it is limited to simple implant geometries because of the constraints of the burnishing tool (Guo et al., 2012).

9.3.1.4 Severe plastic deformation processes

SPD processes are based on metal-forming operations under extensive hydrostatic pressure. Very high strains can be achieved without imposing significant dimensional changes on the processed solid part. The main goal of SPD processes is to produce ultrafine-grained (UFG) bulk metallic components with superior mechanical strength (Valiev et al., 2006). Several different methods have been developed throughout the years, such as equal-channel angular pressing (ECAP), high-pressure torsion, accumulative roll-bonding, cyclic extrusion–compression, and others. An excellent overview of SPD methods is provided by Azushima et al. (2008).

Increased mechanical strength is based on the traditional strengthening mechanism of grain refinement of crystalline metallic materials, which can be described by the Hall–Petch relationship:

$$\sigma_y = \sigma_0 + k \cdot D^{-1/2} \tag{9.6}$$

In this equation, σ_y is the yield stress, σ_0 is the friction force, k is a constant (locking parameter), and D is the grain size (Dieter, 1988). The development of SPD processes has made it possible to achieve grain sizes classified as ultrafine, typically in the range 100 nm $< D <$ 1 μm. In addition to the monotonic mechanical response of SPD-processed metals, their fatigue behavior is also of prime importance to allow for a reliable performance in engineering applications. Mughrabi and Höppel (2010) gave a deep analysis of the fatigue properties of UFG metals and alloys. UFG metals have greater strength and lower ductility than conventional coarse-grained (CG) materials. This leads to differences in high-cycle and low-cycle fatigue regimes according to the total strain fatigue life predicted by Eqn (9.1). These differences can be summarized from the schematic representation shown in Figure 9.9. The parameters shown in Figure 9.9 were defined earlier, in Section 9.2. It is clear that UFG materials are expected to exhibit longer fatigue lives than CG alloys in the high-cycle regime, whereas the opposite occurs in the low-cycle regime.

SPD of magnesium alloys has been developed by several authors (Hamu, Eliezer, & Wagner, 2009; Wang, Chen, Lin, Zhang, & Zhai, 2007). Kim, Lee, and Chumg (2005) showed that the ECAP-processed AZ31 alloy exhibited a greater fatigue threshold and a lower crack growth rate than its CG counterpart as a result of its greater ductility, which increases the ability to accommodate plastic strain during fatigue. Kulyasova et al. (2009) studied the fatigue behavior of the AM60 alloy processed by ECAP at 350 °C, 230 °C, and 150 °C. They observed that the microstructure was characterized

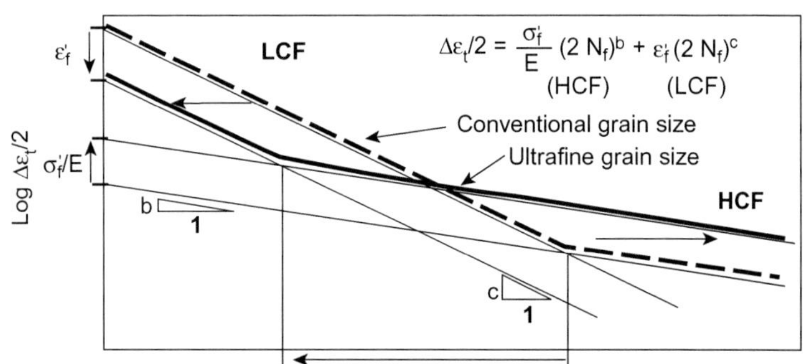

Figure 9.9 Schematic representation of fatigue lives of ultrafine-grained and coarse-grained metallic materials.
From Mughrabi and Höppel (2010).

by a more uniform and smaller grain size with the reduction of ECAP processing temperature, thus leading to an increase of the fatigue limit. Nevertheless, plastic deformation of magnesium alloys at room temperature is difficult because of the reduced ductility of the hexagonal, closely packed crystalline structure. Thus, SPD processes are conducted more easily at progressively increasing temperatures. Recently, Akbaripanah, Fereshtech-Saniee, Mahmudi, and Kim (2013) showed that the fatigue behavior of the AM60 alloy in the low-cycle regime is strongly related to both ductility and texture developed during the ECAP process. The ductility of the alloy increases with the number of passes during ECAP, leading to an improved fatigue life in the low cycle regime. However, texture effects resulting from increased plastic deformation can decrease the fatigue life. The strong relationship between mechanical properties and texture of ECAP-processed magnesium alloys has been recognized by many authors (Balogh, Figueiredo, Ungár, & Langdon, 2010; Figueiredo et al., 2010). Different fiber orientations can be produced depending on the alloying elements, affecting the plasticity and mechanical behavior of magnesium alloys (Agnew, Mehrotra, Lillo, Stoica, & Liaw, 2005). In this regard, the potential of SPD processes to improve fatigue properties of magnesium alloys is still highly unexplored. The correlation between microstructure, texture, plasticity and the fatigue behavior of SPD-processed magnesium alloys and the role of specific alloying elements is not yet understood. The influence of SPD processes on the corrosion resistance of magnesium alloys should be not disregarded. Alvarez-Lopez et al. (2010) showed that the corrosion behavior of the AZ31 alloy in biological fluids can be improved by the grain size reduction attained after ECAP. However, Song et al. (2011) reported that ECAP decreased the corrosion resistance of the AZ91D alloy as a result of the introduction of strain-induced defects, which increased the action of the magnesium matrix. The study of these effects and the interaction of fatigue and corrosion of SPD-processed biomedical magnesium alloys are not seen in the literature.

9.3.2 Coatings

Coatings are traditionally used as a corrosion control method of magnesium alloys. This is a major concern, especially when the alloys are intended for bioabsorbable applications (Staiger et al., 2006). The final performance of the coating layer depends on a complex interaction between microstructure, adhesion, and compactness. This, in turn, is related strongly to the coating technology and processing conditions. The presence of a surface film can also affect the fatigue behavior of the coated substrate. Fatigue performance of coated magnesium alloys can be affected adversely by the coating layer, depending on its structural characteristics (Yerokhin et al., 2004). In this context, control of the degradation rate of the highly active magnesium alloys in the human body by using specific coating layers and the management of their corresponding effect on fatigue properties of the underlying metallic substrate comprise a challenging problem for materials scientists. This section provides an outline of current coating technologies for magnesium alloys and how they affect fatigue performance of these materials.

9.3.2.1 Anodizing

Anodizing is the most common commercial technology for protecting magnesium alloys from corrosion in a variety of environments (Gray & Luan, 2002). It provides a relatively thick, hard, adherent, and abrasion-resistant film. Anodizing is an electrolytic process that converts the surface of the metallic alloy to a film with desirable functional properties. The process can be conducted by using either direct or alternating current in a variety of electrolytes in which the anodized layer does not dissolve faster than it forms. During anodizing, the anions in the electrolyte migrate to the anode where they are discharged. This process generates Mg^{2+}, oxygen, and other species, leading to the formation of magnesium oxide and hydroxide (Blawert, Dietzel, Ghali, & Song, 2006). Anodic films grown by microarc oxidation (MAO) processes have been developed as a method of controlling the degradation rate of biomedical magnesium alloys. Fischerauer et al. (2013) performed *in vivo* studies of microarc-oxidized magnesium implants in rats. They submitted pins of the ZX50 Mg alloy to a commercial, patented MAO process called MAGOXID-COAT®. They observed that the protective layer inhibited corrosion processes during the initial period of implantation, decreasing hydrogen release and enhancing bone formation around the implant. Furthermore, as a consequence of the porosity level of the MAO coating, localized corrosion attack was favored for longer periods of implantation. Thus, the implant was reabsorbed by the body without leaving harmful corrosive products inside the bone tissue that could lead to biocompatibility problems. Other authors confirmed that MAO coatings have a beneficial effect on the bioactivity and biocompatibility of Mg–Ca alloys (Gu et al., 2011). Hiromoto et al. (2008) and Hiromoto and Yamamoto (2010) reported that the precipitation of calcium phosphate on pure magnesium, and its degradation rate can be controlled by the combination of anodizing and autoclaving. The porosity of the anodized layer plays a dominant role in these processes.

In addition to the attractive effects of anodic films with regard to both corrosion resistance and the biological response of magnesium alloys, fatigue performance can also be affected. There are reports indicating that anodized layers can act as stress raisers, contributing as nucleation sites for FCP in aluminum and titanium alloys (Shahzad, Chaussumier, Chieragatti, Mabru, & Rezai-Aria, 2011; Wang et al., 2009). Gu, Xiong, Ning, and Zhang (2012) used MAO to produce anodic films on the biocompatible AZ31 magnesium alloy. Residual stresses of a tensile nature were developed in the anodic films. The stress level depended on the applied voltage, decreasing for higher voltages. This was a consequence of the formation of bigger pores in the anodic film that relieved the residual stresses. Wang et al. (2009) showed that MAO films decreased the fatigue performance of the Ti–Al–Zr alloy. This could be assigned to the tensile residual stresses developed in substrate near the interface because of the compressive residual stresses in the coatings. However, the most important factor contributing to decrease the fatigue life of the MAO-coated titanium alloy was the stress concentrations at notches in the substrate/coating interface, which corresponded to regions where the coating thickness was greater than average values. Fatigue cracks initiated at these notches. Khan, Miyashita, Mutoh, and Koike (2008) studied the effect of anodized layer thickness on FCP behavior of the AM60

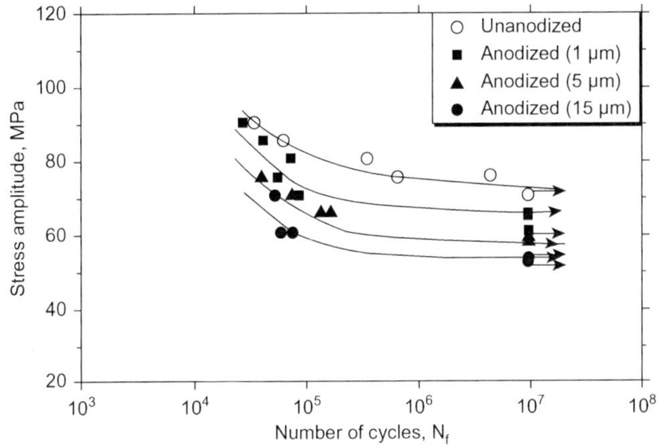

Figure 9.10 *S–N* curves showing the effect of anodized layer thickness on the fatigue performance of the AM60 magnesium alloy.
From Khan et al. (2008).

magnesium alloy. Fatigue strength was found to decrease for progressively thicker anodized layers, as shown in Figure 9.10. The interface between the anodic film and the substrate was uniform when thin films (up to 5 μm) were produced, but contained defects such as pores, and a rough, irregular interface was formed with the substrate. This gave rise to stress concentration, which has shortened fatigue life and accelerated fatigue crack growth.

Metallic implants must keep their mechanical integrity during the healing process. The healing period depends on the tissue and can vary from patient to patient, but it normally takes up to 24 weeks. At a first glance, this could be considered a short period for the onset of fatigue failure. However, for magnesium alloys, even those designed for absorbable purposes, the onset of corrosion can be very rapid, leading to the formation of pits, which are potentially harmful to the fatigue performance of the device, acting as stress raisers that can accelerate fatigue failure (Gu et al., 2011). In this regard, anodic films can be beneficial because they retard the corrosion process of magnesium alloys. Nevertheless, it is of prime importance to design the anodizing parameters properly to avoid the formation of a defective film that would impair fatigue performance instead of enhancing it. Residual stresses and especially the coating thickness must be tailored properly to avoid loss of fatigue performance.

9.3.2.2 Physical vapor deposition

Physical vapor deposition (PVD) technology is traditionally used to produce corrosion- and wear-resistant coatings in a variety of industrial applications (Van Stappen, Stals, Kerhofs, & Quaeyhaegens, 1995). The fatigue behavior of metallic materials can be modified by the deposition of PVD layers (Baragetti, La Vecchia, & Terranova, 2005). As a consequence of the improvements observed in the corrosion, wear, and fatigue properties of PVD-coated materials, their applicability has extended to the biomedical field (Antunes & De Oliveira, 2009). Conventional PVD films are likely to promote galvanic corrosion when deposited on magnesium as a result of the high difference of potentials between the noble ceramic thin films and the active magnesium substrate (Hoche, Blawert, Broszeit, & Berger, 2003). However, this effect can be suppressed, depending on the PVD method and processing parameters, avoiding the development of film pores, which act as preferential sites for the onset of corrosion damage (Hoche, Schmidt, Groß, Trossmann, & Berger, 2011). Several authors have tested different PVD films to protect magnesium alloys against corrosion. Altun and Sen (2006) reported that small structural defects such as pores, cracks, and pinholes are deleterious to the corrosion properties of the AZ91 Mg alloy. In another publication, Altun and Sinici (2008) observed that TiN films produced by cathode arc deposition improved the corrosion resistance of the AZ91 alloy in Na_2SO_4 solution when a dense structure was obtained after deposition. Wu, Wang, Ding, Zhou, and Zeng (2009) and Hollstein, Wiedemann, and Scholz (2003) have also found the same tendency toward increased corrosion resistance of PVD-coated magnesium alloys, especially when duplex or multilayered films are used.

The previously mentioned literature reveals it is possible to use PVD films to protect magnesium alloys against corrosion. Nevertheless, most of the published articles focus on industrial applications, such as automotive or aerospace industries, and the environments in which the corrosion tests are conducted are not adequate to simulate physiological fluids. An exception can be found in the recent report by Rosli et al. (2013). They showed that TiN films produced by magnetron sputtering increased corrosion resistance of the AZ91 alloy in Hanks' solution and can be used to control the degradation rate of this alloy in a physiological medium.

Although the corrosion behavior of PVD-coated magnesium has received much attention, the influence of PVD films on fatigue response is often neglected. Uematsu, Kakiuchi, Teratani, Harada, and Tokaji (2011) studied the corrosion−fatigue behavior of the AZ80A magnesium alloy coated with diamondlike carbon (DLC). However, the DLC film was produced by plasma-enhanced chemical vapor deposition and not by a PVD method. Yet, some conclusions found by these authors are useful to drive the development of PVD layers for improved fatigue strength of magnesium alloys. The major finding is that fatigue cracks nucleate at the substrate under the film defects. Thus, fatigue strength could be improved when the DLC layer is more homogeneous and compact, with less intrinsic defects. Furthermore, the film thickness was also of prime importance. A 3-μm layer could not improve the fatigue strength of the magnesium alloy, whereas a 12-μm film improved it. This concept could be extended to DLC films produced by PVD methods. DLC is recognized as a biocompatible material and has been widely investigated for biomedical purposes (Grill, 2003).

There is a lack of reports in the literature about the corrosion and fatigue behavior of PVD-coated biomedical magnesium alloys. One can hypothesize that this is a result of the following point. The newest developments of magnesium alloys for biomedical applications are concentrated on absorbable alloys (González et al., 2012). Indeed, magnesium alloys are mainly thought of as temporary implants such as pins and screws that are endured only for the necessary time to heal a fractured bone and are then absorbed by the body without the need for a revision surgery (Ma & Xu, 2009).

The presence of a PVD coating would reduce the corrosion rate and increase the mechanical stability of the magnesium alloy during the healing period. However, the coated alloy could endure for an excessive period. Thus, there would be still the need for a revision surgery. One possible way of circumventing this problem would be to deposit a bioactive calcium phosphate film. Calcium phosphate coatings have been deposited successfully by PVD-based methods on biomedical titanium alloys (Jeong, Choe, & Eun, 2011). This approach is not seen in magnesium alloys. However, Tan et al. (2010) showed that a calcium phosphate coating produced by chemical deposition was effective at improving the corrosion resistance of the AZ31 alloy in simulated body fluids (SBFs). This approach indicates that the development of PVD-coated biomedical magnesium alloys has several opportunities for further research.

9.3.3 Ion implantation

Ion implantation has been used widely as a surface modification technique of metals and alloys. The energy of the ions is in the range of several kiloelectron volts to a few megaelectron volts, depending on the desired depth of penetration. Typically, only the subsurface region is implanted and the implanted layer can reach up to 1 μm (Rautray, Narayanan, & Kim, 2011). Unlike common coating methods, ion implantation does not suffer from delamination problems between the coating layer and the substrate. Only the surface properties are modified, whereas the bulk properties of the implanted material remain unaltered (Liu, Chu, & Dinga, 2004). The philosophy of controlling the corrosion rate of magnesium alloys by means of ion implantation has been practiced by many researchers. The increased corrosion resistance of implanted metal surfaces is based on the formation of a protective oxide film and, in the case of magnesium alloys, on the reduction of magnesium content in the near-surface region of the treated component (Xin & Chu, 2010, Chapter 11). Zhao et al. (2013) showed that the corrosion resistance of the WE43 Mg alloy could be improved in a simulated body fluid by dual implantation of titanium and oxygen ions. The formation of a thick TiO_2 film on the surface of the treated alloy provided an effective passive layer that reduced the degradation rate of the substrate. Implantation of titanium ions only can lead to a strong galvanic corrosion effect, accelerating the degradation of the magnesium alloy. The implantation of oxygen ions in a second step after a first titanium implantation eliminated this problem. Pits are known Jamesh, Wu, Zhao, and Chu (2013) used silicon ion implantation to improve the corrosion resistance of the WE43 Mg alloy in SBF. They observed that the treatment was successful without the need of conducting a two-step process because of the high affinity of silicon with oxygen, thus leading to the formation of a stable silicon-rich oxide layer. Aluminum ion implantation has also produced good results resulting from the improvement in the corrosion resistance of magnesium alloys in SBFs (Wu et al., 2012). According to Wu et al. (2012), the aluminum-implanted AZ91 alloy showed fewer corrosion products and pits than its untreated counterpart after a polarization test, as shown in Figure 9.11. Localized corrosion attack has been found to be the source of fatigue cracks during corrosion–fatigue processes of metallic components

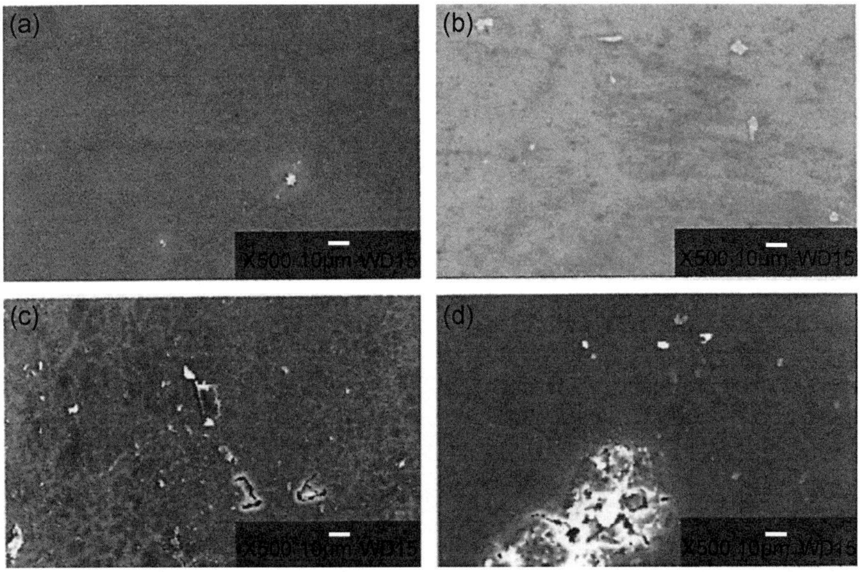

Figure 9.11 Scanning electron micrographs of AZ31 alloy without aluminum ion implantation before polarization test (a), aluminum implanted before polarization test (b), without aluminum ion implantation after polarization test (c), and aluminum implanted after polarization test (d). From Wu et al. (2012).

as a result of the intensified local stress field around the pit (Cerit, Genel, & Edsi, 2009). The crack initiation stage is reduced in the presence of corrosion pits. In this regard, because ion implantation would lead to the formation of less corrosion pits, the stress concentration factor would be lower in the ion-implanted alloy and, therefore, its fatigue performance would be enhanced. However, the fatigue behavior of the implanted alloy was not assessed by Wu et al. Zinc ion implantation of Mg–Ca biodegradable alloys was performed by Wan et al. (2008). The corrosion resistance of the implanted alloy was degraded as a result of galvanic effects. In addition to the corrosion properties, nanoindentation tests were performed to evaluate the surface mechanical properties of the implanted alloy. The results showed that the surface hardness was improved after zinc ion implantation. However, the deterioration of the corrosion resistance would hamper the use of zinc ion-implanted Mg–Ca alloys for biomedical purposes.

Ion implantation can be regarded as a promising surface modification method as a result of the possible improvement in the corrosion resistance of biomedical magnesium alloys, which depends on the type of implanted ions. However, the fatigue properties of ion implantation-treated magnesium alloys have not been investigated in the literature. Shulov and Nochovnaya (2002) reported that the fatigue endurance limit of refractory metals and alloys can be modified greatly by ion beam irradiation. Zhang, Zhang, Yu, and Wang (1993) showed that the endurance limit of a Ti–6Al–4V alloy was increased 12% after boron ion implantation. This effect was explained because of the solid solution and boride precipitation strengthening mechanisms in the near-surface region that acted as obstacles to dislocation movements. Liu, Tang, Wang, Wang, and Jiang (2007) showed that the fatigue properties of a bearing steel were improved significantly by nitrogen ion implantation. Both the increase of microhardness resulting from the formation of nitrides and the compressive residual stresses arising from the implantation process favored the fatigue performance of the treated steel. It is likely that ion implantation can affect the fatigue properties of magnesium alloys markedly because it modifies surface characteristics of the treated material, as reported by Wan et al. (2008) for zinc ion-implanted Mg–Ca alloys. Nevertheless, research in this area is still lacking in the literature. This lack of information allied with the well-established corrosion resistance improvement of ion-implanted magnesium alloys in physiological solutions can be regarded as an opportunity to generate new knowledge in this field.

9.4 Future trends

The fatigue behavior of biomedical magnesium alloys has been little explored in the literature. Basic investigations regarding CFD and FCP approaches are still lacking. The major concerns concentrate on the corrosion resistance of magnesium alloys in physiological fluids because of their highly active behavior, which can lead to accelerated degradation rates. In this regard, the first goal of surface treatments is to provide effective means of controlling the corrosion rate of magnesium alloys; however, their corresponding effects on fatigue response is often disregarded. This could be attributed to the transient nature of the implantable devices based on biodegradable magnesium alloys that are seen as temporary components instead of as permanent prostheses. Nevertheless, the importance of considering the fatigue response of biomedical magnesium alloys even during the relatively short period of the healing of a fractured bone has been highlighted (Gu et al., 2011). Thus, investigations of the effect of surface treatments on fatigue properties of magnesium-based biomaterials should be recognized as a strong area for further research.

In this context, the most promising surface treatments should provide both corrosion and fatigue resistance to magnesium alloys. In the previous sections we outlined the potentialities and limitations of several methods of surface modification regarding mechanical treatments, coatings, and ion implantation. Despite the published literature in this area, several questions remain highly unresolved for a complete understanding of fatigue behavior of biomedical magnesium alloys: How can shot peening be designed properly to increase the fatigue strength in alloys to avoid the overpeening effect? How can the relationship among microstructure, chemical composition, and shot-peening or laser-peening parameters can be explored to optimize fatigue properties? How does surface nanocrystallization by SPD processes affect the corrosion–fatigue behavior of magnesium alloys in physiological media? How are SPD parameters, microstructure, texture, plasticity, and fatigue of magnesium alloys correlated? How do anodizing parameters affect the corrosion–fatigue behavior of magnesium alloys? Can PVD films provide adequate control of degradation rate with improved fatigue properties? How do ion implantation parameters affect the corrosion–fatigue behavior of magnesium alloys?

The answers to these questions are still unknown, even for the most traditional magnesium alloys designed first for automotive or aerospace applications. As the development of new degradable magnesium alloys designed especially for biomedical purposes proceeds, the need for additional studies in this area is increasing, which opens a wide research field for materials scientists.

Effective surface treatments should be based on maintaining the bulk properties of the alloys without introducing defects or impurities that could impair the corrosion and fatigue resistances of treated alloys and their biocompatibility. Based on the reviewed reports, SPD , anodizing, and ion implantation can be considered the most promising surface treatments to improve the corrosion–fatigue properties of biomedical magnesium alloys.

Acknowledgements

The authors are thankful to Elsevier for permitting the reuse of the figures included in this chapter.

References

Agnew, S. R., Mehrotra, P., Lillo, T. M., Stoica, G. M., & Liaw, P. K. (2005). Crystallographic texture evolution of three wrought magnesium alloys during equal channel angular extrusion. *Materials Science and Engineering A, 408*, 72–78.

Akbaripanah, F., Fereshtech-Saniee, F., Mahmudi, R., & Kim, H. K. (2013). The influences of extrusion and equal channel angular pressing (ECAP) processes on the fatigue behavior of AM60 magnesium alloy. *Materials Science and Engineering A, 565*, 308–316.

Altun, H., & Sen, S. (2006). The effect of PVD coatings on the corrosion behavior of AZ91 magnesium alloy. *Materials and Design, 27*, 1174–1179.

Altun, H., & Sinici, H. (2008). Corrosion behavior of magnesium alloys coated with TiN by cathodic arc deposition in NaCl and Na_2SO_4 solutions. *Materials Characterization, 59*, 266–270.

Alvarez-Lopez, M., Pereda, M. D., Del Valle, J. A., Fernandez-Lorenzo, M., Garcia-Alonso, M. C., Ruano, O. A., et al. (2010). Corrosion behaviour of AZ31 magnesium alloy with different grain sizes in simulated biological fluids. *Acta Materialia, 6*, 1763–1771.

Antunes, R. A., & De Oliveira, M. C. L. (2009). Corrosion processes of physical vapor deposition-coated metallic implants. *Critical Reviews in Biomedical Engineering, 37*, 425–460.

Antunes, R. A., & De Oliveira, M. C. L. (2012). Corrosion fatigue of biomedical metallic alloys: mechanisms and mitigation. *Acta Biomaterialia, 8*, 937–962.

Arola, D., & Williams, C. L. (2002). Estimating the fatigue stress concentration factor of machined surfaces. *International Journal of Fatigue, 24*, 923–930.

Azar, V., Hashemi, B., & Yazdi, M. R. (2010). The effect of shot peening on fatigue and corrosion behavior of 316L stainless steel in Ringer's solution. *Surface and Coatings Technology, 204*, 3456–3551.

Azushima, A., Kopp, R., Korhonen, A., Yang, D. Y., Micari, F., Lahoti, G. D., et al. (2008). Severe plastic deformation (SPD) processes for metals. *CIRP Annals, 57*, 716–735.

Balland, P., Tabourot, L., Degre, F., & Moreau, V. (2013). Mechanics of the burnishing process. *Precision Engineering, 37*, 129–134.

Balogh, L., Figueiredo, R. B., Ungár, T., & Langdon, T. G. (2010). The contributions of grain size, dislocation density and twinning to the strength of a magnesium alloy processed by ECAP. *Materials Science and Engineering A, 528*, 533–538.

Baragetti, S., La Vecchia, G. M., & Terranova, A. (2005). Variables affecting the fatigue resistance of PVD-coated components. *International Journal of Fatigue, 27*, 1541–1550.

Barry, N., Hainsworth, S. V., & Fitzpatrick, M. E. (2009). Effect of shot peening on the fatigue behavior of cast magnesium A8. *Materials Science and Engineering A, 507*, 50–57.

Begum, S., Chen, D. L., Xu, S., & Luo, A. A. (2009). Low cycle fatigue properties of an extruded AZ31 magnesium alloy. *International Journal of Fatigue, 31*, 726–735.

Bernard, S., Balla, V. K., Bose, S., & Bandyopadhyay, A. (2011). Rotating bending fatigue response of laser processed porous NiTi alloy. *Materials Science and Engineering C, 31*, 815–820.

Bhuiyan, M. S., Mutoh, Y., Murai, T., & Iwakami, S. (2008). Corrosion fatigue behavior of extruded magnesium alloy AZ61 under three different corrosive environments. *International Journal of Fatigue, 30*, 1756–1765.

Blawert, C., Dietzel, W., Ghali, E., & Song, G. (2006). Anodizing treatments for magnesium alloys and their effect on corrosion resistance in various environments. *Advanced Engineering Materials, 8*, 511–533.

Brar, H., Wong, J., & Manuel, M. V. (2012). Investigation of the mechanical and degradation properties of Mg–Sr and Mg–Zn–Sr alloys for use as potential biodegradable implant materials. *Journal of the Mechanical Behavior of Biomedical Materials, 7*, 87–95.

Carpinteri, A., & Paggi, M. (2009). A unified interpretation of the power laws in fatigue and the analytical correlations between cyclic properties of engineering materials. *International Journal of Fatigue, 31*, 1524–1531.

Cerit, M., Genel, K., & Edsi, S. (2009). Numerical investigation on stress concentration of corrosion pit. *Engineering Failure Analysis, 16*, 2467–2472.

Curtis, S., De los Rios, E. R., Rodopoulos, C. A., & Levers, A. (2003). Analysis of the effects of controlled shot peening on fatigue damage of high strength aluminium alloys. *International Journal of Fatigue, 25*, 59–66.

Denkena, B., & Lucas, A. (2007). Biocompatible magnesium alloys as absorbable implant materials – adjusted surface and subsurface properties by machining processes. *CIRP Annals, 56*, 113–116.

Dieter, G. E. (1988). *Mechanical metallurgy, SI metric edition*. McGraw Hill Book Company.

Figueiredo, R. B., Száraz, Z., Trojanová, Z., Lukác, P., & Langdon, T. G. (2010). Significance of twinning in the anisotropic behavior of a magnesium alloy processed by equal-channel angular pressing. *Scripta materialia, 63*, 504–507.

Fischerauer, S. F., Kraus, T., Wu, X., Tangl, S., Sorantin, E., Hänzi, A. C., et al. (2013). In vivo degradation performance of micro-arc-oxidized magnesium implants: a micro-CT study in rats. *Acta Biomaterialia, 9*, 5411–5420.

Ganesh, P., Sundar, R., Kumar, H., Kaul, R., Ranganathan, K., Hedaoo, P., et al. (2011). Studies on laser peening of spring steel for automotive applications. *Optics and Lasers in Engineering, 50*, 678–686.

Gao, Y. K. (2011). Improvement of fatigue property in 7075-T7451 aluminum alloy by laser peening and shot peening. *Materials Science and Engineering A, 528*, 3823–3828.

Geng, C. J., Wu, B. L., Du, X. H., Wang, Y. D., Zhang, Y. D., Wagner, F., et al. (2013). Low cycle fatigue behavior of the textured AZ31B magnesium alloy under asymmetrical loading. *Materials Science and Engineering A, 560*, 618–626.

González, S., Pellicer, E., Fornell, J., Blanquer, A., Barrios, L., Ibañez, E., et al. (2012). Improved mechanical performance and delayed corrosion phenomena in biodegradable Mg–Zn–Ca alloys through Pd-alloying. *Journal of the Mechanical Behavior of Biomedical Materials, 6*, 53–62.

Gray, J. E., & Luan, B. (2002). Protective coatings on magnesium and its alloys – a critical review. *Journal of Alloys and Compounds, 336*, 88–113.

Grill, A. (2003). Diamond-like coatings as biocompatible materials – an overview. *Diamond and Related Materials, 12*, 166–170.

Gu, X. N., Li, N., Zhou, W. R., Zheng, Y. F., Zhao, X., Cai, Q. Z., et al. (2011). Corrosion resistance and surface biocompatibility of a microarc oxidation coating on a Mg–Ca alloy. *Acta Biomaterialia, 7*, 1880–1889.

Gu, Y., Xiong, W., Ning, C., & Zhang, J. (2012). Residual stresses in microarc oxidation ceramic coatings on biocompatible AZ31 magnesium alloys. *Journal of Materials Engineering and Performance, 21*, 1085–1090.

Gu, X. N., Zhou, W. R., Zheng, Y. F., Cheng, Y., Wei, S. C., Zhong, S. P., et al. (2010). Corrosion fatigue behaviors of two biomedical Mg alloys – AZ91D and WE43 – in simulated body fluid. *Acta Biomaterialia, 6*, 4605–4613.

Guo, Y., Sealy, M. P., & Guo, C. (2012). Significant improvement of corrosion resistance of biodegradable metallic implants processed by laser shock peening. *CIRP Annals – Manufacturing Technology, 61*, 583–586.

Hamu, G. B., Eliezer, D., & Wagner, L. (2009). The relation between severe plastic deformation microstructure and corrosion behavior of AZ31 magnesium alloy. *Journal of Alloys and Compounds, 468*, 222–229.

Hassan, A. M., & Al-Bsharat, A. S. (1996). Improvements in some properties of non-ferrous metals by the application of the ball-burnishing process. *Journal of Materials Processing Technology, 59*, 250–256.

Hatamleh, O., Lyons, J., & Forman, R. (2007). Laser and shot peening effect on fatigue crack growth in friction stir welded 7075-T7351 aluminium alloy joint. *International Journal of Fatigue, 29*, 421–434.

Hill, M. R., DeWald, A. T., Demma, A. G., Hackel, L. A., Chen, H.-L., Dane, C. B., et al. (2003). Laser peening technology. *Advanced Materials and Processing, 161*, 65–67.

Hiromoto, S., Shishido, T., Yamamoto, A., Maruyama, N., Somekawa, H., & Mudai, T. (2008). Precipitation control of calcium phosphate on pure magnesium by anodization. *Corrosion Science, 50*, 2906–2913.

Hiromoto, S., & Yamamoto, A. (2010). Control of degradation rate of bioabsorbable magnesium by anodization and steam treatment. *Materials Science and Engineering C, 30*, 1085–1093.

Hoche, H., Blawert, C., Broszeit, E., & Berger, C. (2003). General corrosion and galvanic corrosion properties of differently PVD treated magnesium die cast alloy AZ91. *Advanced Engineering Materials, 5*, 896–902.

Hoche, H., Schmidt, J., Groβ, S., Trossmann, T., & Berger, C. (2011). PVD coating and substrate pretreatment concepts for corrosion and wear protection of magnesium alloys. *Surface and Coatings Technology, 205*, 5145–5150.

Hollstein, F., Wiedemann, R., & Scholz, J. (2003). Characteristics of PVD-coatings on AZ31hp magnesium alloys. *Surface and Coatings Technology, 162*, 261–268.

Jamesh, M., Wu, G., Zhao, Y., & Chu, P. K. (2013). Effects of silicon plasma ion implantation on electrochemical corrosion behavior of biodegradable Mg–Y–RE alloy. *Corrosion Science, 69*, 158–163.

Jeong, Y.-H., Choe, H.-C., & Eun, S.-W. (2011). Hydroxyapatite coating on the Ti–35Nb–xZr alloy by electron beam-physical vapor deposition. *Thin Solid Films, 519*, 7050–7056.

Jones, R., Molent, L., & Pitt, S. (2008). Similitude and the Paris crack growth law. *International Journal of Fatigue, 30*, 1873–1880.

Jujierm, P., Altenberger, I., & Scholtes, B. (2006). Fatigue and residual stress relaxation of deep rolled differently aged aluminium. *Materials Science and Engineering A, 426*, 4–10.

Khan, S. A., Bhuiyan, M. S., Miyashita, Y., Mutoh, Y., & Koike, T. (2011). Corrosion fatigue behavior of die-cast and shot-blasted AM60 magnesium alloy. *Materials Science and Engineering A, 528*, 1961–1966.

Khan, S. A., Miyashita, Y., Mutoh, Y., & Koike, T. (2008). Effect of anodized layer thickness on fatigue behavior of magnesium alloy. *Materials Science and Engineering A, 474*, 261–269.

Kim, H.-K., Lee, Y.-I., & Chumg, C.-S. (2005). Fatigue properties of a fine-grained magnesium alloy produced by equal channel angular pressing. *Scripta Materialia, 52*, 473–477.

Kulyasova, O., Islamgaliev, R., Mingler, B., & Zehetbauer, M. (2009). Microstructure and fatigue properties of the ultrafine-grained AM60 magnesium alloy processed by equal channel angular pressing. *Materials Science and Engineering A, 503*, 176–180.

Lados, D. A., & Apelian, D. (2004). Fatigue crack growth characteristics in cast Al-Si-Mg alloys Part II. Life predictions using fatigue crack growth data. *Materials Science and Engineering A, 385*, 187–199.

Liao, Y., Suslov, S., Ye, C., & Cheng, G. J. (2012). The mechanisms of thermal engineered laser shock peening for enhanced fatigue performance. *Acta Materialia, 60*, 4997–5009.

Lin, Y. C., Chen, X.-M., Liu, Z.-H., & Chen, J. (2013). Investigation of uniaxial low-cycle fatigue failure behavior of hot-rolled AZ91 magnesium alloy. *International Journal of Fatigue, 48*, 122–132.

Liu, X., Chu, P. K., & Dinga, C. (2004). Surface modification of titanium, titanium alloys and related materials for biomedical applications. *Materials Science and Engineering R, 47*, 49–121.

Liu, W. C., Dong, J., Zhang, P., Korsunsky, A. M., Song, X., & Ding, W. J. (2011). Improvement of fatigue properties by shot peening for Mg-10Gd-3Y alloys under different conditions. *Materials Science and Engineering A, 528*, 5935–5944.

Liu, W., Dong, J., Zhang, P., Zhai, C., & Ding, W. (2009). Effect of shot peening on surface characteristics and fatigue properties of T5-treated ZK60 alloy. *Materials Transactions, 50*, 791–798.

Liu, X., Tang, B., Wang, L., Wang, X., & Jiang, B. (2007). Fatigue life and mechanical behaviors of bearing steel by nitrogen plasma immersion ion implantation. *Surface and Coatings Technology, 201*, 5273–5277.

Ma, E., & Xu, J. (2009). Biodegradable alloys: the glass window of opportunities. *Nature Materials, 8*, 855–857.

Magnissalis, E. A., Zinelis, S., Karachalios, Th, & Hartofilakidis, G. (2003). Failure analysis of two Ti-alloy total hip arthroplasty femoral stems fractured in vivo. *Journal of Biomedical Materials Research, 66B*, 299–305.

Majumdar, P., Singh, S. B., & Chakraborty, M. (2010). Fatigue behavior of boron free and boron containing heat treated Ti–13Zr–13Nb alloy for biomedical applications. *Materials Characterization, 61*, 1394–1399.

Montross, C. S., Wei, T., Ye, L., Clark, G., & Mai, Y.-M. (2002). Laser shock processing and its effects on microstructure and properties of metal alloys: a review. *International Journal of Fatigue, 24*, 1021–1036.

Mughrabi, H., & Höppel, H. W. (2010). Cyclic deformation and fatigue properties of very fine-grained metals and alloys. *International Journal of Fatigue, 32*, 1413–1427.

Nan, Z. Y., Ishihara, S., & Goshima, T. (2008). Corrosion fatigue behavior of extruded magnesium alloy AZ31 in sodium chloride solution. *International Journal of Fatigue, 30*, 1181–1188.

Nascimento, L., Yi, S., Bohlen, J., Fuskova, L., Letzig, D., & Kainer, K. U. (2010). High cycle fatigue behavior of magnesium alloys. *Procedia Engineering, 2*, 743–750.

Niinomi, M. (2007). Fatigue characteristics of metallic biomaterials. *International Journal of Fatigue, 29*, 992–1000.

Paris, P., Gomez, M., & Anderson, W. (1961). A rational analytic theory of fatigue. *Trends in Engineering, 13*, 9–14.

Pazos, L., Corengia, P., & Svoboda, H. (2010). Effect of surface treatments on the fatigue life of titanium for biomedical applications. *Journal of the Mechanical Behavior of Biomedical Materials, 3*, 416–424.

Pugno, N., Ciavarella, M., Cornetti, P., & Carpinteri, A. (2006). A generalized Paris' law for fatigue crack growth. *Journal of the Mechanics and Physics of Solids, 54*, 1333–1349.

Rautray, T. R., Narayanan, R., & Kim, K.-H. (2011). Ion implantation of titanium based biomaterials. *Progress in Materials Science, 56*, 1137–1177.

Rodopoulos, C. A., Curtis, S. A., De los Rios, E. R., & Romero, J. S. (2004). Optimisation of the fatigue resistance of 2024-T351 aluminium alloys by controlled shot peening – methodology, results and analysis. *International Journal of Fatigue, 26*, 849–856.

Rosli, Z. M., Mahamud, Z., Juoi, J. M., Nayan, N., Loon, K. W., Yusuf, Y., et al. (2013). Corrosion behavior of AZ91 Mg-alloy coated with AlN and TiN in NaCl and Hank's solution. *Advanced Materials Research, 626*, 275–279.

Rozali, S., Mutoh, Y., & Nagata, K. (2011). Effect of frequency on fatigue crack growth behavior of magnesium alloy AZ61 under immersed 3.5 mass% NaCl environment. *Materials Science and Engineering A, 528*, 2509–2516.

Rubitschek, F., Niendorf, T., Karaman, I., & Maier, H. J. (2012). Corrosion fatigue behavior of a biocompatible ultrafine-grained niobium alloy in simulated body fluid. *Journal of the Mechanical Behavior of Biomedical Materials, 5*, 181–192.

Salahshoor, M., & Guo, Y. B. (2011). Surface integrity of biodegradable magnesium–calcium orthopedic implant by burnishing. *Journal of the Mechanical Behavior of Biomedical Materials, 4*, 1888–1904.

Sealy, M. P., & Guo, Y. B. (2010). Surface integrity and process mechanics of laser shock peening of novel biodegradable magnesium–calcium (Mg–Ca) alloy. *Journal of the Mechanical Behavior of Biomedical Materials, 3*, 488–496.

Seitz, J.-M., Eifler, R., Stahl, J., Kietzmann, M., & Bach, Fr-W. (2012). Characterization of MgNd2 alloy for potential applications in bioresorbable implantable devices. *Acta Biomaterialia, 8*, 3852–3864.

Shahzad, M., Chaussumier, M., Chieragatti, R., Mabru, C., & Rezai-Aria, F. (2011). Surface characterization and influence of anodizing process on fatigue life of Al 7070 alloy. *Materials and Design, 32*, 3328–3335.

Shiozawa, K., Kitajima, J., Kaminashi, T., Murai, T., & Takahashi, T. (2011). Low-cycle fatigue deformation behavior and evaluation of fatigue life on extruded magnesium alloys. *Procedia Engineering, 10*, 1244–1249.

Shulov, V. A., & Nochovnaya, N. A. (2002). Fatigue strength of metals and alloys modified by ion beams. *Surface and Coatings Technology, 158–159*, 33–41.

Singh, A., & Harimkar, S. P. (2012). Laser surface engineering of magnesium alloys: a review. *Journal of Management, 64*, 716–733.

Song, D., Ma, J. H., Jiang, J. H., Lin, P. H., Yang, D. H., & Fan, J. F. (2011). Corrosion behavior of bulk ultra-fine grained AZ91D magnesium alloy fabricated by equal-channel angular pressing. *Corrosion Science, 53*, 362–373.

Staiger, M. P., Pietak, A. M., Huadmai, J., & Dias, G. (2006). Magnesium and its alloys as orthopedic biomaterials: a review. *Biomaterials, 27*, 1728–1734.

Suresh, S. (2004). *Fatigue of materials* (2nd ed.). Cambridge University Press.

Tan, L.-L., Wang, Q., Geng, F., Xi, X.-S., Qiu, J.-H., & Yang, K. (2010). Preparation and characterization of Ca-P coating on AZ31 magnesium alloy. *Transactions of Nonferrous Metals Society of China, 20*, s648–s654.

Teoh, S. H. (2000). Fatigue of biomaterials: a review. *International Journal of Fatigue, 22*, 825–837.

Tokaji, K., Nakajima, M., & Uematsu, Y. (2009). Fatigue crack propagation and fracture mechanisms of wrought magnesium alloys in different environments. *International Journal of Fatigue, 31*, 1137–1143.

Uematsu, Y., Kakiuchi, T., Teratani, T., Harada, Y., & Tokaji, K. (2011). Improvement of corrosion fatigue strength of magnesium alloy by multilayer diamond-like carbon coatings. *Surface and Coatings Technology, 205*, 2778–2784.

Valiev, R. Z., Estrin, Y., Horita, Z., Langdon, T. G., Zechetbauer, M. J., & Zhu, Y. T. (2006). Producing bulk ultrafine-grained materials by severe plastic deformation. *JOM, 58*, 33–39.

Van Stappen, M., Stals, L. M., Kerhofs, M., & Quaeyhaegens, C. (1995). State of the art for the industrial use of ceramic PVD coatings. *Surface and Coatings Technology, 74-75*, 629–633.

Vojtech, D., Voderova, M., Kubasek, J., Novak, P., Seda, P., & Michalcova, A. (2011). Effects of short-time heat treatment and subsequent chemical surface treatment on the mechanical properties, low-cycle fatigue behavior and corrosion resistance of a Ni–Ti (50.9 at.% Ni) biomedical alloy wire used for the manufacture of stents. *Materials Science and Engineering A, 528*, 1864–1875.

Wagner, L. (1999). Mechanical surface treatments on titanium, aluminum and magnesium alloys. *Materials Science and Engineering A, 263*, 210–216.

Wan, Y. Z., Xiong, G. Y., Luo, H. L., He, F., Huang, Y., & Wang, Y. L. (2008). Influence of zinc ion implantation on surface nanomechanical performance and corrosion resistance of biomedical magnesium–calcium alloys. *Applied Surface Science, 254*, 5514–5516.

Wang, Q. D., Chen, Y. J., Lin, J. B., Zhang, L. J., & Zhai, C. Q. (2007). Microstructure and properties of magnesium alloy processed by a new severe plastic deformation method. *Materials Letters, 61*, 4599–4602.

Wang, Y. M., Zhang, P. F., Guo, L. X., Ouyang, J. H., Zhou, Y., & Jia, D. C. (2009). Effect of microarc oxidation coating on fatigue performance of Ti–Al–Zr alloy. *Applied Surface Science, 255*, 8616–8623.

Webster, G. A., & Ezeilo, A. N. (2001). Residual stress distributions and their influence on fatigue lifetimes. *International Journal of Fatigue, 23*, S375–S383.

Witte, F. (2010). The history of biodegradable magnesium implants: a review. *Acta Biomaterialia, 6*, 1680–1692.

Witte, F., Hort, N., Vogt, C., Cohen, S., Kainer, K. U., Willumeit, R., et al. (2008). Degradable biomaterials based on magnesium corrosion. *Current Opinion in Solid State and Materials Science, 12*, 63–72.

Witte, F., Kaese, V., Haferkamp, H., Switzer, E., Meyer-Lindenberg, A., Wirth, C. J., et al. (2005). In vivo corrosion of four magnesium alloys and the associated bone response. *Biomaterials, 26*, 3557–3563.

Wu, L., Agnew, S. R., Ren, Y., Brown, D. W., Clausen, B., Stoica, G. M., et al. (2010). The effects of texture and extension twinning on the low-cycle fatigue behavior of a rolled magnesium alloy, AZ31B. *Materials Science and Engineering A, 527*, 7057–7067.

Wu, G., Wang, X., Ding, K., Zhou, Y., & Zeng, X. (2009). Corrosion behavior of Ti-Al-N/Ti-Al duplex coating on AZ31 magnesium alloy in NaCl aqueous solution. *Materials Characterization, 60*, 803–807.

Wu, G., Xu, R., Feng, K., Wu, S., Wu, Z., Sun, G., et al. (2012). Retardation of surface corrosion of biodegradable magnesium-based materials by aluminum ion-implantation. *Applied Surface Science, 258*, 7651–7657.

Xin, Y. C., & Chu, P. K. (2010). In H. S. Dong (Ed.), *Surface engineering of light alloys – Aluminium, magnesium and titanium alloys* (p. 393). Abington, UK: CRC Press and Woodhead Publishing (Chapter 11).

Yang, Y., & Liu, Y. B. (2008). High cycle fatigue characterization of two die-cast magnesium alloys. *Materials Characterization, 59*, 567–570.

Ye, C., Suslov, S., Kim, B. J., Stach, E. A., & Cheng, G. J. (2011). Fatigue performance improvement in AISI 4140 steel by dynamic strain aging and dynamic precipitation during warm laser shock peening. *Acta Materialia, 59*, 1014–1025.

Yerokhin, A. L., Shatrov, A., Samsonov, V., Shashkov, P., Leyland, A., & Matthews, A. (2004). Fatigue properties of Keronite® coatings on a magnesium alloy. *Surface and Coatings Technology, 182*, 78–84.

Zeng, R., Han, E., & Ke, W. (2012). A critical discussion on influence of loading frequency on fatigue crack propagation behavior for extruded Mg–Al–Zn alloys. *International Journal of Fatigue, 36*, 40–46.

Zeng, R. C., Xu, Y. B., Ke, W., & Han, E. H. (2009). Fatigue crack propagation behavior of an as-extruded magnesium alloy AZ80. *Materials Science and Engineering A, 509*, 1–7.

Zhang, P., & Lindemann, J. (2005). Effect of roller burnishing on the high cycle fatigue performance of the high-strength wrought magnesium alloy AZ80. *Scripta Materialia, 52*, 1011–1015.

Zhang, P., Lindemann, J., Ding, W. J., & Leyens, C. (2010). Effect of roller burnishing on fatigue properties of the hot-rolled Mg–12Gd–3Y magnesium alloy. *Materials Chemistry and Physics, 124*, 835–840.

Zhang, P., Lindemann, J., & Leyens, C. (2010a). Influence of shot peening on notched fatigue strength of the high-strength wrought magnesium alloy AZ80. *Journal of Alloys and Compounds, 497*, 380–385.

Zhang, P., Lindemann, J., & Leyens, C. (2010b). Shot peening on the high-strength wrought magnesium alloy AZ80 – effect of peening media. *Journal of Materials Processing Technology, 210*, 445–450.

Zhang, D., Zhang, X., Yu, W., & Wang, Z. (1993). Improvement in fatigue lifetime of Ti–6Al–4V alloy by boron implantation. *Surface and Coatings Technology, 58*, 119–123.

Zhao, Y., Wu, G., Lu, Q., Wu, J., Xu, R., Yeung, K. W. K., et al. (2013). Improved surface corrosion resistance of WE43 magnesium alloy by dual titanium and oxygen ion implantation. *Thin Solid Films, 529*, 407–411.

Testing the mechanical properties of surface-modified magnesium and magnesium alloys for biomedical applications

10

T. Imwinkelried
Formerly Synthes GmbH, Oberdorf, Switzerland (Presently at Robert Mathys Foundation)

10.1 Introduction

10.1.1 Biomedical applications of magnesium implants

The concept of biodegradable magnesium implants is not new. The history of biodegradable magnesium implants has been recently reviewed by Witte (2010). He reports a multitude of biomedical applications that were investigated about a century ago, but fell into oblivion after the 1940s.

The goal of a degradable implant is to fulfil a biomedical function and to disappear within a defined duration with acceptable side effects. Degradable implants are particularly attractive if a temporary (nondegradable) implant would be difficult to remove or if a permanent remainder of the implant would present an increased complication risk for the patient.

A use of bioabsorbable materials should allow tissues to regain their natural state and ability to transform and, in the case of children, to continue to grow.

Magnesium implants could offer additional benefits compared to existing implant materials. They might replace permanent metallic implants for indications in which degradation is a significant advantage. They could also allow the treatment of indications in which degradable polymers are limited by their mechanical properties.

There might be additional benefits like the renouncement on antiplatelet medication, which is needed when permanent cardiovascular stents are used (Haude et al., 2013).

10.1.2 The life cycle of a degradable implant

The requirements for a degradable implant can vary considerably during its life cycle – from the implantation through the fulfilment of the desired function until complete degradation.

During implantation, strength and ductility, for example, can be decisive parameters. Strength is important to avoid interoperative failure of the implant (e.g., breaking of a screw head), and ductility is important if the implant is plastically deformed during surgery (e.g., expansion of a stent).

Surface Modification of Magnesium and its Alloys for Biomedical Applications. http://dx.doi.org/10.1016/B978-1-78242-077-4.00010-3

During the fulfilment of the implant function, parameters like strength retention and fatigue strength are important for a safe and effective achievement of that function. An osteosynthesis plate/screw construct, for example, should maintain the reduction of the bone fracture to allow healing and should not break before the healing process is sufficiently advanced (Rüedi & Murphy, 2000).

Finally, during the degradation process, the magnesium implant should disappear within an acceptable time frame without causing significant side effects. The degradation products should be bioabsorbed (i.e., transported away or metabolized).

One should note that these life-cycle phases might overlap, as some degradation could already occur during implantation. In many cases, however, implantation time is short compared to the total life time of an implant.

10.1.3 Goals of mechanical testing

The goal of mechanical testing is to gain knowledge about the performance of a magnesium alloy or of a magnesium implant until fulfilment of its function. It is aimed at simulating worst-case conditions and at exploring the limits of a safe use. With regard to regulatory approval of an implant system, mechanical testing should supply supportive data for proving the safety and efficacy of the medical device.

10.1.4 Assumptions about mechanical anisotropy of magnesium

Magnesium and magnesium alloys have a hexagonal close packed atom lattice. At ambient temperature (i.e., in the operating room), the plastic deformation of magnesium is governed by a limited number of slip systems and the possible formation of twins. As a consequence, yielding in tension or in compression will occur at different stresses for a magnesium single crystal. The degree of anisotropy might be reduced in polycrystalline magnesium by reducing the grain size. Magnesium alloys with grain sizes less than 4 μm have been reported to have very little tension-compression yield stress anisotropy (Gunde, Hänzi, Sologubenko, & Uggowitzer, 2011). In the following, it will be assumed that magnesium behaves as an isotropic material and can thus be described by mechanical values that are not direction-dependant (e.g., single values for the young's modulus, the shear modulus, the yield strength). Anisotropic mechanical properties would make predictions of implant behaviour more difficult.

10.1.5 Chapter outlook and methodology

An accurate simulation of the combined effects of metabolism (i.e., degradation) and of biomechanics (i.e., load condition) by a testing apparatus is a challenging task. As the mechanical testing of a degradable implant system is more demanding than for a permanent implant system, an appropriate and magnesium-specific testing methodology should be adopted.

Table 10.1 Examples of mechanical properties which might be relevant during the life cycle of a biodegradable magnesium implant and potential influence of surface modifications on these properties

Life cycle	Material property	Influence of surface modification (Chapter 10.4)
Implantation related (Chapter 10.2)	Stiffness	No
	Strength	No
	Ductility	No
	Hardness	Yes
	Toughness	No
Degradation related (Chapter 10.3)	Stress corrosion resistance	Yes
	Strength retention	Yes
	Fatigue resistance	Yes
	Abrasion resistance	Yes

Table 10.1 gives an overview of the theoretical part of the chapter, in which the potentially relevant material properties are introduced in Sections 10.2 and 10.3. The influence of surface modifications on these properties is discussed in Section 10.4.

Table 10.2 introduces a general methodology for the testing of magnesium alloys under increasingly demanding test conditions. It relates to examples discussed in the practical Sections 10.5 and 10.6.

10.2 Implantation-related mechanical properties

10.2.1 Material requirements during implantation

The mechanical requirements of magnesium as an implant material depend on their medical indication. During the surgical procedure of implantation, the material might be submitted to peak loads or deformations that can exceed the subsequent values during the function as an implant.

For example, a cardiovascular stent is significantly deformed during the expansion of the balloon catheter. The struts of the stent are stretched, exceeding the limit of elasticity, and will remain permanently deformed. However, this plastic deformation should not negatively affect the subsequent degradation behaviour.

Another example of plastic deformation during implantation is osteosynthesis plates that are used to fix bone fractures. The surgeon usually needs to adapt the shape of the plate to the contour of the bone. Pliers are used to prebend the bone plates and might leave impressions on the surface of the implant. Besides geometrical factors of the plate, the material needs to be ductile enough to allow contouring without cracking.

Table 10.2 Methodology for the testing of magnesium alloys and implants under increasingly severe conditions

Test condition	*In vitro*	*In vivo* (or ex vivo)	Example (Reference)
Without medium at room temperature	Uniaxial or multiaxial tests (tension, compression, shear, torsion, bending)	Ex vivo: Biomechanical tests with implants using harvested tissues (i.e., from cadavers)	*In vitro* 3-point-bending test of magnesium, titanium, and steel wires (Chapter 10.5.1)
With medium at 37 °C, without load	Immersion in medium for various periods, removal and mechanical testing	Implant in animals for various periods, explant and test mechanically	Strength retention test with *in vivo* and *in vitro* degraded specimens (Chapter 10.6.2)
With medium at 37 °C, static load	Elastic pre-tensioning of a device, immersion in medium until failure	*Implant device in healthy tissue with controlled preload*	*In vitro* stress corrosion test of a lag screw fixed with a given torque (Chapter 10.5.3)
With medium at 37 °C, cyclic load	Immersion under cyclic load until failure	*Implant in healthy tissue with known cyclic load*	*In vitro* cyclic bending test of a plate/screw device for bone fixation (Chapter 10.5.5)
With medium at 37 °C, pre-deformation & load	Immersion of previously deformed specimen under load conditions simulating the *in vivo* situation	Implant in pathological or modified healthy tissue to restore function (according to intended use)	Preclinical study of a magnesium stent in a porcine coronary model (Wittchow et al., 2013)

The *in vitro* and *in vivo* columns give an overview of possible procedures for the testing of biomedical magnesium devices. The text in italic refers to *in vivo* test for which no examples could be found. The referenced examples show how these procedures can be concretized.

For bridging the fracture, the osteosynthesis plate needs to be fixed to the bone pieces. This is commonly done using screws. During the insertion of the screw, the friction between bone and the screw threads needs to be overcome. The material needs to withstand multiaxial loads (e.g., torsion, shear, compression) that are transmitted by the screw driver.

These examples show how different the mechanical requirement for a material can be during implantation.

10.2.2 Strength and stiffness

In most applications, the strength and stiffness of the material play an important role for the design of the implant. If degradation can be neglected during the surgical implantation procedure, 'classical' mechanical engineering might be applied to determine a material-specific geometry of the implant.

The compliance of a material towards a mechanical stress that does not permanently deform the material is generally expressed as rigidity or stiffness of the material. For metals, the deformation caused by a mechanical load can be described by a linear stress–strain relation. The proportionality constant reflects the stiffness of the material in tension/compression (Young's modulus E), uniform compression (bulk modulus K), or in shear (shear modulus G). As magnesium is more 'compliant' than steel or nitinol, the struts of a cardiovascular magnesium stent, for example, need to be thicker if a comparable recoil has to be achieved. The AMS-3.0 magnesium stent (Wittchow et al., 2013), for instance, has a cross section of 130 × 120 μm, whereas the stainless steel reference stents have strut thicknesses of 97 and 85 μm.

The limit of elasticity is known as yield strength and usually is defined for a given percentage of permanent deformation. As many metallic alloys harden during the permanent deformation ('strain hardening'), the maximum strength is generally above the yield strength. The characteristic of the stress–strain curve can be important if the implant is deformed during implantation (e.g., for the deployment of a cardiovascular stent).

For the design of magnesium screws, the shear strength of the material might be a limiting factor because it dictates the maximum torque that can be transmitted during implantation. Titanium or steel osteosynthesis screws are often self-tapping, which means that they can be inserted directly into a drilled bone hole (Rüedi & Murphy, 2000). For a magnesium screw with identical diameter, pretapping of the screw thread might be needed to avoid premature failure.

10.2.3 Ductility

The ability to plastically deform is a major benefit of metallic implants. A permanent deformation of the implant might be necessary to adapt to the anatomy of the patient. For example, metallic osteosynthesis plates are often prebent to embrace the shape of the bone pieces that need to be fixed. Cold bending is usually not possible with polymeric implants for which preheated water baths can be used to adjust the shape of the plates.

In case of a metallic stent, ductility is of uppermost importance. During the inflation of the balloon inside a blood vessel, the mesh implant is subjected to plastic deformations that exceed the ductility of many common magnesium alloys. As a consequence, a magnesium alloy used for a stent should be as ductile as possible – a goal that can be achieved by appropriate alloying and thermomechanical processing.

10.2.4 Other properties

Besides stiffness, strength, and ductility, other 'static' mechanical properties can be of importance during and after the implantation.

The hardness of a material is its ability to resist a permanent local deformation of its surface (e.g., by indentation or by scratching). For example, Hänzi et al. (2011) report a fully degradable wound-closing rivet for tissue joining and fixing via endoscopic surgery in the gastrointestinal tract consisting of a fast-degrading magnesium tip for tissue penetration and of a slow-degrading polymeric part.

Similarly, the tip of a Kirschner wire (so-called K-wire used in osteosynthesis) should be harder than cortical bone in order to drill into it. The penetration of a magnesium K-wire into bone will be less easy than for harder materials like stainless steel. However, a hard coating of the magnesium wire might improve the insertion into bone.

The toughness is indicative of the amount of energy a material can absorb before it fails. In fracture mechanics, the resistance towards crack propagation is usually measured using precracked compact tension specimens (Kuhn & Medlin, 2000).

10.3 Degradation-related mechanical properties

10.3.1 Material requirements during degradation

The integrity of an implant is particularly important when mechanical loads are involved. During degradation, the implant should not loose its integrity prior to the fulfilment of its mechanical function. If failure occurs, it should be in a controlled and reproducible manner without causing unwanted side effects. The risk for the patient should always remain acceptable.

To avoid failure of the degradable medical device, a material-specific design has to be adopted. A reliable degradation behaviour is one of the basic material requirements that need to be fulfilled before starting to design an implant. Ideally, the material would degrade in a homogenous way, irrespective of inflicted plastic deformation and elastic load condition. In practice, most magnesium alloys degrade by localized pitting. Designing an implant becomes possible if the pitting does not lead to large preferential degradation and if a threshold for the failure load of the degrading material can be found.

Before determining the strength retention and the fatigue strength during degradation, the premature failure due to a combination of stress and corrosive attack – hereafter referred to as environmentally assisted cracking or stress corrosion cracking – should be avoided.

A magnesium alloy with a high susceptibility towards stress corrosion cracking should not be considered for application as an implant material (even if a coating is applied).

10.3.2 Environmentally assisted cracking

Environmentally assisted cracking is also known as stress corrosion cracking and refers to premature failure of a part due to the combined effects of load and corrosion attack. Its phenomenology and mechanisms are described in Chapter 4.

For the *in vitro* testing, human blood plasma can be considered as a saline solution that contains a variety of ions within certain compositional ranges. Some of these ingredients are more aggressive to magnesium than others. In particular, the content of chlorine ions is among the most important species to consider. Chlorine ions are presumably attacking the passivation layer on the surface of magnesium alloys and can therefore enhance the aggressiveness of the saline solution (Song, 2011).

Besides human blood plasma, proteins, fat, and other metabolic species contained in the respective tissue at the implantation site contribute to the specific environment. A stent, for example, will initially be subject to the blood stream and then become covered by endothelial cells. A osteosynthesis screw inserted into cortical bone will encounter a different environment than if inserted into cancellous bone.

The surface condition is important for the occurrence of environmentally assisted cracking. The surface roughness can have an influence on crack initiation. A smooth surface, for example, can delay the occurrence of cracks and therefore delay the failure of the implant due to environmentally assisted cracking. Ways to prevent or delay stress corrosion cracking, for example, by biocompatible coatings, are discussed in Chapter 4.

10.3.3 Strength retention

Strength retention is an important characteristic of a degradable material, provided that an implant does not fail prematurely and that degradation proceeds in a rather homogenous way. It is concise information for the surgeon about how fast a degradable material loses its holding power and how much strength an implant can retain during the fulfilment of its function. As an example, resorbable polylactide glycolide (PLGA) polymer implants for cranio-maxillofacial surgery are expected to retain at least 60% of their initial strength for up to 8 weeks after implantation, a period long enough to allow for sufficient healing of fractured or osteotomized bones.

In the case of magnesium, degradation proceeds from the surface to the inside and is not a bulk phenomenon as for many polymers. As a consequence, the strength retention depends on the thickness of the implant. Thinner implants will loose their strength much faster than thicker implants. For a rectangular plate with 0.5 mm initial thickness as an example, a degradation depth of 0.1 mm will reduce the bending strength by a much greater extent than for a plate with 1.5 mm thickness. Strength retention is therefore not a material property, but must be evaluated for each implant individually.

10.3.4 Fatigue strength

Beyond strength retention, the fatigue strength becomes important in applications with cyclic loads and might become the limiting factor for the design of an implant.

As an example, a cardiovascular stent is exposed to cyclic blood flow changes that submit blood vessels to swelling/expansion and contraction movements. A strut of a stent has to withstand several million of such load cycles without failing (1 year corresponds to about 30 million heart beats).

Another example to illustrate cyclic loading is to imagine a broken rib that is fixed with osteosynthesis plates and screws. On the one hand, the plated rib would be exposed to the expansion and contraction movement induced by the breathing. On the other hand, less frequent loads from laying down on the side would also have to be considered. As a consequence, a combination of low-cycle and high-cycle fatigue might have to be taken into account for the fatigue testing.

Fatigue testing in a corrosion environment is sometimes also referred to as corrosion fatigue. The effect of surface treatment on the fatigue life of biomedical magnesium devices is the topic of Chapter 8.

10.3.5 Other properties

In addition to the above-mentioned properties, the interaction of magnesium implants with biological tissues could lead to a local increase or decrease of degradation. As a hypothetical example, consider a tendon that is frequently sliding over a magnesium implant. Local abrasion of the metal surface and/or of the degradation products might occur and eventually compromise the performance of the implant system. If such effects would become clinically relevant, the abrasion resistance of a magnesium implant could be improved by an adequate surface coating.

Wear resistance is another property that should be considered when the implant system consists of several parts that are in contact and move with respect to each other. If the relative movements are small, the wear between contact surfaces is also referred to as fretting corrosion. The combination of friction and adhesion can also lead to a wear phenomena called galling (i.e., when sliding surfaces are microscopically roughened and thus strongly adhere to each other).

10.4 Effect of surface modifications on implant performance

10.4.1 Goals of surface modifications

The main goal of surface modification of magnesium implants is the improvement of their performance during the application. Surface modification(s) can influence degradation behaviour, stress corrosion resistance, fatigue strength, wear resistance, optical appearance of the implant, and other properties. Surface modifications can also be used to fulfil additional functions such as local drug delivery.

Wang et al. (2012) have reviewed surface modifications of magnesium alloys for orthopaedic applications. A number of chemical and physical surface modifications as well as combinations of such methods are available.

10.4.2 Effect of surface structure

For titanium implants, increasing the surface roughness can improve or accelerate the adhesion of bone cells (e.g., in dentistry, osteosynthesis). On the other hand, an increased surface roughness might have a negative effect on the fatigue resistance and/or on the stress corrosion cracking tendency of the implant. Surface roughness can locally act as a stress raiser and therefore be responsible for the initiation of surface cracks.

For a degradable implant, the surface roughness might change during the course of degradation.

A polishing treatment can help to avoid or delay the formation of such cracks. Mechanical, chemical or electrochemical polishing techniques might be applied to smooth the surface. However, contact with the polishing media can lead to a contamination of the surface and thus potentially compromise the biocompatibility or the degradation behaviour.

If a coating is applied, the surface roughness might be increased, and the positive effect of delaying the initial degradation might therefore be compromised by the negative effect of increased surface roughness.

10.4.3 Effect of surface chemistry

The surface chemistry of a magnesium implant might be influenced by the shaping operations (e.g., extrusion, machining, rolling, forging, laser cutting) and eventual postprocessing operations (e.g., handling, polishing, storage) used.

Contaminations of the surface can modify the interaction of the implant with the biological environment. For example, fingerprints on a newly machined magnesium surface or the storage in a humid atmosphere can alter the surface chemistry.

Intended surface modifications usually reduce the reactivity of the blank magnesium surface. A coating, for instance, could keep away the corrosive medium from the bulk magnesium material and thus delay the occurrence of stress corrosion.

10.4.4 Effect of internal stresses

Thermomechanical treatments of the surface can lead to local plastic deformation and/or residual internal stresses. A surface modification can modify existing or induce new internal stresses in the surface layer of the implant. Metallurgical factors like grain size, grain boundaries, and dislocation density can therefore differ from the bulk material.

These internal stresses might have positive or negative effects on the properties of the implant as they eventually add to the externally applied stresses.

10.5 Testing mechanical properties *in vitro*

10.5.1 *Mechanical tests without medium*

Mechanical tests without medium are important to characterize magnesium alloys and to compare the performance of medical devices under well-defined load situations, but excluding the influence of corrosion.

For the material characterization in the 'dry' state, basic uniaxial tests for metals according to international standards can be carried out using standardized sample geometries and test configurations (compression: DIN 50,106; tension: ASTM E 8M, EN ISO 527).

The degree of mechanical anisotropy in compression and tension can therefore be checked. As (crystalline) magnesium alloys are not particularly temperature sensitive at 37 °C, it is usually sufficient to carry out these tests at room temperature. Stiffness, yield strength (YS), ultimate tensile or compressive strength (UTS, UCS), and the ductility of the material can be determined.

Additional tests like bending, torsion, or shear might also be used to analyse the performance of a material. A bending test might be a fast and easy way to characterize a material as compressive and tensile stresses occur simultaneously. Figure 10.1 shows a simple 3-point-bending test of metal wires. The example illustrates that a magnesium wire has to be about 1.5 times thicker than titanium or steel to achieve similar flexural properties. The magnesium wire with 3 mm diameter has comparable bending stiffness to the steel wire with 2 mm diameter, i.e., both wires exhibit similar

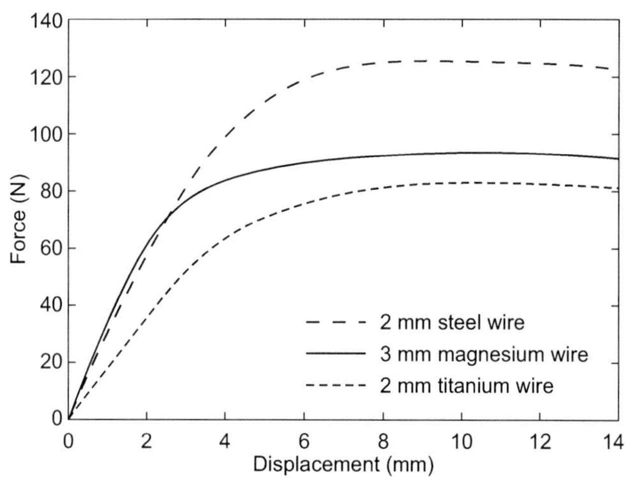

Figure 10.1 Comparison of the bending behaviour of wires with 150 mm length by a three-point-bending test with 60 mm span and a vertical displacement of 5 mm/min. Plunger and supports have a radius of 5 mm. The steel wire (Synthes art. nr. 292.200) and the titanium wire (Synthes art. nr. 492.200) are 2 mm in diameter, whereas the magnesium wire (alloy WE43, as extruded) has a diameter of 3 mm.

elastic behaviour. The maximum bending force for the magnesium wire (alloy WE43) lies between the values of the stainless steel wire (316L) and of the titanium wire (TiAl6V4 alloy).

Flexural properties can be measured according to the standard ISO 178 (for plastics). The deformed specimens can be used to roughly estimate the ductility of the material by a subsequent folding test (EN ISO 7435).

For coated magnesium, further tests are necessary to characterize the mechanical properties of the coating in the dry state (e.g., adherence to the substrate, scratch resistance). For shear testing of calcium phosphate coatings and metallic coatings, the ASTM F1044 can be applied.

For established medical devices, existing standardized tests can be used to evaluate their mechanical behaviour. In the area of osteosynthesis, for example, standards exist for the testing of:

- metallic bone plates (e.g., bending tests according to ASTM F382, ISO 9585)
- metallic bone screws (e.g., torsion test according to ASTM F543)
- intramedullary rods and nails (bending and torsion tests according to ASTM F383, ASTM F1264)
- angled orthopaedic fracture devices (ASTM F384)
- bone staples (pull out and bending according to ASTM F564).

10.5.2 Testing of in vitro degraded samples without load

For degradable polymers, a typical way of *in vitro* testing is described in the ASTM F1635 and ISO 13781 standards. Semifinished components or finished implants are immersed in buffered saline solution at physiological temperature (i.e., PBS at 37 °C) for specific periods of time. The samples are periodically removed and tested for material or mechanical properties (e.g., percent weight loss, inherent viscosity, strength loss).

A limitation of this way of testing is that the used saline solution only roughly simulates the *in vivo* environment. It does not account for the local blood flow, the water or fat content, or the microbiologic interaction with cells in a specific tissue. Another limitation is that the combined effect of stress and degradation is not considered. For example, the polymer might experience a considerable degree of relaxation, i.e., the implant cannot retain the initial load as it gradually yields to the imposed stress due to the imposed temperature (creep) and due to the presence of water (increased mobility of polymer chains).

For magnesium, a similar prolonged (static) immersion in saline solution is not advised because the conditions cannot be kept as constant as in the human body. The release of hydroxide ions increases the pH (as soon as the buffering capacity of the solution is consumed), and the release of magnesium ions increases the osmolality of the solution. The released hydrogen gas might form bubbles and could locally mask the surface. As a consequence, regular medium changes or a dynamic flow system are advised to approximate conditions to a physiological range. Figure 10.2 shows the example of a lag screw that has been degraded in simulated body fluid (SBF) for

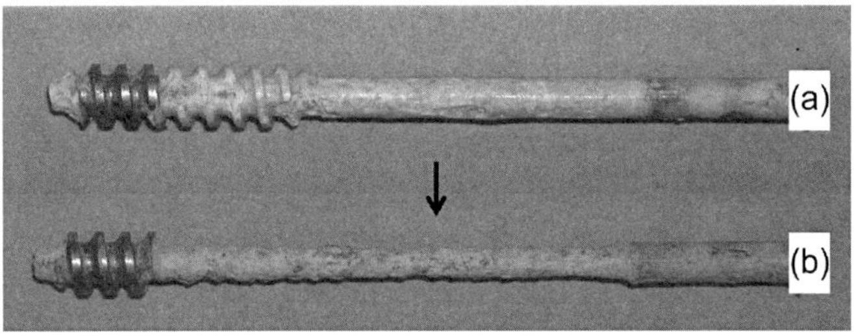

Figure 10.2 Example of a generic magnesium screw after 12 weeks of immersion in 250 ml simulated body fluid with weekly medium change (according to Imwinkelried et al. (2012)). (a) Screw with degradation products. (b) Screw after removal of the degradation products by brushing.

12 weeks with weekly medium changes as described by Imwinkelried et al. (2012). Because the screw was fixed inside a tensioning device, some of the broad threads at the tip are not degraded. The degraded neighbouring threads and the screw shaft have kept their initial shape (Figure 10.2(a)). The outer shape of the screw remained unchanged as the metallic magnesium has been replaced by a thick conversion layer. These white degradation products are then removed with a common hand brush, and the nondegraded magnesium core becomes visible (Figure 10.2(b)). The ductility of the remaining magnesium core can be examined with an additional bending test. Gas release during immersion can be collected and compared to the weight loss of the screw with and without degradation products.

This example shows that magnesium specimens can be degraded *in vitro* for various periods of time, then be removed and tested outside the 'bath' – just as for the tests without medium. Depending on the test, the degradation products might be dehydrated first (e.g., by rinsing with ethanol and subsequent drying) or even removed mechanically or with a chemical agent that does not attack magnesium but dissolves the formed calcium phosphate conversion layer (e.g., hydrofluoric acid).

10.5.3 Testing of in vitro degrading samples with load

After implantation, a medical device might be subject to a certain base load. For example, the above-mentioned lag screws are used in traumatology to put broken bone pieces under compression. The screw itself is experiencing tensile stresses between the screw threads and the screw head. The higher the insertion torque of the screw, the larger are the tensile stresses along the screw axis. The screw shown in Figure 10.2 was tested at a low insertion torque and did not fail within 12 weeks in SBF. With higher insertion torques, the combined effect of tensile stress and corrosion attack might lead to premature failure due to stress corrosion cracking. A coating could be helpful in preventing stress corrosion cracking because it might keep the corrosive medium away from the magnesium surface. However, it is important to make sure that the coating is also effective when the implant is under load.

10.5.4 Testing of in vitro degrading samples after plastic deformation

If implants are submitted to plastic deformation during the implantation, it is necessary to test the magnesium samples *in vitro* after deforming them permanently. This kind of test is particularly important if coatings or other surface modifications are used. Due to the plastic deformation of the base material, a coating might locally crack or even delaminate and thus induce a completely different degradation behaviour (e.g., with more pronounced localized corrosive attack). In addition to the plastic deformation, a given load might be applied to test worst-case conditions.

An example for *in vitro* testing of plastically deformed samples with applied load is shown in Figure 10.3. A rectangular plate is first deformed in a bending template with a defined curvature (Figure 10.3(a)). In a second step, the plate is elastically deformed to fit inside a polymer implant holder. Figure 10.3(b) shows how the exerted force can be measured before immersing the floating device into SBF. This measurement can be repeated at each medium change to track the force decrease over time and to characterize the strength retention of the alloy. Figure 10.3(c) shows the tensioned magnesium specimen with the implant holder floating at the (SBF) surface of a gas-collecting tube.

Premature failure of the sample might occur before the sample has degraded to a point where no load is left (Figure 10.3(d)). In addition, the released gas can be collected. The amount of released gas can be compared to the weight loss of the samples.

10.5.5 Fatigue testing under cyclic stress

Besides a given base load as mentioned above, implants will, in most cases, be submitted to cyclic and/or dynamic loads. As an example, a cardiovascular stent is deployed by the expansion of the balloon and exerts a radial pressure on the artery wall to keep the vessel open. In addition to this base load, the pulsating blood flow puts the stent under cyclic stress. An adequate *in vitro* test is therefore needed to determine the

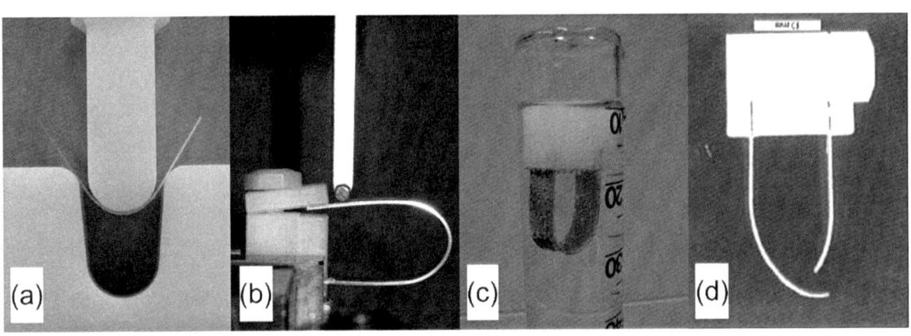

Figure 10.3 Immersion test of a rectangular magnesium specimen after plastic deformation and tensioning. (a) Controlled deformation in a bending template. (b) Force measurement of the spring type specimen before fixation in the implant holder. (c) Immersion in simulated body fluid with a floating PE polymer holder. (d) Example of a broken specimen after a immersion.

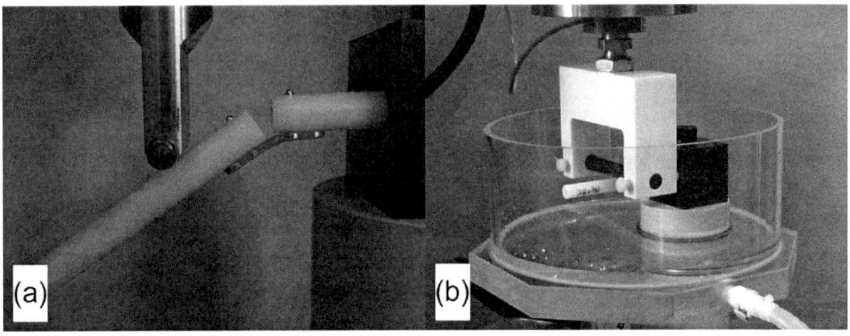

Figure 10.4 *In vitro* testing of a magnesium plate/screw construct with a simulated fracture gap. (a) Static bending test. (b) Cyclic bending test in distilled water.

failure behaviour of the stent in such harsh conditions (i.e., corrosive environment, plastic deformation, cyclic stresses).

Another example can be found in osteosynthesis. When a bone fracture is fixed using plates and screws, the plate is usually prebent to follow the contour of bone, and the screws are fixed to the bone in such a way that the fracture is well reduced and eventually put under compression (Rüedi & Murphy, 2000). As a consequence, the plastically deformed plate is submitted to a basic load. Cyclic loads come in addition as early mobilization of the patient is sought to accelerate bone healing.

A major difficulty in fatigue testing is to define adequate testing conditions. The forces occurring *in vivo* are often not well known and can be multiaxial. The aggressiveness of the used medium and exposure time are important parameters to consider in defining adequate cycle times.

For uncoated magnesium, distilled water might be a good medium to start with. On one side, distilled water does not give deposits on the magnesium surface. On the other side, the magnesium ions that are leached out of the material usually do not change the pH as $Mg(OH)_2$ is formed. The important effect of distilled water is that it can alter the occurrence and growth of cracks. Premature failure therefore will occur at lower load levels and/or at shorter periods than if no medium is used.

Figure 10.4(a) shows a static bending test of a magnesium plate that is fixed to two polymer tubes with screws to simulate a bone fracture. For cyclic bending tests in an aqueous medium, it is important to avoid contact with other metals to prevent galvanic corrosion, which could falsify the degradation conditions. In the example shown in Figure 10.4(b), a carbon fibre rod (black) is used instead of the metal plunger used without the medium.

10.6 Testing implant performance *in vivo*

10.6.1 Suitability of animal models

The suitability of animal models in testing the performance of an implant *in vivo* depends on how well the human environment of a particular indication can be simulated in an animal tissue. Small animals such as mice, rats, or rabbits are often used for first screening tests of degradable implant materials. For testing the performance of a human implant, however, small animal testing might not be sufficient. Frink et al. (2011) recommends the use of large animals if physiological processes are to be investigated, which is the case for magnesium degradation.

The metabolism of an animal is of considerable importance when degradation is involved. Degradation rate might depend on the size of the animal. Witte et al. (2008) report that the blood flow in the human body, for instance, is an order of magnitude higher than in a small animal (rat, rabbit). In a rat study with the slow-degrading magnesium alloy WE43, Tschegg et al. (2011) revealed a significant increase of push out force of a pin implanted in the distal femur compared to titanium and/or PLGA polymer pins of same geometry. Because the degradation-related gas release was slow enough to avoid the formation of gas bubbles, it did not interfere with bone apposition, and the receding implant was replaced by bone. The same material implanted on top of the nasal bone of a minipig, however, lead to the formation of gas bubbles (Imwinkelried, Beck, Iizuka, & Schaller, 2012).

The blood flow in miniature pigs is more similar to humans than that in small animals. Because the physiology of the miniature pig's heart resembles the human heart, these animals are often used to test the *in vivo* performance of cardiovascular stents.

Wang, Liu, Fang, and Shi (2007) consider the miniature pig to be a useful large animal model for dental and orofacial research. Minipigs are not ruminants (like sheep) and have a temporomandibular joint similar to the humans. If bone healing is studied, the use of skeletally mature animals is advised.

10.6.2 Testing of in vivo degraded samples

One method to test the *in vivo* degradation behaviour of magnesium is to place an implant inside a given tissue of an animal and to let it degrade for a defined period. After explantation, mechanical testing of the partially degraded implants can be done.

Thomann et al. (2010) implanted 25-mm-long McCa0.8 pins with 2.5 mm diameter in the intramedullar cavity of rabbit tibiae for 3, 6 and 12 months. After explantation, three-point-bending tests of the degraded pins were carried out, and the strength retention of the material could be determined.

Imwinkelried et al. (2012) used large rectangular plates that were implanted on top of the nasal bone of miniature pigs for up to 24 weeks. Figure 10.5(a) shows the removal of magnesium plate after 12 weeks of implantation. Figure 10.5(b) illustrates the three-point-bending testing of the *in vivo* degraded rectangles. To compare the strength retention of *in vivo* and *in vitro* degradation, identical plates were immersed in SBF (SBF27) for up to 12 weeks. *In vitro* degradation was found to occur about four times faster than *in vivo*.

10.6.3 Extrapolation of mechanical properties from residual volumes

To avoid the sacrifice of an animal for the *in vivo* degradation of an implant, the residual volume of an implant might be determined using *in vivo* imaging

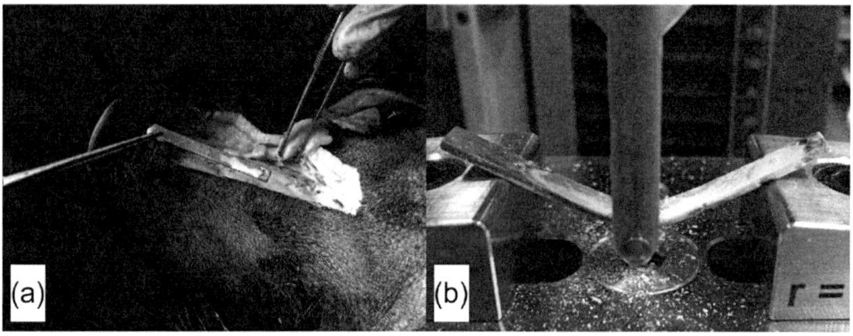

Figure 10.5 Mechanical testing of *in vivo* degraded magnesium plate. (a) Removal of plate from the nasal bone of a miniature pig after 3 months. (b) Three-point-bending test of *in vivo* degraded plate.

techniques (e.g., live computed tomography (CT)). Numerical modelling might then be used to estimate the remaining strength and stiffness of the implants. In practice, it may not be possible to determine the residual volume of an implant due to the limited image resolution. Furthermore, the conversion zone (i.e., the volume of the implant that has been transformed into degradation products) supposedly has a similar density to the magnesium because it can hardly be distinguished from nondegraded material.

Witte et al. (2010) were able to distinguish between metallic magnesium and the corrosion layer by using synchrotron-radiation-based micro-CT: (SRμCT) on explanted rat femurs. They were also able to determine the corrosion rate based on the pit depths of coated and noncoated cylinders. However, they did not extrapolate the remaining mechanical properties from these residual volumes.

Cho et al. (2012) have used conventional postmortem micro-CTs (μCTs) on explanted bone blocks from rabbit femurs containing magnesium screws. On strongly degraded samples, they were able to determine the residual equivalent diameter of the screws and to determine the remaining bending and torsion strength using numerical modelling.

As can be seen by these two examples, the determination of residual magnesium volumes either needs sophisticated imaging techniques or a sufficiently thick conversion zone. An *in vivo* determination of residual implant volumes remains very challenging, even if image resolution could be further improved.

10.6.4 Testing of implants using functional animal models

The registration process of a medical device requires the proof of its safety and performance. As a consequence, the intended use of an implant system has to be tested using a functional animal model. The results gained from such preclinical tests are the basis for the regulatory approval of an implant system or for the respective human clinical trial required for getting such approval from the authorities.

Cardiovascular magnesium stents, for instance, have been implanted into the coronary arteries of miniature pigs exactly as in the human procedure. Wittchow et al.

(2013) showed that the efficacy and healing characteristic of a magnesium stent was comparable to permanent metallic stents in the porcine model used.

For the testing of osteosynthesis implants, few functional animal models are available because the conflict among clinical comparability, standardization, and reproducibility remains a major problem (Frink, Andruszkow, Zeckey, Krettek, & Hildebrand, 2011).

A common way to simulate a bone fracture is to apply an osteotomy (i.e., to cut a bone in two) and fix the bone pieces with the implant system being tested. The use of large animals is important not only because a similar metabolism should be sought, but also has the advantage that human-size implants can be used.

Furthermore, biomechanics in the animal often represent a worst case, but static and dynamic forces occurring are often unknown and are difficult to measure.

10.6.5 Likely future trends

The future in testing of mechanical properties of magnesium alloys will depend on which biomedical applications will be successful. The requirements of the regulatory bodies for the registration of magnesium implants will also influence the amount and the quality of the test data.

The requirements on test planning, documentation, and statistical analysis as well as for the validation of experimental methods and setups are likely to increase. A standardization of the *in vitro* testing of mechanical properties of magnesium alloys might be reached.

Accelerated *in vitro* degradation might become a way to simulate the *in vivo* behaviour of magnesium implants and to reduce the amount of animal testing.

The use of sophisticated *in vivo* imaging techniques, the implantation of sensors, and numerical modelling of the biomechanics will improve our knowledge of the animal models used.

10.7 Sources of further information

The following books by Woodhead Publishing can provide further information about:

- Fundamentals of magnesium alloy metallurgy (Pekguleryuz, Kainer, & Kaya, 2013)
- Metals for biomedical devices (Niinomi, 2010)
- Corrosions of magnesium alloys (Song, 2011)
- Coatings for biomedical applications (Driver, 2012)

Further information about mechanical testing can be found in:

ASM Handbook Volume 08: Mechanical Testing and Evaluation (Kuhn & Medlin, 2000)

Since 2009, a dedicated symposium on biodegradable metals for biomedical applications has taken place every year (Berlin 2009, Maratea 2010 and 2012, Quebec City 2011, Umang Island 2013). The related abstracts and publications can provide further information about testing media and methods.

Acknowledgements

All figures in this chapter are a courtesy of Synthes GmbH, Oberdorf, Switzerland, and have not been published previously. I thank my colleagues and superiors from the biomaterials department for their support and in particular Dr Andrea Montali for the revision of the text. I also acknowledge the support of the prototype shop and of the materials testing department for manufacturing and testing of the magnesium specimens.

References

Cho, S. Y., Chae, S.-W., Choi, K. W., Seok, H. K., Kim, Y. C., Jung, J. Y., et al. (2012). Biocompatibility and strength retention of biodegradable Mg–Ca–Zn alloy bone implants. *Journal of Biomedical Materials Research. Part B, 101B*(2), 201–212.

Driver, M. (2012). *Coatings for biomedical applications*. Cambridge, UK: Woodhead Publishing Limited.

Frink, M., Andruszkow, H., Zeckey, C., Krettek, C., & Hildebrand, F. (2011). Experimental trauma models: an update. *Journal of Biomedicine and Biotechnology, 2011*. http://dx.doi.org/10.1155/2011/797383. Article ID 797383, 15 pages, 2011.

Gunde, P., Hänzi, A. C., Sologubenko, A. S., & Uggowitzer, P. J. (2011). High-strength magnesium alloys for degradable implant applications. *Material Science and Engineering A, 528*, 1047–1054.

Hänzi, A. C., Metlar, A., Schinhammer, M., Aguib, H., Lüth, T. C., Löffler, J. F., et al. (2011). Biodegradable wound-closing devices for gastrointestinal interventions: degradation performance of the magnesium tip. *Materials Science and Engineering C, 31*, 1098–1103.

Haude, M., Erbel, R., Erne, P., Verheye, S., Degen, H., Böse, D., et al. (2013). Safety and performance of the drug-eluting absorbable metal scaffold (DREAMS) in patients with de-novo coronary lesions: 12 month results of the prospective multicentre, first-in-man BIOSOLVE-I trial. *Lancet, 381*, 836–844.

Imwinkelried, T., Beck, S., Iizuka, T., & Schaller, B. (2012). Effect of plasmaelectrolytic coating on the strength retention of in vivo and in vitro degraded magnesium implants. *Acta Biomaterialia, 9*, 8643–8649.

Kuhn, H., & Medlin, D. (2000). *ASM handbook volume 08: Mechanical testing and evaluation*. Materials Park, Ohio 44073-0002: ASM International.

Niinomi, M. (2010). *Metals for biomedical devices*. Cambridge, UK: Woodhead Publishing Limited.

Pekguleryuz, M., Kainer, K., & Kaya, A. (2013). *Fundamentals of magnesium alloy metallurgy*. Cambridge, UK: Woodhead Publishing Limited.

Rüedi, T. P., & Murphy, W. M. (2000). *AO principles of fracture management*. Davos, Switzerland: AO Publishing.

Song, G. L. (2011). *Corrosion of magnesium alloys*. Cambridge, UK: Woodhead Publishing Limited.

Thomann, M., Krause, C., Angrisani, N., Bormann, D., Hassel, T., Windhagen, H., et al. (2010). Influence of a magnesium-fluoride coating of magnesium-based implants (MgCa0.8) on degradation in a rabbit model. *Journal of Biomedical Materials Research. Part A, 93*(4), 1609–1619.

Tschegg, E. K., Lindtner, R. A., Doblhoff-Dier, V., Stanzl-Tschegg, S. E., Holzlechner, G., Castellani, C., et al. (2011). Characterization methods of bone-implant-interfaces of bioresorbable and titanium implants by fracture mechanical means. *Journal of the Mechanical Behavior of Biomedical Materials, 4*(5), 766–775. http://dx.doi.org/10.1016/j.jmbbm.2010.08.004.

Wang, S., Liu, Y., Fang, D., & Shi, S. (2007). Review article – the miniature pig: a useful large animal model for dental and orofacial research. *Oral Diseases, 13*, 530–537.

Wang, J., Tang, J., Zhang, P., Li, Y., Wang, J., Lai, Y., et al. (2012). Surface modification of magnesium alloys developed for bioabsorbable orthopedic implants: a general review. *Journal of Biomedical Materials Research. Part B, 100B*(6), 1691–1701.

Wittchow, E., Adden, N., Riedmüller, J., Savard, C., Waksman, R., & Braune, M. (2013). Bioresorbable drug-eluting magnesium-alloy scaffold: design and feasibility in a porcine coronary model. *EuroIntervention, 8*, 1441–1450.

Witte, F. (2010). The history of biodegradable magnesium implants: a review. *Acta Biomaterialia, 6*, 1680–1692.

Witte, F., Fischer, J., Nellesen, J., Vogt, C., Vogt, J., Donath, T., et al. (2010). In vivo corrosion and corrosion protection of magnesium alloy LAE442. *Acta Biomaterialia, 6*, 1792–1799.

Witte, F., Hort, N., Vogt, C., Cohen, S., Kainer, K. U., Willumeit, R., et al. (2008). Degradable biomaterials based on magnesium corrosion. *Current Opinion in Solid State and Materials Science, 12*, 63–72.

Biocompatibility testing and marketing authorisation of degradable magnesium implants

11

Anneke Loos
Hannover Medical School, Hannover, Germany

11.1 Introduction

An implant that disappears after its function has been fulfilled has fascinated visionary inventors for more than 150 years and will probably lead to revolutionary changes in the field of surgery once the technical and regulatory hurdles are overcome. Besides polymers, metal alloys based on magnesium, and to a lesser extent iron, have been the focus of attention of scientists and the medical device industry. Like with many other innovations on the way from bench to bedside, the properties of new materials pose a challenge with respect to the current marketing authorisation procedures. The lack of knowledge about the possible (unwanted) negative consequences that new materials might have on the patient reduces the innovative potential of new implant materials. In the case of degradable magnesium implants, their ability to dissolve causes difficulties when standard biological tests are applied. With regard to safety evaluations, regulatory authorities ask questions concerning the degradation kinetic as well as the frequency of occurrence and the nature of intermediates released by the bulk material or toxic alloy components (e.g., aluminium, yttrium, or neodymium) and their localisation in or transport in the body.

This chapter discusses the biological evaluation of degradable magnesium implants with respect to the marketing authorisation of new implants. It starts with the classification of implants, which has become increasingly complicated because manufacturers have started modifying them with drugs or cells to create additional properties (Section 2). Since standards are a key aspect of the marketing authorisation process of implants, Section 3 gives insight into the meaning, development, and application of these documents. Section 4 gives an overview of the different aspects of biological safety that have to be considered for marketing authorisation. In Section 5, biocompatibility testing of degradable magnesium implants and the necessary modifications of standard methods for dealing with their intrinsic characteristics are discussed. Section 6 gives a review of the current approaches and ideas of the scientific community that are used to gain information about the toxicity of various magnesium alloys in *in vitro* and *in vivo* assays.

Surface Modification of Magnesium and its Alloys for Biomedical Applications. http://dx.doi.org/10.1016/B978-1-78242-077-4.00011-5

11.2 Classification – which directive applies?

If the aim is to obtain marketing authorisation for a new implant, one of the first considerations should be the classification of the product. The reason for this is that the classification can have an impact on many factors, like the requirements concerning the quality management system of the manufacturer, the marketing authorisation process, the necessary tests for safety assessment, and the responsible regulatory authority. This section gives a general overview of the current regulatory situation, the various ways that an implant can be classified, and the implications of the classification.

One thing that makes the way from bench to bedside for a new medical product time-consuming is that the requirements and procedures for marketing authorisation differ from country to country. The only exemptions are the countries of the European Union (EU), where a uniform EU legislation applies for all the member countries. Besides national marketing authorisation, it is also possible for the manufacturer to request approval for internal European trade, which is then accepted in all of the 28 member countries. Every other country in the world has its own laws, procedures, and regulatory authorities. Besides Europe, the two biggest markets for implants are currently Asia and the United States. In the United States, the Food and Drug Administration (FDA) is the competent authority for marketing authorisation of all medical products. In other countries (e.g., within the European Union), the responsibilities are divided between different authorities, like the European Medicines Agency (EMA) and the inspection authorities ("notified bodies" like Technischer Überwachungsverein (TÜV) or Deutscher Kraftfahrzeug-Überwachungs-Verein e.V. (DEKRA), www.dekra.de). Some effort has been made to harmonise the procedures in the different countries, e.g., through internationally accepted standards by the International Medical Device Regulators Forum (IMDRF), which carries out the mission of the Global Harmonization Task Force (GHTF), or by the International Conference on Harmonization of Technical Requirements for Registration of Pharmaceuticals for Human Use (ICH) (www.ich.org).

Medicinal products can be divided into three groups: drugs (or pharmaceuticals), medical devices, and advanced therapy medicinal products (ATMPs, Figure 11.1). Drugs can be differentiated from medical devices by their primary mode of action. Pharmaceuticals act immunologically, pharmacologically, or metabolically, whereas a medical device does not. Medical devices usually replace or modify the anatomy or a physiological process. They are used to recognise, prevent, monitor, treat, compensate, or heal diseases, injuries, and handicaps and for contraception. The group of ATMPs includes products for cell therapy, gene therapy, and products based on methods of tissue engineering. Detailed definitions of the product classes can be found in the legislation of the respective country. Please bear in mind that these definitions might differ from country to country!

Implants typically belong to the medical devices or ATMP group. However, an implant can also be classified as a pharmaceutical if it is combined with a drug and if the only function of the implant is as a carrier for the pharmaceutical (e.g., contraceptive hormone implants). Combinations of medical devices and drugs

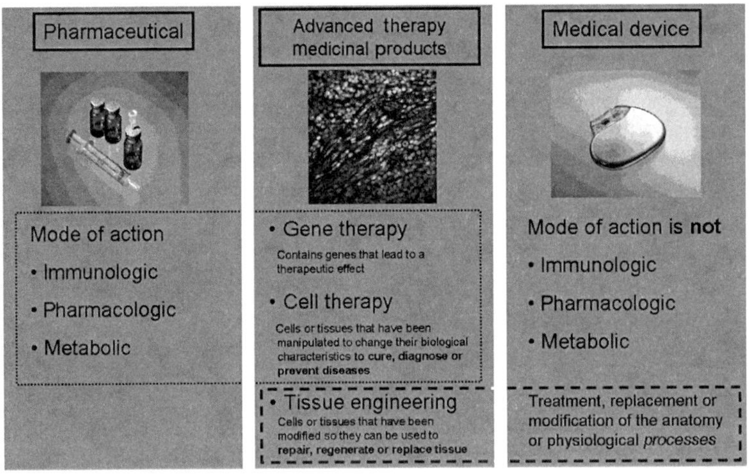

Figure 11.1 Medicinal products are divided into pharmaceuticals, advanced-therapy medicinal products (ATMPs), or medical devices.

(e.g., drug-eluting stents) or ATMPs and medical devices (e.g., a collagen matrix combined with cells) are also possible. Product classification is essential for the marketing authorisation process, because it indicates which directive applies, the tests that have to be performed, the quality management system needed for manufacturing, and the responsible regulatory authority for marketing authorisation (Figure 11.2). For example, a typical feature of the marketing authorisation procedure of medical devices is the use of a standard like the International Organization for Standardization ISO 10993, which does not apply to the preclinical testing of pharmaceuticals. The classification also has an impact on developmental costs and time. For example, the marketing authorisation process for a pharmaceutical is 10 times more expensive (US$ 500 million to US$ 2 billion) than and takes twice as long (approximately 10–12 years) as the marketing authorisation of a medical device (Adams and Brandnter, 2006; DiMasi, Hansen, and Grabowski, 2003; Kaplan et al., 2004). In case of doubt, it is highly recommended to seek legal advice for the product classification.

In general, the classification of implants made of degradable materials follows the rules described above. Degradable implants are commonly classified as medical devices and tested accordingly. Changes in the product classification can occur when the implant surface is modified (e.g., coated with a drug) or when it is combined with cells. In these cases, aside from the testing of the final product, the medical device part also has to be tested separately.

To summarise, the proper classification of a new implant should always be one of the first steps because basics like the required quality management system, test procedures, and regulatory authority might differ depending on the classification as medical device, pharmaceutical, ATMP, or combinations thereof. Implant classification can differ from country to country because besides the EU, every country has its own classification rules and marketing authorisation procedures.

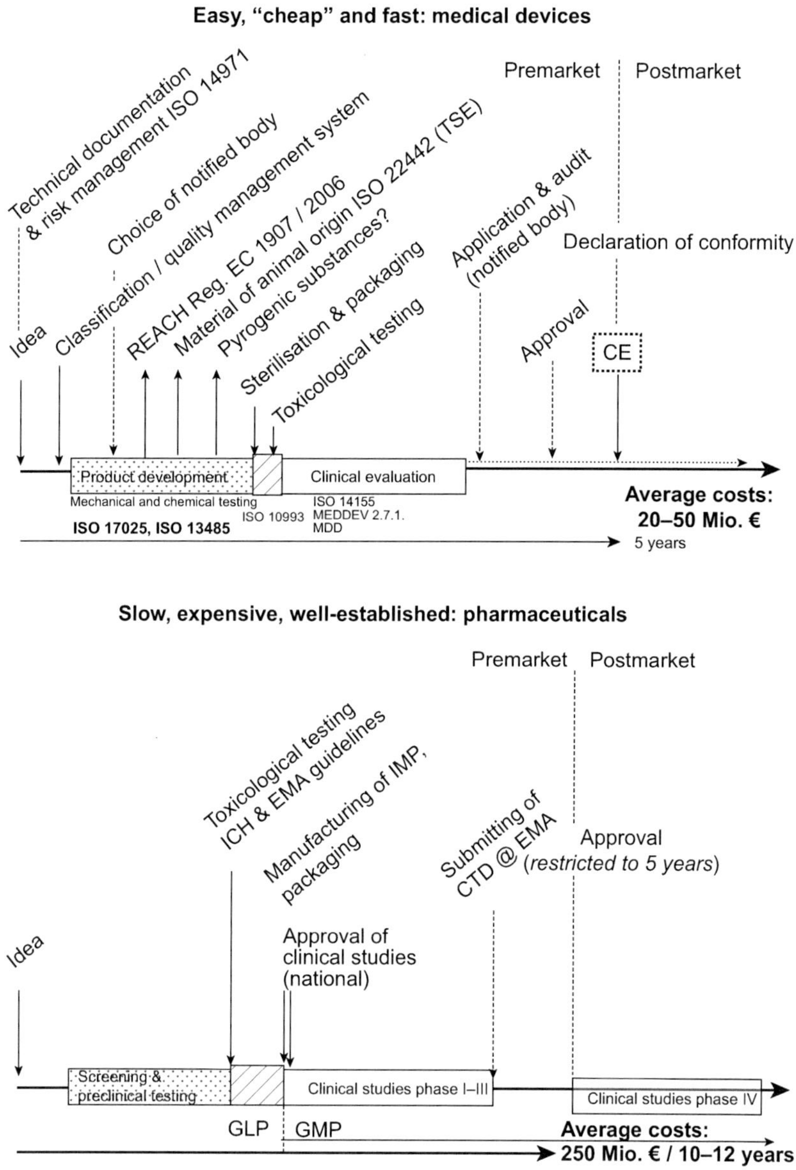

Figure 11.2 Comparison between the market approval process of pharmaceuticals, medical devices, and ATMPs in the European Union. CTD = Common Technical Document; IMP = investigational medicinal product, ICH = International Conference on Harmonization of Technical Requirements for Registration of Pharmaceuticals, AMNOG = Arzneimittelmarktneuordnungsgesetz (Germany), CAT = Committee for Advanced Therapies, CHMP = Committee for Medicinal Products for Human Use, EMA = European Medicines Agency (EMEA), GLP/GMP = Good Laboratory Practice/Good Manufacturing Practice, TSE = transmissible spongiform encephalopathy. RM = risk management

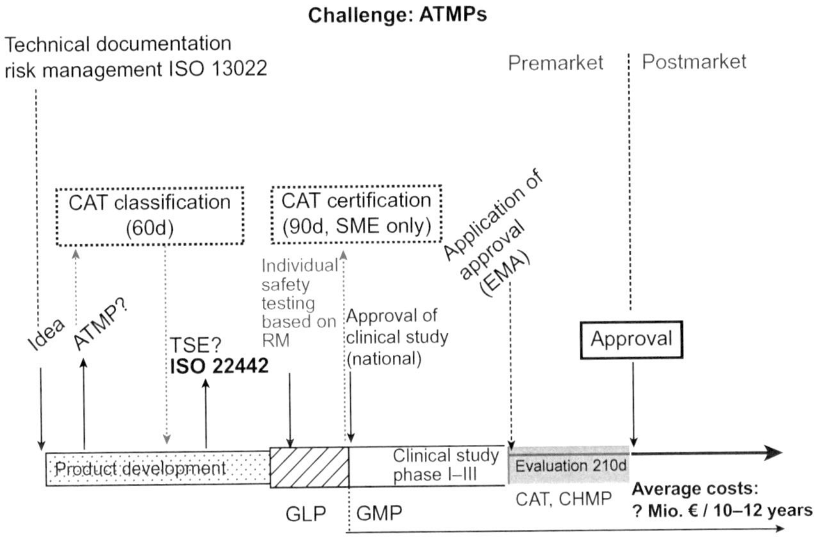

Figure 11.2 Continued.

11.3 Testing according to international standards

We are surrounded by standardised language, safety requirements, and technical equipment in our daily lives; we have become so used to this that the only time we think about standards is when they are missing. One example is the lack of compatibility of power supplies for mobile phones. Every mobile phone needs its own power supply because an internationally acknowledged uniform, interchangeable connector plug does not exist at present. The following section covers basic information about the meaning, classification, development, and use of standards.

11.3.1 *What is a standard?*

According to ISO (www.iso.org), a standard is defined as a "document approved by a recognised body, that provides, for common and repeated use, rules, guidelines or characteristics for products or related processes and production methods, with which compliance is not mandatory. It may also include or deal exclusively with terminology, symbols, packaging, marking or labelling requirements as they apply to a product, process or production method" (ISO Central Secretariat, 2007). In contrast, the GHTF definition is slightly different and more general: "document, established by consensus and approved by a recognised body, that provides, for common and repeated use, rules, guidelines or characteristics for activities or their results, aimed at the achievement of the optimum degree of order in a given context" (GHTF/SG1/N044:2008). The GHTF definition additionally refers to the following note to ISO/IEC Guide 2:2004, definition 3.2: "Standards should be based on the consolidated results of science, technology and experience, and aimed at the promotion of optimum community benefits."

Standards make our lives safer by defining minimal safety requirements and providing a basis for quality assessment (e.g., in the field of construction, food, traffic, electronics, and medicine). The aim of international standardisation organisations is to publish international agreements in the form of standards to facilitate the international trade of goods and services and to promote international cooperation in all fields of technical, scientific, economic, and intellectual activity. Standards provide definitions that support communication; they help to eliminate technical barriers and reduce the overall costs, which results in a positive impact on the economy (Blind, 2004).

11.3.2 What kinds of standards exist?

Standards can be subdivided into national or international standards according to their origin and into horizontal or vertical standards according to their scope.

On the international level, all standards are developed at the ISO except for electronics and telecommunications standards, which are provided by the International Electrotechnical Commission (IEC) and the International Telecommunication Union (ITU). The European analogues are the European Committee for Standardization (CEN), the European Committee for Electrotechnical Standardization (CENELEC), and the European Telecommunications Standards Institute (ETSI). Besides the international standardisation organisations, almost every country has its own national organisation: e.g., BNSI (Barbados), SA (Australia), NC (Cuba), EVS (Estonia), DIN (Germany), COSQC (Iraq), JISC (Japan), ILNAS (Luxembourg), SNZ (New Zealand), SSMO (Sudan), and ANSI (USA). Scientific or technical societies also publish guidelines and standards on specific topics. Relevant standardised testing protocols for the toxicological evaluation of implants are also provided by the Organization for Economic Cooperation and Development (OECD) and the American Society for Testing and Materials (ASTM International). In general, the OECD focus is on the assessment of the potential toxicity of environmental substances, whereas ASTM International is more concerned with the safety assessment of materials for medical applications. Sometimes, the ISO standards refer to OECD or ASTM methods.

Additionally, a hierarchy of different levels exist for standards (Figure 11.3). Level 1 is the basic or horizontal standard. Documents in this group describe fundamental concepts, principles, or general requirements. They usually cover general safety and performance aspects that are applicable to a wide range of products or processes, e.g., risk or quality management, or clinical investigation. With respect to implants, examples of horizontal standards are ISO 14971 "Medical devices—Application of risk management to medical devices," IEC 62366 "Medical devices—Application of usability engineering to medical devices," ANSI/AAMI HE 74 "Human factors design process for medical devices," or ISO 14630 "Non-active surgical implants[a]—General requirements." Level 2 is a group standard or a semihorizontal standard that is applicable to

[a] Non-active in this regard means that the implant does not have an externally driven energy source.

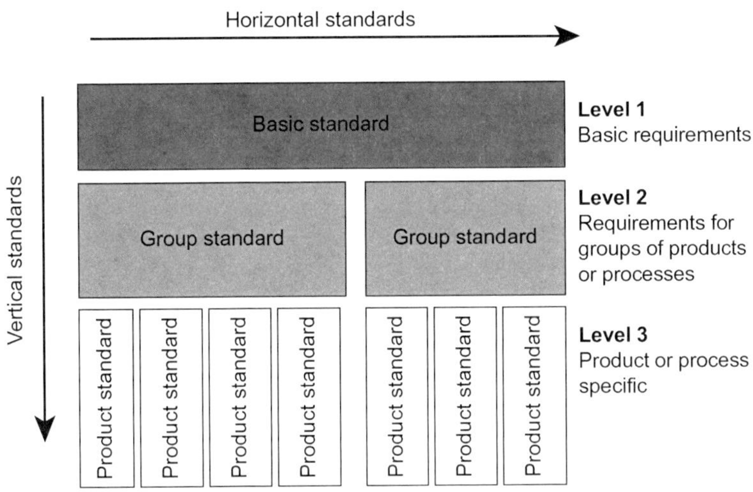

Figure 11.3 According to ISO and GHTF, standards can be classified as basic, group, or product standard depending on their scope.

families of products or processes. Examples of semihorizontal standards are the ISO 11135 series "Sterilization of health care products Ethylene oxide" or the ISO 10993 series "Biological evaluation of medical devices." Level 3 covers product-specific standards that generally deal with the required safety and performance aspects of a specific product or process, such as ISO 25539-2 "Cardiovascular implants–Endovascular devices–Part 2: Vascular stents" for coronary stents.

11.3.3 Development of (international) standards

One important feature of standards mentioned in the GHTF definition is that they are consensus documents. This means that a more or less randomly mixed (international) group of approximately 5–20 people, who may have very different opinions and backgrounds, tries to reach a mutual consensus on wording. On an international level, most of the members of the groups are from industry; however, scientists are rarely part of this group. Finding a consensus can be very time-consuming and usually takes several years. As a result, standards are sometimes not very detailed and in general not state of the art. The procedures of the ISO, which are described in the following section, are an example of the development of an international standard.

ISO standards are developed by an ISO committee (a technical committee or TC). The ISO website gives an overview of its committee structure and the standards the committees are working on. Relevant standards for implants are being developed, for example, in TC 150 "Implants for surgery," TC 198 "Sterilization of health care products," TC 194 "Biological evaluation of medical devices," TC 168 "Prosthesis and orthotics," TC 106 "Dentistry," and TC 210 "Quality management and corresponding general aspects for medical devices." When committees have many members and several topics to cover, subcommittees are formed to concentrate on specific issues.

The experts within the ISO committees are assigned by their corresponding ISO member bodies, which are usually national standardisation organisations (e.g., ANSI, DIN, or BSI; please refer to the ISO website for more information). Only one national standardisation organisation per country is a member of the ISO. The ISO member bodies can choose to be participating members or observing members. This decision depends on the national interests and has implications for the influence on the development and the contents of a standard. Participating member bodies organise the national committees, which are open to all stakeholders (mirror committees). If a person wants to contribute actively to the development of an ISO standard for a specific topic, the first step is to identify and contact the national ISO member body. After admission to the national mirror committee, it is possible to be officially assigned as an expert to the ISO committee that is responsible for writing the standard.

During development, an ISO standard has to pass through different project stages (Table 11.1, Figure 11.4). Every stage is connected to a project deliverable. The intervals between the different stages are fixed. Each stage is terminated with a ballot; however, the conditions for further maturing the document might differ at the various stages. For example, at the preliminary stage, a simple majority vote decides the fate of a preliminary work item (PWI). In contrast, at the proposal stage, at least five participating members have to declare their active support of the new work item proposal (NWIP) and nominate experts. Provided the latter is the case, a simple majority of the participating members is sufficient for the NWIP to become a working draft (WD). After publication of a new standard, it will be subjected to periodic review cycles during which whether the standard is still needed and whether changes are required or not will be discussed.

Table 11.1 Stages of ISO project deliverables

Stage	Name of the document	Acronym
Preliminary	Preliminary work item	PWI
Proposal	New work item proposal	NWIP
Preparatory	Working draft	WD
Committee	Committee draft	CD
Enquiry	Draft international standard	DIS
Approval	Final draft international standard	FDIS
Publication	International standard	IS

Adapted from www.iso.org.

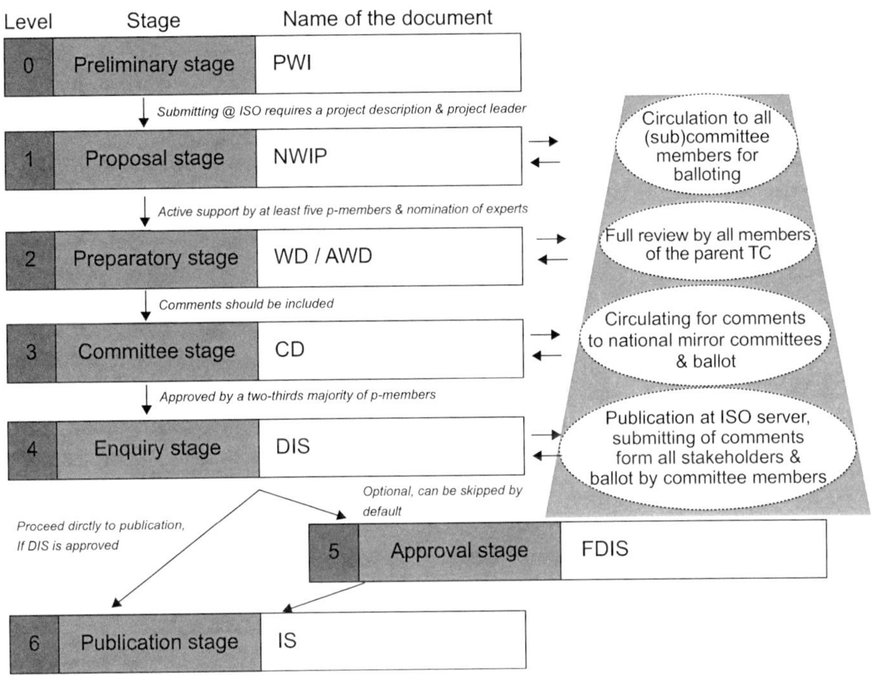

Figure 11.4 ISO standard development process. Standard development is driven by consensus—the document matures by including subsequently increasing circles of stakeholders. For some steps time limits exist. The overall process requires approximately 4–5 years. PWI = preliminary work item, NWIP = new work item proposal, WD = working draft, AWD = approved working draft, CD = committee draft, DIS = draft international standard, FDIS = final draft international standard, TC = technical committee, ISO = International Organization for Standardization, p-member = participating member.

11.3.4 Using standards

The use of standards is voluntary, and they do not have to be followed exactly. If a standard is not followed (in whole or in part), a comprehensible explanation/justification in the context of a marketing authorisation application is required. All relevant ISO standards have to be taken into account for the safety evaluation of an implant (if it is classified as a medical device). If no ISO standard exists for a certain topic, any national standard may be used. Manufacturers may use other methods if they meet the relevant essential principles implemented by law, such as industry agreed methods, internal standard operating procedures of the manufacturer, or other sources representing the current state of technology or science. Nevertheless, the acceptability of these methods has to be reviewed by the regulatory authorities or conformity assessment bodies. However, the advantage of using standards is that a recognised standard "is deemed to offer the presumption of conformity to specific Essential Principles of Safety and Performance" (GMHTF/SG1/N044:2008). This means that every time a standard is applied during the development of a product or process (e.g., biological safety testing), the regulatory authorities will not ask further questions regarding the validity and applicability of the test method covered by the corresponding standard. Therefore, using standards is usually the easiest and quickest way to prove the conformity of the product with the Essential Principals. However, standards are often not applicable (in whole or in part) to the product in question. In this case, the manufacturer has to explain why the standard was not applicable. This has to be done even if the reasons are obvious.

Taken together, standards are international agreements to facilitate the international trade of goods and services and to promote international cooperation in all fields of technical, scientific, economic, and intellectual activity. Standards can be subdivided into national or international standards according to their origin and horizontal or vertical standards according to their scope. Standards are consensus documents that make them sometimes very superficial and the development of which usually takes several years. Because of this, they are in general technically not state of the art. The use of standards is voluntary, and they do not have to be followed exactly. Due to a regular review process, the content can change. Therefore, it is important to work always with the latest version of a standard.

11.4 Biocompatibility – one aspect of biological safety

In addition to durability and mechanical stability, ensuring biological safety is mandatory for each implant. This does not mean that the implant has to be safe in every regard (there are no risk-free treatments); it means that the advantages of having the implant outweigh the possible risks connected with the operation procedure, the implant function, and the implant itself. Of course, this has to be verified for every implant. Risk management is the method of choice to ensure that the beneficial effects of an implant will offset the possible risks ("fit for purpose"). Several ISO standards exist that provide guidance in this regard (e.g., ISO 14971 "Application of risk management to medical devices" or the ISO 22442 series regarding the risks connected with materials of animal origin). All potential dangers should be taken into account; however, they do not necessarily have to be tested. A convincing argument with references to the literature might also be sufficient in some cases.

The biological safety of implants can be impaired by various biological and chemical substances (Figure 11.5). For example, it is very important for implants to be free of microorganisms such as bacteria, fungi, and other parasites to avoid the risk of infection. This can be achieved with different sterilisation methods such as heat, radiation, or chemical inactivation. The sterility requirement seems obvious and easy to achieve. However, the existence of more than 20 ISO standards regarding sterilisation methods for medical devices and their validation, for sterilisers, and for sterile packaging (please refer to Table 11.2 for an overview) indicates, that this is a challenge. Sterilisation can also have an impact on the implant material, e.g., through the creation of free radicals by disrupting the carbon chains in polymers. Therefore, it is necessary to choose a sterilisation method that is suitable for the implant material (and packaging). For example, a sterilisation method using water steam would not be suitable for magnesium alloys.

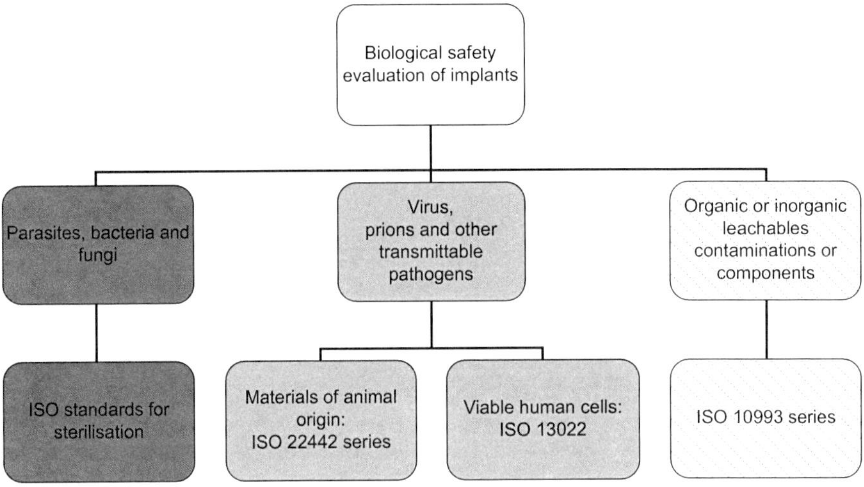

Figure 11.5 Biological safety evaluation of implants involves the assessment of the risk of pathogens and chemical contaminations (leachables). Various ISO standards give guidance on risk evaluation and minimisation.

Implants should also be free of transmittable pathogens such as viruses or prions. Contamination with these pathogens may occur if they have been made with or consist of materials of animal origin, e.g., bovine serum albumin, collagen, trypsin or pepsin, amino acids, collagen, or deoxycholate. Furthermore, human cells can also transmit pathogens if they are part of the implant. A strategy of how to evaluate and control the risk of infection with respect to materials of animal origin or human cells is described in the ISO standard series 22442 or standard 13022 (Figure 11.5). Biological materials are difficult and sometimes impossible to sterilise with conventional methods using heat or radiation because they become inactivated or are destroyed during the process. In these cases, it is common practice to work under sterile conditions. Guidance concerning this topic will be given by new ISO standard 18362 (Table 11.2), which will be published probably in 2015.

In addition to biological contamination, the chemical composition of the starting material, production residues (e.g., organic solvents, separating agents), leachables, degradation products, wear debris, ageing, or surface effects have to be taken into account for the biological evaluation of implants. The material-related aspects are tested and evaluated according to the ISO 10993 series (Figure 11.5). The ISO 10993 series consists of 20 parts, and a new section (Part 22) for nanomaterials is currently under development (Table 11.3). The set of tests that should be considered for evaluating the safety of a medical device depends on the manner and duration of contact with the body. A re-evaluation is necessary for the following types of changes: a change of supplier for the materials; specification changes; an alteration of the composition, processing, packaging, or sterilisation process; or if the material is used for a new application. Other reasons for re-evaluation could be the indication of adverse effects on humans after the product is on the market or if the product is altered during storage.

Table 11.2 Exemplary list of standards that cover the topic of sterilisation of medical devices

Topic	Number	Title
General requirements	EN 556-1	Sterilisation of medical devices – Requirements for medical devices to be designated "STERILE," part 1: Requirements for terminally sterilised medical devices
	EN 556-2	Sterilisation of medical devices – Requirements for medical devices to be designated "STERILE," part 2: Requirements for aseptically processed medical devices
	ISO 17664	Sterilisation of medical devices – Information to be provided by the manufacturer for the processing of resterilisable medical devices
Technique & equipment	EN 285	Sterilisation – Steam sterilisers, large sterilisers
	EN 13060	Small steam sterilisers
	EN 1422	Sterilisers for medical purposes – Ethylene oxide sterilisers – Requirements and test methods
	EN 14180	Sterilisers for medical purposes – Low-temperature steam and formaldehyde sterilisers – Requirements and testing

Table 11.2 Exemplary list of standards that cover the topic of sterilisation of medical devices – cont'd

Topic	Number	Title
Methods	EN ISO 11135-1	Sterilisation of health care products – Ethylene oxide, part 1: Requirements for development, validation, and routine control of a sterilisation process for medical devices
	EN ISO 11137-1	Sterilisation of health care products – Radiation, part 1: Requirements for development, validation, and routine control of a sterilisation process of medical devices
	EN ISO 11137-2	Sterilisation of health care products – Radiation, part 2: Establishing the sterilisation dose
	EN 13824	Sterilisation of medical devices – Aseptic processing of liquid medical devices – Requirements
	EN ISO 14160	Sterilisation of single-use medical devices incorporating materials of animal origin – Validation and routine control of sterilisation by liquid chemical sterilants
	EN ISO 14937	Sterilisation of health care products – General requirements for characterisation of a sterilising agent and the development, validation, and routine control of a sterilisation process for medical devices

Continued

Table 11.2 Exemplary list of standards that cover the topic of sterilisation of medical devices – cont'd

Topic	Number	Title
	EN 15424	Sterilisation of medical devices – Low-temperature steam and formaldehyde– Requirements for development, validation and routine control of a sterilisation process for medical devices
	EN ISO 17665-1	Sterilisation of health care products – Moist heat, part 1: Requirements for the development, validation, and routine control of a sterilisation process for medical devices
	ISO 18362	Processing of cell-based health care products

The set of tests described in the ISO 10993 series are often referred to as "biocompatibility testing." Alternatively, if an implant passes all necessary tests mentioned in ISO 10993, it is described as "biocompatible." The word comes from the Greek bios (life) and compatible (compliant). The term biocompatibility was first defined by David F. Williams in the late 1980s as "the ability of a material to perform with an appropriate host response in a specific application" (Williams, 1987). In 2008, he redefined the earlier version of his definition as follows: "Biocompatibility refers to the ability of a biomaterial to perform its desired function with respect to a medical therapy, without eliciting any undesirable local or systemic effects in the recipient or beneficiary of that therapy, but generating the most appropriate beneficial cellular or tissue response in that specific situation, and optimising the clinically relevant performance of that therapy" (Williams, 2008). Later, the International Union of Pure and Applied Chemistry (IUPAC, www.iupac.org) defined biocompatibility as the "ability of a material to perform with an appropriate host response in a specific application" and defined biomedical therapy as the "ability to be in contact with a living system without producing an adverse effect" (Vert et al., 2012).

Biocompatible materials can also be broken down by type as "bioinert," "biotolerant," or "bioactive." A bioinert material shows virtually no chemical or biological interaction with the surrounding tissue, whereas a bioactive material interacts with

Table 11.3 List of all currently available parts of ISO 10993 "Biologic evaluation of medical devices"

10993-1	Evaluation and testing
10993-2	Animal welfare requirements
10993-3	*Test for genotoxicitiy, carcinogenicity and reproductive toxicity*
10993-4	*Selection of tests for interactions with blood*
10993-5	*Test for in vitro cytotoxicity*
10993-6	Test for local effects after implantation
10993-7	Ethylene oxide sterilisation residuals
10993-8	Selection and qualification of reference materials for biological tests
10993-9	Framework for identification and quantification of potential degradation products
10993-10	*Test for irritation and delayed-type hypersensitivity*
10993-11	*Test for systemic toxicity*
10993-12	Sample preparation and reference materials
10993-13	Identification and quantification of degradation products from polymeric medical devices
10993-14	Identification and quantification of degradation products from ceramics
10993-15	Identification and quantification of degradation products from metals and alloys
10993-16	Toxicokinetic study design for degradation products and leachables
10993-17	Establishment of allowable limits for leachable substances
10993-18	Chemical characterisation of materials
TS 10993-19	Physicochemical, morphological and topographical characterisation of materials
TS 10993-20	Principles and methods for immunotoxicology testing of medical devices
10993-22	Guidance on nanomaterials (in development)

Parts of ISO 10993 that include test procedures using implant extracts are in italics

the surrounding tissue and is typically used where a tight connection between tissue and implant material is required. In contrast, a biotolerant material is not bioactive and, in the long run, not bioinert either. However, it can be safely used for a certain time (from several months to several years) before compatibility with the surrounding tissue starts to fade/diminish. Interestingly, the term biocompatibility or biocompatible is not defined in the European or American laws on the requirements and procedures of marketing authorisation. It is also not defined in any of the ISO standards mentioned above.

In summary, the term "biocompatibility" has various definitions. In the context of marketing authorisation, it refers to the property of an implant (medical device) that is mainly confirmed by passing a set of tests documented in ISO 10993. Biocompatibility is only one aspect of the biological evaluation of implants.

11.5 Preclinical biocompatibility testing of biodegradable magnesium implants for marketing authorisation

11.5.1 General principles of biocompatibility testing for marketing authorisation

The aim of biocompatibility testing for marketing authorisation is to demonstrate that the implant is safe for its intended use ("fit for purpose"), which is determined by the manufacturer. To make sure that the implant is fit for purpose, all potential dangers should be taken into account; however, they do not necessarily need to be tested. It is also possible to replace testing by an evaluation of the relevant literature, especially to avoid unnecessary animal experiments. All of the available literature can be used for the evaluation as long as the evaluation process is explained in a comprehensive manner so that the regulatory bodies can follow the argumentation easily and are able to evaluate the suitability of the implant for its intended use.

In vitro and *in vivo* testing should always be conducted with the final product (or a test item that has been treated like the final product), which means that it has to already be packaged and sterilised (original packaging). This is done to ensure that no changes have occurred to the implant and that toxic substances have not been released from the packaging and come into contact with the implant during the sterilisation process. Given that ageing can also induce changes to the sterilised implant in its primary packaging, tests of the final aged product should be considered. The extent to which the final aged product needs to be tested has to be evaluated by risk management.

The test items to be used for biocompatibility testing should be traceable and marked accordingly (e.g., with their designations and batch numbers). Tests for marketing authorisation should be performed by certified laboratories (EU: ISO 17025, USA: Good Laboratory Practice, GLP). Some companies do the preclinical testing in-house; in that case, their laboratories will be included in the product audit for approval by the regulatory authorities or regulatory competent bodies. In the interest of reducing animal testing, animal experiments performed at universities or other noncertified laboratories can also be used for marketing authorisation if they meet the ISO 10993 requirements. Currently, there is discussion about whether these data should be excluded from studies for marketing authorisation; therefore, it might be possible that only data collected by certified laboratories are accepted by the regulatory bodies in the future. In any case, all the test results have to be documented in a test report, including the raw data. Finally, the manufacturer has to summarise its conclusions based on the tests performed.

11.5.2 Standards for biocompatibility testing of magnesium implants

Currently, there are no specific standards for testing the biocompatibility of biodegradable magnesium implants. However, some efforts have been made to adapt ISO 10993 "biological evaluation of medical devices" to issues related to degradable (magnesium) implants. As a result of these efforts, TR 37137, titled "Biological evaluation of medical devices Guidance for absorbable implants," has been published recently. The document is based on a gap analysis and lists changes to ISO 10993, which are required for biocompatibility testing of absorbable implants made of polymers or metals. The contents of the document will be discussed within the working groups of the ISO technical committee 194 "Biological and clinical evaluation of medical devices," which develops the 10993 series. If the experts within the working groups agree to the proposed changes, they will be included gradually in the next revision. Complementary to TR 37137, the ASTM document F1983-99 (2008) "Standard Practice for Assessment of Compatibility of Absorbable/Resorbable Biomaterials for Implant Applications" gives guidance on the examination of tissue reactions to absorbable biomaterials for implants that will reside in bone or soft tissue longer than 30 days and less than 3 years.

There are also few application-specific standards available. On the international level, ancillary to ASTM F3036-13 "Standard Guide for Testing Absorbable Stents," an ISO document dealing with absorbable cardiovascular implants either based on polymers or metals (ISO/TS 17137) has been recently published. For aspects related to biocompatibility, this document refers to ISO 10993. Also an ASTM standard is available that gives guidance to mechanical testing of orthopaedic absorbable implants (plates and screws for internal fixation, ASTM F2502-11).

11.5.3 Biocompatibility testing of implants according to ISO 10993

ISO 10993, "Biological evaluation of medical devices," is a semihorizontal level 2 standard that consists of 20 parts (Table 11.3) and is applicable to all kinds of medical devices as well as implants. Examples of medical devices are hospital bed linens, plaster, wound dressing materials, ultrasound gels, dialysis equipment, surgical instruments, condoms, contact lenses, catheters, artificial hips, and stents. Part 1 of ISO 10993 provides guidance on the biological evaluation and use of the standard. Besides biological tests, several chemical analyses of degradation products and leachables, which are also are part of the biocompatibility assessment of a product, have to be considered (Figure 11.6). Annex A in ISO 10993-1 gives an overview of the tests required for biocompatibility assessment. Since the standard is relevant for many different products, the manner and duration of body contact of a medical device is a decisive factor for the toxicological topics, which has to be addressed in the biocompatibility evaluation either as a test result or in relation to the existing literature. For implants, usually all tests mentioned in ISO 10993 have to be taken into account. The only exception applies to implants that are not exposed to blood, in which case the tests for hemocompatibility (ISO 10993-4) may be omitted.

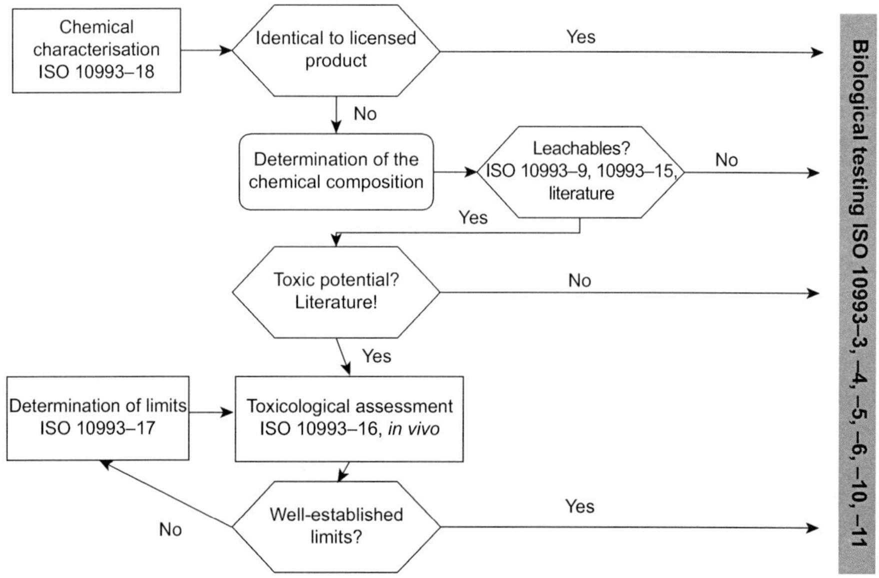

Figure 11.6 Biocompatibility evaluation of medical devices adapted from ISO 10993-1.

11.5.4 Specific considerations for testing magnesium-based implants

Due to the general character of ISO 10993, not all medical devices can be tested in accordance with this standard without encountering some difficulties. In the case of degradable magnesium implants, their ability to dissolve causes problems when some parts of ISO 10993 are applied. This is because many of the tests require the preparation of an aqueous extract of the implant, for example "Genotoxicity" (Part 3), "Hemocompatibility" (Part 4), "Cytotoxicity" (Part 5), "Irritation, Sensitization" (Part 10), and "Systemic toxicity" (Part 11). If a magnesium implant is extracted in an aqueous solution, it degrades to magnesium, hydroxide ions, and hydrogen gas. The degree of degradation is reflected in the increase in osmolality and the pH of the extract, both of which can also lead to positive results for toxicity. The *in vitro* degradation of magnesium alloys is enhanced compared to the *in vivo* situation (Bobe et al., 2013; Witte et al., 2006) depending on the extraction media (Bobe et al., 2013; Feyerabend et al., 2012; Willumeit et al., 2011; Xin, Hu, Chu, 2011). This has to be taken into account for evaluation of the results from the extraction-based testing methods according to ISO 10993. In cases of a very fast *in vitro* degradation of the magnesium implant, the tests have to be adapted or they might not be suitable for biological evaluation. Besides the extraction-based tests (originally designed to detect toxic production residues, leachables, and degradation products of permanent polymer or metallic implants), animal experiments can also lead to false results if unsuitable test items, implantation sites, or animal models are chosen. Since vascularisation, gas, and fluid exchange are factors that play an important role in in vivo degradation (reviewed in Witte (2010)), it is strongly advisable to choose the implantation site according to the intended use of the implant.

11.5.4.1 Necessary modifications to biological tests of ISO 10993 for magnesium-based implants

For the extraction-based tests (Parts 3, 4, 5, 10, and 11), it is recommended that cell culture media or a medium similar to body fluids (Xin et al., 2011; Willumeit et al., 2011; Feyerabend et al., 2012) be used for extraction and that the extraction be performed at a temperature of 37 °C (Xin et al., 2011). With regard to the *in vitro* tests (ISO 10993 Parts 3, 4, and 5), it is generally a good idea to test the final product at various stages of degradation to check for the toxicity of by-products. The osmolality and pH of the extract should be controlled to avoid misinterpretation of the test results. However, ISO 10993-12:2012, "Sample preparation, controls and reference materials," states that the pH of the extract "shall not be adjusted unless a rationale is provided." Adjusting the pH for extraction-based indirect cytotoxicity testing according to ISO 10993 Part 5 can be justified given that basic pH is cytotoxic itself and the increase in pH might interfere with test components, e.g., XTT, MTT, or MTS (Fischer et al., 2010). High pH values, high osmolality, and a high cytotoxic potential of the product extract could also affect genotoxicity testing. In addition, testing of various extract dilutions might also be necessary.

For *in vivo* experiments, a pH adjustment could be necessary to avoid misleading results if the implantation site does not reflect the actual *in vivo* cell exposure conditions (e.g., in terms of physiological perfusion) (Hoffheinz and Dimitroff, 1928) in irritation tests (ISO 10993 Part 10) or tests for systemic toxicity (Part 11). If an adverse pathology is monitored, an assessment of the reversibility of the observed effects after a certain time should be considered. During implant retrieval and tissue sample collection, care should be taken to minimise artefacts due to histological processing and handling (e.g., further degradation of the material or shrinkage).

The development of hydrogen gas can be problematic in *in vivo* tests. The misinterpretation of *in vivo* results can be reduced by selecting an implantation site according to the intended use of the implant.

Cellular attachment to fast degrading magnesium implants might not be possible to be studied *in vitro*, because the fast changes to the surface resulting from the release of hydrogen gas prevent cellular attachment. However, this functional test is not part of ISO 10993; therefore, it is not relevant for marketing authorisation.

11.5.4.2 Additional options for the modification of standard protocols in ISO 10993-5

A simple way to avoid adjustment of the pH could be to use a test system not affected by a high concentration of hydroxide ions (e.g., neutral red staining). Some other alternatives could be counting the cell number, estimating the total protein, or using a qualitative live/dead assay based on cell membrane integrity. Fischer et al. (2010)

proposed using a BrdU (5-bromo-2′-deoxyuridine) incorporation assay, which measures cell proliferation and is not disturbed by the pH. However, the latter is not advisable for cytotoxicity testing for marketing authorisation, because according to ISO 10993-5 the output of the test has to demonstrate cell viability. Cell proliferation can contribute to viability; however, nonproliferating cells can also be viable. XTT, MTT, or MTS tests can also be applied with a slight modification of the protocol: One example is to exchange the cell culture medium prior to XTT, MTT, or MTS application (Witte et al., 2006). Another possibility is to perform the test in parallel without cells to determine the absorption baseline. The latter can be further subtracted from the measured values to obtain the corrected baseline values (Fischer et al., 2010). In addition to the problem of pH and osmolality changes, the released magnesium ions (or other ions) might also affect the test by interacting with test components. Therefore, compatibility of the magnesium alloy with the test method should be tested before.

Since the composition of the extraction media has a huge impact on the degradation of magnesium alloys and ISO 10993 does not specify the use of a specific medium, the choice of cell culture medium might also be worth considering. Commonly used cell lines for cytotoxicity testing, such as L929 or NIH 3T3 fibroblasts, are relatively undemanding and grow in many standard cell culture media such as RPMI 1640, DMEM, or MEM. All of the media mentioned are commercially available with different compositions with regard to their glucose, salt, and buffer concentration. Some companies are also willing to produce individualised cell culture media according to the needs of the customer.

Another factor to consider is the cell type used for the *in vitro* experiments. According to ISO 10993 Part 5, several cell lines are proposed for cytotoxicity testing, and it is possible to use primary human cells as well. Current publications indicate that the results of cytotoxicity testing might differ depending on the cell line used (e.g., Bobe et al., 2013; Feyerabend et al., 2010). Therefore, it might support the biological evaluation to use a cell type that reflects the actual *in vivo* cell exposure.

To summarise, the aim of biocompatibility testing for marketing authorisation is to demonstrate that the implant is safe for its intended use (“fit for purpose”), which is determined by the manufacturer. Currently, there are no specific standards for testing the biocompatibility of biodegradable magnesium implants. The standards of the ISO 10993 series can be applied, but the protocols might need slight modifications and additional controls, e.g., pH and osmolality measurements. Adaption of the ISO 10993 series according to the requirements of degradable magnesium implants will be pursued in a stepwise manner during the periodic revisions of the documents.

11.6 Nonstandardised *in vitro* and *in vivo* techniques for biocompatibility evaluation of magnesium alloys

A survey of the available literature demonstrates that most of the groups use *in vitro* viability tests such as MTT or XTT for biocompatibility assessment. Some groups have also published results using neutral red live/dead assays (two fluorescent dyes: one passes through the intact cell membrane, whereas the other does not) or viable cell counting. However, there are little data available on the genotoxicity and hemocompatibility of magnesium (Di Virgilio, Reigosa, and de Mele, 2011; Gu, Zhen, Cheng, Zhong, and Xi, 2009). Besides the *in vitro* data, animal experiments have been performed that focus on *in vivo* degradation and the local effects after implantation. Witte et al. (2007) also tested sensitisation using the Magnusson–Kligman test (GMPT) according to 10993-10 for comparison of the various magnesium alloys. One publication focuses on the fate of hydroxide gas developed during the degradation of magnesium (Bartsch, Willbold, Yarmolenko, and Witte, 2012).

Some scientists try to explain the effects of magnesium and alloying components on the cellular level to reveal the mechanisms, unknown dangers, and potential markers for the biocompatibility of magnesium implants (Feyerabend et al., 2010; Purnama, Mantovani, and Couet, 2013). In recent years, it has become increasingly apparent that *in vitro* corrosion of magnesium alloys is highly variable depending on the solution used in the test set-up. Ions such as chloride, biological buffers, proteins, and amino acids and temperature can influence *in vitro* degradation (reviewed in Xin et al. (2011)). In a new approach using artificial neural networks, the influence of the CO_2 content is highlighted (Willumeit, Feyerabend, and Huber, 2013). Dynamic conditions involving sheer stress, continuous mass transport, and the exchange of fluids also have an impact on degradation (Levesque, Hermawan, Dube, and Mantovani, 2008).

To compare the results of the *in vitro* data in the future, it would be worthwhile to develop a standardised procedure for measuring *in vitro* degradation with a defined corrosion (or extraction) medium, temperature, and specific atmospheric conditions (CO_2, O_2, N_2) in addition to reference materials. In would also be helpful to establish a definition of the different phases of degradation to facilitate compatibility of the results of the toxicity tests of degradation intermediates. However, despite the many efforts to find a predictable model for an *in vitro/in vivo* correlation, it seems that we will continue to depend on animal experiments for information about the time course and intermediates of degradation.

References

Adams, C. P., & Brandnter, V. V. (2006). Estimating the cost of new drug development: is it really 802 million dollars? *Health Affairs (Millwood), 2*, 420–428.

ANSI/AAMI HE 74. Human factors design process for medical devices.

ASTM F3036-13. Standard guide for testing absorbable stents biological evaluation.

ASTM F2502-11. Standard specification and test methods for absorbable plates and screws for internal fixation implants.

Bartsch, I., Willbold, E., Yarmolenko, S., & Witte, F. (2012). In vivo fluorescence imaging of apoptosis during foreign body response. *Biomaterials, 33*, 6926–6932.

Blind, K. (2004). *The economics of standards. Theory, evidence, policy*. Cheltenham: Edward Elgar.

Bobe, K., Willbold, E., Morgenthal, I., Andersen, O., Studnitzky, T., Nellesen, J., et al. (2013). In vitro and in vivo evaluation of biodegradable, open-porous scaffolds made of sintered magnesium W4 short fibres. *Acta Biomaterialia, 10*, 8611–8623.

Di Virgilio, A. L., Reigosa, M., & de Mele, M. F. L. (2011). Biocompatibility of magnesium particles evaluated by in vitro cytotoxicity and genotoxicity assays. *Journal of Biomedical Materials Research Part B, 99B*, 111–119.

DiMasi, J. A., Hansen, R. W., & Grabowski, H. G. (2003). The price of innovation: new estimates of drug development costs. *Journal of Health Economics, 2*, 151–185.

Feyerabend, F., Drücker, H., Laipple, D., Vogt, C., Stekker, M., Hort, N., et al. (2012). Ion release from magnesium materials in physiological solutions under different oxygen tensions. *Journal of Materials Science: Materials in Medicine, 23*, 9–24.

Feyerabend, F., Fischer, J., Holtz, J., Witte, F., Willumeit, R., Drücker, H., et al. (2010). Evaluation of short-term effects of rare earth and other elements used in magnesium alloys on primary cells and cell lines. *Acta Biomaterialia, 6*, 1834–1842.

Fischer, J., Prosenc, M. H., Wolff, M., Hort, N., Willumeit, R., & Feyerabend, F. (2010). Interference of magnesium corrosion with tetrazolium-based cytotoxicity assays. *Acta Biomaterialia, 6*, 1813–1823.

GHTF/SG1/N044:2008. (2008). *Role of standards in the assessment of medical devices*. Global Harmonization Task Force. Study Group 1.

Gu, X., Zheng, Y., Cheng, Y., Zhong, S., & Xi, T. (2009). In vitro corrosion and biocompatibility of binary magnesium alloys. *Biomaterials, 30*, 484–498.

Hoffheinz, S., & Dimitroff, N. (1928). Experimentelle Studie über die resorption parental zugeführten metallischen Magnesiums. *Deutsche Zeitschrift fuer Chirurgie, 208*, 346–353.

IEC 62366. Medical devices – Application of usability engineering to medical devices.

ISO 10993. Biological evaluation of medical devices.

ISO 11135. Sterilization of health care products – Ethylene oxide.

ISO 13022. Medical products containing viable human cells – Application of risk management and requirements for processing practices.

ISO 14630. Non-active surgical implants – General requirements.

ISO 14971. Medical devices – Application of risk management to medical devices.

ISO 22442 series. Medical devices utilizing animal tissues and their derivatives.

ISO 25539-2. Cardiovascular implants – Endovascular devices – Part 2: Vascular stents.

ISO Central Secretariat. (2007). *Joining in – Participating in international standardization*, ISBN 978-92-67-10453-9. http://www.din.de/sixcms_upload/media/2896/joining_in_2007.pdf.

ISO/CD 18362. Processing of cell-based health care products.

ISO/IEC 17025. General requirements for the competence of testing and calibration laboratories.

ISO/IEC Guide 2. Standardization and related activities – General vocabulary.

ISO/TR 37137. Biological evaluation of medical devices – Guidance for absorbable implants.

ISO/TS 17137. Cardiovascular implants and extracorporal systems – Cardiovascular absorbable implants.

Kaplan, A. V., Baim, D. S., Smith, J. J., Feigal, D. A., Simons, M., Jefferys, D., et al. (2004). Medical device development–from prototype to regulatory approval. *Circulation, 109*, 3068–3072.

Levesque, J., Hermawan, H., Dube, D., & Mantovani, D. (2008). Design of a pseudo-physiological test bench specific to the development of biodegradable metallic biomaterials. *Acta Biomaterialia, 4*, 284–295.

Purnama, A., Mantovani, D., & Couet, J. (2013). Caveolin: a possible biomarker of degradable metallic materials toxicity in vascular cells. *Acta Biomaterialia, 10*, 8754–8760.

Vert, M., Doi, Y., Hellwich, K.-H., Hess, M., Hodge, P., Kubisa, P., et al. (2012). Terminology for biorelated polymers and applications (IUPAC Recommendations 2012). *Pure and Applied Chemistry, 84*(2), 377–410.

Williams, D. F. (1987). *Definitions in biomaterials*. Amsterdam: Elsevier.

Williams, D. F. (2008). On the mechanisms of biocompatibility. *Biomaterials, 29*, 2941–2953.

Willumeit, R., Feyerabend, F., & Huber, N. (2013). Magnesium degradation as determined by artificial neural networks. *Acta Biomaterialia, 10*, 8722–8729.

Willumeit, R., Fischer, J., Feyerabend, F., Hort, N., Bismayer, U., Heidrich, S., et al. (2011). Chemical surface alteration of biodegradable magnesium exposed to corrosion media. *Acta Biomaterialia, 7*, 2704–2715.

Witte, F. (2010). The history of biodegradable magnesium implants: a review. *Acta Biomaterialia, 6*, 1680–1692.

Witte, F., Abeln, I., Switzer, E., Kaese, V., Meyer-Lindenberg, A., & Windhagen, H. (2007). Evaluation of the skin sensitizing potential of biodegradable magnesium alloys. *Journal of Biomedical Materials Research Part A*, 1041–1047.

Witte, F., Fischer, J., Nellesen, J., Crostack, H. A., Kaese, V., Pisch, A., et al. (2006). In vitro and in vivo corrosion measurements of magnesium alloys. *Biomaterials, 7*, 1013–1018.

Xin, Y., Hu, T., & Chu, P. K. (2011). In vitro studies of biomedical magnesium alloys in a simulated physiological environment: a review. *Acta Biomaterialia, 7*, 1452–1459.

Index

Note: Page numbers followed by "f" and "t" indicate figures and tables, respectively.

D

E

J

K

L

Printed in the United States
By Bookmasters